AF566595

Fatigue Crack Growth Threshold Concepts

Fatigue Crack Growth Threshold Concepts

Proceedings of the International Symposium on Fatigue Crack Growth Threshold Concepts sponsored by the Mechanical and Physical Metallurgy Committees of The Metallurgical Society of AIME and the Corrosion and Environmental Effects Committees of The Metallurgical Society and American Society for Metals held at the Fall Meeting of The Metallurgical Society in Philadelphia, Pennsylvania, October 3-5, 1983.

Edited by

D.L. Davidson
Southwest Research Institute
San Antonio, Texas 78284

S. Suresh
Brown University
Providence, Rhode Island 02912

A Publication of 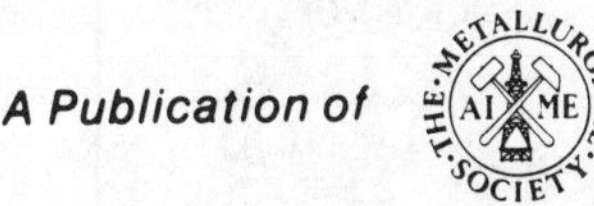 The Metallurgical Society of AIME

A Publication of The Metallurgical Society of AIME
420 Commonwealth Drive
Warrendale, Pennsylvania 15086
(412) 776-9000

Printed in the United States of America.
Library of Congress Catalogue Number 84-70400
ISBN NUMBER 089520-475-4

Foreword

In recent years, the mechanics and mechanisms associated with fatigue crack growth at very low temperatures have been recognized as topics of considerable engineering importance and academic interest, and it has become important to explain the origin of the growth threshold. Research activity dealing with the influence of various factors on the threshold for growth of "long" and "short" fatigue cracks has accelerated as there is increasing evidence that crack growth near the threshold is one of the important factors in the development of materials and in the design and lifetime assessment of structure.

The papers included in this volume were presented at the International Symposium on Fatigue Crack Growth Threshold Concepts held in Philadelphia, Pennsylvania, October 3-5, 1983, during the Fall Meeting of The Metallurgical Society of AIME. This symposium was sponsored by the Mechanical and Physical Metallurgy Committees of TMS and the Corrosion and Environmental Effects Committee of TMS and American Society for Metals. The five sessions were well-attended by scientists and engineers from many countries.

This symposium was organized specifically to concentrate on the concepts underlying the fatigue crack growth threshold. The First International Conference on Fatigue Thresholds was held in Stockholm, Sweden, in June 1981. Research in the period between these two conferences has led to an improved understanding of the many mechanisms influencing the near-threshold behavior of fatigue crack growth in engineering materials.

These proceedings deal with many aspects of near-threshold fatigue crack growth including the roles of composition, microstructure, environment, temperature, crack closure, crack size, and variable amplitude loading, The current level of knowledge of fatigue thresholds is accurately reflected by the contents of the manuscripts.

The papers published in this book were reviewed by individuals working in the field. We would like to express our appreciation first to the authors for their participation in the symposium and for their cooperation in the preparation of the manuscripts. Our sincere thanks also go to the many reviewers for their sacrifice of time and effort in helping to ensure the quality of the papers.

We are grateful to Southwest Research Institute, University of California, Berkeley, and Brown University for providing us with the use of their typing, printing, copying, mailing, and telex and telephone facilities. Many have helped support us in the organization of the symposium and the editing of this volume, but we are especially appreciative of the cheerful and efficient assistance of Mrs. Rose Fontana, Miss Madeleine Penton, and Mrs. Louise Gray.

D.L. Davidson
Southwest Research Institute
San Antonio, Texas

S. Suresh
Brown University
Providence, Rhode Island

January 1984

Contents

Foreword v

MICROSTRUCTURAL AND ENVIRONMENTAL INFLUENCES

Some Aspects of Near-Threshold Crack Growth: Microstructural and Environmental Effects 3
J. Petit

Near-Threshold Fatigue Crack Growth Behavior in 7XXX and 2XXX Alloys: A Brief Review 25
A.K. Vasudevan and P.E. Bretz

The Use of the Cyclic Stress Strain Curve and A Damage Model for Predicting Fatigue Crack Growth Thesholds 43
E.A. Starke, Jr., F.S. Lin, R.T. Chen, and H.C. Heikkenen

Environmental Effects on Theshold Stress Intensity Factor in 70-30 Alpha Brass and 2024-T351 Aluminum Alloy 63
J.P. Bailon, M. El Boujaini, and J.I. Dickson

Effect of Microstructure and Load Ratio on Theshold Stress Intensity Factor in Titanium Alloys 83
J.C. Chesnutt and J.A. Wert

Environmental and Frequency Effects on Near-Threshold Fatigue Crack Propagation in a Structural Steel 99
A. Bigonnet, D. Loison, R. Namdar-Irani, B. Bouchet, J.H. Kwon, and J. Petit

Near-Threshold Fatigue Crack Propagation of Hy80 and HY130 Steels 115
J.L. Horng and M.E. Fine

The Effect of Microstructure on Fatigue Crack Path and Crack Propagation Rate 131
G.T. Gray, III, A.W. Thompson, and J.C. Williams

Origin of Crack Closure in the Near-Threshold Fatigue Crack Propagation of Fe and Al-3% Mg 145
D.H. Park and M.E. Fine

The Effects of Grain Size and Stress Ratio on Fatigue Crack Growth in 7091 Aluminum Alloy 163
P.E. Bretz, J.I. Petit, and A.K. Vasudevan

Influence of Temperature on the Near-Threshold Fatigue Crack Growth Behavior of a Nickel Base Superalloy 185
J.L. Yuen and P.Roy

Influence of Temperature and Load Ratio
on Near-Threshold Fatigue Crack Growth Behavior of CrMoV Steel......... 205
P.K. Liaw, A. Saxena, V.P. Swaminathan, and T.T. Shin

CRACK CLOSURE

Near-Threshold Fatigue Crack Propagation:
A Perspective on the Role of Crack Closure............................ 227
S. Suresh and R.O. Ritchie

Near-Threshold Fatigue Crack Growth and Crack Closure
in 17-4 PH Steel and 2024-T3 Aluminum Alloy........................... 263
A.F. Blom

Influence of R. Ratio and Orientation on the Fatigue Crack
Threshold and Subsequent Crack Growth of a Low-Alloy Steel............ 281
A.J. Cadman, C.E. Nicholson, and R. Brook

An Assessment of Internal Hydrogen Versus Closure Effects
on Near-Threshold Fatigue Crack Propagation........................... 299
K.A. Esaklul, A.G. Wright, and W.W. Gerberich

Fatigue Crack Closure and the Fatigue Threshold....................... 327
C.J. Beevers, K. Bell, and R.L. Carlson

Influence of Stress Ratio on Fatigue Threshold and
Structure Sensitive Crack Growth in Ni Base Superalloys............... 341
R.A. Venables, M.A. Hicks, and J.E. King

APPLICATION AND SPECIAL TECHNIQUES

Application of Fatigue Threshold Concepts
to Variable Amplitude Crack Propagation............................... 361
S. Suresh and A.K. Vasudevan

On the Interaction of Overload and Metallurgical Variables
of Fatigue Crack Growth in the Threshold Regime....................... 379
C.H. Newton, T.S. Vecchio, R.W. Hertzberg, and R. Jaccard

Concepts of Fatigue Crack Growth Threshold Gained
by the Ultrasound Method.. 399
S.E. Stanzl and H.M. Ebenberger

An Assessment of the Fatigue Threshold
as a Design Parameter... 417
R. Brook

SMALL CRACKS

The Effects of Texture and Grain Size
on the Short Fatigue Crack Growth Rates in Ti-6Al-4V................... 433
C.W. Brown and D. Taylor

Near-Threshold Crack Tip Strain and Crack Opening
for Large and Small Fatigue Cracks.................................... 447
J. Lankford and D.L. Davidson

Propagation of Small Surface Cracks in Titanium Alloys................ 465
C. Gerdes, A. Gysler, and G. Lutjering

Investigation of the Growth Threshold for Short Cracks................ 479
W.L. Morris and M.R. James

Mechanics of Growth Threshold of Small Fatigue Cracks................. 497
K. Tanaka and Y. Nakai

Crack Closure and the Conditions
for Factigue Crack Propagation.. 517
A.J. McEvily and K. Minakawa

Dislocation Models for Threshold Fatigue Crack Growth................. 531
T. Mura and J. Weertman

CONCLUDING REMARKS

D.L. Davidson and S. Suresh... 553

R.O. Ritchie.. 555

J. Lankford... 559

Subject Index... 561

Author Index.. 565

Microstructural and Environmental Influence

SOME ASPECTS OF NEAR-THRESHOLD CRACK GROWTH :

MICROSTRUCTURAL AND ENVIRONMENTAL EFFECTS

J. PETIT

Laboratoire de Mécanique et de Physique des Matériaux
E.R.A. N°123 DU C.N.R.S.
E.N.S.M.A. - 86034 POITIERS CEDEX
FRANCE

A study of crack growth in the low rate range and at threshold level has been performed on several high strength Al alloys - i.e. 2618-T651, 2024-T351, 7020-T651 and 7075 in four aged conditions - in air, vacuum and nitrogen with traces of water vapour. Using a ΔK^*_{eff} concept taking into account an effect of mixed-mode crack opening, it is shown that crack growth in vacuum is proportional to $(\Delta K^*_{eff})^4$, with the microstructural influence corresponding essentially to two mechanisms related to precipitate conditions. In nitrogen the environmental effect observed in the mid rate range is related to the decrease in surface energy induced by water vapour adsorption, and in the low rate range to an "embrittling" effect of hydrogen generated by water molecules dissociation and leading to a characteristic crack growth behaviour and to the lowest threshold levels. In air the crack growth process is analysed as the result of a complex and competitive interaction of different factors : adsorption, oxide thickening and microstructure.

Introduction

Various studies have been conducted on fatigue crack growth (F.C.G.) in low rate range and on the threshold in aluminium alloys. A large set of information is now available on the role of environment, microstructure, crack closure, slip and strain distributions at the crack tip, strength parameters and other related subjects.

This paper presents new results from an analysis of complimentary F.C.G. tests and crack closure measurements performed on aluminium alloys. These results are discussed on the basis of a review of available literature so as to get a better understanding of the threshold concept and of the near-threshold F.C.G. behaviour.

Review

A few years ago different studies conducted on the influence of environment on fatigue crack growth (1,2) had shown that the fatigue resistance of metals is generally lower in atmospheric than in vacuum environment. In the case of aluminium alloys, this atmospheric effect has been attributed to water vapour adsorption on freshly cracked surfaces and subsequent hydrogen embrittlement (3,5-9) which is observed for a crack growth rate lower than a critical value, depending on the partial pressure of water vapour and on the test frequency. The analysis presented (10) has shown that the fatigue behaviour in vacuum depends on the fatigue mechanism itself which is characteristic of the alloy under given loading conditions, and underlined that changes in this mechanism are related to microstructure as shown in an Al-4 % Cu alloy (11).

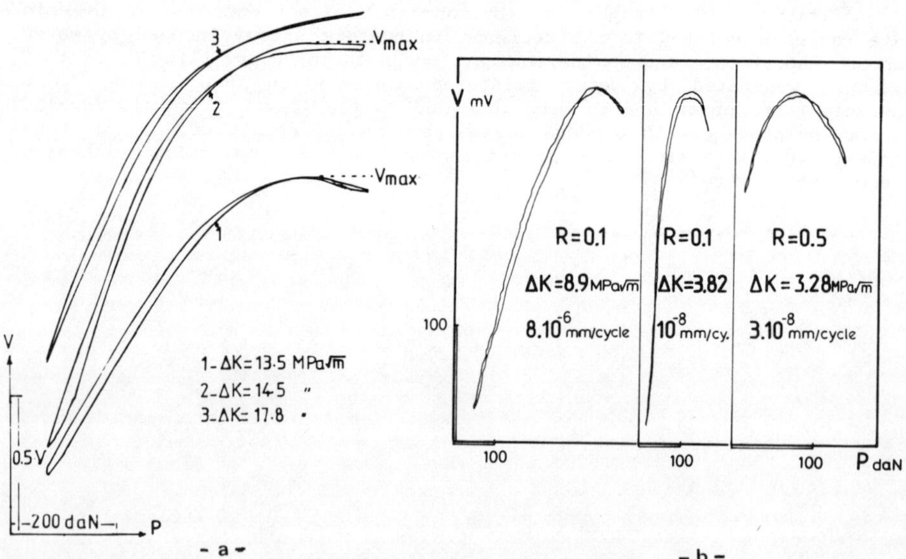

Figure 1 - Potential drop measurements in vacuum. Examples of diagrams obtained on the 2618 T651 alloy (3,4) :
a) Constant ΔP test in the mid ΔK range ; b) Threshold tests.

More recently, studies have been conducted in the low rate range and near-threshold level on high strength Al alloys (12,13) i.e. in conditions where environmental and microstructural influences are predominant. A major role of crack closure has been underlined (4) and threshold has been related to a change in the predominant mode of crack opening from mode I to a mixed mode as the stress intensity magnitude decreases (14). The effect of an aggressive environment has been shown to move the predominant mode transition down to lower values of ΔK (15). A large decrease observed in closure potential drop measurements in vacuum during the unloading cycle (4) has been related to the existence of discrete contact points between the two cracked surfaces when the crack closes (Figure 1a). At low F.C.G. rate an additional decrease in potential is observed near P_{max} (Figure 1b) which gives evidence of such contacts ; this phenomenon can be related to mode II displacements parallel to the crack plane which lead to contacts between the two cracked surfaces when the size of the surface irregularities becomes of the same order as the CTOD (Figure 2). In such conditions one can expect the existence of a second roughness-induced phenomenon (the first one being the "roughness-induced closure" (16)) reducing the effective stress intensity factor in the upper part of the cycle.

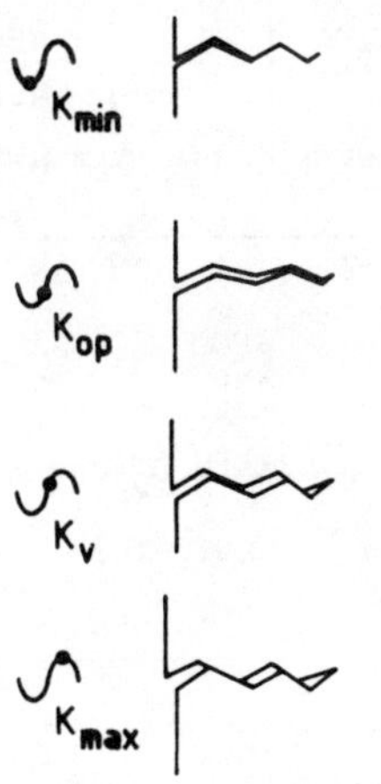

Figure 2 - Schematic illustration of the roughness induced effect near P_{max} induced by Mode II displacement in the low ΔK range.

In a first approximation the F.C.G. behaviour in vacuum has been described with a ΔK^*_{eff} concept which supposes an inhibition in loading effect at the crack tip above P_v as defined in figure 3a. A propagation law in $\Delta K^*_{eff})^4$ has been obtained which is consistent with the models based on damage accumulation (15, 17-19) (Figure 3b).

The influence of corrosion deposits on near threshold F.C.G. has been studied (20-22) in some Al alloys on the basis of the "oxide induced closure" model first proposed for steels (23) ; it has been shown that, at low load ratios, environmentally influenced F.C.G is controlled by three concurrent processes :

- a crack closure due to the wedge effect of corrosion deposits which tends to increase the threshold level ; such enlarged oxide formation has been reasoned as resulting from fretting contact induced by mode II rubbing observed at low stress intensities and load ratios (23).

- a strong environmental effect resulting from water vapour adsorption and which considerably accelerates near- threshold F.C.G. rate even in the presence of traces of water vapour in an inert gas (3) (Figure 4),

- a limitation in aggressive environment transport up to the crack tip due to the reduction in crack opening and to trapping of water vapour by oxide deposits, which tend to decrease the F.C.G. rate and to raise threshold levels (3, 20, 24) at low R ratios.

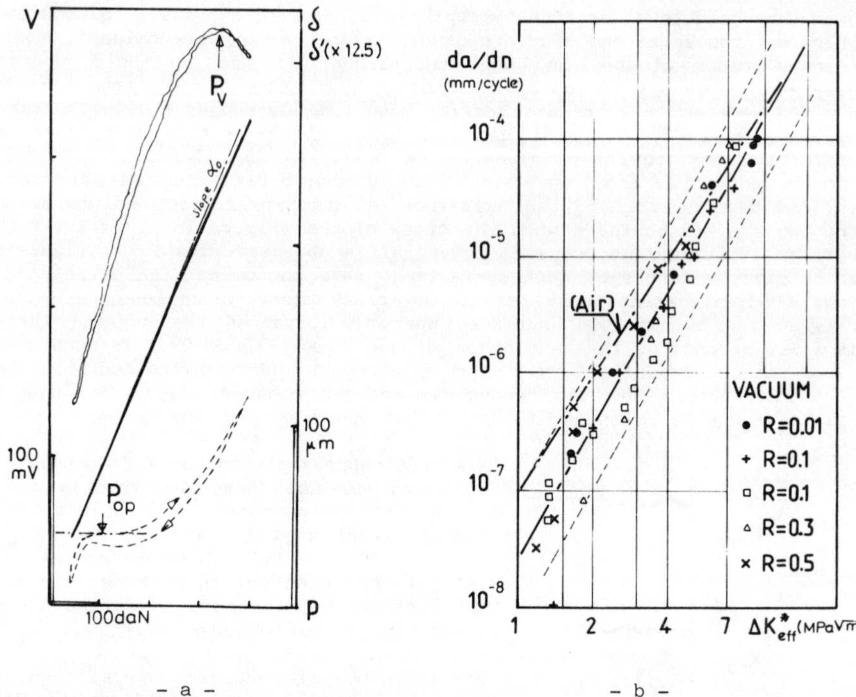

- a - - b -

Figure 3 - a) Definition of $\Delta K^*_{eff} = K_v - K_o$ (K_v corresponds to P_v and K_o to P_{op} as defined by Elber (4).
b) Example of diagram (da/dN) vs ΔK^*_{eff} on the 2618-T651 alloy. Test frequency = 40 Hz (4).

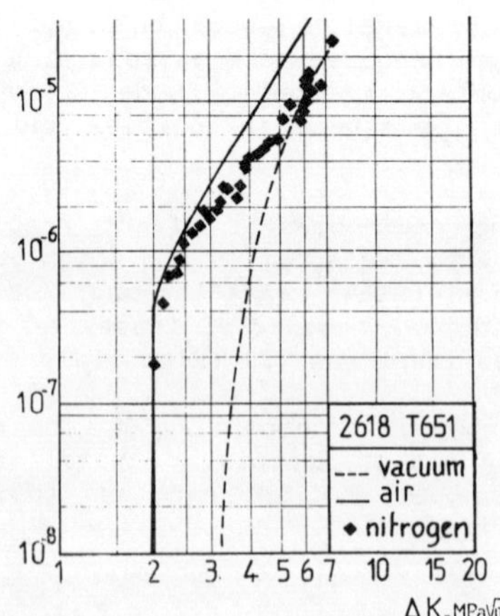

Figure 4 - Fatigue threshold data on the 2618 T651 alloy in industrial nitrogen (50 ppm $H_2O + O_2$) compared to air and vacuum data (3).
(R = 0.5 ; 40 Hz).

At high load ratios only the second of the above mentioned process occurs (Figure 4).

The influence of microstructure on F.C.G. has been studied essentially on Al-Zn-Mg-Cu alloys (20-22, 25-27). Similar observations have been made on a high purity alloy (28).

In vacuum, tests on a 7075 alloy in four aged conditions (20) have shown a higher threshold level and a higher F.C.G. resistance for structures including shearable precipitates as compared to overaged structures with non-shearable precipitates (Figure 5).

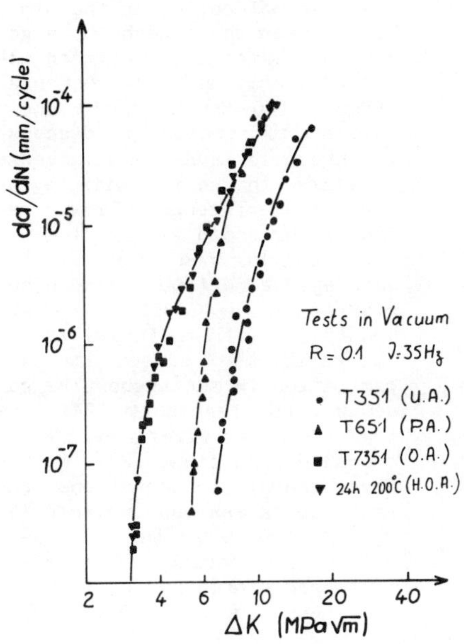

Figure 5 - Threshold data on the 7075 alloy in four aged conditions in vacuum (20).

Such differences have been related to the existence of two major crack growth mechanisms (20, 26, 28) :

- a crystallographic transgranular mode of failure in underaged alloys (typically in 7075-T351) containing shearable hardening GP zones of very small size (about 10 A diameter) ; in addition, Cu and Mg in solid solution lower S.F.E. and restrict cross-slip (26,30). In such cases the damage process results from the to and fro motion of dislocations (very easy in the absence of dispersoïd trapping points) which progressively induces dislocation jogging and cutting up. So a predominantly planar slip mode is observed leading to very slow F.C.G.

- a smooth mode of failure in overaged alloys containing incoherent precipitates (typically 7075-HOA); the large precipitates (about 200 A) which would resist shearing are sufficiently widely spaced for the dislocations to pass between them. In addition, lesser contents in Cu and Mg in solid solution lead to easier cross-slip (26). A faster damage accumulation localised around the precipitates is in agreement with a higher F.C.G. resulting from a predominantly wavy slip mode and with a flat growth path (20, 26).

A mixed behaviour is observed in the peak aged conditions (7075 T651) corresponding to GP zones and coherent dispersoïd plates (15x50 Å) (20). In the low rate range, i.e. where crack advances step by step and the strain is localised to one shear direction (30),the planar slip mode can occur, while in the mid-rate propagation range the wavy slip mechanism controlled by the dispersoïd precipitation is operative.

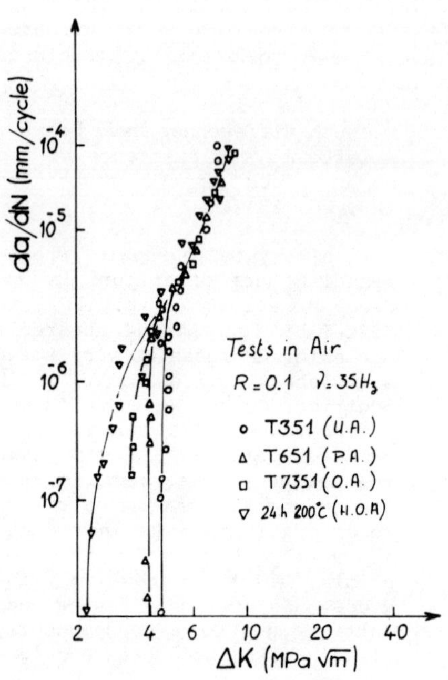

Figure 6 - Threshold data on the 7075 alloy in four aged conditions in air (20).

In air (Figure 6), in the mid rate range a similar behaviour has been observed for all the aged conditions (20) corresponding to a transgranular F.C.G. mechanism resulting from a gliding process occuring on several slip systems which are assumed to be activated by adsorption of reactive molecules. On the other hand, near the threshold range, a strong interaction between microstructural and environmental effects (20-22) has been shown to lead to substantial decrease in threshold level and increase in oxide thickening with aging (22). The existence at low R ratio of threshold values close to the ones obtained in vacuum for each aged conditions is more con sistent in the -T 351 and -T 651 conditions with an inhibition of the environment effect due to a limitation in the aggressive gaseous phase transport (24) instead of wedge effect as in the overaged conditions (22). Microfractographic observations and oxide thickness measurements (20 21, 22) support this analysis which corroborates the afore mentioned microstructural influence near threshold (20).

The experimental results presented below will be discussed on the basis of this review.

Experimental conditions

The chemical compositions of the alloys studied are given in table I and typical properties are listed in table II.

Table I - Chemical composition

Alloy	Si	Fe	Cu	Mn	Mg	Cr	Ni	Zn	Ti	Zr
2024	0.10	0.22	4.46	0.66	1.50	0.01	-	0.04	0.02	
2618	0.20	1.10	2.60	-	1.55	-	1.15	0.10	0.10	-
7020	0.35	0.40	0.20	0.25	0.90	0.35	-	4.40	0.03	0.13
7075	0.07	0.16	1.52	0.04	2.44	0.20	-	6.00	0.04	-

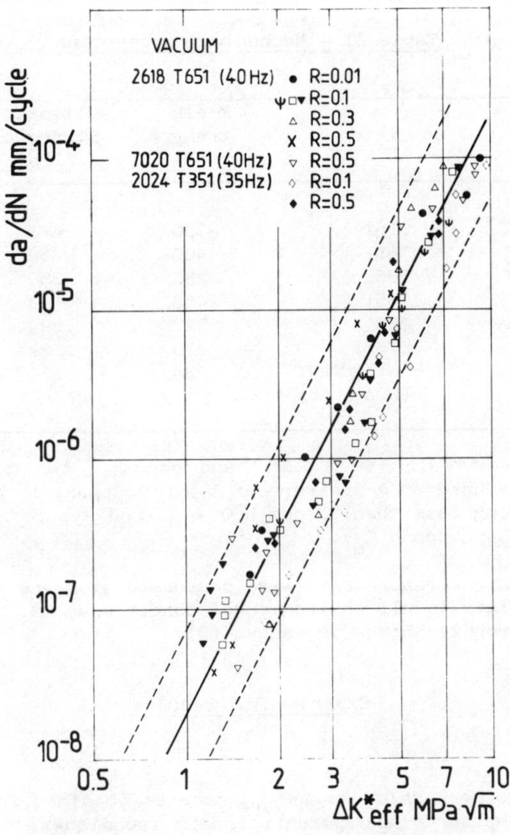

Figure 7 - da/dN vs ΔK^*_{eff} diagram for tests in vacuum for different aluminium alloys.

The specimens used were of 10 mm thick CT75 type, machined so that the stress axis upon testing would be parallel to rolling direction.

Tests were performed in an environmental chamber mounted on an electro-hydraulic machine and providing a vacuum better than 5.10^{-4} Pa (36). Crack growth was optically monitored with a travelling microscope, and F.C.G. measurements were made by using a shedding method.

Table II - Mechanical properties

Alloy	Condition	Yield Strength (MPa)	Ultimate Strength (MPa)	Elongation %
2024	T351	320	473	17.5
2618	T651	400	460	10.0
7020	T651	280	350	11.0
7075	T351	458	583	10.6
7075	T651	527	590	11.0
7075	T7351	470	539	11.7
7075	24 h at 200 °C (HOA)	234	338	14.4

In addition to air (50 % R.H.) and vacuum, two controlled gaseous environments were used at a pressure of 1.2 atmosphere : an industrial dry nitrogen containing less than 50 ppm $H_2O + O_2$ and a pure nitrogen (< 3 ppm H_2O, < 1 ppm O_2, < 1 ppm H_nC_m).

Crack closure measurements were performed by means of a differential compliance technique in all environmental conditions, (3, 32), complemented by a potential drop technique in vacuum (3).

Experimental results

Tests in vacuum

Figure 7 gathers da/dN vs ΔK^*_{eff} results obtained from crack closure measurements (compliance and potential drop techniques) on the 2024-T351, 7020-T651 and 2618-T651 alloys for different R ratios.In these three alloys F.C.G. in vacuum is essentially controlled by non-shearable dispersoïd precipitates over the entire range of growth rates. It might be possible to detect some slight differences from one alloy to another which could lead to several closely parallel data, but all the measurements show that whatever be the alloy and R ratio, these data can be considered to be based on a same growth process which can be described by a law of the form $(\Delta K^*_{eff})^4$.

Figure 8 shows results obtained on the 7075 alloy in four aged conditions (corresponding F.C.G. tests are presented in Figure 5a). These data correspond to two typical behviour of the form $(\Delta K^*_{eff})^4$:

- a first one observed on the 7075 HOA (R = 0.1 and 0.5) which is in good agreement with the data presented in figure 7 corresponding to a F.C.G. controlled by the micro-precipitation with a very flat growth path (figure 9a).

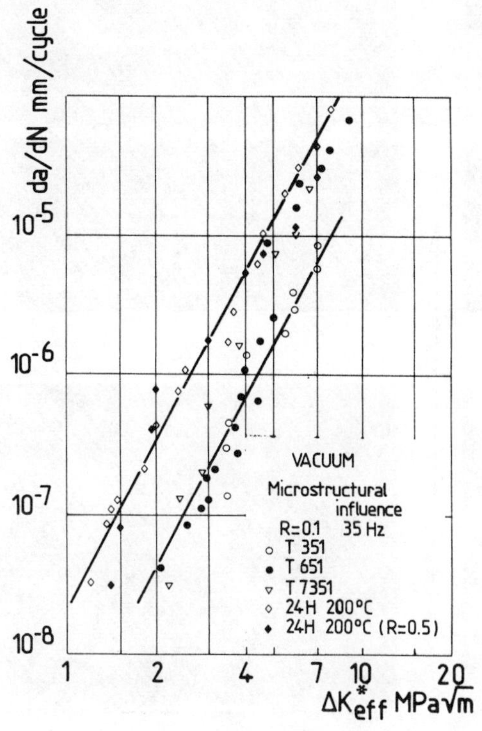

Figure 8 - da/dN vs ΔK^*_{eff} diagram for tests in vacuum at 35 Hz on the 7075 alloy in four aged conditions.

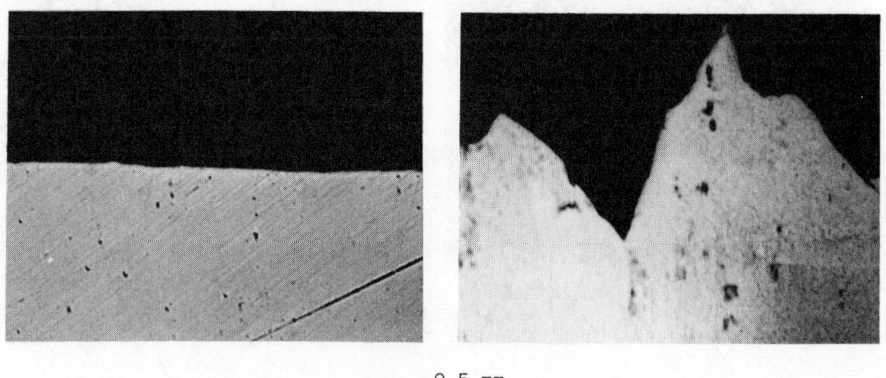

0.5 mm

a) 7075 24 h 200 °C in air and vacuum.

b) 7075-T351 in vacuum.

Figure 9 - Cross section of the crack growth path near threshold.

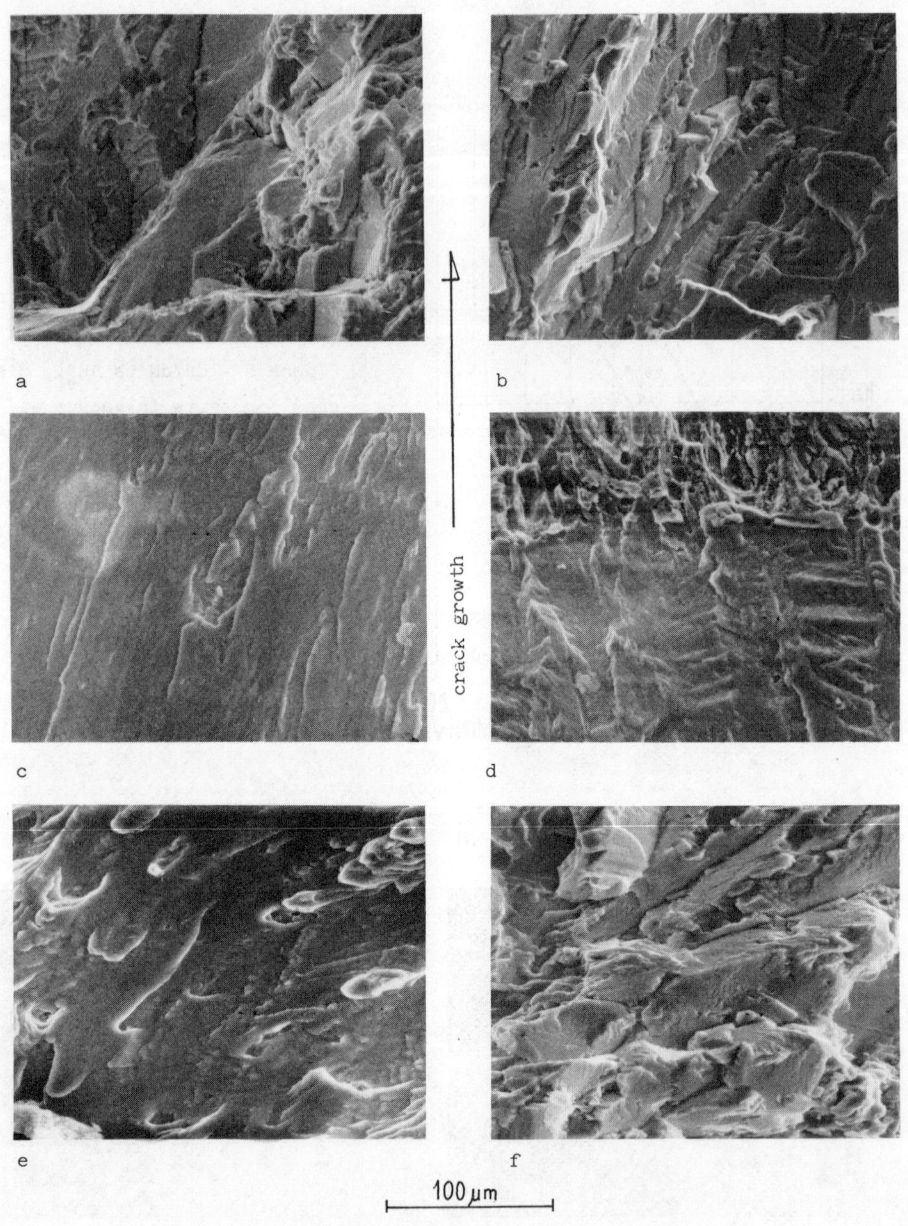

Figure 10 - Microfractographic aspect of fracture surfaces on the 7075 alloy in vacuum (35 Hz, R = 0.1)

10a - T351, 6×10^{-5} mm/cycle   10b - T351, 6×10^{-8} mm/cycle

10c - T7351, 5×10^{-5} mm/cycle   10d - T7351 at threshold

10e - T651, 3×10^{-6} mm/cycle   10f - T651, 2×10^{-7} mm/cycle

- a second one observed on the 7075-T351 which corresponds to a very **crystallographic growth** path (figure 9b) involving a planar slip mode and leading to slower crack growth and higher threshold level (20, 33).

The results obtained on the 7075 alloy show that in the range 10^{-5} to 10^{-6} mm/cycle the change from the first crack growth data in the mid rate range to the second data at low growth rates occurs (figure 5). Early beginning of such a change can be detected close to threshold on the 7075 T7351 alloy which correspond to the observation in some localized grains of a crystallographic mode of failure (figure 10b).

Tests in nitrogen

Figure 11 presents the diagram da/dN versus ΔK determined on the 7075 T351 and 7075 T7351 alloys tested in the industrial nitrogen (50 ppm H_2O + O_2) in comparison with previous tests in air and vacuum (figure 5) performed at 35 Hz and R = 0.1.

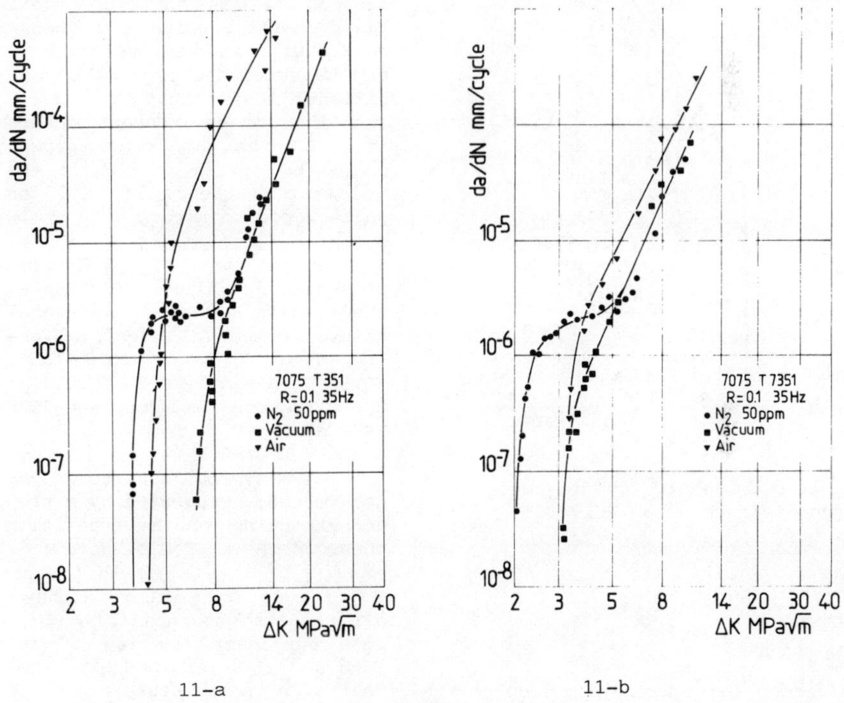

11-a 11-b

Figure 11 Threshold data in industrial nitrogen on the 7075 alloys in two aged conditions.

The crack growth behaviour is similar to the one observed at 40 Hz on the 2618-T651 alloy (figure 4) and on the 7020-T651 alloy (3) and consists of three typical phases as decribed previously (3) :

- in the mid-rate range, a behaviour close to the one in vacuum ;

- in the range 2×10^{-6} to 6×10^{-6} mm/cycle a transition where F.C.G. rate is independant of ΔK ;

- near the threshold, a substantially different behaviour with the fastest F.C.G. and the lowest threshold level.

Figure 12 presents a complementary test performed on the 2618 T651 alloy in the pure nitrogen environment (< 3 ppm H_2O,<1 ppm O_2,<1 ppm $C_n H_m$).

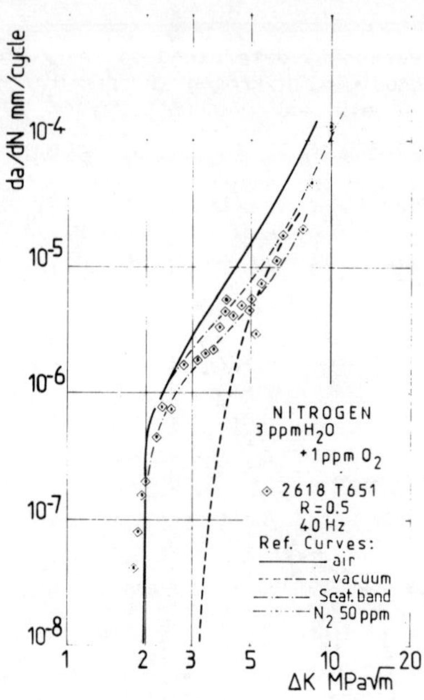

Figure 12 - Threshold test in pure nitrogen environment on 2618 T651 alloy.

The same growth behaviour as in the industrial nitrogen is observed. To eliminate a possible effect of outgassing pollution during the test in pure nitrogen, a third test was performed with a continuous admission of gas through a cell fixed on the specimen. No change in F.C.G. data was observed. It can be concluded from these additional tests that the transition due to environment occurs at about the same crack growth range for given load ratio (R) and test frequency (f) and for water vapour contents < 50 ppm.

da/dN vs ΔK_{eff} diagram presented in figure 13 compares the results obtained on several alloys tested in industrial nitrogen at 40 Hz : 2618-T651, 7075-T351 and 7075-T7351 at R = 0.1, and 7020-T651 and 7075-T351 at R = 0.5

Two typical regimes can be determined, separated by a step corresponding to the transition observed in the F.C.G. data :

- in the mid rate range all results obtained fit with a good agreement the law of the form ΔK^4_{eff} determined here above in vacuum and taking into account the effect of microstructure except on the 2618-T651 alloy tested in nitrogen at 1 Hz and R = 0.5 which corresponds to a line with a slope m = 4 slighty shifted up to the higher rates.

In the low rate range down to threshold the F.C.G. data can be described by a law of the form $\Delta K^{2.8}_{eff}$ for the studied alloys.

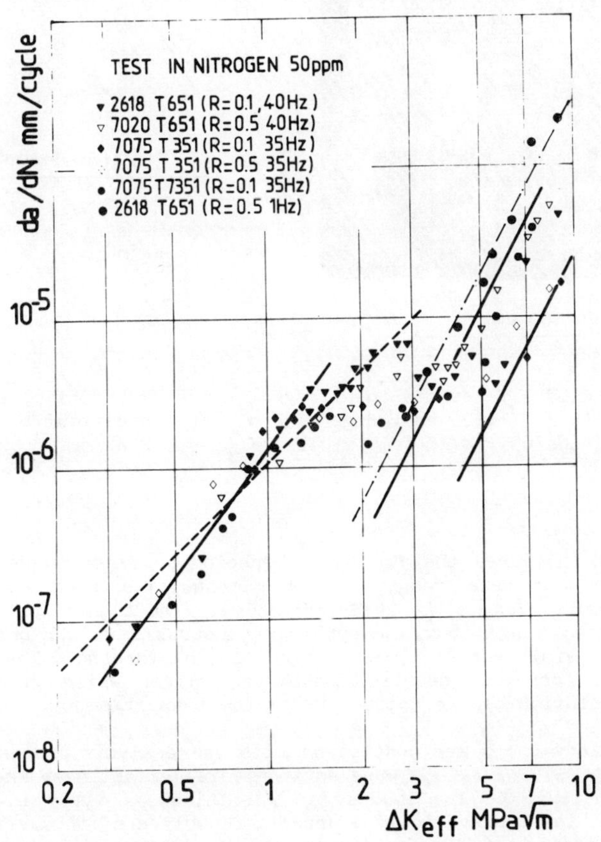

Figure 13 - da/dN vs ΔK_{eff} for tests in nitrogen

The straight full lines corresponding to ΔK^4_{eff} plotted in the mid-rate range are reported from figure 8. The dotted line corresponds to the reference slope m = 2 (CTOD model).

Microfractographic observations show that the transition corresponds to a change from a crystallographic sharply-angled growth path to a flatter and faceted fracture surface (Figure 14).

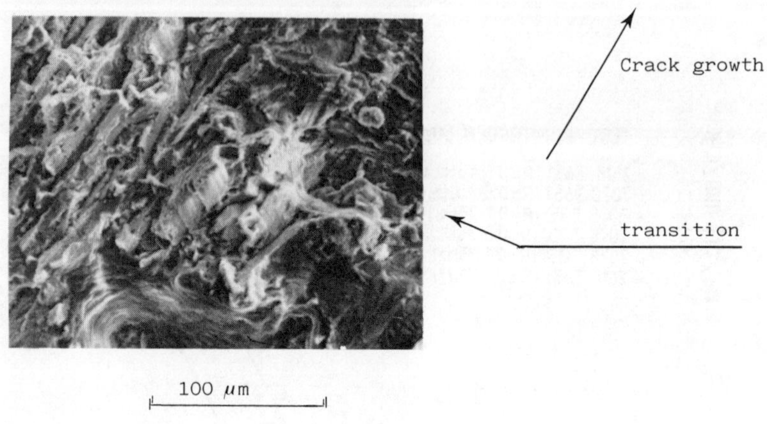

Figure 14 - Threshold in nitrogen on the 7075-T351 alloy ($R = 0.1$, 35 Hz). Beginning of the transition ($da/dN = 4.10^{-6}$ mm/cycle, $\Delta K_{eff} = 6$ MPa m).

Figure 15 compares the fracture morphologies in nitrogen, air and vacuum in the mid-rate range, in the transition zone and near the threshold, as observed on the 2618-T651 alloy. The transition in nitrogen corresponds to a change from a morphology comparable to the one in vacuum to a morphology similar to the one in air. But at threshold level, a thick oxide layer of corrosion debris is observed in air while in nitrogen no significant evolution can be noticed below the transition.

These observations are consistent with the analysis proposed for the 2618-T651 alloy (3), i.e. : a pure environmental effect near the threshold in nitrogen whatever be the load ratio which leads to very low threshold levels and, at low load ratio,to a drastic reduction of this effect in air due to the presence of the oxide layer which considerably reduces the access to the crack tip for the active gas by geometrical lowering of the rms molecular free path and trapping of water vapour molecules.

Tests in air

Figure 16 shows the da/dN vs ΔK_{eff} diagram determined in atmospheric environment (about 50 % R.H.) on different Al alloys. It is interesting to notice the following remark : independantly of R, all the results fit a straight line with a slope $m = 3.3$ with an acceptable scatter, except in the very low F.C.G. rate range where the deflexion from the mean line could be related to the occurence of mode II opening. These results are in accordance with previous observations made on the 2618 alloy alone (3).

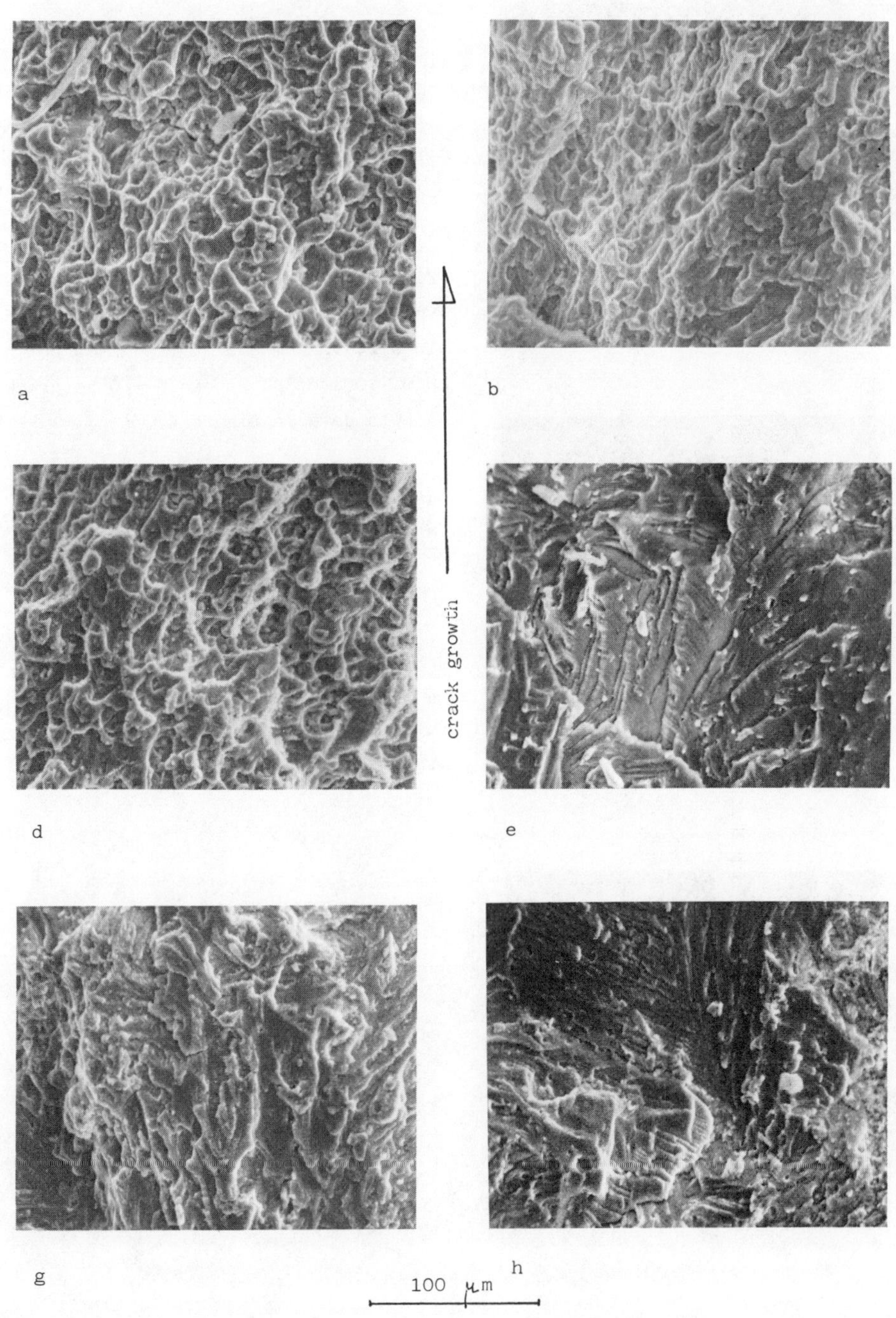
a
b
d
e
g
h
crack growth
100 μm

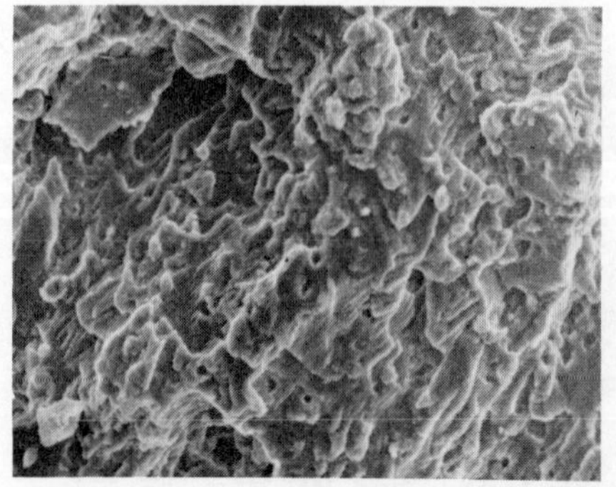

c

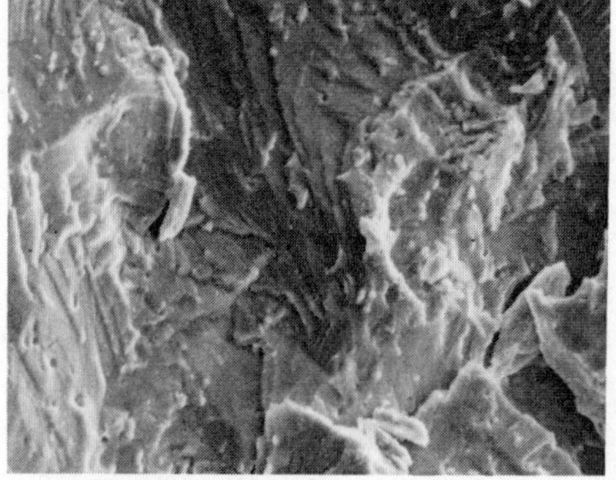

f

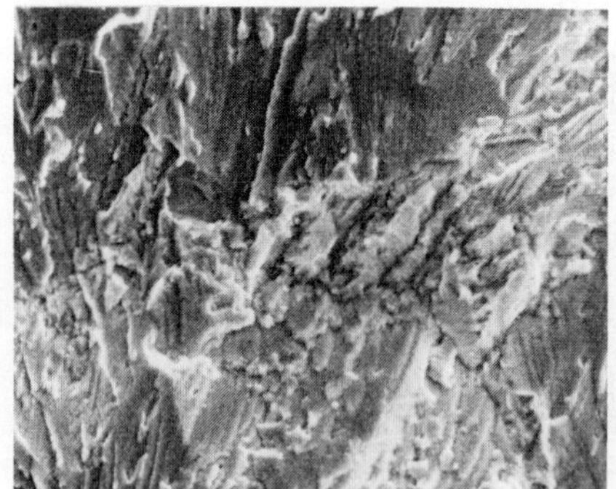

i

Figure 15 - Microfractographic observations on the 2618-T651 alloy tested in air, vacuum and nitrogen (50 ppm $H_2O + O_2$) at 40 Hz and R = 0.5 (crack growth data in figure 4).

da/dN mm/cycle	3×10^{-5}	10^{-6}	4×10^{-7}
vacuum	a	b	c
nitrogen	d	e	f
air	g	h	i

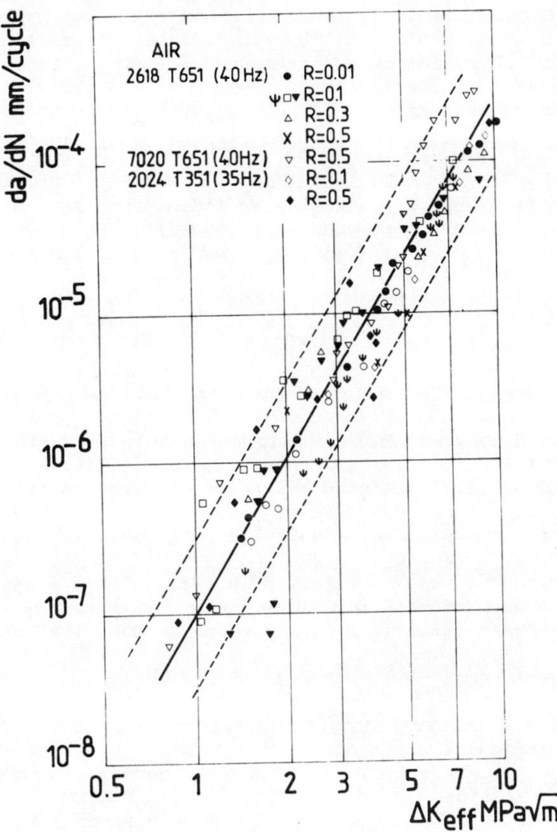

Figure 16 - da/dN vs ΔK_{eff} data in air for different alloys.

Discussion

Tests in vacuum

In a review of the crack growth models (27) it has been shown that most reliable data taken in an inert environment or in vacuum in the mid-rate range give laws of the form $\Delta K^{4 \pm 0.5}$.

A simple relationship for F.C.G. rate as reported in (21-24) is of the form :

$$da/dN = A \, \Delta K^4 / \mu \sigma^2 U \tag{1}$$

(A dimensionless constant, μ shear modulus, σ strength parameter, U plastic work required for the creation of a unit area for fatigue crack).

The analysis in ΔK^*_{eff} in vacuum shows that laws of the form $(\Delta K^*_{eff})^4$ can describe the propagation down to threshold by taking into account an effect of mode II opening.

As it has been postulated (20), the low ΔK region exhibits a mixed mode of crack opening (mode I + mode II) and then the crack being driven forward mainly in mode I opening, the F.C.G. rate decreases and the ΔK^4 law is not verified. The correction introduced by ΔK^*_{eff} consists in a first approach to eliminate the effect of mode II on F.C.G. rate.

So one can expect that :

$$U = U_I + U_{II} \qquad (2)$$

where U_I = plastic work corresponding to mode I opening

U_{II} = plastic work corresponding to mode II displacements

U_I would correspond to γ , energy for creation of a unit area of new free surface.

Then, the relation between da/dN and ΔK^*_{eff} can be written as :

$$da/dN = A\ (\Delta K^*_{eff})^4/\mu\sigma^2\gamma \qquad (3)$$

The tests on the 7075 alloy above presented the precise influence of microstructure.

Conditions T351 and T7351 lead to comparable strength resistance but to very different F.C.G. data and threshold levels, while overaged conditions (T7351 and H.O.A.) present significant differences in strength resistance but very close threshold levels. So, change in F.C.G. and threshold with aging cannot be explained by variations of σ alone.

The experimental results reported in figure 7 and 8 suggest the existence of two typical γ levels depending upon microstructure and corresponding to the two characteristic F.C.G. mechanisms described above.

Tests in nitrogen

The results obtained indicate the existence of two F.C.G. mechanims separated by a typical transition.

- in the mid-rate range, the F.C.G. behaviour can be taken into account by the relation (4) with $\Delta K^*_{eff} = \Delta K_{eff}$ (mode II opening effect shifted towards low ΔK (20). The influence of environment can be analysed as a lowering in γ due to water vapour adsorption (Rehbinder effect (38)).

The experimental data show that at 40 Hz γ (N_2) is lower than $\gamma(v)$ (figure 13). Such a decrease in γ has been shown to be dependant upon test frequency i.e. upon time. Hence, this behaviour suggests a dependance upon the transport mechanism of water vapour molecules to the crack tip. A similar mechanism has been analysed by WEI et al (34) who propose that the crack growth in an agressive environment can be considered as the sum of two components :

$$(da/dN)_e = (da/dN)_v + \theta\,(da/dN)_{e,s} \qquad (4)$$

with suffixes e = environment, v = vacuum, e,s = saturated environment, and θ = fractional physical surface coverage.

To make $\theta = 1$ for $(da/dN)_e = (da/dN)_{e,s}$, (5)

the relation (4) can be written as :

$$(da/dN)_e = (da/dN)_v + \theta\,[(da/dN)_{e,s} - (da/dN)_v] \qquad (6)$$

From (3) and (6) θ is given as :

$$\theta = \frac{1/ (N_2) - 1/ (V)}{1/^{\gamma} (e,s)-1/^{\gamma}(v)} \qquad (7)$$

Considering a single loading cycle WEI et al (34) proposed for θ the following expression :

$$\theta = F\ P_o/SN_o\ RT\ 2f \qquad (8)$$

F = Knudsen flow parameter
S = Physical surface
N_o = Density of surface sites
R = Gas constant
T = Temperature
P_o = Surrounding partial pressure of water vapour
f = Test frequency

From the experimental results, the relations (7) and (8) show that at a low P_o, one gets $\theta \simeq 0$ at 40 Hz while a change in $\gamma\ (N_2)$ as compared to $\gamma(v)$ observed at 1 Hz, corresponds to $0 < \theta < 1$.

- In the low rate range, the present study suggests that an "embrittling" process in an inert environment containing traces of water vapour is observed in the very low rate range under a critical rate which is about 5×10^{-6} mm/cycle for water vapour contents lower than 50 ppm.

In a recent accurate microfractographic study LANKFORD and DAVIDSON (37) show that within the threshold to a Paris law transition region, crack extension does not occur for many consecutive cycles and they introduce a concept of intermittent crack growth .These results are in accordance with previous observations performed in the near-threshold region for 7075 T651 (13, 20).

In addition, these authors (37) consider that an "embrittling" effect of hydrogen can introduce stress intensity/environmental conditions under which crack growth rate can be CTOD controlled in the mid-rate range ($\Delta K > 8$ MPa $\sqrt{m}$).

ACHTER (5) and BOWLES (6) have developped a model based on hydrogen diffusion and embrittlement ; they determined a critical F.C.G. rate above which no effect can be observed :

$$(da/dN)_{cr} = 3\times6\ 10^{-6} \frac{P}{f(MT)^{1/2}} \qquad (9)$$

(M atomic mass of the gaseous phase, P = partial pressure at the crack tip)

Using $P = P_o$ give a maximum for $(da/dN)_{cr}$: at 40 Hz one gets 4.10^{-6} mm/cycle in the industrial nitrogen and 5×10^{-7} mm/cycle in the pure nitrogen. The effective pressure at the crack tip being much lower than the surrounding pressure P, $(da/dN)_{cr}$ should be lower than the rate of the observed transition.

But these authors (5,6) have considered a continuous advance of the crack instead of an intermittent crack growth. In the latter conditions the time for hydrogen diffusion is the time where the crack is temporarily stationary and the concept of a critical rate loses its significance in the low rate range.The test performed in the pure nitrogen (3 ppm H_2O) support this analysis. Hence, the experimental results presented here above in the low rate range can be considered to be consistent with an intermittent crack growth allowing hydrogen assisted fatigue, and the apparent ΔK exponent higher than the theoritical m = 2 (corresponding to the dotted line in figure 13) can result from an increase in the stationary number of cycles with decreasing F.C.G. rates.

Tests in air

In the case of air environment, water vapour and oxygen contents are respectively at about 0.75 % and 20 % instead of ≃ 25 ppm for both of them in the industrial nitrogen, i.e. 300 times more of water vapour and 8000 times more of oxygen. In such conditions one can expect the existence of an environmental effect through the studied rate range but also a competition between water vapour and oxygen adsorption processes, the high reactivity of water vapour as compared to oxygen (33) being partly offset by the higher partial pressure of oxygen.

So, the observed F.C.G. data in $\Delta K_{eff}^{3.3}$ can be interpreted as a medium behaviour between the two typical ones described above.

Conclusions

1 Crack growth in vacuum can be described by a law of the form

$$\frac{da}{dN} = A \frac{(\Delta K^*_{eff})^4}{\mu \sigma^2 \gamma}$$

where ΔK^*_{eff} is a new concept for the effective stress intensity range determined from crack closure through potential drop measurements.

2 Microstructural influence on F.C.G. rate in vacuum corresponds to change in the energy γ for creation of a unit area of new free surface.

3 In a gaseous atmosphere with traces of water vapour and oxygen:

- for F.C.G. rate higher than a critical value depending upon test frequency and partial pressure of water vapour, the F.C.G. rate increase observed in comparison with F.C.G. in vacuum is associated to a lowering in γ depending upon the fractional physical surface coverage θ, and hence depends upon R ratio and test frequency ;

- for F.C.G. rate lower than the critical value or when crack growth occurs by a step by step advance process (da/dN $< 10^{-5}$ mm/cycle), hydrogen diffusion and embrittlement at the crack tip subsequent to dissociation of adsorbed water molecules, accelerate the propagation and lower the threshold level. Crack growth rate could be CTOD controlled.

4 In ambient air, the observed $(\Delta K_{eff})^{3.3}$ propagation law could be related to a competition between water vapour and oxygen adsorption process. At low R ratio the oxide formed on cracked surfaces acts essentially as a trap of water vapour and hence reduces drastically the environmental effect. In some overaged conditions which enhance oxide thickening, the threshold can result from crack wedging (7075-T7351).

ACKNOWLEDGEMENT

The author is grateful to Professor J. de Fouquet for the fruitful discussion and critical comments on this paper.

References

1 - C.M. HUDSON, "Problems of fatigue of metal in a vacuum environment" NASA TN D-2563 (1965).

2 - C.M. HUDSON and S.K. SEWARD, "A litterature Review and Inventory of the Effects of Environment on the Fatigue behavior of Metals", Engng Fract. Mech., 8 (1976) 315-322.

3 - J. PETIT and A. ZEGHLOUL, "Gaseous Environmental Effect on Threshold Level in High Strength Aluminium Alloys", Proc. ISFT Fatigue Treshold, Stockholm, J. Bäcklund, A. Blom and C.J. Beevers eds, EMAS, Warley (U.K.), A (1981) 563-579.

4 - J. PETIT, N. RANGANATHAN and J. de FOUQUET, "Vacuum effect on fatigue crack propagation at low rate", Proc. E.C.F. 3 "Fracture and Fatigue", ed. J.C. Radon, Pergamm Press, London (1980) 329-337.

5 - M.R. ACHTER, "The adsorption model for environmental effects in fatigue crack growth", Scripta Met. 2 (1968) 525-528.

6 - C.Q. BOWLES, "The role of environment, frequency and wave shape during fatigue crack growth in aluminium alloys", Report L.R. 270, Delft University of Technology, The Netherlands (1978).

7 - D.G. CHAKRAPANI and E.N. PUGH, "Hydrogen Embrittlement in a Mg-Al alloy", Met. Trans. 7A (1976) 173-178.

8 - R.J. JACKO and D.J. DUQUETTE, "Hydrogen Embrittlement of a Cyclically Deformed High Strength Al Alloy", Met. Trans.8A (1977) 1821-1827

9 - J. ALBRECH, B.J. McTIERNAN, I.M. BERNSTEIN and A.W. THOMPSON, "Hydrogen Embrittlement in a High Strength Al alloy" Scripta Met. 11, (1977), 893-897.

10 - B.I. VERKIN and N.M. GRINBERG, "The Effect of Vacuum on the Fatigue Behaviour of Metals and Alloys", Mat. Sci and Engng, 41 (1979) 149.

11 - B. BOUCHET, "Influence de l'environnement sur le comportement en Fatigue d'un alliage léger type AU4", Doc. es Sc. Thesis (1976) Poitiers, France.

12 - J. PETIT, B. BOUCHET, C. GASC, J. DE FOUQUET, "A contribution to the study of the influence of environment on the crack growth rate of high strength aluminium alloys in fatigue", Proc. ICF4 "Fracture, 1977" University of Waterloo, Press Canada M 201 (1977) 867-871.

13 - R.B. KIRBY and C.J. BEEVERS, "Slow fatigue crack growth and threshold behaviour in air and vacuum of commercial Aluminium Alloys", Fat. Engng and struct.1 (1979) 203-215.

14 - D.L. DAVIDSON and J. LANKFORD, "Fatigue Crack Tip Plastic Strain in High Strength Aluminium Alloys", Fat. Engng Mat. and Struct. 3 (1980) 289-303.

15 - D.L. DAVIDSON, "Incorporationg threshold and environmental effects into the damage accumulation model for fatigue crack growth, Fat. Engng Mat. and Struct. 3 (1980) 229-236.

16 - R.O. RITCHIE and S. SURESH, "Some considerations on Fatigue Crack Closure at Near-Threshold Stress Intensities due to Fracutre Surface Morphology, Metalurgical Transactions A, 13A (1982) 937-940.

17 - J. WEERTMAN, "Theory of fatigue crack growth based on a BCS theory work hardening", Int. J.Fracture Mech. 9 (1973) 125.

18 - T. MURA and C.T. LIN, "Theory of fatigue crack growth for work hardening materials", Int. J. Fracture Mech. 10 (1974) 284.

19 - J.R. RICE, "Fatigue and Propagation", ASTM STP 415 (1967) 247.

20 - J. PETIT, P. RENAUD and P. VIOLAN, "Effect of microstructure on crack growth in a high strength aluminium alloys", Proc. ECF4, Leoben,Austria, K.L. Maurer, F.E. Matzer eds, EMAS, Warley (U.K.) 2 (1982) 426-433.

21 - P. RENAUD, P. VIOLAN, J. PETIT and D. FERTON, "Microstructural Influence on Fatigue Crack Growth Near Threshold in 7075 Alloy", Scripta Met. 16 (1982) 1311-1316.

22 - A.K. VASUDEVAN and S. SURESH, "Influence of Corrosion Deposits on Near Threshold Fatigue Crack Growth Behavior in 2XXX and 7XXX Series Aluminium Alloys", Met. Trans. 13 A (1982) 2271--2279.

23 - S. SURESH, G.F. ZAMISKI and R.O. RITCHIE, "Oxide-Induced Crack Closure An Explanation for Near-Threshold Corrosion fatigue Crack Grouth Behavior", Met. Trans. 12 A (1981) 1435-1443.

24 - R.P. WEI, P.S. PAO, R.G. HART, T.W. WEIR and G.W. SIMMONS, "Fracture Mecahnics and Surface Chemistry Studies of Fatigue Crack Growth in Aluminium Alloys", Met.Trans. 11A (1980) 151-158.

25 - G.R. YODER, L.A. COLLEY and T.W. CROOKER, "On microstructural control of near threshold fatigue crack growth in 7000 series Aluminium Alloys", Scripta Met. 16 (1982) 1021.

26 - R.J. SELINES, "The fatigue behavior of high strength aluminium alloys" Doc. of. Sc. M.I.T. (1971).

27 - P.E. IRWING and L.N. Mc CARTNEY, Metal Sci., 11 (1977) 351-361.

28 - J. LINDIGKEIT, A. GYSLER and G. LUTJERING, "The effect of microstructure on the fatigue crack propagation behavior of an Al-Zr-Mg-Cu alloy Met. Trans. 12 A (1981) 1613-1619.

29 - G. THOMAS and J. NUTTING, "The plastic deformation of aluminium and aluminium alloys", J. Inst. of Metals 85 (1956) 1.

30 - B. TOMKINS, Metal. Sc. 13 (1979) 387.

31 - J. PETIT, A. NADEAU, M.C. LAFARIE, N. RANGANATHAN, "Realisation d'un caisson hermétique pour essais dynamiques", Rev. Phys. Appl. 15 (1980) 919-923.

32 - A. OHTA and E. SASAKI, "A method for determining the stress intensity threshold level for fatigue crack propagation", Int. J. of Fracture,11 (1975) 1040-1050.

33 - J. OUDAR, "La Chimie des Surfaces", P.U.F. ed., Vendôme, France (1973)

34 - R.P. WEI and G.W. SIMMONS, Recent progress in understanding environment assisted fatigue crack growth, Int. Journ. of Fracture, 17 (1981) 235-247.

35 - R. M. PELLOUX, Trans. of ASM, 62 (1969) 281-285.

36 - F.A. McCLINTOCK, "Fatigue and Propagation", ASTM STP 415 (1967) 169.

37 - J. LANKFORD and D.L. DAVIDSON, "Fatigue crack micromechanisms in ingot mowder metallurgy 7XXX aluminium alloys in air and vacuum", Acta Met. 8 (1983) 1273-1284.

38 - A.R.C. WESTWOOD, Adsorption-sensitive flow and fracture of solids, 7ème Réunion Internationale de la Société de Chimie Physique, Paris 19-23 Sept. 1983 - Elsevier ed. Proceedings in press.

NEAR-THRESHOLD FATIGUE CRACK GROWTH BEHAVIOR OF 7XXX AND 2XXX ALLOYS: A BRIEF REVIEW

A. K. Vasudévan, P. E. Bretz
ALUMINUM COMPANY OF AMERICA
Alcoa Laboratories
Alcoa Center, PA 15069
USA

Summary of near-threshold fatigue crack growth characteristics of 7XXX and 2XXX alloys is presented in terms of aging, composition and fabricating practice. In general, the variations in near-threshold fatigue crack growth behavior can be attributed to the synergistic interaction of microstructure and environment that result in several competing mechanisms involving moisture-induced embrittlement, deformation mode, crack closure and crack-branching processes. Dominance of these processes strongly depend on composition, aging and fabrication treatment.

Literature survey indicates that large body of information on the near-threshold fatigue crack growth (FCG) behavior is available on Fe-base and Ti-base alloys. However, compared to these alloys, not much work is known in aluminum alloys despite its extensive use in fatigue-critical applications. Few studies have emerged in the last several years concerning the near-threshold crack growth behavior in aluminum alloys.

It has been observed that both microstructure and environment can play an important role on the behavior of fatigue cracks at near-threshold stress intensity range (ΔK_o). However, the nature of the underlying interactions between microstructure and environment is not clearly understood. Wei and co-workers (1, 2) have indicated that FCG in aluminum alloys can be explained based on hydrogen embrittlement phenomenon. Schmidt and Paris (3) have interpreted their 2024-T351 results mainly in terms of plasticity-induced closure concepts. In addition, mode II shear displacements are observed to be important at low stress intensities and load ratios (4). Recently, Vasudevan and Suresh (5) have put forth a mechanistic argument that the near-threshold FCG rates are influenced by mutually competitive processes involving moisture-induced embrittlement, crack closure and mode of slip. This has been further extended, more recently, to include crack deflection by Suresh, Vasudevan and Bretz (6), as an additional mechanism contributing to near-threshold fatigue resistance.

The present paper describes the importance of the microstructural and environmental interactions that can strongly contribute to the near-threshold FCG resistance in aluminum alloys. In particular, the dominant role of aging, composition and fabrication process in 7XXX and 2XXX alloys on the ΔK_o will be described. For the sake of brevity, most of the experimental data in the text will be presented in terms of the threshold stress intensity ΔK_o dependence on the load ratio (R), for environment such as moist air.

Materials

Large body of the data presented in this paper has been taken from the published literature. Most of the data represent 7XXX and 2XXX alloys. We have recently published aging effects in 7075 alloys (5, 6), where measurements of fracture surface oxide thicknesses and CTOD are given. In order to enable us to interpret the data in a coherent package, we present some newer information pertaining to: (a) effect of test frequency on 7075-T7351, (b) effect of copper content on 7X75-T7 alloys (7), (c) effect of grain size in P/M 2024-T351 and 7091-T7E69 alloys (8, 9), and (d) low density, high modulus alloys containing Li (10, 11).

Phenomenalogical Observations

Figure 1 represents the fatigue crack growth behavior in terms of (da/dN) vs. ΔK for various commercial aluminum alloys tested in moist air at a load ratio of $R \sim 0.33$. The 1100-0 alloy data, taken from McKittrick, et al. (12), was tested at $R \sim 0.05$ in laboratory air environment. The range of yield stress varied from $\sim$ 30 MPa (1100-0 alloy) to $\sim$ 500 MPa (7091 P/M alloy). The contributing strengthening mechanisms in these alloys differ significantly: subgrain strengthening in 1100; solid solution strengthening in 5182; precipitation strengthening in 7050, 7091, 2124 and 6061 alloys. Thus, these alloys represent the response of various substructural differences to the fatigue crack growth behavior. The effect of these all alloys in Figure 1 can be clearly seen to markedly affect the crack growth rates below about 10^{-8} m/cycle, compared to higher growth rates ($\gtrsim 10^{-8}$ m/cycles). The threshold ΔK_o values ranged from about 1.2 MPa$\sqrt{m}$ to $\sim$ 3.5 MPa$\sqrt{m}$.

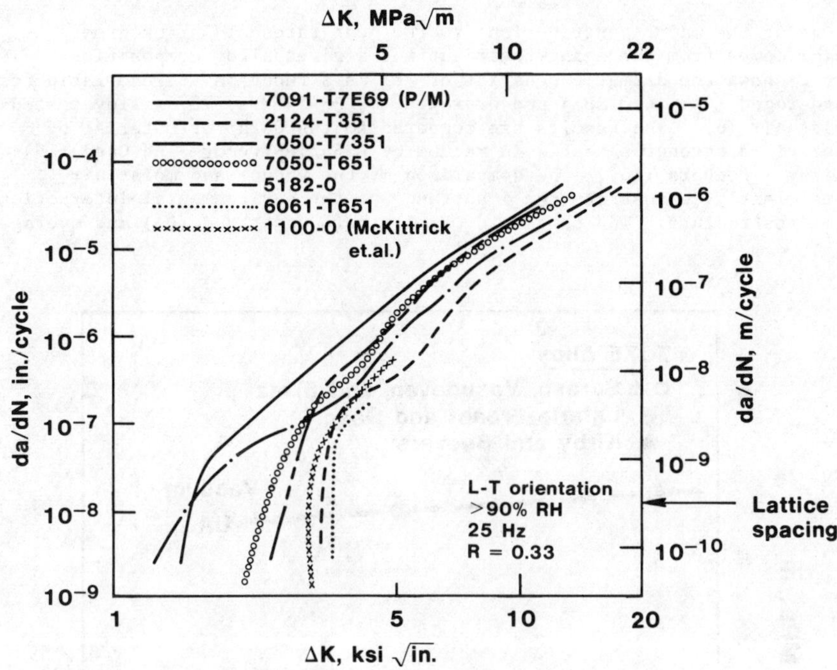

Figure 1 - Fatigue crack growth behavior of several aluminum alloys at R $\sim$ 0.33, 25 Hz and > 90% RH air. 1100-0 alloy taken from McKittrick, et al. (12), was tested at R $\sim$ 0.05 in laboratory air.

One can broadly categorize that composition and microstructure have major effects on the magnitude of threshold ΔK_o. For a given alloy like 7050, the precipitate morphology in the peak-aged temper T651 has a higher ΔK_o value compared to that developed in the overaged T7351 temper which has a higher volume fraction of incoherent precipitates. In the peak aged condition, compositional differences such as 7050 vs. 6061 show strong effects on ΔK_o. Also, fabricating process, plate I/M vs. extruded P/M alloy (7050 vs. 7091) can alter the ΔK_o levels.

Minor influences on threshold behavior in 7XXX and 2XXX-type alloys can be attributed to the alloy purity levels (Fe + Si %) and dispersoid type (Cr vs. Zr). Reference on these effects on threshold can be made to Bretz, et al. (7), Filler (13) and Zedalis and Fine (14).

7X75 Alloys

One of the major contributions to the near-threshold crack growth behavior comes from the aging treatment for a given alloy composition. Figure 2 shows the dramatic behavior of ΔK_o as a function of load ratio for the underaged ($\sigma_y \sim 450$ MPa) and overaged ($\sigma_y \sim 454$ MPa) 7075 alloy tested in moist air (6). The results are compared to the data for material of similar yield strengths tested in vacuum by LaFarie-Frenot and Gasc (15) and Kirby & Beevers (16). The comparison of the vacuum and moist air results clearly demonstrate the importance of the environmental interactions with microstructure. The threshold ΔK_o for both underaged (UA) and overaged

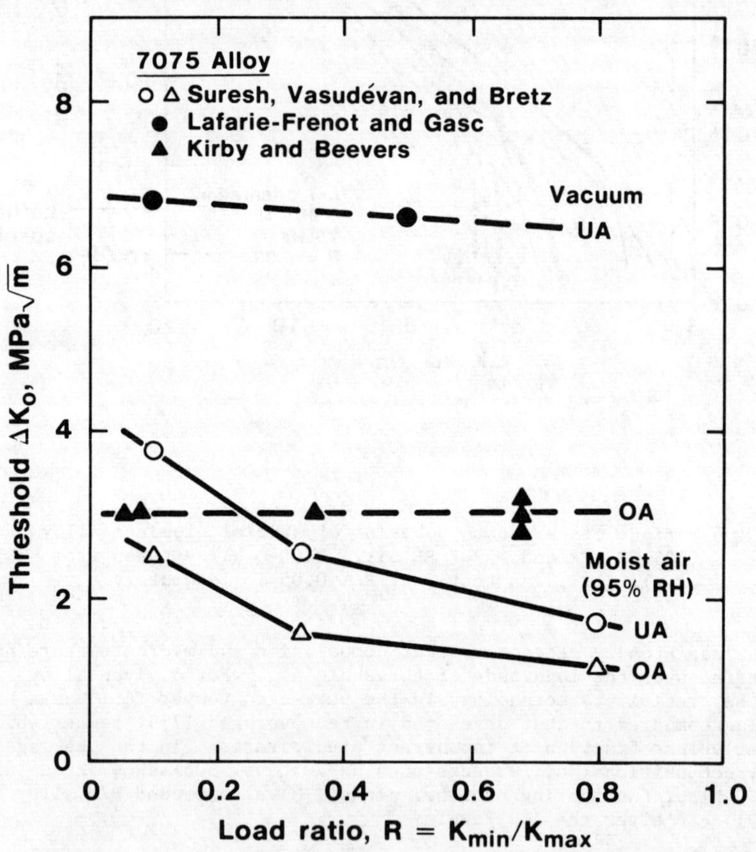

Figure 2 - Variation in threshold stress intensity ΔK_o with load ratio R for 7075 alloy in the underaged (UA) and overaged (OA) conditions tested in vacuum (15, 16) and 95% RH moist air (6).

(OA) tempers, tested in moist air, decreased with increasing load ratio. This trend, however, was completely absent for both tempers when tested in vacuum. The magnitude of the differences in ΔK_o between UA and OA was more significant in vacuum than in moist air. For the moist air condition, the measured oxide thicknesses on the fracture surfaces of material in the UA temper were insignificant compared to that of material in the OA temper (5). This overall difference in ΔK_o for the two aging conditions can be attributed to the complex interplay of competing mechanisms involving moisture-induced embrittlement, mode of deformation, crack closure and crack deflection due to the interaction of microstructure with the environment (5, 6).

Recently, Suresh (17) has quantitatively estimated changes in the crack-tip stress intensities due to crack deflection and growth rates for non-linear fatigue cracks. For near-threshold ΔK_o region in aluminum alloys, studies by Suresh, Vasudevan and Bretz (6) and Carter, et al. (19), have shown that such crack deflection mechanisms can significantly alter the fatigue crack growth characteristics. To illustrate this, crack profiles for UA and OA tempers, at low stress intensities ($\Delta K \sim 4.4$ MPa$\sqrt{m}$ and $R \sim 0.1$ and 0.8) are shown in Figures 3(a) through (d). Qualitatively, UA material seems to exhibit irregular crack profile compared to the OA temper.

Substantial fracture surface oxidation and oxide-induced closure were observed in 7075-T7351 alloy (5, 6), at 25 Hz frequency and $R \sim 0.33$. Since the condition for this oxide formation is fretting, the applied test frequency should play an important role in the amount of the oxide thickness. It takes about a millisecond to form a monolayer of oxide on aluminum. Hence, if the test frequency applied is such that the fretting can be induced at a rate comparable to rate of oxide formation, then the oxide thickness would be greater and, hence, the amount of closure. This was achieved by applying a 200 Hz test frequency to 7075-T7351 alloy at $R \sim 0.33$ in 95% RH air. Figure 4 shows such a result in terms of da/dN vs. ΔK_o. For sake of comparison, vacuum data (16) and 25 Hz data (5, 6) for the 7075-T7351 alloy are shown. Application of nearly an order of magnitude higher test frequency to the alloy resulted in a decrease in the crack growth rates below 10^{-8} m/cycle, when compared to the 25 Hz result. Measured oxide thickness at ΔK_o for 200 Hz case was ~ 0.15 μm compared to ~ 0.1 μm (25 Hz). Comparing this to the maximum crack tip opening displacement (CTOD) ~ 0.09 μm suggests that oxide-induced closure is partly responsible for the increase in the closure levels due to oxide; the 200 Hz curve exhibited significantly higher growth rates at low ΔK compared to the vacuum result. This suggests that moisture-induced embrittlement could be playing an important role in addition to the contribution from the oxide-induced closure mechanism. Such frequency effects, relating to the enhancement of oxide formation on fracture surfaces, was also observed in tempered steel (21), wherein the oxide thicknesses increased with test frequency.

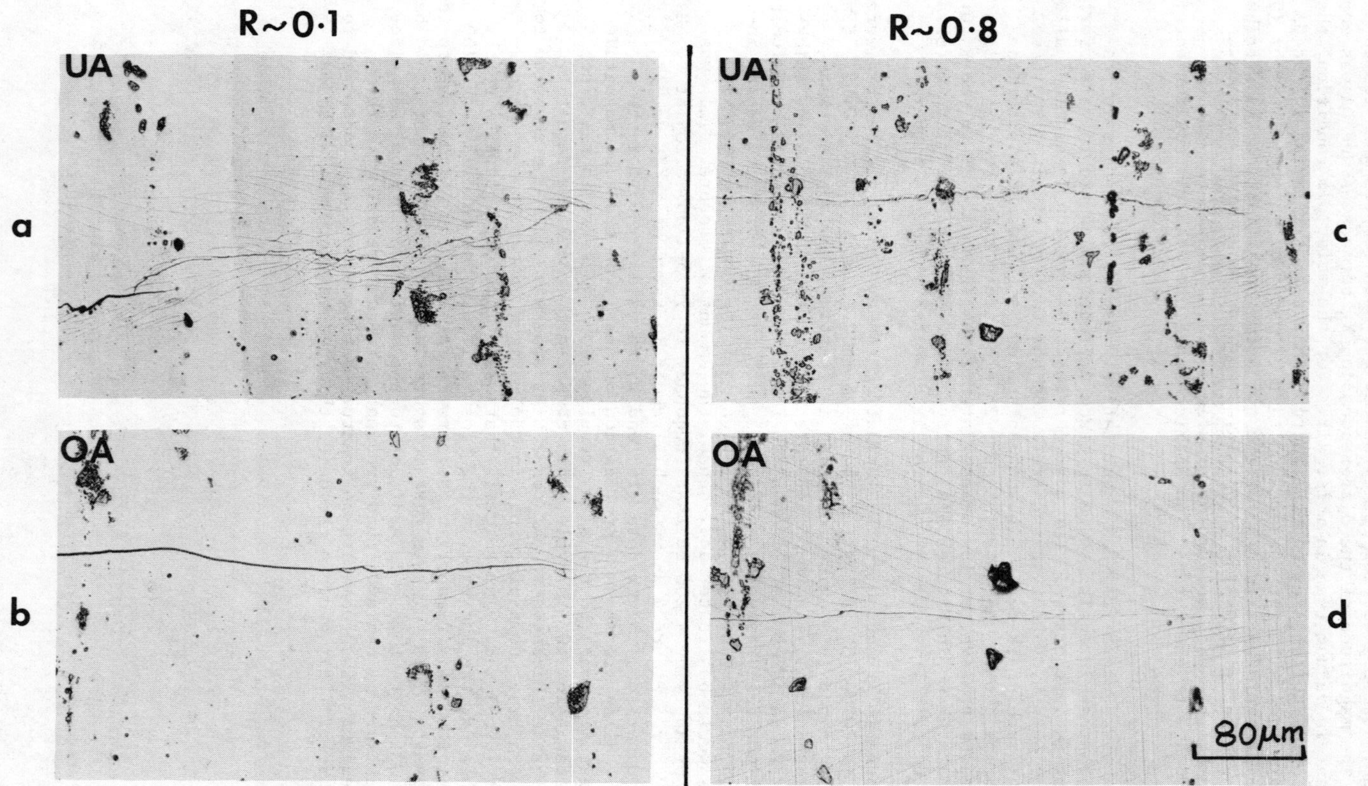

Figure 3 - Profiles of fatigue cracks in 7075-UA and 7075-OA alloys at 25 Hz in moist air environment. (a) UA at R = 0.1, (b) OA at R = 0.1, (c) UA at R = 0.8, and (d) OA at R = 0.8.

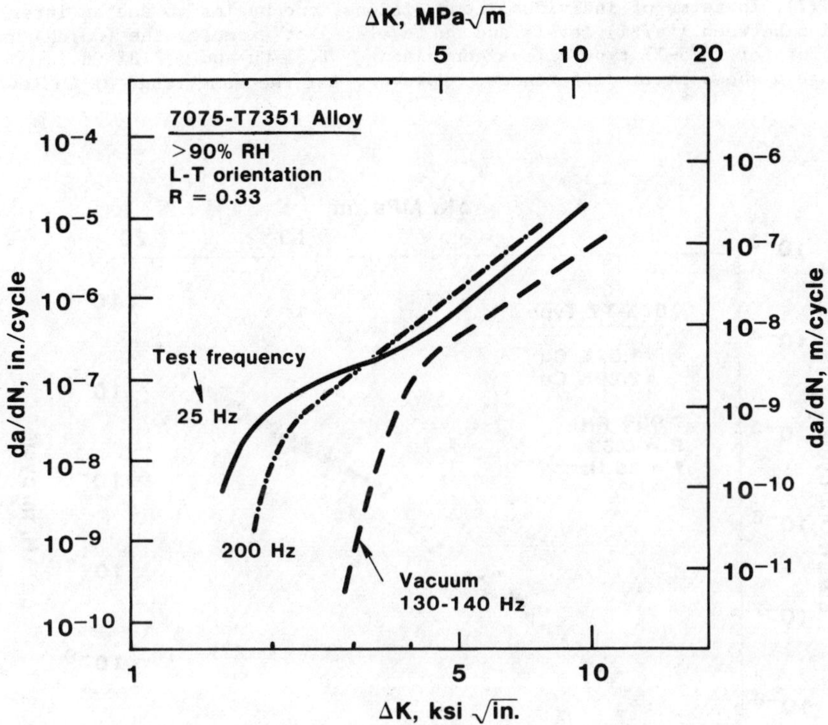

Figure 4 - Fatigue crack propagation rates (da/dN) with stress intensity range ΔK of 7075-T7351 alloy tested in moist air at 25 Hz and 200 Hz frequencies and R ∿ 0.33. For comparison, vacuum data from Kirby & Beevers (16) are plotted.

Cu Effect in 7X75 Alloy

Copper has been intentionally added to commercial 7XXX alloys not only to improve strength levels but also to enhance the resistance to stress-corrosion cracking (21, 22). While the precipitation reactions in 7XXX alloys are still not clearly understood, in general, Cu seems to affect the precipitation kinetics and grain boundary precipitate-free zone (23). Lin and Starke (24) performed detailed evaluation of FCG rate behavior of the Al-6% Zn-2% Mg alloys containing 0.01% Cu to 2.1% Cu, in T651 temper, under dry air and distilled water environments. They observed that the FCG rates decreased systematically with decreasing Cu content in dry air. However, in distilled water, 0.01% Cu exhibited the lowest growth rates with an un-systematic behavior for alloy containing higher Cu levels. Their study was not extended to the threshold regions. Bretz, et al. (9), observed that Cu effect was small in the range of 1.5% to 2.3%, at low ΔK region for all

7XXX-T7 alloys, since the (da/dN) vs. ΔK data fell within a scatter band for all alloys. However, if one plots the (da/dN) vs. ΔK data from Bretz, et al. (7), in terms of individual compositions, one begins to see an interaction between (Fe/Si) levels and Cu levels. For example, the (da/dN) vs. ΔK plot for 7075-T7 type alloys containing ∿ 1.5% Cu and ∿ 2.3% Cu in Figure 5 shows small differences. However, for the same range of Cu levels

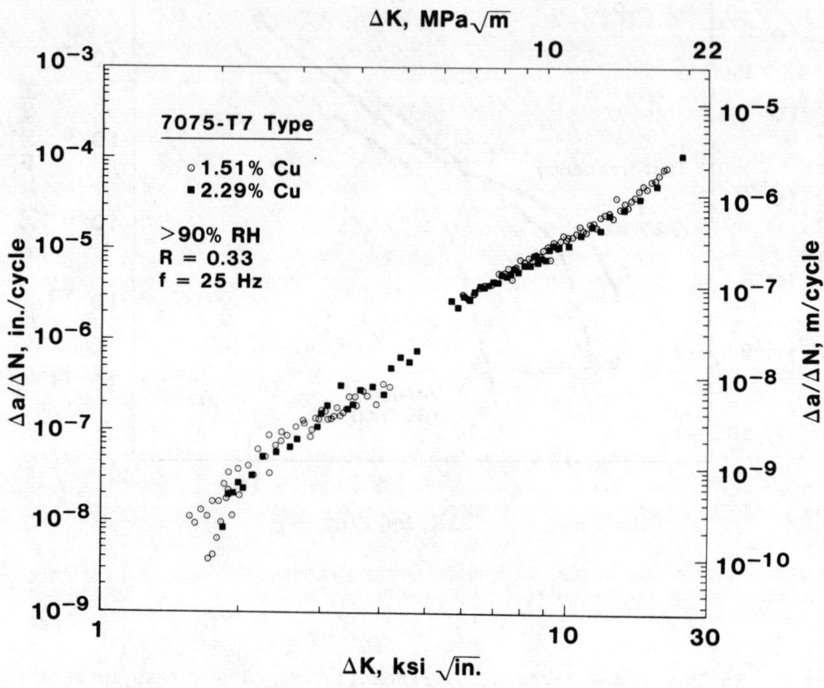

Figure 5 - Effect of Cu content on fatigue crack growth resistance of 7075 -T7 alloy in moist air at R ∿ 0.33.

a higher purity version of 7075 (i.e., 7475), exhibits large differences at near-threshold region (Fig. 6). A 7050-type alloy also showed trends similar to 7075 alloy. Measured oxide thickness for 1.5% Cu was ∿0.1 μm compared to 0.04 μm for 2.3% Cu in 7475 system. A slightly lower oxide thickness was observed in 7075 alloys. The reduction in oxide thickness at high Cu levels seems to indicate that the oxidation resistance in these alloys for T7 tempers could depend on Cu. While the optical metallography indicated similar grain structures, the difference between the alloy 7075 vs. 7475 seems to lie in the amount of constituents (Al_7Cu_2Fe and Mg_2Si). The calculated total volume fraction of the constituents is ∿ 2% for 7075 alloy compared to ∿ 0.6% for 7475 alloy. The calculated total amount of Cu that could be tied up in Al_7Cu_2Fe is about 0.6% in 7075. This would indicate that the actual Cu content in 7075 would be lower by ∿ 0.6% compared to 7475 alloy. Even though the oxide thickness variations and the absolute differences in Cu content in both the alloys are nearly the same, the near-threshold FCG

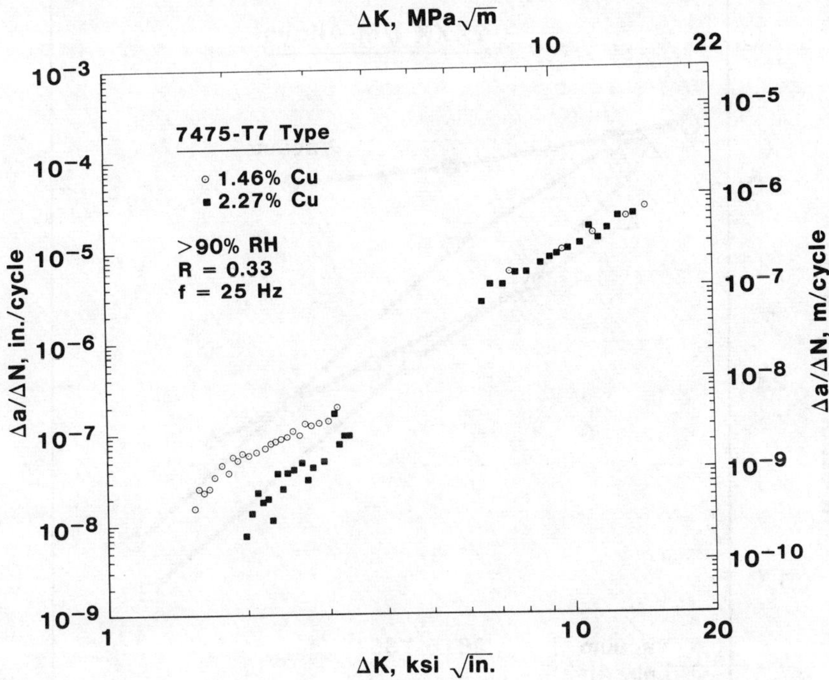

Figure 6 - Effect of Cu content on the fatigue crack growth resistance of 7475-T7 tested in moist air at R ∿ 0.33.

behavior differ significantly. This observation seems to differ from that of Lin and Starke (24). As a result, the effect of Cu in 7X75 alloys (in T7 temper) at ΔK_o regions needs further clarification.

2XXX Alloys

Compositional variations in 2XXX alloys (16, 25-27) affecting threshold ΔK_o are shown in Figure 7 as a function of R-ratio. One can infer from the trend in Figure 7 that the 2XXX alloys seem to behave similarly to the 7XXX alloys shown previously in Figure 2. Garrett and Knott (28) observed aging effects on ΔK_o in 2024 alloy. The trend in the results was observed to be similar to that in 7075. Suresh, et al. (25), and Vasudevan and Suresh (5) measured insignificant oxide thicknesses in 2021-T6 and 2024-T351 alloys. This indicates that oxide-induced closure effects do not play an important role in 2XXX-type alloys (5). In spite of this observation and being a completely different alloy system, the behavior of 2XXX alloys seems to be similar to 7075 alloy. This may imply that the complex concurrent mechanisms, without oxide-induced closure, described for 7075 alloys (5, 6) could still be applicable to the 2XXX alloys.

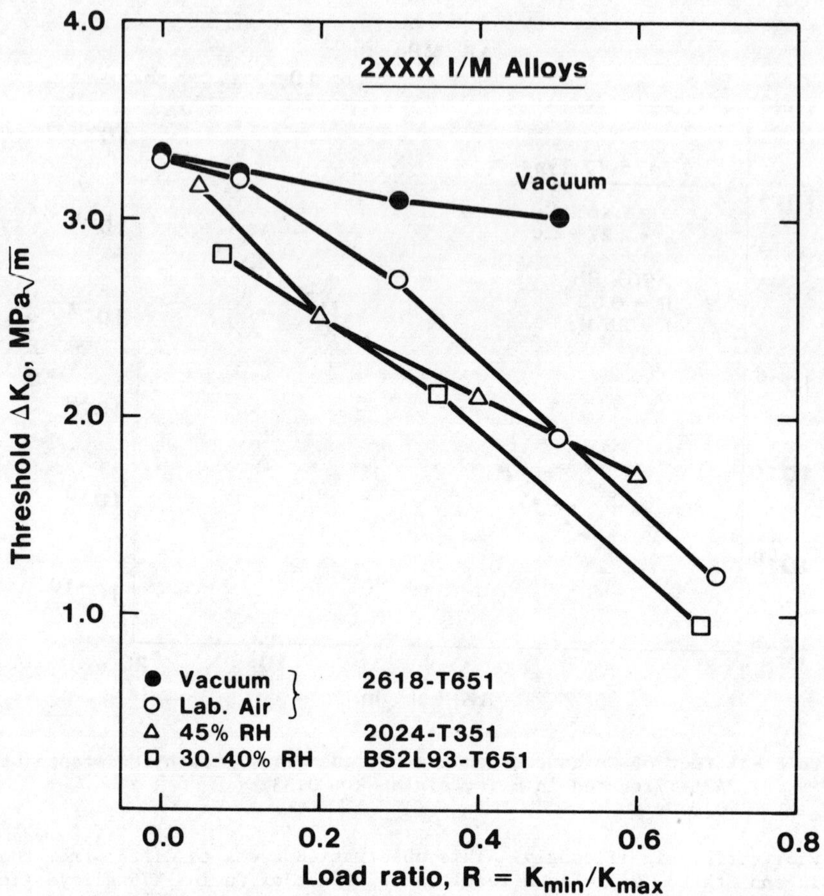

Figure 7 - Variation of ΔK_o with load-ratio R for fatigue tests conducted in vacuum and moist air for several 2XXX alloys: 2618-T651 (26), 2024-T351 (27) and B52L93-T651 (16).

Al-Li Alloys

Very little information is available on these alloys at near-threshold regions. To date, Coyne, et al. (29), observed at growth rates of $\sim 10^{-8}$ m/cycle, $\Delta K \sim 7$ MPa$\sqrt{m}$ in dry air compared to ~ 9 MPa$\sqrt{m}$ for distilled water in Al-3% Li-1.25% Mn alloy in the underaged condition. In the peak aged temper, for same growth rates, the difference in ΔK for the two environments was negligible. They attributed the FCG resistance to high modulus of Al-3% Li alloys.

Recently, Bretz, Mueller and Vasudevan (10) and Vasudevan, et al. (11), have studied the near-threshold FCG behavior of Al-4.5% Cu-1.0% Li-0.5% Mn-0.2% Cd (Alcoa 2020 alloy) in detail. They concluded that the near-threshold behavior of this alloy in T651 temper is predominantly controlled by the degree of planar slip that resulted in severe crack branching. Figure 8 compares recent results on the same alloy composition (2020 type) that has different substructures. Zr-containing alloy was completely unrecrystallized compared to coarse recrystallized structure of 2020 alloy.

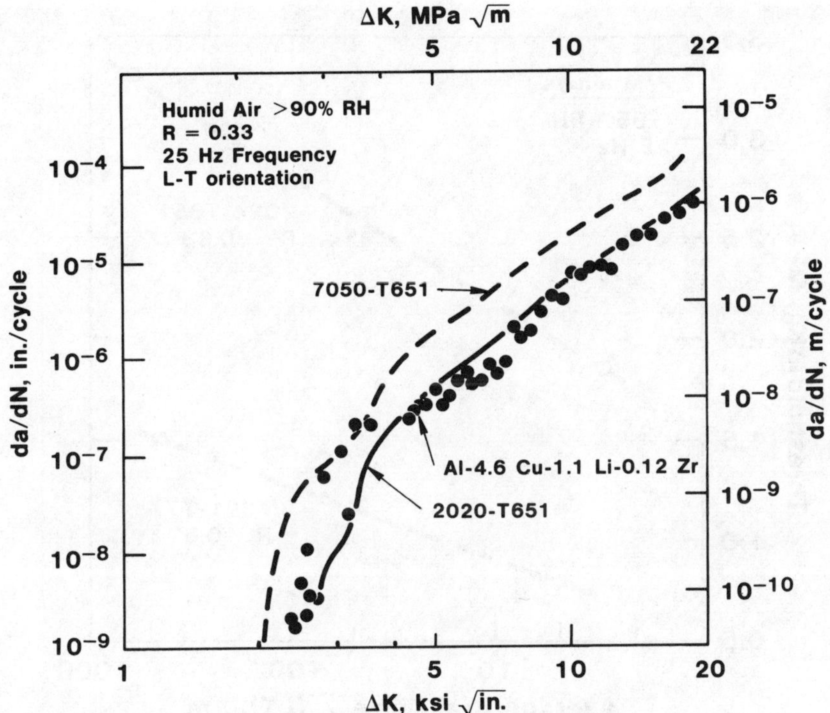

Figure 8 - Comparison of fatigue crack growth resistance of 2020-type (Al-Li-Cu-X) alloys with 7050-T651.

Both alloys showed comparable yield strengths at T651 ($\sigma_y \sim 518$ MPa) temper. Compared to the conventional 7050-T651 alloy having the same yield strength as the 2020-type alloys, both the Al-4.5% Li-1.0% Cu-X alloys exhibited better FCG resistance at all crack growth rates.

Effects of Fabrication

Grain Size

Effect of grain size on FCG resistance has been examined by several investigators (8, 9, 30-32). In all cases, an increase in grain size resulted in decrease in FCG rates in near-threshold region. Several authors (25, 33-35) have attempted to model the relationship of ΔK with grain size. Fracture surface roughness and crack closure (18, 34, 36) levels with grain size have been argued to influence the ΔK. However, the underlying mechanism for the improvement in near-threshold crack growth rates due to grain size alone is not clear.

Figure 9 shows the ΔK_o dependence on average grain size for P/M 2024-T351 (8) and 7091-T7X (10) alloys at R ∿ 0.33 and ∿ 0.8, respectively, in moist air environment. Clearly, for both alloy compositions, ΔK_o increased

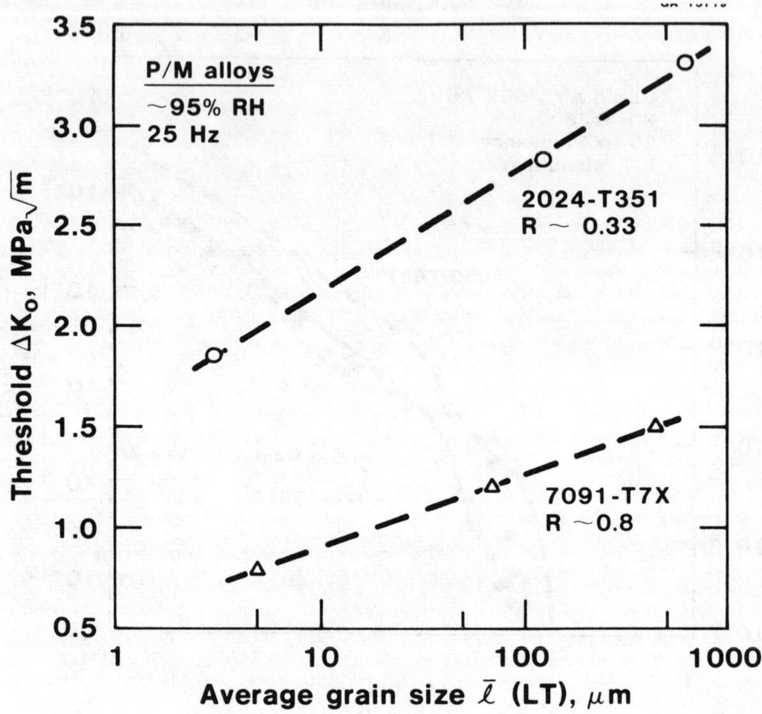

Figure 9 - Variation of average grain size with ΔK_o for 7091 and 2024 P/M alloys tested in moist air.

with grain size. In the 7091-T7X alloy, similar trend was observed at lower load ratio R ∿ 0.1 (10). This increase in ΔK_o with grain size has been attributed to increase in crack branching (17) with grain size (8, 10). Effect of grain size on FCG at low ΔK levels was also observed in Al-6% Zn-2.6% Mg-1.7% Cu alloy tested in vacuum at R ∿ 0.1 (31). They explained their results on the basis of reversed slip of dislocations within the plastic zone of a moving fatigue crack. Limited available data in vacuum do not permit comparison of the trend on the effect of grain size on ΔK_o between moist air and vacuum.

Grain size variations in a given alloy are, in general, obtained through thermomechanical treatment, prior to heat treatment. While grain sizes do vary with such processing, other microstructural changes (such as texture) do occur simultaneously. In addition, grain size (and shape) distribution are also important issues to be accounted for the changes in ΔK_o. Furthermore, aluminum alloys are known to be quench sensitive, particularly

P/M alloys (38, 39). As a result, aging kinetics could differ for different grain size material. Due to these complex microstructural changes occurring during fabrication and heat treatment, and subsequent interaction with fatigue crack growth in moist air environment, it becomes difficult to assess the underlying mechanisms controlling the near-threshold crack growth behavior for materials having different grain sizes. In addition, closure concept alone cannot explain the behavior, since the ΔK_o is found to increase with grain size at both low and high load ratios (Figure 9). These are some of the important competing effects that enter into the study of grain size effects on ΔK_o in 2XXX and 7XXX alloys.

Powder Metallurgy Alloys

The powder metallurgy fabrication approach, from rapid solidification, results in fine metallurgical structures which cannot be produced by conventional ingot metallurgy techniques. In heat-treatable P/M alloys, such fine microstructures give rise to significant improvements in strength, toughness, corrosion and S-N fatigue behavior, compared to I/M alloys. However, it has been observed that the constant amplitude fatigue crack growth behavior of P/M alloys is relatively poor compared to I/M alloys, particularly at near-threshold region. The following paragraphs discuss the near-threshold behavior of 7XXX-type P/M alloy, where most of the work has been conducted.

Figure 10 compares the P/M 7091 alloy with I/M 7075 alloy in the overaged temper. Over the entire load ratio range, ΔK_o values for the P/M alloy are lower than I/M alloy. Similar trends in the result are observed in the

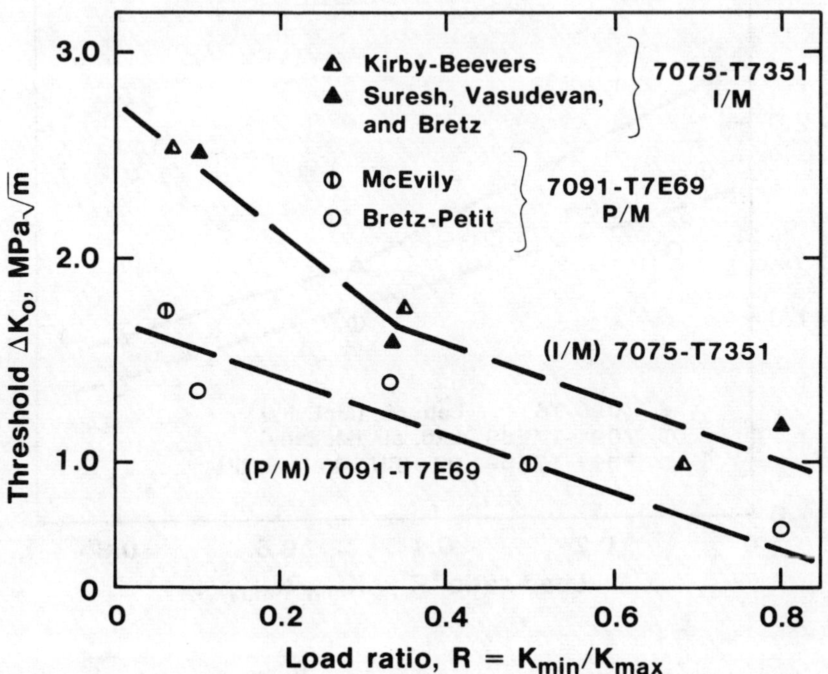

Figure 10 - Comparison of ΔK_o variation with load ratio for 7075 I/M and 7091 P/M alloys in the overaged temper in moist air (6,9,16,37).

peak aged temper for the same alloys. The major alloying elements (Zn, Mg, and Cu) are nearly the same in both alloys, the yield strength of 7091 alloy is about 10% greater than the 7075 alloy. The important difference between these two alloys lies in the fabrication process, extruded P/M 7091 compared to plate I/M 7075. The grain structure of 7091 alloy is extremely fine compared to the unrecrystallized structure of 7075 alloy. Qualitatively, surface roughness differences could be small between the two alloys. Measured oxide thickness of 7091 alloy is about 50% lower than that of 7075 alloy (5). Hence, oxide-induced closure is relatively negligible (5). Since 7075-T7351 alloy showed smooth fracture surface at both low and high load ratios [Figures 3(a) through (d)], one may assume similar crack profiles in 7091-T7E69 alloy (9). This may suggest that moisture-induced embrittlement could be a viable mechanism in 7091 P/M alloys, assuming less contribution from oxide-induced closure and crack branching. To date, systematic experiments to compare the fatigue crack growth behavior of P/M and I/M alloys have not been conducted; consequently, mechanistic implications of the observed trends in the experimental data would be speculative.

The effect of aging (T6 vs. T7E69) on the threshold ΔK_o with load ratio is shown in Figure 11 (9, 37). Strict comparison between 7090 and 7091

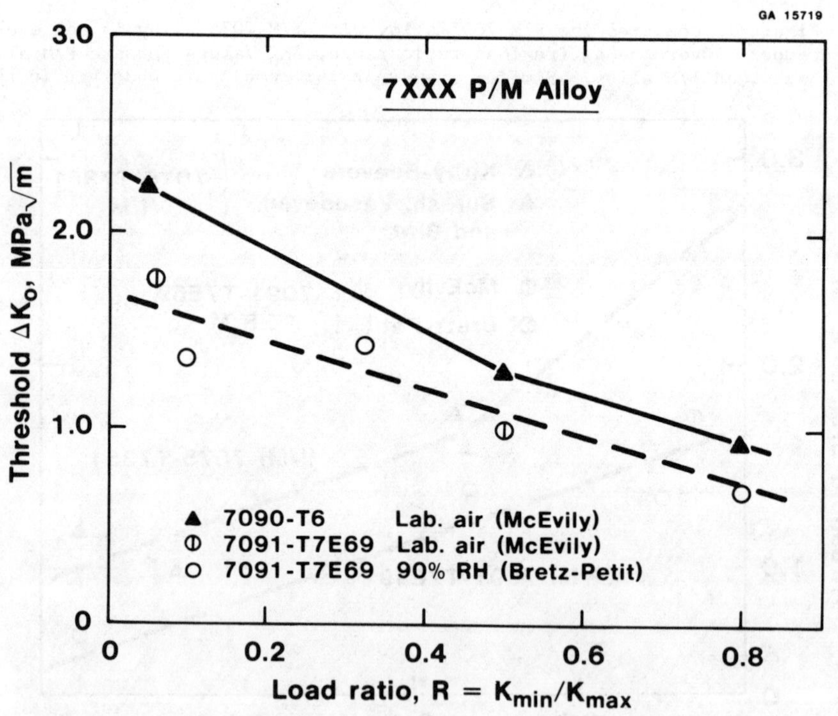

Figure 11 - Comparison of composition and temper on ΔK_o variation with load ratio in 7XXX P/M alloys.

alloys cannot be made, as they do differ in composition. Since no T6 data on 7091 alloy were available, 7090-T6 data were used instead. This would enable us to qualitatively observe the trends due to aging. The general variation in the data is similar to that of I/M 7075 alloy (Figure 2), implying that aging treatment affects the ΔK_o behavior in P/M alloys. Peak aged temper seems to exhibit relatively higher ΔK_o values compared to the overaged temper at all load ratios. Oxide thickness and surface roughness measurements are not available in these alloys in order to explain the behavior.

In order to observe the overall differences between the alloy composition, aging and fabrication treatments, one can describe all of the available data from 7XXX (I/M), 2XXX (I/M), and 7XXX (P/M) on a single ΔK_o vs. load ratio plot (Figure 12) for moist air environment at 5-50 Hz frequency. While there is some overlapping trend between the range of 2XXX and 7XXX alloys, in general, the ΔK_o values decrease with increase in R-ratio for all aluminum alloys. Interestingly, P/M 7XXX alloy results seem to be displaced significantly to lower ΔK_o values from I/M 7XXX alloys.

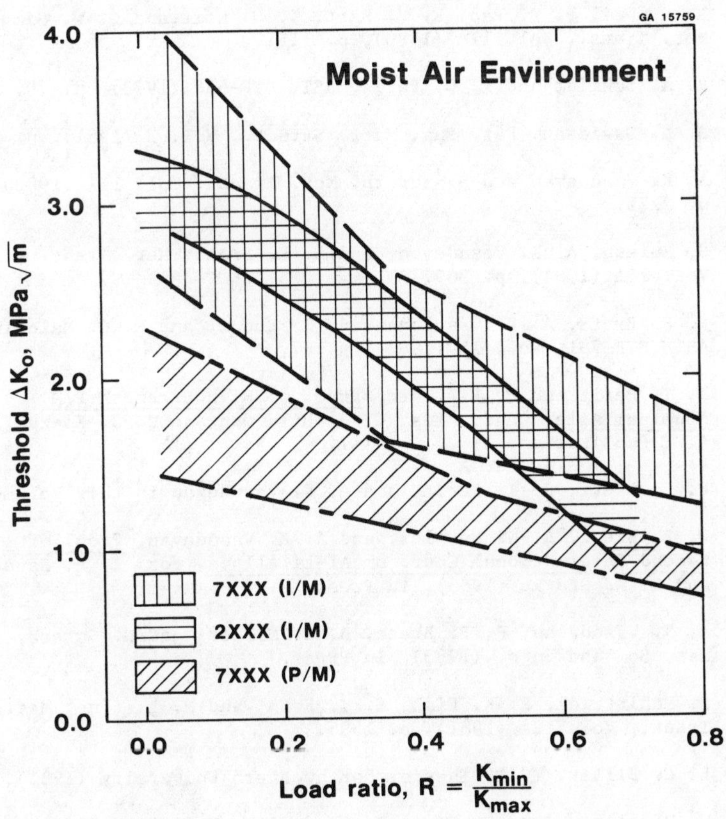

Figure 12 - Relationship of ΔK_o variation with load ratio for 7XXX (I/M), 2XXX (I/M), and 7XXX (P/M) alloys in moist air environment, at 5-50 Hz frequency.

Summary

In this paper, a brief review has been presented on the near-threshold effects on microstructure and environment for aluminum alloys. It is observed that composition, fabrication and aging treatments strongly affect the near-threshold fatigue crack growth behavior. In general, this behavior is governed by several mutually competitive mechanisms involving moisture-induced embrittlement, crack closure, deformation characteristics and crack deflection processes.

Acknowledgments

This work was supported by Alcoa Internal Research funds. The authors wish to acknowledge A. C. Miller, R. A. Petri, D. Emert, R. C. Malcolm for their assistance to this work and J. T. Staley and R. R. Sawtell for their encouragement. Thanks to Mrs. Antoinette Perkins for typing the manuscript.

References

1. R. P. Wei, Eng. Fract. Mech., Vol. 1 (1970), p. 633.

2. R. P. Wei, P. S. Pao, R. G. Hart, T. W. Weir and G. W. Simmons, Met. Trans., Vol. 11A (1980), p. 1151.

3. R. A. Schmidt and P. C. Paris, ASTM STP-536 (1973), p. 79.

4. D. L. Davidson, Fat. Eng. Matr. Struct., Vol. 3 (1981), p. 229.

5. A. K. Vasudevan and S. Suresh, Met. Trans., Vol. 13A (1982), p. 2271.

6. S. Suresh, A. K. Vasudevan, and P. E. Bretz, Met. Trans. A, Vol. 15A (1984), p. 369.

7. P. E. Bretz, A. K. Vasudevan, R. J. Bucci, and R. C. Malcolm, ASTM STP-791, Vol. II (1983), p. 67.

8. J. I. Petit and P. E. Bretz, Proc. of High Strength P/M Aluminum Alloy Conf., Eds. G. J. Hildeman and M. J. Koczak, AIME Publication (1982), p. 147.

9. P. E. Bretz, J. I. Petit, and A. K. Vasudevan in this Volume.

10. P. E. Bretz, L. N. Mueller, and A. K. Vasudevan, Proc. of Second International Conf. on Al-Li Alloys, Eds. T. H. Sanders and E. A. Starke (1983), in Press.

11. A. K. Vasudevan, P. E. Bretz, A. C. Miller, and S. Suresh, Mat. Sc. and Engg. (1983), in Press.

12. J. McKittrick, P. K. Liaw, S. I. Kwun, and M. E. Fine, Met. Trans., Vol. 12A (1981), p. 1535.

13. L. C. Filler, M. S. Thesis, Northwestern University (1981).

14. M. Zedalis and M. E. Fine, Scripta Met., Vol. 16 (1982), p. 1411.

15. M. C. LaFarie-Frenot and C. Gasc, Fat. Eng. Matl. and Structures, Vol. 6 (1983), p. 329.

16. B. R. Kirby and C. J. Beevers, Fat. Eng. Mat. and Struct., Vol. 1 (1979), p. 203.

17. S. Suresh, Met. Trans., Vol. 14A (1983), p. 2375.

18. S. Suresh, Engg. Fract. Mech., Vol. 18 (1983), p. 577.

19. R. D. Carter, E. W. Lee, C. J. Beevers, and E. A. Starke, Submitted to Met. Trans. A.

20. A. Bignonet, R. Namdar-Irani, and M. Truchon, Scripta Met., Vol. 16 (1982), p. 795.

21. J. Herenguel and G. Chaudron, Metaux et Corrosion, Vol. 16 (1941), p. 33.

22. W. A. Anderson, Precipitation from Solid Solution, ASM Publication (1959), p. 150.

23. T. H. Sanders and J. T. Staley, Fatigue and Microstructure, ASM Publication (1978), p. 467.

24. F. S. Lin and E. A. Starke, Jr., Matls. Sc. and Engg., Vol. 43 (1980), p. 65.

25. S. Suresh, I. G. Palmer, and R. E. Lewis, Fat. Engg. Matl. Struct., Vol. 5 (1982), p. 133.

26. J. Petit and J. L. Maillard, Scripta Met., Vol. 14 (1980), p. 163.

27. T. J. Mackay, Engg. Fract. Mech., Vol. 11 (1979), p. 753.

28. G. G. Garrett and J. F. Knott, Acta Met., Vol. 23 (1975), p. 841.

29. E. J. Coyne, Jr., T. H. Sanders and E. A. Starke, Jr., Proc. of First International Conference on Al-Li Alloys, Eds. T. H. Sanders and E. A. Starke, Jr., AIME Publications (1980), p. 293.

30. G. R. Yoder, L. A. Cooley, and T. W. Crooker, Scripta Met., Vol. 16 (1982), p. 1021.

31. J. Lindikeit, G. Terlinde, A. Gysler, and G. Lutjering, Acta Met., Vol. 27 (1979), p. 1717.

32. V. W. C. Kuo and E. A. Starke, Jr., Met. Trans., Vol. 14A (1983), p. 435.

33. Y. Mutoh and V. M. Radhakrishnan, J. Engr. Mat. Tech., Vol. 103 (1981), p. 229.

34. K. Minakawa and A. J. McEvily, Scripta Met., Vol. 15 (1981), p. 633.

35. J. Masounave and J. P. Bailon, Scripta Met., Vol. 10 (1976), p. 165.

36. S. Suresh, G. F. Zamiski, and R. O. Ritchie, Met. Trans., Vol. 12A (1981), p. 1435.

37. A. J. McEvily, Annual Report, AFOSR-81-0046 (1983 January).

38. D. S. Thomson, Met. Trans., Vol. 6A (1975), p. 671.

39. J. P. Lyle and W. S. Cebulak, Met. Trans., Vol. 6A (1975), p. 685.

THE USE OF THE CYCLIC STRESS STRAIN CURVE AND A DAMAGE MODEL FOR PREDICTING FATIGUE CRACK GROWTH THRESHOLDS

E.A. Starke, Jr.,* F.S. Lin,** R.T. Chen,* and H.C. Heikkenen***

*Department of Materials Science
University of Virginia
Charlottesville, VA 22901

**Reynolds Metals Company
Richmond, VA 23261

***Georgia Institute of Technology
Atlanta, GA 30332

Fatigue crack initiation and fatigue crack propagation both involve the concept of cyclic accumulated damage. The details of the damage structure can be related to a material's cyclic stress strain response (CSSR). The CSS curve determined from the CSSR can be used to establish the onset of persistent slip bands (PSB's) which have been associated with the fatigue limit. In the present paper, the critical strain necessary to form PSB's is equated to the strain at ΔK_{th}. The PSB concept is then used to modify the Chakrabortty fatigue crack propagation equation and to predict the intrinsic fatigue crack growth threshold. It is shown that this concept allows an accurate prediction of ΔK_{th} and near threshold fatigue crack propagation when slip reversibility is minimal.

I. Introduction

It is the goal of the fatigue researcher to develop an understanding of fatigue mechanisms to the degree that would allow an accurate prediction of the life of a component under a given set of experimental conditions. This includes a conceptional awareness of fatigue as an evolutionary process involving both micro- and macro-instabilities which lead to final failure. Classically, the fatigue process is divided into two stages, crack initiation and subsequent crack propagation, and researchers normally focus on one aspect or the other with little or no attempt to wed the two. This approach may be partly due to the increased emphasis on damage tolerant design which most often assumes a preexisting flaw or crack, and thus eliminates the initiation stage.

Both stages of fatigue involve the concept of "damage" which leads to the initiation of a crack and which develops in the plastic zone at the crack tip and ultimately results in fracture. There are different views as to the definition of this "damage" (1), and some of these have been expressed at a recent ASTM conference on the subject (2). However, in the context of this paper fatigue damage will be associated with the dislocations, substructure, slip bands, etc., that develop under cyclic loading. Consequently, knowledge of the basic mechanisms in the cyclic deformation of metals and alloys is a necessary prerequisite in elaborating a fatigue damage concept. The details of the damage structure that develops during cyclic loading has been extensively studied in the past 25 years and discussed in a number of review articles (3-5). The objective of this paper is to use the fundamental concepts established by these studies to develop a model which predicts fatigue crack growth thresholds.

II. Background

A. Cyclic Stress Strain Response and Damage

Since plastic deformation is not completely reversible, modifications to the microstructure occur during cyclic straining. These modifications result in changes in the flow properties. The cyclic stress strain response (CSSR) of a material is conveniently represented as a plot of saturation stress (if saturation occurs) versus the plastic strain amplitude. If saturation does not occur the stress amplitude at half the fatigue life is used. For fcc single crystals this so-called CSS-curve exhibits three regimes (6,7) consistent with a model described by Winter (8). A semi-log plot is usually necessary to adequately differentiate the three regimes, Fig. 1, which are due to differences in deformation behavior that occur as the strain amplitude is varied (6,7).

The plateau, regime II, is of special interest since its onset is related to the formation and development of persistent slip bands (PSB's) which are known to be the sites of crack initiation and early crack propagation (9). The plateau extends from the smallest plastic strain amplitude value giving rise to PSB formation up to the plastic strain amplitude where the entire crystal is filled (7). Although plastic deformation and some strain accumulation occurs in regime I, no PSB's are formed at strain amplitudes below γ_{II}, Fig. 1. This strain amplitude and its associated threshold stress have been related to the fatigue limit of metals (10,11). In this regard, the initial formation of PSB's may be considered the critical threshold damage necessary to nucleate a transgranular fatigue crack.

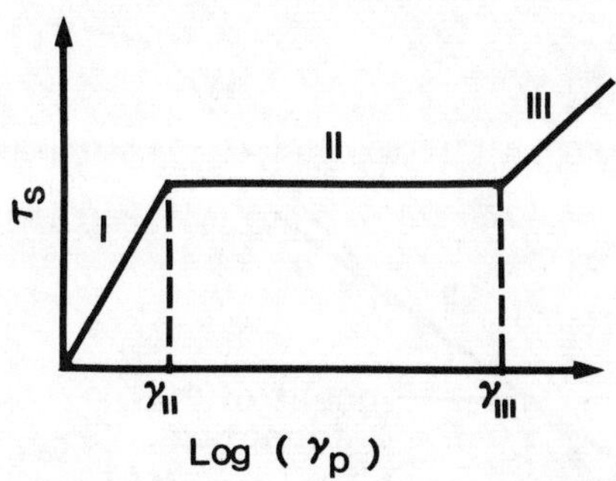

Fig. 1. Schematic representation of the cyclic stress strain curve for single crystals.

The cyclic deformation behavior of two phase alloys often differs from that of single phase materials. In alloys containing shearable precipitates, saturation may be limited or nonexistent, and softening may occur. The strengthening effects associated with coherent precipitates are reduced when they are sheared by dislocations (13). This leads to localized deformation and the development of PSB's. Although they are much narrower and different in appearance from the ladder-structure of PSB's in single phase materials (14-16), they are known to be the sites for crack nucleation (17) and early crack extension (18). The CSSC's of age-hardened single crystals are similar to those of single phase alloys (14,19,20), and there exists a critical strain below which PSB's do not form and the fatigue life appears to be infinite.

Mughrabi and Wang (21) recently compared the cyclic deformation of fcc polycrystals with single crystals. They showed that the CSS data of polycrystals could be converted to that of single crystals by the Sachs factor (22) in the low plastic strain range of regime I but diverged as the single crystal curve entered the plateau regime II. This is not surprising since surface observations usually indicate that most grains show a preference for single slip at low $\Delta\varepsilon_p/2$, where $\Delta\varepsilon_p$ is the axial plastic strain range. The dislocation arrangements and PSB's in fatigued polycrystalline copper were also found to be very similar to those of fatigued single crystals for $\Delta\varepsilon_p/2 \leq 10^{-3}$.

Lee (19) observed a plateau in the CSSC of polycrystalline Al-Cu alloys but most results (21,23) seem to indicate that this is the exception rather than the rule. The absence of a plateau is related to differing slip behavior from grain to grain and to the lack of correlation between PSB activity in different grains (21). However, when a double logarithmic plot is used for the CSS data, the PSB threshold for polycrystalline materials corresponds to the beginning of the linear regime II, as indicated in Fig. 2. This has been verified by surface observations. Since the CSSC's of single and polycrystalline alloys can be made coincident up to regime II by using the Sachs factor ($\Delta\varepsilon_p/2 = \gamma_p/M$ where M = 2.24), a close approximation of $\Delta\varepsilon_p/2$ for PSB formation can be obtained from single crystal studies.

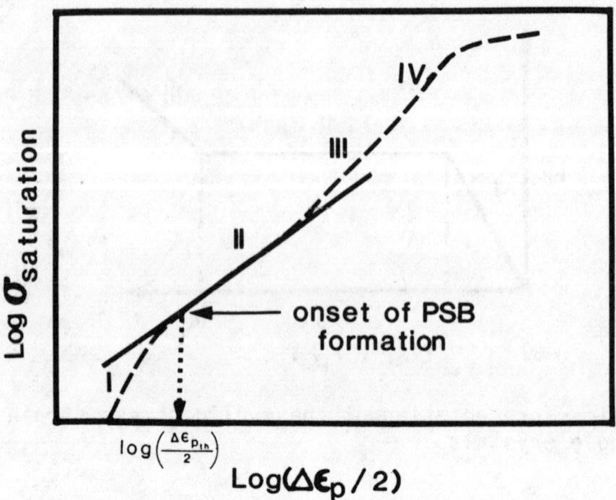

Fig. 2. Schematic representation of the cyclic stress strain curve for polycrystals.

B. Fatigue Crack Propagation and Damage

There are a number of fatigue crack growth models that employ the damage concept (24-28) and some of these have been recently reviewed (29). In a later section we will utilize the model developed by Chakrabortty (27) which employs LCF data to describe the damage at the crack tip and a microstructural parameter to predict crack growth rates without the use of adjustable constants. Following Coffin, Chakrabortty assumes that a small test specimen can be located in the critically strained region and subjected to the same strain history as that in the plastic zone, Fig. 3. This specimen can be considered to be a standard low cycle fatigue specimen used in the laboratory. The following is a brief outline of the Chakrabortty model.

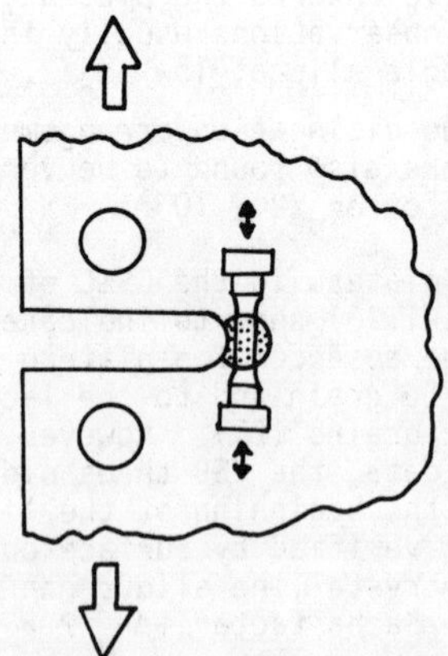

Fig. 3. Schematic showing the method used for modeling the reversed plastic zone using a LCF specimen.

Chakrabortty assumes that for a ductile material and proper ΔK range fatigue crack growth occurs by exhaustion of ductility due to cyclic plastic strain in the reverse plastic zone ahead of the crack tip. Exhaustion of ductility is described by the Coffin-Manson low cyclic fatigue law

$$\Delta\varepsilon_p = 2\varepsilon_f'(2N_f)^c \qquad [1]$$

where ε_f' is the fatigue-ductility coefficient and c is the fatigue ductility exponent. Consequently, the FCGR is a function of ε_f' and $|c|$. The cyclic plastic strain depends on the CSSR, which is normally represented by a simple power-law relationship of the form

$$\sigma' = k'(\Delta\varepsilon_p)^{n'} \qquad [2]$$

where σ' is the saturation stress, k' the cyclic strength coefficient, and n' the cyclic strain hardening exponent. It is evident from the previous section that the CSSC cannot be represented by such an expression over the total strain range, but for the near threshold and Paris regions of the da/dN-ΔK curve the parameters derived from regime II are most applicable. From eq. [2] we see that the FCGR is also a function of k' and n'.

The cyclic strain amplitude is controlled by the stress intensity range, ΔK, and the distance from the crack tip. Majumdar and Morrow (25) use continuum mechanics concepts to show that

$$\Delta\sigma \cdot \Delta\bar{\varepsilon} = \frac{\Delta K^2}{(1+n')\pi E x} \qquad [3]$$

where E is Youngs' modulus and $\Delta\sigma$ and $\Delta\bar{\varepsilon}$ are the cyclic stress and strain ranges at a distance x from the crack tip. Using

$$\Delta\bar{\varepsilon} = \Delta\sigma/E + \Delta\bar{\varepsilon}_p \qquad [4]$$

one obtains

$$x = \frac{\Delta K^2/[(1+n')\pi\ Ek']}{(\Delta\bar{\varepsilon}_p)^{n'+1} + \frac{k'}{E}(\Delta\bar{\varepsilon}_p)^{2n'}} \qquad [5]$$

which predicts the approximate cyclic plastic strain for a continuum and represents strain as a continuously decreasing function of distance from the crack tip. However, a very important aspect of the Chakrabortty model is the inclusion of the effect of slip distance. For a real material, the plastic strain ahead of the crack tip will not decrease monotonically with distance as predicted by Majumdar and Morrow, but should be more uniform within a zone which is dependent on microstructure, as shown schematically in Fig. 4. For most materials the zone size will be equivalent to the distance between major deformation barriers which is often equal to the grain size for single phase and age-hardenable alloys. Consequently, the plastic strain at a given distance from the crack tip also depends on this microstructural parameter, defined as $\bar{\rho}^{\prime}$. Chakrabortty developed a strain averaging method to obtain $\Delta\varepsilon_{pn}$,

$$\Delta\varepsilon_{pn} = \int_{x=r_{n-1}}^{r_n} x^{\frac{1}{2}} \Delta\bar{\varepsilon}_p dx \Big/ \int_{x=r_{n-1}}^{r_n} x^{\frac{1}{2}} dx \qquad [6]$$

where $r_n = r_o + \sum_{i=1}^{n} \bar{\rho}_i^{\prime}$, and $2r_o$ is the crack tip opening displacement for a particular ΔK. The zones are assumed to approach a stationary crack tip with the speed of da/dN determined when the sum of the fractions of fatigue life spent in the various zones is equal to unity. The fraction of fatigue life spent in a zone depends on $\bar{\rho}^{\prime}$ and the position of the zone n. The resulting expression is

$$\frac{da}{dN} = 2\sum_n \bar{\rho}_n^{\prime} (\Delta\varepsilon_{pn}/2\varepsilon_f^{\prime})^{-1/c} \qquad [7]$$

which can be solved numerically.

Comparisons of predicted crack growth curves using the Chakrabortty equation with FCP data obtained experimentally show good agreement for intermediate and high values of ΔK for several alloy systems (30,31). However, at low values of ΔK the Chakrabortty equation overestimates the FCGR and does not accurately indicate ΔK_{th}.

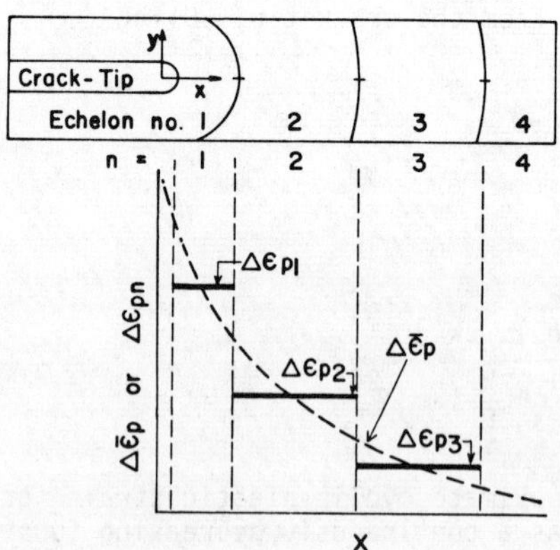

Fig. 4. Schematic of average microstructural deformation zones and their cyclic plastic strain ranges. The estimated cyclic plastic strain for a continuum is indicated by the dashed line. After Chakrabortty (27).

III. Modeling Fatigue Crack Growth Thresholds

The fatigue thresholds stress intensity can be considered to consist of a closure component, ΔK^c_{th}, and a component related to the material's resistance to crack extension, ΔK^i_{th}, giving (32)

$$\Delta K_{th} \text{ (measured)} = \Delta K^c_{th} + \Delta K^i_{th}. \qquad [8]$$

ΔK^c_{th} is the stress intensity range which must be exceeded to overcome the extrinsic influence on crack extension and depends on a closure associated with residual stresses (33), oxides (34), surface roughness (35,36), etc. ΔK^i_{th} is the stress intensity range which must be exceeded to overcome the intrinsic resistance of the material to crack extension and depends on elastic modulus, cyclic hardening exponent, cyclic yield strength, slip length (grain size), slip mode, slip reversibility, and grain orientation/deflection. Models describing the crack closure contribution to ΔK_{th} have been described elsewhere in this Proceedings (37,38). We will be concerned with the intrinsic component of the fatigue threshold, but will exclude the contribution due to grain orientation/deflection which has been treated in detail by Suresh (39). Any comparison of the model developed here with experimentally measured ΔK_{th} must be made after the measured value has been corrected for both closure and deflection.

Intrinsic effects on FCP resistance are treated in the Chakrabortty model described in the previous section. Our extension of that model to include the threshold concept is based on the premise that there is a plastic strain amplitude, $(\Delta\varepsilon_p/2)_{th}$ below which deformation is reversible and fatigue damage is not accumulated (40). For plastic strain amplitudes greater than or equal to $(\Delta\varepsilon_p/2)_{th}$, fatigue damage is accumulated and crack nucleation and extension can occur. We equate $(\Delta\varepsilon_p/2)_{th}$ to the critical strain necessary to form PSB's, as described in the Background section and incorporate $(\Delta\varepsilon_p/2)_{th}$ in eq. [7] giving

$$\frac{da}{dN} = 2\sum_n \bar{\rho}_n{}' [(\Delta\varepsilon_{pn} - \Delta\varepsilon_{p_{th}})/2\varepsilon_f{}']^{-1/c}. \qquad [9]$$

Our assumptions are supported by the evidence that cracks nucleate in PSB's and early crack extension occurs crystallographically along PSB's. These PSB's reform and develop continuously ahead of the crack tip (18). Near-threshold crack extension in deeply-notched samples, e.g., compact tension specimens, has also been observed to occur by the crystallographic stage I shear mode (30,41,42), with a transition to the noncrystallographic stage II mode at higher da/dN and ΔK's (Paris regime) (43,44). An example of such crystallographic facets in the threshold regime of an age-hardened Al-Zn-Mg alloy (45) is shown in the scanning electron fractograph of Fig. 5.

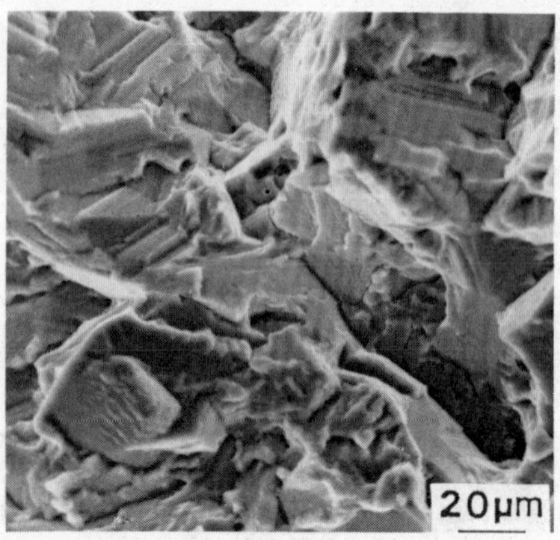

Fig. 5. Scanning electron fractograph near ΔK_{th} for an underaged Al-6Zn-2Mg-0.2Zr alloy tested in vacuum.

Figure 6 shows the effect of variations in $(\Delta\varepsilon_p)_{th}$ on the crack growth rates, as a function of ΔK. There is little or no effect for crack growth rates greater than 10^{-7} m/cycle for this example. The curves were obtained using equation [9] and LCF data from an underaged Al-Zn-Mg alloy having only shearable precipitates (46). Consequently, the slip length was controlled by the grain size and $\bar{\rho}'$ was taken to equal the grain diameter, D = 15 µm, for Fig. 6a. Figure 6b was obtained similarly using a larger grain size to illustrate the interactive effects of $(\Delta\varepsilon_p)_{th}$ and $\bar{\rho}'$. By equating ΔK_{th} to the stress intensity factor corresponding to a crack growth rate, da/dN, of 10^{-10} m/cycle, and plotting ΔK_{th} versus $(\Delta\varepsilon_p)_{th}$, we can schematically show the influence of $(\Delta\varepsilon_p)_{th}$ on ΔK_{th} for two different grain sizes, Fig. 7.

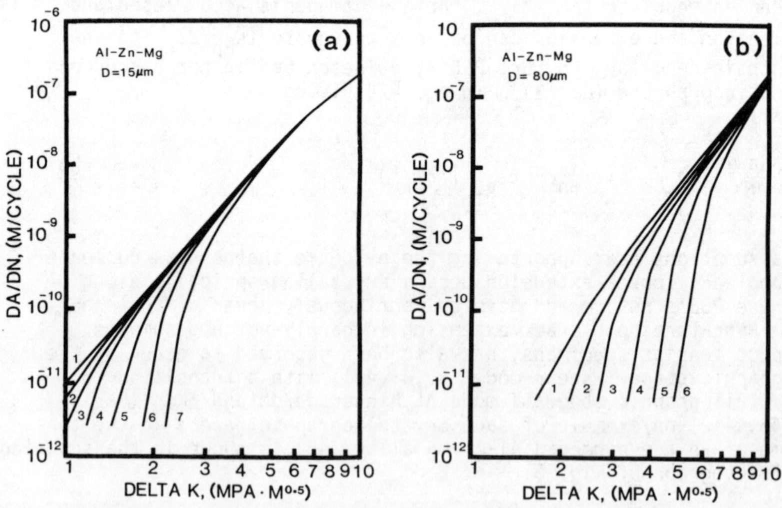

Fig. 6. Schematic of the crack growth rates as a function of ΔK for various $(\Delta\varepsilon_p)_{th}$ beginning with 2.5 x 10^{-5} and increasing in multiples of 2 to 10^{-3}. (a) for a slip length of 15 µm and (b) for a slip length of 80 µm.

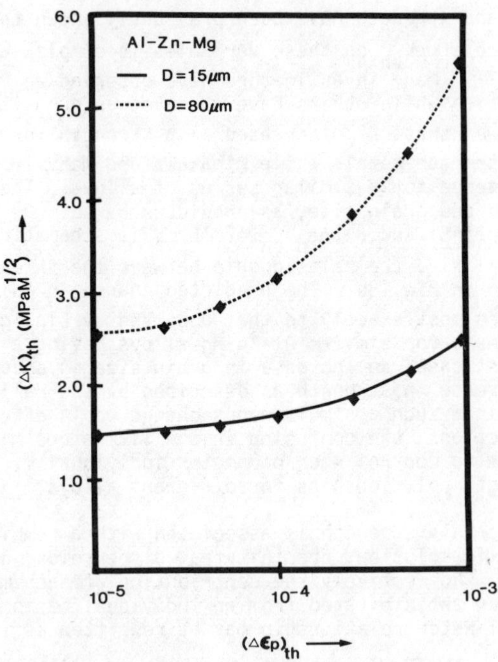

Fig. 7. Schematic representation of the predicted change in ΔK_{th} with $(\Delta\varepsilon_p)_{th}$ for two different slip distances.

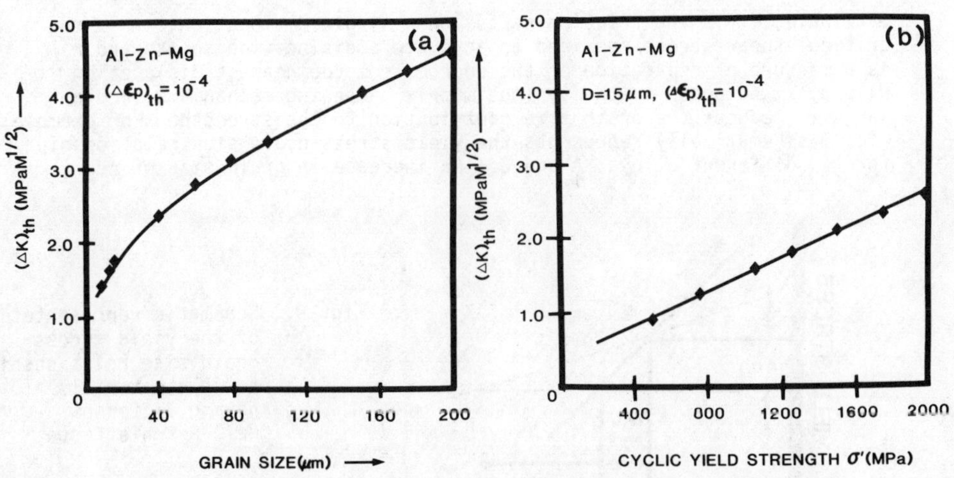

Fig. 8. Schematic representation of the predicted change in ΔK_{th} with (a) grain size and (b) cyclic yield strength.

Grain size and strength have been previously shown to effect ΔK_{th}, but the dependence of ΔK_{th} on these variables is complex and somewhat controversial (47). Many investigators have observed an increase in ΔK_{th} with grain size while others have observed an opposite effect (48). Beevers (32) showed that ΔK_{th} increased with strength for a range of medium to high strength metals while Minakawa and McEvily (49) showed an inverse dependence for a similar series of alloys. The relationship between threshold and grain size, as predicted by eq. [9] using the LCF data of Heikkenen (46) and a $(\Delta\varepsilon_p)_{th}$ of 10^{-4}, is schematically represented in Fig. 8a. Similarly, the relationship between the threshold and yield strength is shown in Fig. 8b. The predicted change in ΔK_{th} with grain size corresponds almost exactly to that observed by Lindigkeit et al. (50) in vacuum tests for similar Al-Zn-Mg alloys having a constant yield strength. In most cases an increase in grain size is accompanied by a concurrent decrease in strength as described by the Hall-Petch relationship (51,52). Since such a simultaneous change would effect the threshold in opposite directions, the confusion in the literature may be associated both with failure to control each parameter individually, and different grain size-strength relationships for different alloys.

In most materials strength is associated with a combination of mechanisms, e.g., solid solution, precipitates, dispersoids, grain and subgrain boundary effects. Collectively the contribution of each mechanism is somewhat less than anticipitated from an individual basis (53). Consequently, the Hall-Petch relationship may be rewritten as (54)

$$\tau_y = m\tau_o + (m^2\tau^* r^{\frac{1}{2}})d^{\frac{1}{2}} \qquad [10]$$

where m is an orientation factor related to the number of slip systems available, τ_o is the friction stress on a single dislocation, τ^* is the critical shear stress required to activate a dislocation source and r is a measure of separation of the source from the nearest dislocation pileup. τ_o depends on the various matrix hardening mechanisms, and a large τ_o reduces the grain size contribution to the strength. For example, Fig. 9 schematically represents the yield stress-grain size relationship for two different alloys, A and B. An increase in grain size of Δd_A

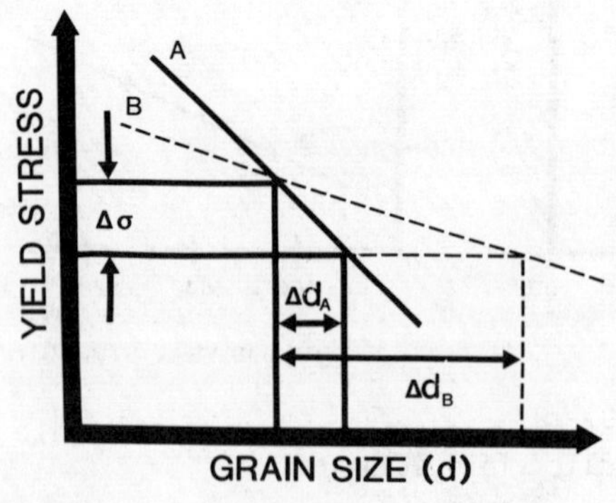

Fig. 9. Schematic representation of the yield stress-grain size relationship for two materials having different Hall-Petch effects.

for alloy A results in a decrease in strength equal to $\Delta\sigma$. However, a much larger increase in grain size, Δd_B, is required for the same decrease in strength for alloy B. Consequently, an examination of the strength-ΔK_{th} relationship may show a positive effect for alloy A (strength dominance) and a negative effect for alloy B (grain size dominance).

IV. Comparisons With Experimental Measurements

Low cyclic fatigue data was obtained for a polycrystalline Al-6Zn-2Mg-0.2Zr alloy having a recrystallized grain size of 15 μm and a random texture. This composition was selected so only shearable precipitates would be present in the underaged condition. Measurements were made in laboratory air (relative humidity ~60%), dry air (maximum H_2O content, 3 ppm), and vacuum (10^{-6} torr) environments for underaged specimens and dry air for overaged specimens. FCP tests were performed on compact tension specimens for the same aging conditions in similar environments. FCGR's were experimentally corrected for closure which was measured at 0.2 Hz from plots of load versus displacement using a clip-on displacement gage. The details of sample preparation and experimental procedures are given elsewhere (46).

Table 1. Cyclic and Microstructural Parameters Used for ΔK_{th} and FCGR Prediction

Alloy	Test Environment	$\lvert c \rvert$	ε_f' (EFP)	n' (RNP)	K' (MPa) (RKP)	ρ' (m) (RHO)	E (MPa)
Underaged	Lab. Air	0.56	0.15	0.046	1058	15×10^{-6}	7×10^4
Underaged	Dry Air	0.58	0.20	0.048	1058	15×10^{-6}	7×10^4
Underaged	Vacuum	0.58	0.22	0.048	1058	15×10^{-6}	7×10^4
Overaged	Dry Air	0.65	0.32	0.065	1100	15×10^{-6}	7×10^4

The cyclic and microstructural parameters used for calculated FCGR's and ΔK_{th} prediction are given in Table 1. There was no significant difference in the LCF results for the different environments for the underaged alloy. The CSS curve for this aging condition is given in Fig. 10, as saturation stress versus the log of the plastic strain amplitude, and compared to the results obtained by Renard et al. (23) for a similar alloy and aging treatment. Equipment limitations prevented us from making measurements below a plastic strain amplitude of 10^{-4} but Renard et al.'s data covers the range 8×10^{-7} to 10^{-2}. There is very good agreement between the region of overlap with the stress amplitudes from our measurements being slightly higher. This is probably associated with the presence of zirconium, a small grain size and a slightly higher Hall-Petch effect for our alloy. There is a significant change in slope at $\Delta\varepsilon_p/2$ of 10^{-4} which may be associated with the regime I-regime II transition. This contention is supported by Wilhelm's (14) studies on underaged Al-Zn-Mg single crystals. He found that PSB's did not form at $\gamma_p < 10^{-4}$ but did form at slightly higher values. Consequently, we will equate $(\Delta\varepsilon_p/2)_{th}$ to $10^{-4}/2.24$ for this alloy system. Closure-compensated FCGR's of the underaged alloy were corrected for crack deflection using the procedures described by Suresh (39) and given in Table 2. No deflection correction was required for the overaged alloy.

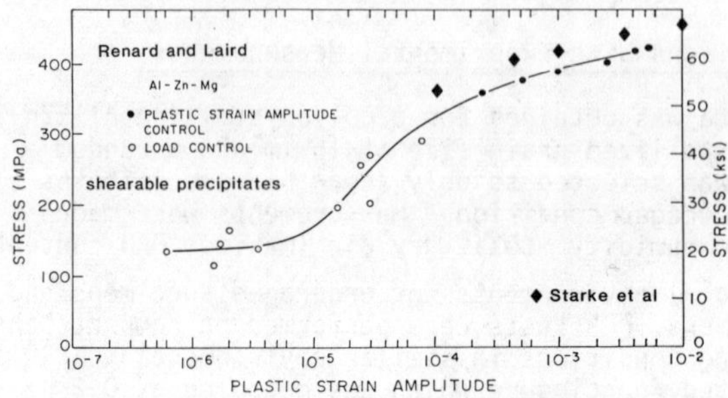

Fig. 10. CSSC for underaged Al-Zn-Mg alloys.

Table 2. Crack Deflection Corrections for the Underaged Al-Zn-Mg Alloy

Environment	Deflection (Θ)	Weight $(\frac{D}{D+S})$	$(d\bar{a}/dN)/(da/dN)$	$\Delta K_{eff}/\Delta K_I$
Lab Air	5°	0.5	0.998	1
Dry Air	20°	0.5	0.97	0.98
Vacuum	40°	0.5	0.88	0.94

single kinked crack

$$\left(\frac{d\bar{a}}{dN}\right) = \left(\frac{D\cos\Theta + S}{D + S}\right)\left(\frac{da}{dN}\right), \quad \overline{\Delta K_I} \approx \Delta K_I\left[\frac{D\cos^2(\frac{\Theta}{2}) + S}{D + S}\right]$$

$\frac{d\bar{a}}{dN}$: measured FCG rate (apparent)

$\frac{da}{dN}$: FCGR for a straight crack at the same ΔK

ΔK_I: nominal ΔK for mode I crack

$\overline{\Delta K_I}$: the average stress intensity range in each segment

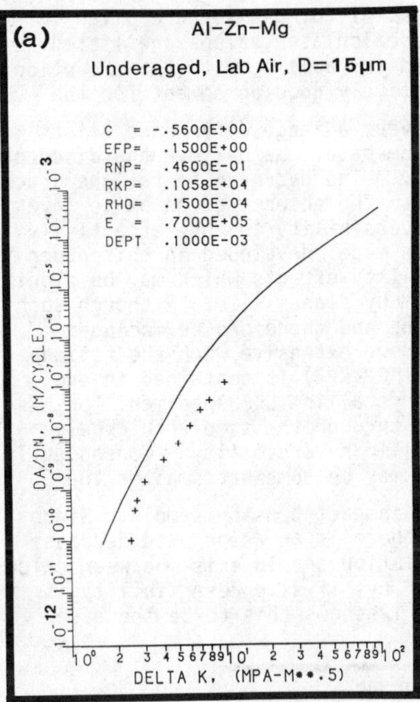

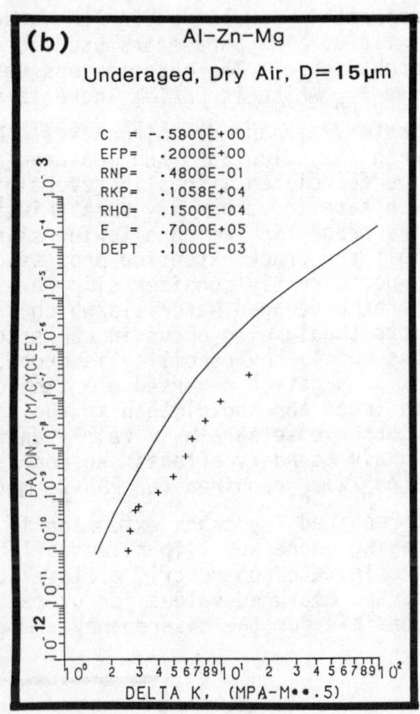

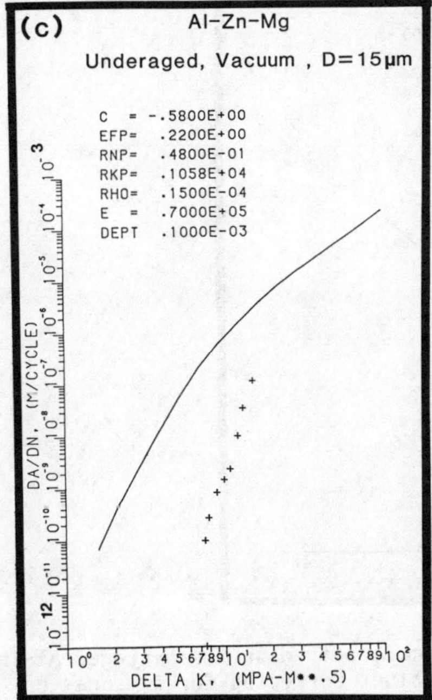

Fig. 11. Predicted and measured (+) crack growth data for underaged Al-Zn-Mg tested in (a) lab air, (b) dry air and (c) vacuum. Measured values have been corrected for closure and crack deflection.

Comparisons of the calculated intrinsic and closure-deflection-corrected experimental FCGR curves for the underaged Al-Zn-Mg alloy are given in Fig. 11a-c. The parameters used for the calculated values are listed on each Figure. The notation was defined in Table 1, with the exception of $(\Delta\varepsilon_p)_{th}$ which is DEPT. There is relatively good agreement for the laboratory air and dry air curves. However, a large difference exists between the calculated and measured vacuum data. We believe the discrepancy may be associated with slip reversibility. The decrease in fatigue crack growth rate in vacuum may be attributed to the absence of an oxide layer in the crack tip region allowing slip reversibility to more effectively inhibit the crack extension process. The model developed in this paper does not directly consider slip reversibility effects which may be significant for underaged materials which deform by planar slip. Although such effects should also occur in LCF specimens and therefore be included in the model, reversibility is probably more extensive when the fatigue crack propagation reversed plastic zone (FCP-RPZ) is contained in one grain (near threshold) than in the polycrystalline LCF specimen. For the latter case strain is fairly uniform through the sample thickness and grain boundary effects can decrease slip reversibility. Consequently, the $(\Delta\varepsilon_p/2)_{th}$ required for PSB formation may be somewhat smaller than that required for crack extension in the compact tension sample. Since overaging decreases slip reversibility (there is an associated decrease in precipitate coherency), a close correlation should exist between calculated and measured values for overaged alloys if slip reversibility is responsible for the discrepancy. Figure 12 shows this to be the case.

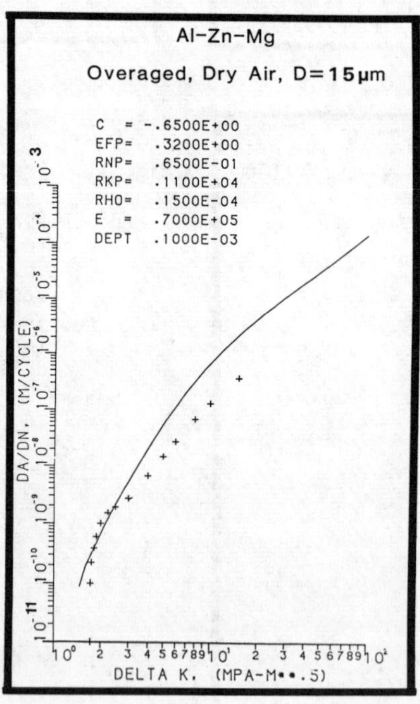

Fig. 12. Predicted and measured (+) crack growth data for overaged Al-Zn-Mg tested in lab air. Measured values corrected for closure--no deflection correction was necessary.

Recent studies by Carter et al. (55) offered the opportunity of comparing calculated and measured values for the commercial alloy 7475 whose nominal composition is Al-6Zn-2.4Mg-1.6Cu-0.2Cr. Figure 13 shows the results of this analysis for vacuum FCP tests. Measured and calculated ΔK_{th} values are tabulated in Table 3 along with those obtained for the Al-Zn-Mg alloy. There is a smaller discrepancy between measured and calculated FCG's for the underaged alloy, Fig. 13a, than observed for the underaged Al-Zn-Mg alloy of Fig. 11. This is probably due to the presence of the incoherent Cr-rich dispersoid in 7475 which should decrease slip reversibility. The calculated and measured values were coincident, however, for the overaged condition. As mentioned previously, slip reversibility should be minimal for this condition.

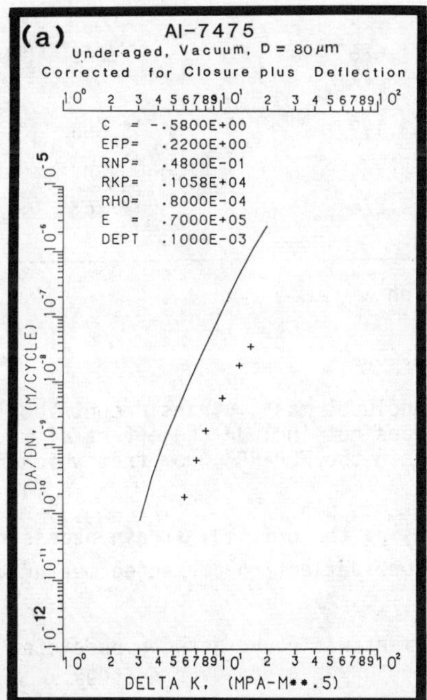

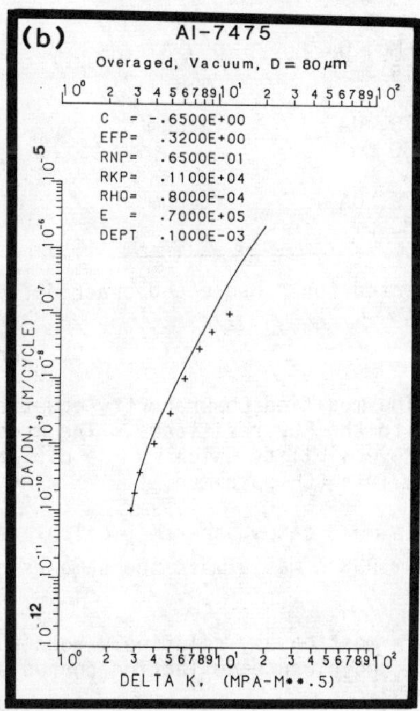

Fig. 13. Predicted and measured (+) crack growth data for 7475 (a) underaged and (b) overaged. All data corrected for closure and deflection.

Table 3. Comparison of Calculated Intrinsic ΔK_{th} and Experimentally Measured Values*

Alloy Condition	Environment	ΔK_{th} (MPa $m^{\frac{1}{2}}$) calculated	ΔK_{th}* (MPa $m^{\frac{1}{2}}$) experimental
Al-Zn-Mg, UA D = 15 μm	Lab Air	1.640	2.453
Al-Zn-Mg, UA D = 15 μm	Dry Air	1.648	2.586
Al-Zn-Mg, UA D = 15 μm	Vacuum	1.677	7.503
Al-Zn-Mg, OA D = 15 μm	Dry Air	1.626	1.745
Al-7475, UA D = 80 μm	Vacuum	3.169	5.655
Al-7475, OA D = 80 μm	Vacuum	2.876	2.863

*corrected for closure and crack deflection

V. Conclusions

1. The modified Chakrabortty equation includes most intrinsic contributions to the FCP resistance. The model does not include the effect of slip reversibility which may be different in the FCP-RPZ from that which occurs in a LCF specimen.

2. In most cases, the ΔK_{th} calculated using the critical strain necessary to form PSB's was almost the same as closure-deflection-corrected measured values.

3. In most cases, relatively good agreement was found between predicted FCGR's and closure-deflection-corrected measured values over a large ΔK range.

4. Poor agreement existed between predicted and measured values for those conditions that allowed extensive slip reversibility, i.e., underaged specimens tested in vacuum.

VI. Acknowledgements

This research was sponsored by the Air Force Office of Scientific Research, U.S. Air Force Systems Command, under Grant AFOSR-83-0061, Dr. Alan H. Rosenstein, program manager. We would like to thank Dr. S. Suresh for furnishing us with a copy of his paper on crack deflection prior to publication.

References

1. J.T. Fong, in Damage in Composite Materials, ASTM STP 775, American Society for Testing Materials, 1982, p. 243.

2. J. Lankford, D.L. Davidson and R.P. Wei, eds., Quantitative Assessment of Fatigue Damage, ASTM STP 811, American Society for Testing Materials, 1983.

3. J.C. Grosskreutz, Phys. Stat. Sol. (b), 1971, Vol. 47, p. 11.

4. Campbell Laird, in Fatigue and Microstructures, M. Meshii, ed., ASM, Metals Park, Ohio, 1979, p. 149.

5. H. Mughrabi, in Proc. of 5th Inter. Conf. on the Strength of Metals and Alloys, Pergamon Press, Oxford, 1980, Vol. 3.

6. H. Mughrabi, K. Herz and F. Ackermann, in Proc. of 4th Inter. Conf. on the Strength of Metals and Alloys, Pergamon Press, Oxford, 1976, Vol. 3, p. 1244.

7. H. Mughrabi, Mat. Sci. Eng., 1978, Vol. 33, p. 207.

8. A.T. Winter, Phil. Mag., 1974, Vol. 30, p. 719.

9. C. Laird and D.J. Duquette, in Corrosion Fatigue, National Association of Corrosion Engineers, Houston, Texas, 1972, p. 88.

10. C. Laird, Mat. Sci. Eng., 1976, Vol. 22, p. 231.

11. H. Mughrabi, Scripta Met., 1979, Vol. 13, p. 479.

12. P.M. Kelly, Int. Met. Rev., 1973, Vol. 18, No. 172C, p. 31.

13. E.A. Starke, Jr. and Gerd Lütjering, in Fatigue and Microstructures, M. Meshii, ed., ASM, Metals Park, Ohio, 1979, p. 205.

14. M. Wilhelm, Mat. Sci. Eng., 1981, Vol. 48, p. 91.

15. C. Calabrese and C. Laird, Mat. Sci. Eng., 1974, Vol. 13, p. 141.

16. W. Vogel, M. Wilhelm and V. Gerold, Acta Met., 1982, Vol. 30, p. 21.

17. C.Y. Kung and M.E. Fine, Met. Trans. A, 1979, Vol. 10A, p. 603.

18. W. Vogel, M. Wilhelm and V. Gerold, Acta Met., 1982, Vol. 30, p. 31.

19. J.K. Lee, Cyclic Deformation and Strain Localization in Al-4%Cu Containing Θ'' Precipitates, Ph.D. Thesis, University of Pennsylvania, 1980.

20. A.S. Cheng, J.C. Figueroa, C. Laird and J.K. Lee, in Deformation of Polycrystals; Mechanisms and Microstructures, Proc. 2nd Risø Int. Symp. on Metallurgy and Materials Science, September, 1981, Risø National Laboratory, Roskilde, Denmark, 1981, p. 405.

21. H. Mughrabi and R. Wang, in Deformation of Polycrystals: Mechanisms and Microstructures, Proc. 2nd Risø Int. Symp. on Metallurgy and Materials Science, September, 1981, Risø National Laboratory, Roskilde, Denmark, 1981, p. 87.

22. G. Sachs, Z. VDI, 1928, Vol. 72, p. 734.

23. Alain Renard, A.S. Cheng, R. de La Vraux and C. Laird, Mat. Sci. Eng., 1983, Vol. 60, p. 113.

24. H.W. Liu and N. Iino, Fracture (ICF2), Chapman and Hall, 1969, p. 812.

25. S. Majumdar and J.D. Morrow, in Fracture Toughness and Slow-Stable Cracking, ASTM STP 559, American Society for Testing Materials, 1974, p. 159.

26. S.D. Antolovich, A. Saxena and G.R. Chanani, Engr. Fract. Mech., 1975, Vol. 7, p. 649.

27. S.B. Chakrabortty, Fat. Engr. Mat. and Struct., 1979, Vol. 2, p. 331.

28. D.L. Davidson and J. Lankford, in Quantitative Assessment of Fatigue Damage, ASTM STP 811, American Society for Testing Materials, 1983, p. 371.

29. J.P. Bailon and S.D. Antolovich, in Quantitative Assessment of Fatigue Damage, ASTM STP 811, American Society for Testing Materials, 1983, p. 313.

30. S.B. Chakrabortty and E.A. Starke, Jr., Met. Trans. A, 1979, Vol. 10A, p. 1901.

31. E.J. Coyne, Jr. and E.A. Starke, Jr., Int. J. Fract., 1979, Vol. 15, p. 405.

32. C.J. Beevers, in Fatigue Thresholds: Fundamentals and Engineering Applications, Vol. I, J. Bäcklund, A. Blom and C.J. Beevers, eds., Engineering Materials Advisory Services, Ltd., West Midlands, UK, 1982, p. 257.

33. W. Elber, Eng. Fract. Mech., 1970, Vol. 2, p. 37

34. R.O. Ritchie and S. Suresh, Met. Trans. A, 1978, Vol. 9A, p. 291.

35. N. Walker and C.J. Beevers, Fat. of Eng. Mat. and Struct., 1979, Vol. 1, p. 135.

36. R.O. Ritchie, S. Suresh and C.M. Moss, J. Eng. Mat. and Tech., Trans ASME Series H, 1979, Vol. 102, p. 293.

37. S. Suresh and R.O. Ritchie, "Near Threshold Crack Propagation: A Perspective on the Role of Crack Closure," This Proceedings.

38. C.J. Beevers, K. Bell and R.L. Carlson, "Fatigue Crack Closure and Fatigue Threshold," This Proceedings.

39. S. Suresh, Met. Trans. A, 1983, Vol. 14A, p. 2375.

40. H.C. Heikkenen, E.A. Starke, Jr. and S.B. Chakrabortty, Scripta Met., 1982, Vol. 16, p. 571.

41. R.W. Hertzberg and W.J. Mills, in Fractography-Microscopic Cracking Processes, ASTM STP 600, American Society for Testing Materials, 1976, p. 220.

42. C.J. Beevers, Metal Sci., 1977, Vol. 11, p. 362.

43. A. Otsuka, K. Mori and T. Miyata, Eng. Fract. Mech., 1975, Vol. 7, p. 429.

44. A.J. McEvily, Metal Sci., 1977, Vol. 11, p. 274.

45. H.C. Heikkenen, F.S. Lin and E.A. Starke, Jr., in Environmental Degradation of Engineering Materials in Hydrogen, M.R. Louthan, Jr., R.P. McNitt, and R.D. Sisson, Jr., eds., Virginia Polytechnic Institute, Blacksburg, VA, 1981, p. 459.

46. Herman C. Heikkenen, "A Study of the Fatigue Behavior of an Al-6Zn-2Mg-0.1Zr Alloy," M.S. Thesis, Georgia Institute of Technology, Atlanta, GA, November, 1981.

47. J. McKittrick, P.K. Liaw, S.I. Kwun and M.E. Fine, Met. Trans. A, 1981, Vol. 12A, p. 1535.

48. R.O. Ritchie, Int. Metals Rev., 1979, Vol. 24, p. 205.

49. K. Minakawa and A.J. McEvily, in Fatigue Thresholds: Fundamentals and Engineering Applications, Vol. I, J. Bäcklund, A.F. Blom and C.J. Beevers, eds., Engineering Materials Advisory Services, Ltd., West Midlands, UK, 1981, p. 373.

50. J. Lindigkeit, G. Terlinde, A. Gysler and G. Lütjering, Acta Met., 1979, Vol. 27, p. 1717.

51. E.O. Hall, Proc. Phys. Soc., 1951, Vol. 64, p. 747.

52. N.J. Petch, J. Iron Steel Inst., 1953, Vol. 197, p. 25.

53. Edgar A. Starke, Jr., in Strength of Metals and Alloys (ICSMA 6), R.C. Gifkins, ed., Pergamon Press, New York, 1983, Vol. 3, p. 1025.

54. R.B. Nicholson, in Strengthening Methods in Crystals, A. Kelly and R.B. Nicholson, eds., Applied Science Pub., Ltd., London, 1971, p. 535.

55. R.D. Carter, E.W. Lee, E.A. Starke, Jr. and C.J. Beevers, "The Effect of Microstructure and Environment on Fatigue Crack Closure of 7475 Aluminum Alloy," Met. Trans. A, 1983, in press.

ENVIRONMENTAL EFFETCS ON ΔK_{th}

IN 70-30 α-BRASS AND 2024-T351 AL ALLOY

J.P. Baïlon, M. ElBoujdaini and J.I. Dickson

Department of Metallurgical Engineering
Ecole polytechnique
P.O. Box 6079, Station "A",
Montréal (Québec)
Canada, H3C 3A7

In an attempt to elucidate the exact role played by plasticity-induced crack closure and crack closure induced by corrosion product wedging during fatigue crack propagation in the threshold region, the values of ΔK_{th} for 70-30 α-brass and 2024-T351 aluminium alloy were measured at two R-ratios (0.1 and 0.5) and for four different environments, namely, laboratory air ($\simeq$ 35% R.H.), dry argon, wet argon (100% R.H.) and silicone oil. Continuous recording of the load versus CMOD allowed the measurement of K_{op} and therefore of ΔK_{eff} at the threshold. While the environment caused variations in ΔK_{th} values for a material of up to 40%, $(\Delta K_{eff})_{th}$ remained almost constant independent of the environment and of the R-ratio. The results, supported by fractographic observations, indicate that both plasticity-induced and corrosion product-induced crack closure occurred in the near-threshold region, and that their relative importance depended on the amount of water vapour in the atmosphere and on the plasticity of the materials.

Oxide-induced crack closure effects were particularly important for the tests in wet argon on the Al alloy at R = 0.1. Plasticity-induced crack closure effects were particularly important for the tests in inert environments on α-brass at R = 0.1. At R = 0.5, the influence of these plasticity-induced and oxide-induced crack closure effects essentially disappeared in both materials.

The amount of intergranular cracking in the near-threshold region of α-brass increased with increasing humidity suggesting the possibility of hydrogen embrittlement effects.

The similarity between the transgranular crystallographic facets produced in the near-threshold region with those produced by stress-corrosion cracking is also discussed.

Introduction

Since the first experimentally measured value of the threshold stress intensity factor, ΔK_{th}, for fatigue crack growth was reported by Linder [1] in 1965, a large amount of effort has been expended on improving our knowledge of the fatigue crack growth threshold and on elucidating the mechanisms by which fatigue crack propagation occurs at the ultra low growth rates obtained at or near ΔK_{th}. Different parameters which have been found to have an important influence on the threshold regime include the load or stress ratio, $R = \sigma_{min}/\sigma_{max}$, microstructural parameters such as the grain size, the strength level, which is largely determined by the microstructure, and the fatigue testing environment. The specific effects of these parameters in different materials (low and high strength steels, aluminum alloys, titanium alloys, copper and α-brass, etc.) have been summarized in recent articles [e.g. 2, 3] and presented in conference proceedings [e.g. 4]. A fundamental understanding of the crack extension mechanisms which operate at these ultra-low fatigue crack growth rates, however, is still lacking and many questions remain unanswered concerning the combined role of the environment and the stress ratio R.

The present state of knowledge concerning the coupled effect of the R-ratio and the environment on ΔK_{th} and on the ultra-low fatigue crack growth rates can be briefly summarized by noting that four basic mechanisms are most frequently invoked in proposed explanations. These are the following:

Plasticity-induced crack closure. This phenomenon proposed by Elber [5], has proved to be very useful in rationalizing the influence of the R-ratio and overloading effects on regime II (the Paris regime) of the log da/dN - vs - log ΔK fatigue crack growth curve. The extension of this mechanism to the near-threshold regime (regime I) at first glance appears straightforward. In regime I, however, the crack propagation rates become so low that the kinetics of the reaction between the environment and the freshly produced fracture surfaces can no longer be ignored. Moreover, the small values of the crack tip opening displacement (CTOD) characteristic of the near-threshold regime are now of the same order of magnitude as the size-scale of the fracture surface roughness. Consequently, other or additional mechanisms are often invoked to explain the near-threshold crack growth behaviour.

Roughness-induced crack closure: A salient feature of fracture surfaces produced in the near-threshold regime is their strongly crystallographic aspect [6]. The deviations of individual crystallographic facets from the macroscopic crack plane can be quite important and a component arising from this surface roughness can now be expected to make a significant contribution to crack closure behaviour. This effect, termed roughness-induced crack closure, was first proposed by Cooke and Beevers [7] and is often considered as making a probable contribution to closure related behaviour. An extensive bibliography on this type of crack closure may be found in reference [8].

Oxide-induced crack closure: Fracture surfaces produced in laboratory air at ultra-low fatigue crack growth rates generally present a film of oxide or corrosion product, easily visible with the naked eye. The shade of this oxide film becomes darker or more pronounced as the crack growth value at which it was produced decreases. The appearance of the near threshold fracture surfaces produced in vacuum or in an inert environment generally suggests much less oxidation of the fracture surfaces.

A number of experimental studies [11-17] have shown that in environments containing a sufficient (but not necessarily large) quantity of water vapour, the thickness of the oxide film formed in the near-threshold regime could be of the same order of magnitude as the calculated CTOD [13]. In such a case, this oxide film acts as a wedge which tends to increases the value of K_{op}, the stress intensity factor at which the crack becomes fully opened, which consequently results in an increase in the measured value of ΔK_{th}. Reference [8] also contains a good bibliography concerning this type of fatigue crack closure.

These different contributions to the occurrence of crack closure are not necessarily independent. In fact, it appears that the marked microscopic roughness of the fracture surfaces produced in the near-threshold regime may [11, 17] assist the formation of oxide debris which can then act as a wedge giving rise to oxide-induced crack closure.

It can also be difficult to differentiate between plasticity-induced crack closure and the other types of crack closure since the mere effect of the opposite fracture surfaces meeting together during crack closure will result in local plasticity. These different types of crack closure cannot be considered as truly independent. The wedging effect caused by corrosion product debris can be expected to become more important as the R-ratio decreases, which favours a large plasticity-induced crack closure effect and which can favour the formation of fretting oxide. This oxide wedging effect can also be expected to increase in importance as the surface roughness increases, which favours a large roughness-induced crack closure effect, and as the susceptibility of the material to oxidation or to fretting oxidation increases. The probable interdependence of different crack closure effects is indicated by the experimental observations showing that the oxide wedging effect is much less important, even for low values of the R-ratio, in the case of high strength steels (yield strength ≥ 700 MPa) as demonstrated by Suresh and Ritchie [18, 19] and in the case of strongly work-hardened copper, as shown by Liaw et al [15].

For aluminum alloys, the experimental results are scarcer and the situation is less clear than for steels. For these alloys, the oxide wedging effect does not appear to be the only mechanism which can satisfactorily explain the combined influence of the environment and of the R-ratio. Vasudevan and Suresh [20] and Renaud et al [21] mention the possible role that hydrogen embrittlement, associated with the dissociation of water vapour present in the atmosphere, as a mechanism which can play a non-negligeable role in the near-threshold regime for some aluminum alloys.

Hydrogen embrittlement. The fourth mechanism invoked for explaining fatigue crack behaviour in the near-threshold regime is hydrogen embrittlement. This mechanism, which has frequently been employed to explain stress corrosion cracking or corrosion fatigue results, probably has been given, until recently, insufficient consideration in interpreting fatigue propagation results in ordinary gaseous environments. Ritchie [22] proposed a hydrogen embrittlement mechanism to interpret the influence of water vapour and of hydrogen on ultra low fatigue crack growth rates for high strength steels. A similar explanation was also later employed to explain results obtained for medium strength steels [e.g., 17] and for certain aluminum alloys [11, 20, 21]. Hydrogen embrittlement is also an attractive mechanism for explaining the frequent observation of intergranular cracking at the very low crack propagation rates obtained in the region of the knee in the log da/dN-log ΔK fatigue curves. It has recently been proposed [23] to explain the intergranular cracking observed in this portion of the crack propagation curve in the case of OFHC copper. While the experimental results available are not yet sufficiently complete to conclude that hydrogen embrittlement effects are definitely responsible, a contribution of such a mechanism in the near-threshold region cannot a priori be ignored. If hydrogen embrittlement effects are indeed important, the influence of the cycling frequency which determines the crack propagation rate with respect to the rate of the formation and absorption in the region ahead of the crack tip of atomic hydrogen should become an important test variable.

In order to shed further light on the possible role of each of the mechanisms mentioned in the preceding as being able to influence ΔK_{th} and the near-threshold fatigue crack propagation behaviour, a testing program was initiated at Ecole Polytechnique. In the present paper, the results of one of the first studies in this program will be presented.

Experimental procedure

For the present study, two face-centred cubic materials of rather different characteristics were chosen. One was aluminum alloy 2024-T351, a material of moderate ductility, quite sensitive to the presence of water vapour and more characteristic of a wavy slip than of a planar slip material. The second was 70-30 α-brass, a more ductile material, which is less sensitive to the presence of water vapour and which is very characteristic of a planar slip material. The gain size of this α-brass (after recrystallisation) was ≃ 12μm.

The monotonic and cyclic mechanical characteristics measured for these two materials are presented in Table I.

Table I. Mechanical characteristics of the tested materials

	Monotonic					Cyclic		
Material	YS (MPa)	UTS (MPa)	ε_f (%)	K (MPa)	n	YS (MPa)	K' (MPa)	n'
2024-T351	362	458	21	665	0.11	380	530	0.18
70-30 α-brass	142	505	36	555	0.31	200	988	0.21

YS = yield strength; UTS = ultimate tensile strength;
ε_f = tensile elongation; K, n, K' and n' = parameters of the work hardening curves $\sigma = K\varepsilon^n$ or $\sigma_c = K'\varepsilon_c^{n'}$

CT specimens, conforming to ASTM specification E-399 with width W = 50.8 mm and thickness B = 12.7 mm were machined from rolled plates 16-25 mm thick for a TL crack propagation orientation. The tests were performed on an MTS servohydraulic machine interfaced to a desk-top computer, which was employed for controlling the test and for data acquisition. The tests were performed at R-ratios of 0.1 and 0.5 and at a frequency of 20 Hz.

Two methods were employed to perform the ΔK_{th} measurements, the conventional method and a rapid method. In the conventional method, ΔK was decreased in a step of 10% or less and the crack allowed to propagate over a distance equal to at least three times the calculated plane strain plastic zone size before a new stepped decrease in ΔK was effected. The underlying principles of the rapid method employed have been previously described [24]. This method has subsequently been improved and completely automatized [25].

The crack lengths were calculated based on the CMOD measurements, however, a travelling microscope was also employed to verify that these calculated values agreed with those measured optically.

Tests were performed in four environments: laboratory air (approximately 35 to 40% relative humidity), dry argon, wet argon and silicone oil. For the tests other than those in air, a plexiglass environmental chamber was attached to both sides of the fracture surface employing a silicone glue which does not release acetic acid vapour during setting. For the dry argon, a commercial argon containing 10 ppm or less of water vapour and oxygen was passed successively over copper chips and zirconium foil at 550°C before being admitted at room temperature to the environmental chamber. The same commercial argon was passed through boiling water and cooled to room temperature to produce the wet argon environment. The silicone oil employed was Dow Corning Oil no 550F, which was continuously pumped into the environmental chamber using a peristatic pump.

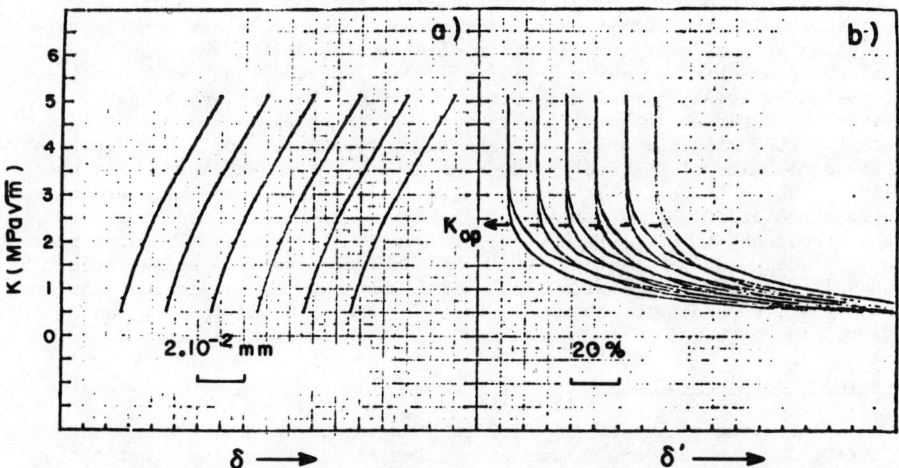

Figure 1 - Curves of a) K vs δ_m and b) K vs δ' for α-brass tested at R = 0.1 in lab air. The value of K_{op} is taken as that corresponding to δ' = 0.02 (2%).

During the tests, average hysteresis loops of K versus δ_m, the instantaneous CMOD measurement, were obtained by taking measurements during 200 cycles of 50 pairs of experimental points (K, δ_m) per cycle. The method of Elber [5] was employed in order to obtain the value of K_{op}, the stress intensity factor at which the crack becomes completely opened, from these average loops. In the near-threshold region, the truly elastic portion of the hysteresis loop was small; therefore this portion was determined optically for each recorded loop and typically extended from approximately 0.75 to 0.95 K_{max}. As well, since the sample compliance depends on the crack length, it was found preferable for improved precision, to base the K_{op} measurement on the relative value of CMOD. Accordingly, K was plotted versus δ' where $\delta' = (\delta_m - \delta_e)/\delta_e$ and δ_e is the elastic CMOD obtained from the linear portion of the K versus δ_m (or experimentally measured CMOD) curve. The value of K_{op} was then taken as the K corresponding to $\delta' = 0.02$ on the K vs δ' curves. An example is given in figure 1.

The fracture surfaces obtained were studied by scanning electron microscopy (SEM).

Experimental results

Comparison between the results obtained by the two methods

Typical examples of experimental log da/dN -vs- log ΔK curves obtained employing the conventional (manual) and the accelerated (automatic) methods of obtaining ΔK_{th} are presented in figure 2. The accelerated method can be seen to give da/dN values considerably lower than those obtained by the conventional technique for $\Delta K > \Delta K_{th}$. This technique, however, was not proposed [24] as a technique for the obtention of precise crack propagation rate curves. The technique employs an exponential decrease of the applied load which obliges the fatigue crack to arrest. If the damping coefficient [24] for this load decrease is too high, the propagation rates measured are underestimated since the crack continually propagates in the presence of overloading effects. After each crack arrest, however, a new load decreasing stage with a lower damping coefficient is imposed. As a result, the overloading effects become progressively less important and crack arrests occur at ΔK values progressively closer to ΔK_{th}. Eventually the damping coefficient employed is sufficiently low that crack arrest occurs at ΔK_{th}. Crack arrest during the next load decreasing procedure should then again occur at the same value of ΔK, indicating that ΔK_{th} has been attained. For the ΔK values at which crack arrest occurs in these last load-decreasing procedures, the measured da/dN values corresponding to the last measurement prior to crack arrest and to ΔK_{th} are low, $\simeq 1$ to 3×10^{-8} mm/cycle. From the present and previous results [24, 25], it can be concluded that the accelerated method is reliable and permits the determination of ΔK_{th} employing a testing period that, depending on the sensitivity of the material to overloading effects, is 30 to 60% of that required with the conventional method.

Experimental values of ΔK_{th}

The measured values of ΔK_{th} for the two materials and R-ratios and the different environments are summarized in Table II, which also lists the values measured at the threshold for K_{op} and for the effective stress intensity factor, $(\Delta K_{eff})_{th} = (K_{max} - K_{op})_{th}$.

For the 2024-T351 alloy tested at R = 0,1, the ΔK_{th} value obtained in wet argon (5.28 $MPa.m^{1/2}$) is considerably higher than that obtained in air (3.70 $MPa.m^{1/2}$), while the threshold values obtained in dry argon and in silicone oil (3.83 and 3.82 $MPa.m^{1/2}$) respectively, are only slightly higher than the lab air value. For α-brass cycled at R = 0.1, testing in wet argon compared to air causes a significant increase in ΔK_{th} (5.83 $MPa.m^{1/2}$ compared to 4.74 $MPa.m^{1/2}$), which increase, however, is relatively less important than that obtained for the aluminum alloy (23% compared to 43%). On the other hand, the tests performed in the two inert environments results in an increase of ΔK_{th} of 30% compared to the lab air value, a considerably more important increase than for 2024.

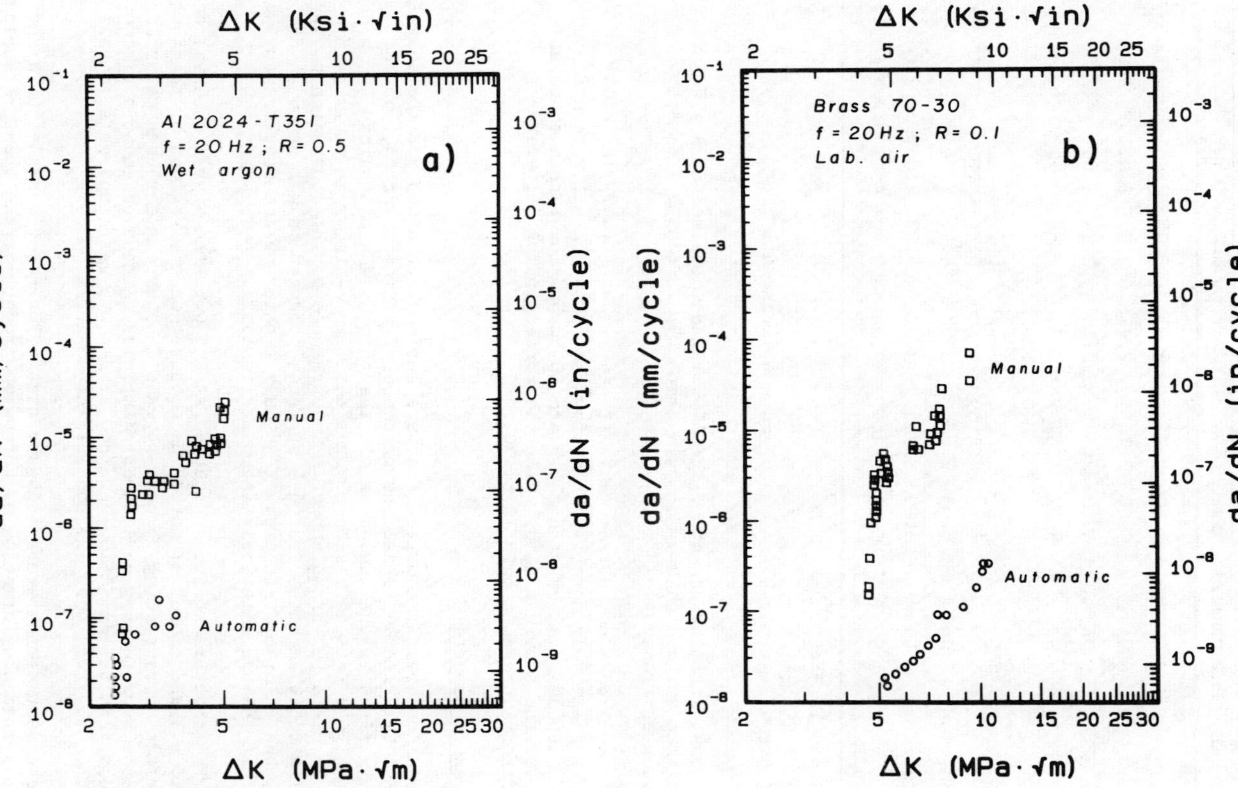

Figure 2 - Fatigue crack growth curves da/dN - vs - ΔK for a) Al alloy 2024-T351 at R = 0.5 in wet argon b) 70-30 α-brass at R = 0.1 in laboratory air (≃ 35% R.H.)

Table II - Experimental values of ΔK_{th}, K_{op}, $(\Delta K_{eff})_{th}$

Material	Environment	R = 0.1			R = 0.5		
		ΔK_{th}	K_{op}	$(\Delta K_{eff})_{th}$	ΔK_{th}	K_{op}	$(\Delta K_{eff})_{th}$
2024-T351	LA	3.70	1.70	2.41	2.30	2.48	2.12
	WA	5.28	2.90	2.97	2.45	2.81	2.09
	DA	3.83	2.34	2.10	2.34	2.65	2.03
	SO	3.82	1.87	2.37	2.68	2.94	2.42
70-30 α-brass	LA	4.74	2.41	2.88	4.60	6.05	3.15
	WA	5.83	2.60	3.88	4.45	5.00	3.90
	DA	6.30	3.95	3.05	4.33	4.80	3.86
	SO	6.13	3.44	3.00	4.60	5.52	3.68

All K values in $MPa(m)^{1/2}$.
LA = laboratory air　　WA = wet argon
DA = dry argon　　SO = silicone oil

The most striking aspect of the ΔK_{th} values obtained at R = 0.5 is that, at this quite high R-ratio, the threshold is little influenced by the testing environment.

It can also be noted from Table II that the intrinsic influence of the R-ratio is relatively more important for the 2024-T351 alloy than for the α-brass. The ΔK_{th} values at R = 0.5 are 30 to 50% lower than those at R = 0.1 for the aluminum alloy while the corresponding difference in ΔK_{th} values for the brass is only 10 to 30%.

Fractographic aspects

Visual observations of the fracture surfaces obtained for the test in air at R = 0.1 clearly reveal the presence of corrosion product on the fracture surface in the regions where the threshold was attained. For the aluminum alloy, this corrosion product has a grey-black aspect, while for the α-brass it has a red-brown aspect. The presence of this corrosion product is equally visible on the fracture surfaces produced in wet argon at R = 0.1; however, the color of this product is much lighter on the fracture surfaces produced in the two relatively inert environments.

On the fracture surfaces produced at R = 0.5, the visual observations indicated a light amount of corrosion product on the regions of the fracture surface corresponding to near-threshold behaviour for both materials and for the four testing environments employed. These observations tended to suggest that the oxidation obtained at R = 0.5 was comparable to that obtained in the two inert environments at R = 0.1.

These results therefore indicate that a low R-ratio and a humid environment favours the formation of corrosion product on the fracture surfaces produced in the near-threshold region and therefore increases the possibility of significant oxide crack closure effects.

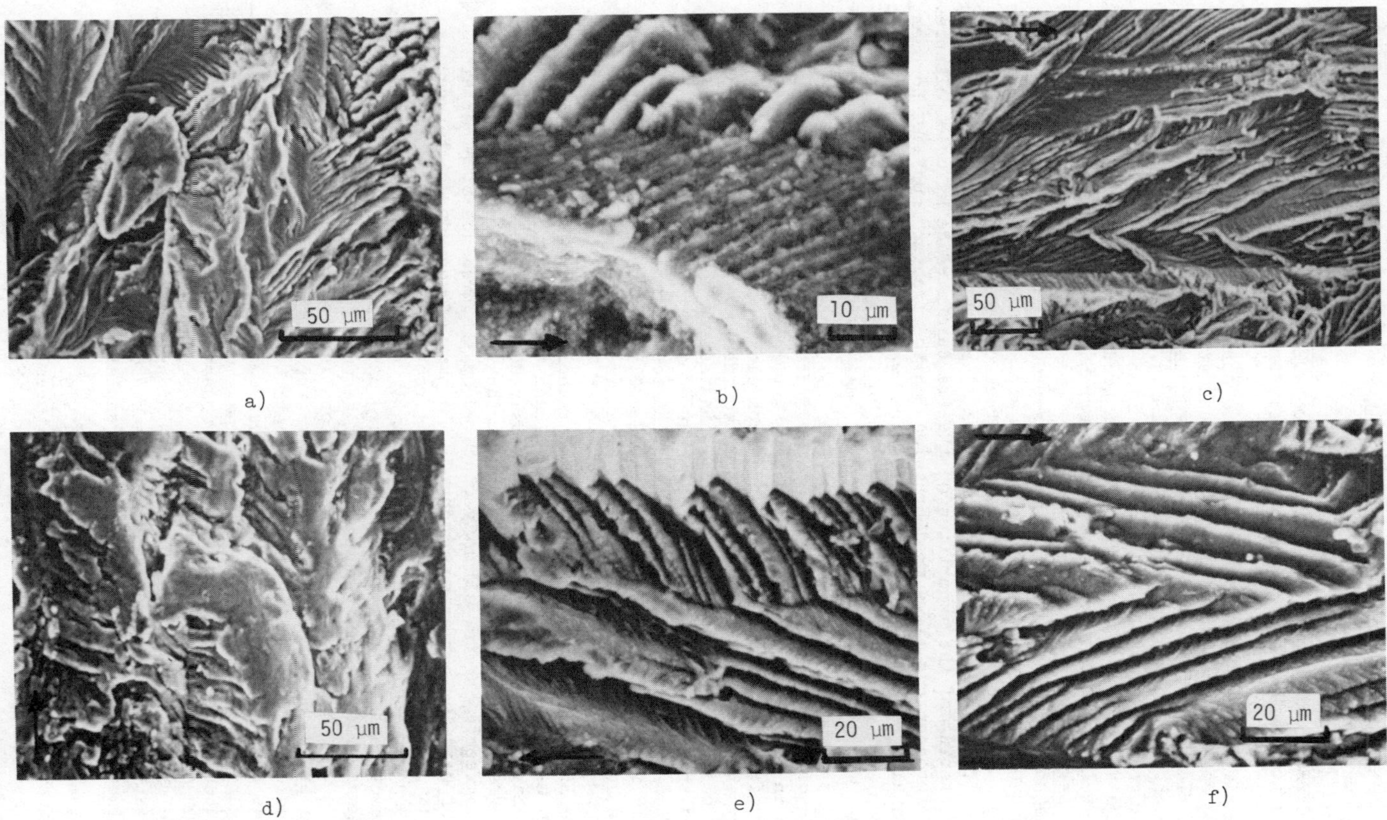

<u>Figure 3</u> - Fractographic aspects of Al alloy 2024-T351 for $da/dN \leq 10^{-7}$ mm/cycle a) Lab. air (R=0.1), b) Wet Ar (R=0.1), c) Dry Ar (R=0.1), d)e) Silicone oil (R=0.1) f) Silicone oil (R=0.5). The arrows indicate the direction of crack propagation.

The SEM fractographic observations were carried out keeping in mind that to obtain a general view a large number of different microscopic regions must be studied.

The SEM fractographic observations on 2024-T351 showed that the fracture surface at the threshold was generally transgranular and strongly crystallographic. As previously reported [6], and reverified in the present study employing etch-pitting, the fracture surfaces correspond to {100} planes. For the fracture surfaces produced at R = 0.1 in lab air, microregions flattened by crack closure can be observed (figure 3a), as well as the presence of corrosion product, especially visible on the edges of the crystallographic facets. The fracture surfaces produced at R = 0.1 in wet argon are strongly covered with corrosion product which partially obscures the cristallographic transgranular fractographic features (figure 3b).

The near-threshold fracture surfaces produced in dry argon for R = 0.1 were less oxidized than that produced in air. A few microregions where the fracture surface had been flattened by crack closure were observed (figure 3c). The fracture surfaces produced in silicone oil indicated considerably more flattening of the asperities on the fracture surface by crack closure. In the microscopic valleys or depressions on the fracture surface, however, the crystallographic fractographic features were particularly clear (figure 3d,e), since these were relatively unoxidized and were not or little influenced by the occurrence of crack closure effects.

One of the striking features of the fracture surfaces produced at R = 0.5 in 2024-T351 for all the testing environments employed was the absence of microregions flattened by crack closure. The crystallographic fractographic features produced in the two inert environements were particularly clearly observable (figure 3f), indicating that these were less oxidized than the fracture surfaces produced in the two humid environments (lab. air and wet argon). The amount of corrosion product on the near-threshold fracture surface produced at R = 0.5 in wet argon was sufficient to often completely obscure the fractographic features.

The SEM observations on the near-threshold fracture surfaces produced in 70-30 α-brass showed that for R = 0.1 and testing in lab air, the fracture surface was similar to that previously described [6], i.e., a mixture of intergranular facets (≃ 50% of the surface) and transgranular highly crystallographic facets, slightly obscured by the presence of a certain amount of oxide or corrosion product. A number of secondary intergranular cracks can be observed (figure 4a). The fracture surfaces produced at R = 0.1 in wet argon present a significantly greater percentage of intergranular fracture facets (≃ 75%) and are more oxidized than those produced in air (figure 4b). As well, the amount of secondary intergranular cracking indicated by the fractographic observations is more important. The fracture surfaces produced at this same R-ratio in the two inert environments present considerably less intergranular facets (< 20%) than those obtained in air. The crystallographic facets are less oxidized and more clearly observed (figure 4c,e). Three sets of facets which on stereofractographs appear to be mutually perpendicular were often observed suggesting {100} orientations. One of the important features of the fracture surfaces produced in silicone oil at R = 0.1 was the presence of an important amount of microregions flattened by the occurrence of crack closure (figure 4d), which at first glance can be mistaken for intergranular facets.

The fracture surfaces produced at R = 0.5 in α-brass in the different environments presented fractographic features (amount of intergranular facets, secondary cracking, oxidized nature of the fracture surfaces,

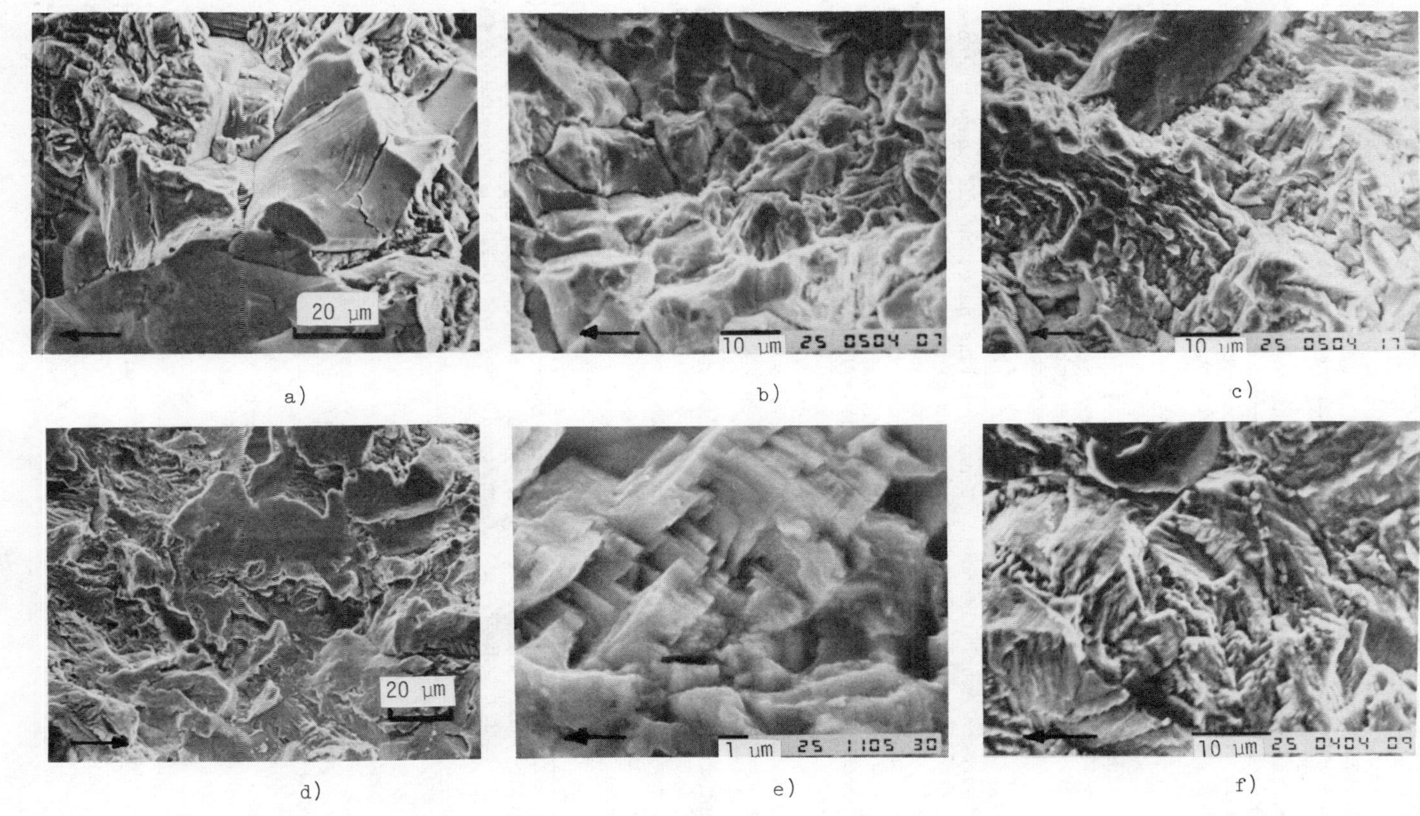

Figure 4 - Fractographic aspects of 70-30 α-brass for $da/dN \leq 10^{-7}$ mm/cycle a) Lab. air (R=0.1), b) Wet Ar (R=0.1), c) Dry Ar (R=0.1), d, e) Silicone oil (R=0.1) f) Lab. air (R=0.5). The arrows indicate the direction of crack propagation.

clarity of the transgranular facets) similar to those described for R = 0.1, with the notable exception that there was an almost complete absence of microregions flattened by crack closure for the higher R-ratio.

A last experimental observation related to the fractographic results was that the introduction of silicone oil or, to a lesser extent, of dry argon into the environmental chamber while cycling above ΔK_{th} in air very quickly resulted in an expansion of the plastic zone size. Although the plastic zone size was difficult to measure employing the travelling microscope (magnification 400X) to view the crack tip region, the rapid increase in plastic zone size as the inert environment was introduced could be easily followed, particularly for tests performed on α-brass.

Discussion

The correct explanation of the experimental dependence of the threshold values on the environment and on the R-ratio must take into account the influence of these testing variables on the values of ΔK_{th}, K_{op} and $(\Delta K_{eff})_{th}$ as well as on the fractographic features observed for the near-threshold region. As well, these explanations must be consistent with the basic properties of the tested materials (plasticity, sensitivity to oxidation and to hydrogen embrittlement). Considerations based on all these factors should permit a qualitative determination of the relative importance of the contributions made by the four mechanisms generally considered as having an influence on ΔK_{th}. These mechanisms, which are hydrogen embrittlement and plasticity-induced, roughness-induced and oxide-induced (or corrosion product-induced) crack closure, have been briefly described in the introduction.

For ease of comparison, the influence of the environment and of the R-ratio on the threshold values, ΔK_{th}, K_{op} and $(\Delta K_{eff})_{th}$ are summarized in Figure 5. The more important aspects of these results as well as the more important fractographic observations, with respect to the basic properties of each material are also summarized in Tables III and IV, for 2024-T351 and α-brass respectively. As well, these tables summarize the conclusions that have been arrived at concerning the relative importance of the four mechanisms considered.

Al alloy 2024-T351. This material is highly sensitive to oxidation in a humid environment. Consequently, the fracture surfaces for tests produced in laboratory air and in wet argon appear strongly oxidized when observed by SEM for both of the R-ratios employed. For R = 0.1, the presence of this oxide efficiently acts as a wedge propping the crack open during a portion of the load - δ_m cycle. The greater amount of oxide formed in wet argon compared to air results in a significant increase of K_{op}. For R = 0.5, the values of K_{op} are close to those of K_{min} and there is no or little contact between the opposite fracture surfaces in any of the environments employed. This lack of contact is also indicated by the apparent difference from the visual observations of the amount of oxide layer present for the two R-ratios which is not indicated from the SEM observations where the fracture surfaces produced in wet argon at R = 0.5 appeared to be the most heavily-oxidized. It appears that the darker oxide produced at R = 0.1 is a fretting-corrosion oxide, while no or little fretting but considerable oxidation occurred at R = 0.5.

In the inert environments of dry argon or silicone oil and for tests at R = 0.1, the increase in plastic zone size of this alloy of moderate plasticity gives rise to a moderate combined effect of plasticity-induced

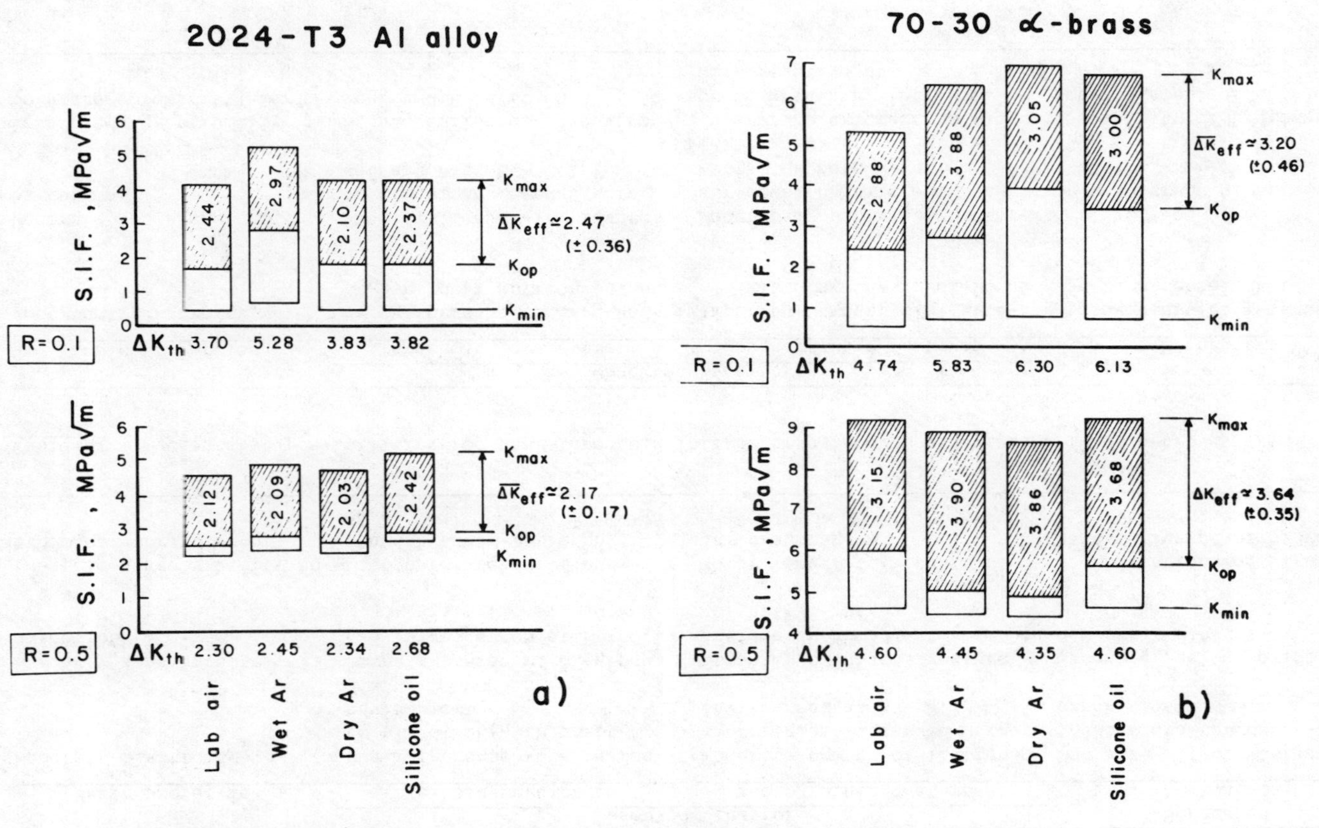

Figure 5 - Threshold values, ΔK_{th}, K_{op} and $(\Delta K_{eff})_{th}$ for
a) Al alloy 2024-T351, b) 70-30 α-brass
The $(\Delta K_{eff})_{th}$ values are given with ± one standard deviation

Table III - Mechanistic possibilities for near-threshold fatigue crack growth behavior in aluminum alloy 2024-T351

Basic properties	Observed behavior: Threshold parameters	Observed behavior: Fractographic features	Suggested Mechanisms
Moderate plasticity	Small increase with respect to air of ΔK_{th} in inert environment	Moderate amount of flattened microregions observed in inert environment at R = 0.1	Moderate effect of plasticity-induced and roughness-induced crack closure
Strongly sensitive to oxidation	Large increase of ΔK_{th} and K_{op} in wet argon at R = 0.1	Strongly oxidized fracture surface in humid environments for both R-ratios	Strong effect of oxide induced crack closure
Possible susceptibility to hydrogen embrittlement	Little influence of environment and R-ratio on ΔK_{eff}	No or very little intergranular cracking	No indications of significant influence of hydrogen embrittlement.

Table IV - Mechanistic possibilities for near-threshold fatigue crack growth behavior in 70-30 α-brass

Basic properties	Observed behavior: Threshold parameters	Observed behavior: Fractographic features	Suggested Mechanisms
Important plasticity	Large increase of ΔK_{th} and K_{op} in inert environment at R = 0.1	Important amount of flattened microregions in silicone oil at R = 0.1	Strong plasticity-induced and roughness-induced crack closure
Moderate resistance to oxidation	Small increase with respect to air of ΔK_{th} and K_{op} in wet argon at R = 0.1	Moderately oxidized fracture surfaces in humid environments for both R-ratios	Small effect of oxide-induced crack closure
Possible susceptibility to hydrogen embrittlement	Little influence of environment and R-ratio on ΔK_{eff}	Intergranular cracking and crack branching increase with relative humidity	Possibility of hydrogen embrittlement effects

and roughness-induced crack closure, as indicated by the flattened microregions at asperities on the fracture surface. This effect does not result in a significant increase in ΔK_{th} with respect to the laboratory air value, indicating that the increased contribution of plasticity-induced crack closure in these environments are probably offset by a decreased contribution of oxide-induced crack closure. The fractographic observations indicate significantly more flattening of microregions at asperities by plasticity-induced and roughness-induced crack closure on the fracture surfaces produced at R = 0.1 in silicone oil than in dry argon, suggesting that the latter environment is less inert and that the ΔK_{th} value measured in this dry argon environment is influenced by a less important plasticity-induced crack closure effect and a more important oxide-induced crack closure effect than that in silicone oil. When the R-ratio is increased to 0.5, both types of crack closure effects essentially or largely disappear.

It should be noted that since the fracture surface flattening in silicone oil is confined to asperities on the fracture surfaces, the deformation of these surfaces should not be due to hydrodynamic wedging effects.

The possibility of a hydrogen embrittlement effect at ΔK_{th} for the 2024-T351 alloy, as suggested for aluminum alloys by some authors [12, 20, 21], is not supported by the lack of influence of the environment and of the R-ratio on $(\Delta K_{eff})_{th}$. There was a notable absence of intergranular facets on the fracture surfaces and while stress-corrosion cracking of 2024-T351 is generally intergranular [26] for cracking in the ST or SL orientations, the pancake-shaped grains of this microstructure tend to prevent the occurrence of intergranular cracking in the TL orientation. As previously pointed out [3, 6], the crystallographic facets observed in the near-threshold region in f.c.c. metals, including in relatively inert environments, very strongly resemble those produced during transgranular stress-corrosion cracking of f.c.c. alloys, including of Al alloys [27]. The present fractographic observations, including the lack of intergranular facets, therefore, cannot be considered as precluding a possible hydrogen embrittlement effect.

It can therefore be concluded that, at a low R-ratio, 2024-T351 is especially sensitive to oxide-induced crack closure effects. The other effects appear moderate, with the exception of the tests in silicone oil, where the fractographic observations suggest a quite strong combined plasticity-induced and roughness-induced crack closure effect at R = 0.1.

70-30 α-brass. This material can be characterized as highly plastic and moderately sensitive to oxidation in a humid environment. Accordingly, an important influence of plasticity-induced and roughness-induced crack closure can be expected at a low R-ratio, especially in the inert environments, where an expansion of the plastic zone size was observed on introducing the inert environment. Similar increases in plastic-zone size have been reported by other authors [28, 29]. From slip line observations, Ait Bassidi [29] noted an approximate doubling of zone size on a duplex stainless steel, following the introduction of silicone oil.

The greater amount of flattened microregions on the fracture surfaces produced in silicone oil at R = 0.1 in comparison to that produced in dry argon is in agreement with the particularly observable increase in plastic zone size noted on introducing silicone oil into the environmental chamber during fatigue cycling. The influence of both these inert environments on significantly increasing ΔK_{th} and K_{op} at R = 0.1 with respect to the laboratory air values is certainly an indication of strong plasticity-induced and roughness-induced crack closure effects.

The higher value of ΔK_{th} at R = 0.1 in wet argon with respect to laboratory air tends to suggest a moderate oxide-induced crack closure effect; however, the measured K_{op} values were similar in both these humid environments. One possibility for explaining the higher ΔK_{th} and $(\Delta K_{th})_{eff}$ values but similar K_{op} value for wet argon with respect to laboratory air is the greater amount of intergranular and secondary cracking observed in the more humid environment. This secondary cracking or crack branching can be expected to cause the true effective ΔK_{th} value to be lower than the measured $(\Delta K_{th})_{eff}$, since this effect is not expected to influence K_{op}.

The important amount of intergranular cracking in the near-threshold region of α-brass and the increase in the percentage of intergranular facets with increasing humidity of the environment suggests the possibility of a significant effect on ΔK_{th} of hydrogen embrittlement. As for the Al alloy, such a possibility is not supported by the observed lack of influence of the environment and the R-ratio on $(\Delta K_{eff})_{th}$, whose value appears to remain essentially constant for each material, within the expected experimental accuracy. While it can be expected that an increasing importance of hydrogen embrittlement effects should tend to decrease $(\Delta K_{eff})_{th}$, this decrease in the case of intergranular cracking with crack branching can be expected to be at least partially offset by the increasing relaxation of the stress intensity factor associated with this crack branching. This tendency to overestimate $(\Delta K_{eff})_{th}$ tends to make it difficult to detect hydrogen embrittlement from the measured mechanical parameters. This relative independence of $(\Delta K_{eff})_{th}$ on the environment and R-ratio, nevertheless, tends to suggest that the possible decrease of the true $(\Delta K_{eff})_{th}$ as a result of hydrogen embrittlement is small.

As already mentioned, even the transgranular facets obtained in the near-threshold region in the different environments strongly resemble those obtained during stress-corrosion cracking. It appears improbable that in the inert silicone oil, that these could be produced by an environmentally-assisted fracture mode. Rather, it appears that this strong resemblance is associated with both types of cracking being produced by small, discontinuous cleavage-type crack bursts, as experimentally demonstrated in the case of stress-corrosion cracking of Admiralty metal [30], but with such crack bursts being able to occur in the absence of environmental assistance in the fatigue threshold region.

Constancy of $(\Delta K_{eff})_{th}$

Let us now consider whether the differences in the $(\Delta K_{eff})_{th}$ values obtained for each material are significant. In the case of carefully carried out measurements, an experimental dispersion of the order of ±10% can be expected for ΔK_{th} results from different samples. The dispersion on the measurements of K_{op} should be no less than ±10%. That on $(\Delta K_{eff})_{th}$, which is obtained as the difference between ΔK_{th} and K_{op}, should therefore be no better than ±20%. Figure 5 gives the average $(\Delta K_{eff})_{th}$ value ± a standard deviation for each material and R-ratio. It should be noted that one standard deviation is always less than 15% of $(\Delta K_{eff})_{th}$ and that all values are within 1.5 standard deviations of $(\Delta K_{eff})_{th}$. For the influence of environment on ΔK_{th}, it is clear from Table II that significant differences exist for both materials at R = 0.1, but that the differences at R = 0.5, where $K_{op} \approx K_{min}$, are not statistically significant for both materials. It is also clear that a significant influence of R-ratio on ΔK_{th} also exists. Looking at the influence of environment of $(\Delta K_{eff})_{th}$ for each R-ratio, one observes in all cases good to very good agreement for tests in three environments. The fourth environment giving poorer but still reasonable agreement is generally different suggesting that this lesser agreement is only the result of the quite high experimental dispersion possible. The average of the sets of three $(\Delta K_{eff})_{th}$ values that show good agreement are, for 2024-T351, 2.30 and 2.08 $MPa(m)^{1/2}$ and for α-brass 2.97 and 3.81 $MPa(m)^{1/2}$, for R = 0.1 and 0.5 respectively. The agreement for 2024-T351 is very good. For α-brass the indication of a higher $(\Delta K_{eff})_{th}$ at a higher R is the opposite of that which can be expected, which again indicates that this difference is only the result of the quite high dispersion possible on the $(\Delta K_{eff})_{th}$ values. This is further indicated by the average of all four $(\Delta K_{eff})_{th}$ values (Figure 5) at the two R-ratios being in better agreement for this material.

In considering the experimental dispersion possible in the $(\Delta K_{eff})_{th}$, values and the sporadic nature of this dispersion, the conclusion that can be drawn is that the value of $(\Delta K_{eff})_{th}$, within the expected experimental accuracy, is independent of the R-ratio and testing environment. This important result indicates that the ΔK_{th} value can be decomposed, according to the following equation:

$$\Delta K_{th} = (\Delta K_{eff})_{th} + (K_{op} - K_{min}),$$

with the value of $(\Delta K_{eff})_{th}$ being essentially a material constant. This constant value should represent a limiting value of effective ΔK below which there is insufficient damage accumulated ahead of the crack tip to result in the fracture of this damaged zone.

It can be expected that $(\Delta K_{eff})_{th}$ should depend on the bulk material properties and should be influenced as these properties in the crack tip region are changed, for example, by a change in microstructure or by the occurrence of hydrogen embrittlement. In contrast, the $(K_{op} - K_{min})$ contribution to ΔK_{th} should be strongly influenced by the extrinsic mechanisms that could influence in the near-threshold region, such as the formation of oxide or corrosion product or a change in crack tip plastic zone size associated with a change in environment. This contribution to ΔK_{th} can also be expected to be influenced by bulk properties, particularly the material's inherent plasticity, which can be varied by heat treatment or by cold-working.

As found in the present study, for a material highly susceptible to oxidation, e.g., 2024-T351 Al alloy, $(K_{op} - K_{min})$ strongly increases in the case of testing in a very humid environment at a low R-ratio. For a material that is less susceptible to oxidation but more susceptible to plasticity-induced crack closure, e.g., α-brass, this contribution to ΔK_{th} is more strongly influenced by the presence of an inert environment (dry argon or silicone oil) which results in an increase of the crack tip plastic zone size.

The combined influence of the environment and of the R-ratio on ΔK_{th} manifests itself by effects which are, at least to some extent, interrelated and which depend on the basic properties of the material (plasticity, susceptibility to oxidation, susceptibility to hydrogen embrittlement). As a result, it should often be difficult to predict correctly a priori the net effect of this combined influence in the case of a particular material tested in a particular environment at a particular (low) R-ratio.

Conclusions

From the present study, the following conclusions can be drawn:

1. Testing at a low R-ratio in a very humid environment results in an important increase of ΔK_{th} and of K_{op} for 2024-T351 Al relative to air.

2. For the same alloy, oxide-induced crack closure effects can have a strong influence on ΔK_{th} and on the near-threshold fatigue behaviour.

3. Testing at a low R-ratio in an inert environment results in an significant increase of ΔK_{th} and of K_{op} for 70-30 α-brass relative to air.

4. For this material, plasticity-induced and roughness-induced crack closure effects can have a strong influence on ΔK_{th} and on the near-threshold fatigue behaviour.

5. The amount of intergranular cracking and crack-branching observed in the near-threshold region of α-brass increases with increasing humidity, suggesting a possible role played by hydrogen embrittlement.

6. For each of these materials, while the crack closure effects influence K_{op}, the value of $(\Delta K_{eff})_{th}$ remains essentially independent of the R-ratio and of the testing environment, for the environments employed.

7. It can therefore be proposed that ΔK_{th} is the sum of two contributions, $(\Delta K_{eff})_{th}$, a material constant, and $(K_{op} - K_{min})$. The former corresponds to a critical value of ΔK_{eff}, below which there is insufficient accumulation of fatigue damage ahead of the crack tip to result in a crack advance. The second contribution, $(K_{op} - K_{min})$ characterizes the net effect of the different crack closure mechanisms, and is strongly influenced by the environment and by the R-ratio.

8. The strong similarity between transgranular fractographic facets produced during stress-corrosion cracking and those produced near the fatigue threshold including in inert environments can be explained by proposing that both types of crack propagation are produced by small, discontinuous cleavage-type crack bursts.

Acknowledgements

The authors are particularly indebted to Mr Louis Handfield for his considerable assistance in automatizing the fatigue tests. The technical assistance of Jean Claudinon and Jacques Desrochers was much appreciated. Financial support from the Natural Sciences and Engineering Research Council of Canada, (grants A-6694 and A-6330) and from the Quebec "Formation des Chercheurs et Actions Concertées" Program (grant CRP 336-74) is gratefully acknowledged.

References

[1] Linder, B.M. "Extremely slow crack growth rates in Aluminum alloy 7075-T6", M.Sc. Thesis, Lehigh University, 1965.

[2] Ritchie, R.O., "Near-threshold fatigue-crack propagation in steels", Intl. Metals Review, (1979), 20 (516) pp. 205-230.

[3] Dickson, J.I., Baïlon, J.P. and Masounave, J. "A review on the threshold stress intensity range for fatigue crack propagation", Can. Metall. Quart. (1981), vol. 20, pp. 317-329.

[4] Bäcklund, J., Blom, F.A. and Beevers, C.J. "Fatigue Thresholds", EMSA Publ. Ltd, Warley, U.K., (1982), Volumes 1 and 2.

[5] Elber, W., "The significance of fatigue crack closure" in "Damage tolerance in aircraft structures", ASTM-STP 486, Am. Soc. Test. Mater., Philadelphia (1971), pp. 230-240.

[6] Bailon, J.P., Chappuis, P., Masounave, J. and Dickson, J.I. "Fractographic aspects near the threshold in several alloys", in ref [4], vol. 1, pp. 277-291.

[7] Cooke, R.J. and Beevers, C.J. "Slow fatigue crack propagation in pearlitic steels", Mater. Sci. Eng. (1974), vol. 13, pp. 201-210.

[8] Gray, G.T., Williams, J.C. and Thompson, A.W. "Roughness-induced crack closure: an explanation for microstructurally sensitive fatigue crack growth", Metall. Trans. A(1983), vol. 14A, pp. 421-433.

[9] Minakawa, K. and McEvily, A.J. "On near-threshold crack growth in steels and aluminum alloys", in ref [4], pp. 373-390.

[10] Minakawa, K. and McEvily, A.J. "On the crack closure in the near-threshold region", Scripta Metall. (1981), vol. 15, pp. 633-636.

[11] Benoit, D., Lieurade, H.P., Namdar-Irani R. and Tixier, R. "Oxydation des surfaces de rupture par fatigue des aciers aux basses vitesses de fissuration", Mém. Sci. Rev. Metall., (1981), pp. 569-583.

[12] Petit, J. and Zeghoul, A. "Gaseous environmental effect on threshold level in high strength aluminum alloys", in ref [4], vol. 1, pp. 563-579.

[13] Ritchie, R.O. "Environmental effects on near-threshold fatigue crack propagation in steels: a re-assessment", in ref [4], vil. 1, pp. 503-526.

[14] Suresh, S., Zarniski, G.F. and Ritchie, R.O. "Oxide-induced crack closure: an explanation for near threshold corrosion fatigue crack growth behaviour", Metall. Trans A(1981), vol. 12A, pp. 1435-1443.

[15] Liaw, P.K., Leax, T.R., Williams, R.S. and Peck, M.G. "Influence of oxide-induced crack closure on near-threshold fatigue crack growth behaviour", Acta Met.(1982), vol. 30, p. 2071-2078.

[16] Liaw, P.K., Saxena, A., Swatimathan, V.P. and Shih, T.T. "Effects of load ratio and temperature on the near-threshold fatigue crack propagation behavior in a CrMoV steel", Metall. Trans.A(1983), vol. 14A, pp. 1631-1640.

[17] Stewart, A.T. "The influence of environment and stress ratio on fatigue crack growth at near-threshold stress intensities in low alloy steels", Eng. Fract. Mech. (1980), vol. 13, pp. 463-478.

[18] Suresh, S. and Ritchie, R.O. "Mechanistic dissimilarity between environmentally influenced fatigue crack propagation at near-threshold and higher crack growth rates in lower strength steels", Metal Sci., vol. 16, (1982) pp. 529-538.

[19] Ritchie, R.O. and Suresh, S. "A comparison of environmentally-influenced near-threshold fatigue crack growth in high and lower strength steels at conventional frequencies", in "Ultrasonic fatigue", Proceedings of the 1st Intl. Conf. on Fatigue and Corrosion Fatigue up to Ultrasonic Frequencies, Champion (Penn.) Oct. 25-30, 1981, J.M. Wells et al. ed., ASTM Publications, Philadelphia, pp. 443-460.

[20] Vasudevan, A.K. and Suresh, S. "Influence of corrosion deposits on threshold fatigue crack growth behaviour in 2XXX and 7XXX Series aluminum alloys", Metall. Trans. A(1982), vol. 13A, pp. 2271-2280.

[21] Renaud, P., Violan, P., Petit, J. and Ferton, D. "Microstructural influence on fatigue crack growth near threshold in 7075 Al alloy" Scripta Metall. (1982), vol. 16, pp. 1311-1316.

[22] Ritchie, R.O. "Influence of microstructure on near-threshold fatigue crack propagation in ultra-high strength steels", Metal Science (1977), vol. 11, pp. 368-381.

[23] Marchand, N., Baïlon, J.P. and Dickson, J.I., to be published.

[24] Baïlon, J.P., Chappuis, P. and Masounave, J. "A rapid experimental method for measuring the threshold stress intensity factor", in ref. [4], pp. 77-98.

[25] El Boujdaini, M. "Influence de l'environnement sur le seuil de propagation en fatigue", M.Sc. thesis, Ecole polytechnique, Montréal, (1983).

[26] Speidel, M.O. and Hyatt, M.V., "Stress-corrosion cracking of high strength aluminum alloys" in Advances in Corrosion Science and Technology, edited by M.G. Fontana and R.W. Staehle, Plenum Press, New York (1972), vol. II, p. 115-335.

[27] Dorward, R.C. and Hasse, K.R., "The fractography of long-transverse stress-corrosion cracking in Al-Zn-Mg-Cu Alloys", Corrosion Science (1982), vol. 22, pp. 251-257.

[28] De Fouquet, J. "Effet d'environnement" in "La fatigue des matériaux et des structures", C. Bathias and J.P. Baïlon, ed., Maloine S.A., Paris (1980), pp. 291-312.

[29] Aït Bassidi, M. "Etude des mécanismes de fissuration par fatigue-corrosion de différents types d'aciers inoxydables". Ph.D. thesis, Ecole polytechnique, Montréal, (1983).

[30] Pugh, E.N. "On the propagation of transgranular stress-corrosion cracks", Proceedings of NATO Advanced Research Institute Conference on Atomistics of Fracture, (Corsica, May 1981), Plenum Press, New York, NY, pp. 997-1010.

EFFECT OF MICROSTRUCTURE AND LOAD RATIO ON ΔK_{th} IN TITANIUM ALLOYS

James C. Chesnutt and John A. Wert

Rockwell International Science Center
Thousand Oaks, California 91360

Both microstructure and load ratio were found to have a significant effect on near-threshold fatigue crack propagation in Ti-6Al-4V. The effect of microstructure on fatigue crack propagation at near-threshold growth rates was evaluated for three distinct microstructures: recrystallization annealed, beta-annealed and solution treated and overaged. At a constant load ratio, the crack growth rate and ΔK_{th} for these microstructures correlated well with microstructural dimensions and low cycle fatigue parameters.

Beta-transformed, especially beta-annealed, microstructures of Ti-6Al-4V are of interest as a result of their excellent crack propagation resistance. Holding time above the beta transus and cooling rate were both shown to affect the crack propagation resistance of these microstructures. For the beta-transformed microstructures, the effect of load ratio was strongly dependent on microstructure. An increase in load ratio from R = 0.1 to R = 0.75 resulted in considerable increases in crack growth rate in a beta-quenched microstructure, but only moderate increases in a beta-annealed microstructure.

Threshhold ΔK is seen, for titanium alloys, to affect the durability of components designed using fracture mechanics design and spectrum loading conditions.

Introduction

The application of titanium alloys to aircraft structures designed to damage tolerant requirements has prompted considerable interest in fatigue crack propagation in these alloys, especially Ti-6Al-4V. While microstructure has been shown to have considerable effect on growth rate in the Paris Law region (1,2), the effect at near-threshold growth rates has not been as extensively characterized.

Methods for predicting fatigue crack propagation behavior from low cycle fatigue properties and microstructural parameters are currently receiving considerable interest. Charkrabortty (3) and Charkrabortty and Starke (4) have utilized low cycle fatigue properties and microstructural parameters to verify fatigue crack propagation behavior of three titanium-vanadium alloys in the crack growth rate range of 10^{-6} to 10^{-3} mm/cycle (4×10^{-8} to 4×10^{-5} in./cycle). The microstructures which Chakrabortty et al used were single phase beta and beta with precipitated α and/or ω phases. For their study, the fatigue crack propagation behavior could be characterized in terms of the beta grain size for the single phase material or the α particle spacing for the multiphase conditions. An objective of this program is to extend the predictive method to more complex microstructures such as those produced in Ti-6Al-4V and to evaluate the applicability of the method to lower crack growth rates (approximately 10^{-7} mm/cycle), a growth rate region in which the crack growth rate (da/dN) vs stress intensity (ΔK) relationship may deviate significantly from linearity.

There is a continuing interest in beta-processing of alpha-beta titanium alloys for improved shape definition and for cost reduction associated with lower forging loads. Fracture toughness and fatigue crack propagation resistance improvements are often associated with the coarse, colony microstructures produced by such beta processing (2,5). These microstructures may also exhibit considerable scatter in the fatigue crack growth rate data compared to alpha-beta processed microstructures.

The importance of fatigue crack growth behavior to modern high performance aircraft design is exemplified by the damage tolerant design methodology embodied in military standard MIL-STD-1530A, Aircraft Structural Integrity Program, Airplane Requirements (6) and implemented by Chang et al in B1-B design (7).

Material and Experimental Procedure

The material used for this study was in two forms, plate and pancake forgings. Chemistries for both forms are given in Table I. Plate material was used for the study of three microstructural conditions (RA, BA and STOA) at constant R. The plate had been purchased by Rockwell International to the fracture toughness plate specification for the B-1 aircraft. The plate had a moderate texture, with basal poles two to four times random in both the rolling direction (L) and the transverse direction (T).

Specimen blanks were removed from the plate such that the loading axis of the tensile specimens would be parallel to the T direction and the FCP specimen would have a TL orientation. The blanks were then heat treated and finished machined. The heat treatments based on previous experience (8) and adjusted to the beta transus temperature of this plate are given in Table II.

The material used for the study of the effect of beta processing and R ratio was from a pancake forging used in a previous study (8). The pancake forging had been forged above the beta transus and air cooled; the resulting condition is referred to as beta-forged (BF). The forging was weakly textured. Specimen blanks were removed such that the loading axis of the tensile specimens would be parallel to the C direction of the forging and the FCP specimens would have a CR orientation. The blanks were then heat treated and finish machined. The heat treatments are given in Table III.

Table I. Alloy Composition, wt %

Condition	Al	V	Fe	C	O	N	H
Plate	6.1	4.0	0.19	0.020	0.122	0.018	0.009
Forging	6.2	4.1	0.22	0.010	0.122	0.012	0.003

Table II. Heat Treatment Schedules for Plate Material

Condition	Heat Treatment
Recrystallization Anneal (RA)	927°C/4h/cool @ 50°C/h to 760°C/AC
Beta Anneal (BA)	1038°C/0.5h/AC + 704°C/4h/AC
Solution Treat and Overage (STOA)	955°C/1h/WQ + 593°C/4h/AC

Table III. Heat Treatment Schedules for Forging Material (1)

Condition	Heat Treatment
Beta Forge (BF)	As-forged plus 704°C/2h/AC
Beta Anneal (BA)	1038°C/0.5h/AC + 704°C/2h/AC
Beta Quench (BQ)	1038°C/0.5h/WQ + 704°C/2h/AC

Key: AC = air cool
WQ = water quench
(1) Forged at 42°C above the beta transus temperature and air cooled.

Optical metallography specimens were prepared by electropolishing using a Buehler recirculating pump cell. Specimens were etched in Kroll's etchant (95% H_2O, 3% HNO_3, 2% HF) or an etchant composed of equal parts of 10% aqueous oxalic acid and 1% aqueous HF solutions.

Tensile tests were conducted in accordance with ASTM E 8 using an Instron test machine with a 90kN load cell. Specimens with a 6.4 mm diameter x 25.4 mm long gauge length were tested at a strain-rate of $1.67 \times 10^{-4}\ s^{-1}$.

Fatigue crack propagation tests were conducted in accordance with the proposed ASTM standard test method (9) using the load shedding technique. All other test conditions conformed to ASTM E 647 using a compact type (CT) specimen with B = 12.7 mm and W = 50 mm. Tests were conducted in laboratory air (approximately 50% relative humidity) at a frequency of 20-30 Hz.

Experimental Results

Microstructure and Mechanical Properties

The microstructures produced for this study are shown in Fig. 1. The plate microstructures (Fig. 1a, b and c) are three which previously have been shown to have significantly different fatigue crack propagation behavior. The transformed beta microstructures (Fig. 1d, e and f) each have distinctive characteristics. The beta-forged microstructure, Fig. 1d, comprises Widmanstatten alpha with little or no grain boundary alpha. The beta-annealed microstructure, Fig. 1e, is similiar to the beta-forged microstructure, but has smaller Widmanstatten alpha colonies, and nearly continuous grain boundary alpha. The beta-quenched microstructure, Fig. 1f, comprises a fine martensitic alpha structure without discernable colonies or grain boundary alpha. It should be noted that a previous study has shown that the microstructure resulting from quenching thick sections (greater than about 12 mm) of Ti-6Al-4V from temperatures above the beta transus, results in a combination of martensitic and nucleation and growth alpha, the latter in the specimen interior.

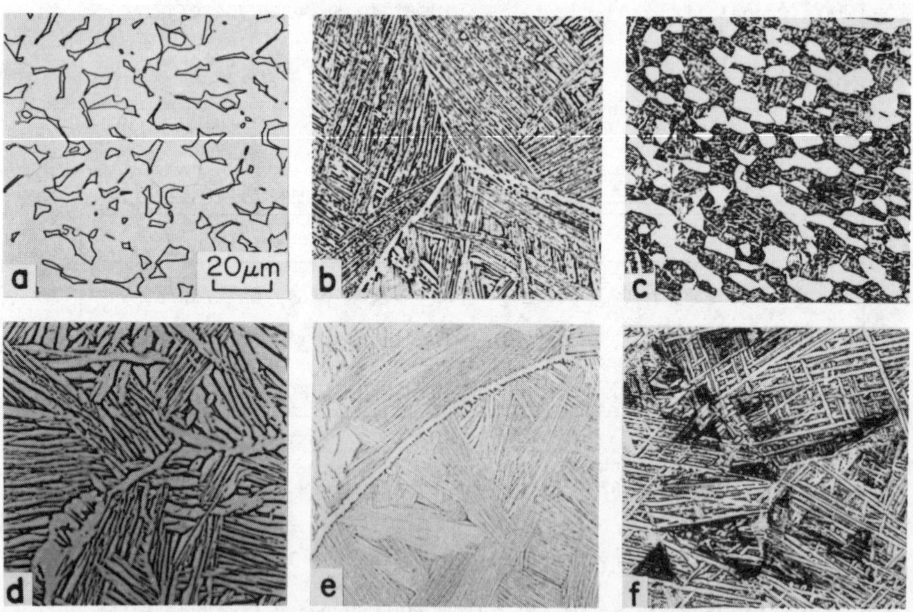

Figure 1 - Microstructures studied: (a) recrystallization annealed (RA) plate, (b) beta-annealed (BA) plate, (c) solution treated and overaged (STOA) plate, (d) beta-forged (BF) forging, (e) beta annealed (BA) forging, and (f) beta-quenched (BQ) forging.

Room temperature tensile properties for the six microstructures studied are given in Table IV.

Table IV. Tensile Properties of the Microstructural Conditions Studied

Condition	$\sigma_{0.2}$(MPa)	σ_u(MPa)	e(%)	RA(%)
		Plate Material		
RA	862	910	---	30
BA	862	927	---	21
STOA	1038	1081	---	35
		Forging Material		
BF	774	852	11.2	23
BA	784	874	9.4	13
BQ	848	928	5.4	7

Effect of Microstructure on FCP at Constant Load Ratio

Three microstructural conditions were tested in laboratory air. Test data were compared to fatigue crack propagation curves predicted from a characteristic microstructural parameter and low cycle fatigue parameters.

Fatigue Crack Propagation All three microstructural conditions were tested in laboratory air; the fatigue crack propagation behavior is shown in Fig. 2. For the RA condition, a fairly smooth curve was obtained with some scatter in the data at ΔK levels between 10 and 20 MPa•m$^{1/2}$ (Fig. 3b). The threshold stress intensity (ΔK_{th}) for this condition is found, by the least squares method of the proposed ASTM test method (9) to be 5.92 MPa•m$^{1/2}$. Ritchie (10) provides a different definition for the threshold stress intensity (ΔK_o) which he defines as the stress intensity below which

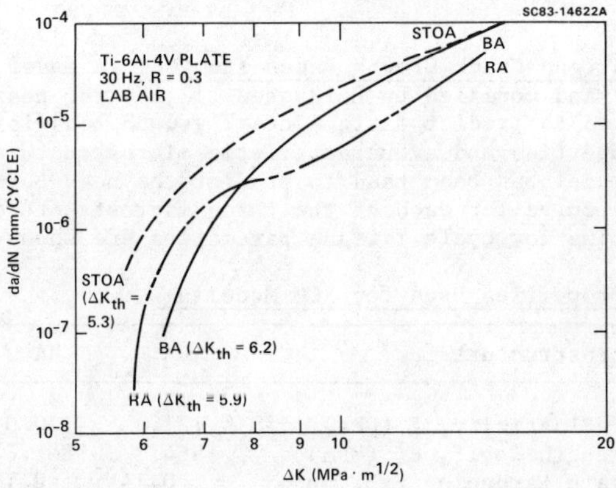

Figure 2 - FCP of Ti-6Al-4V as a function of microstructure, laboratory air, R = 0.3, 30 Hz.

no measurable growth (> 0.1 mm) occurs in 10^7 cycles. For the data plotted in Fig. 3, this criterion was also applied, resulting in a value for ΔK_o of 5.71 MPa•$m^{1/2}$, a difference of less than 5%. For the remainder of the

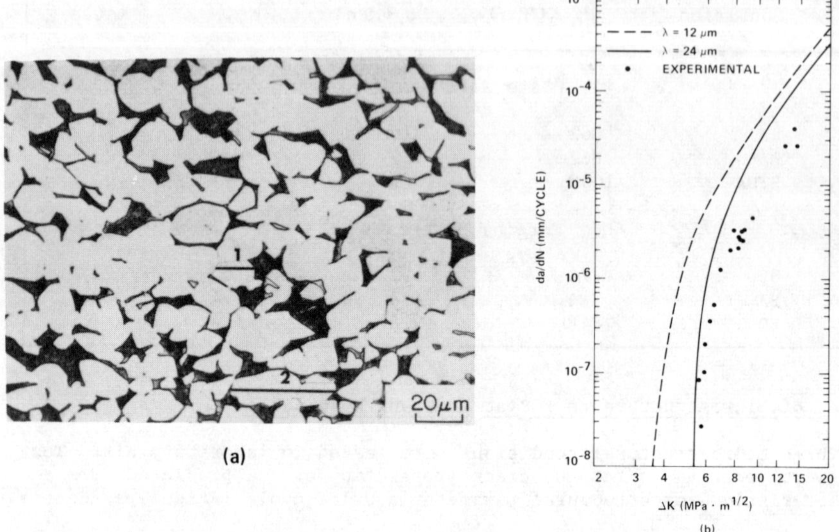

Figure 3 - Predicted FCP for Ti-6Al-4V, RA plate; (a) microstructure, (b) predicted curve and experimental data.

tests, the least squares method was used, a choice which is further justified by the discontinuous nature of crack propagation at near threshold growth rates in titanium alloys, especially for transformed beta microstructures. For these microstructures, a least squares fit of the data in the lowest decade of crack growth should give the most consistent results. FCP data for the BA and STOA microstructures are shown in Figs. 4 and 5, respectively.

Prediction of Fatigue Crack Growth Rates from LCF A model, developed by Chakrabortty (11) and modified by Heikkenen (12,13) for near-threshold growth, can be used to predict fatigue crack growth behavior from low-cycle fatigue (LCF) properties and a characteristic microstructural parameter. This predictive model has been used to predict the near-threshold fatigue crack growth rate curve for each of the three microstructures tested in laboratory air. The low cycle fatigue parameters are shown in Table V.

Table V. Properties Used for FCP Modeling

Microstructure	RA	BA	STOA
Modulus of Elasticity, E (GPa)	123.4	120.0	126.2
Cyclic Strength Coeff, K' (MPa)	1964.3	1801.6	2060.8
Cyclic Strain Hardening Exp., n'	0.14	0.14	0.14
Fatigue Ductility Coeff., ε_f'	0.40	0.27	0.52
Fatigue Ductility Exp., c	-0.76	-0.79	-0.75

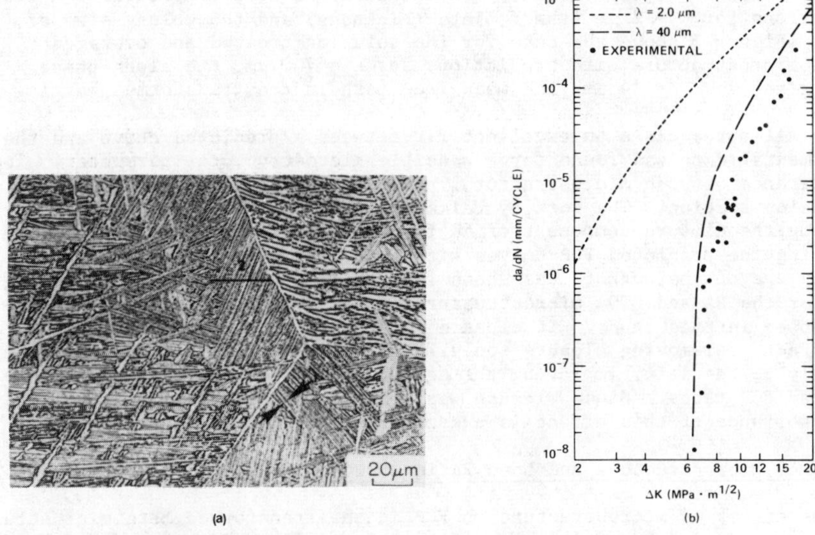

Figure 4 - Predicted FCP for Ti-6Al-4V, BA plate; (a) microstructure, (b) predicted curve and experimental data.

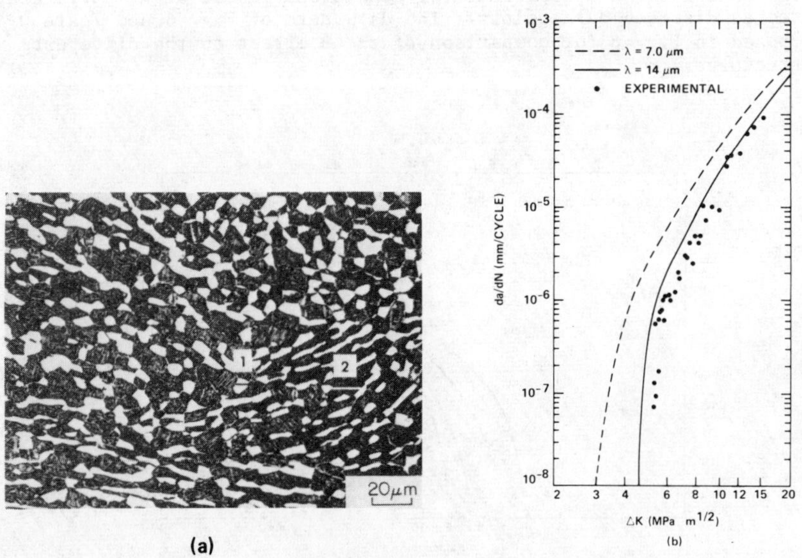

Figure 5 - Predicted FCP for Ti-6Al-4V, STOA plate; (a) microstructure, (b) predicted curve and experimental data.

The results of the predictions along with the experimental data are shown in Figs. 3, 4 and 5. Figure 3 shows the data and prediction for two different microstructural parameters in RA microstructure, 1) the mean alpha phase grain size (λ = 12 μm) and 2) the alpha phase mean free path (λ =

24 μm). Figure 4 shows the case for the beta annealed (BA) condition with predictions for λ = 2 μm (the α plate thickness) and the colony size of 40 μm. Figure 5 shows the case for the solution-treated and overaged (STOA) microstructure with predictions for λ = 7.0 μm, the alpha-phase grain size, and λ = 14 μm, the mean free path in the transformed matrix.

In all three cases an excellent fit between a predicted curve and the experimental data was found for a sensible microstructural parameter. The significance of such microstructural parameters will be addressed in the discussion section. The work by Heikkenen (12,13) showed the importance of removing the closure component of ΔK from the experimental results before comparing the predicted FCP curves with the experimental data. Previous work by one of the authors has shown that the closure component is negligible for the RA and STOA microstructural conditions when R = 0.3, the R value used in this study. If closure is important for the BA condition, the effect of removing closure would be to shift the experimental data slightly to the left, improving the agreement between the predicted and measured FCP rates. Since closure was not measured for the BA condition, the importance of this effect is unknown.

Effect of Microstructure and Load Ratio on FCP

The effect of microstructure on FCP in the transformed beta microstructures is shown for R = 0.1 in Fig. 6, and R = 0.75 in Fig. 7. The data for beta-forged and beta-quenched microstructures tested at R = 0.1 are shown as scatter bands indicative of the scatter observed in crack growth rates. The data for all three microstructural conditions tested at R = 0.75 are presented as discrete data points. The data sets of Fig. 6 and 7 are superimposed in Fig. 8 for comparison of the R effect on the different microstructures.

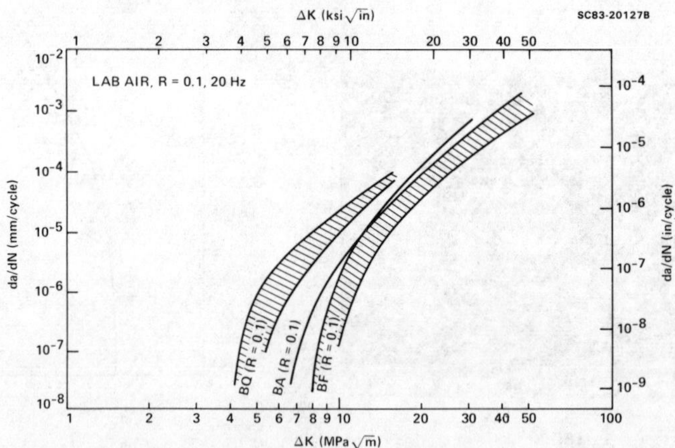

Figure 6 - FCP of Ti-6Al-4V forging as a function of microstructure, laboratory air, R = 0.1, 20 Hz.

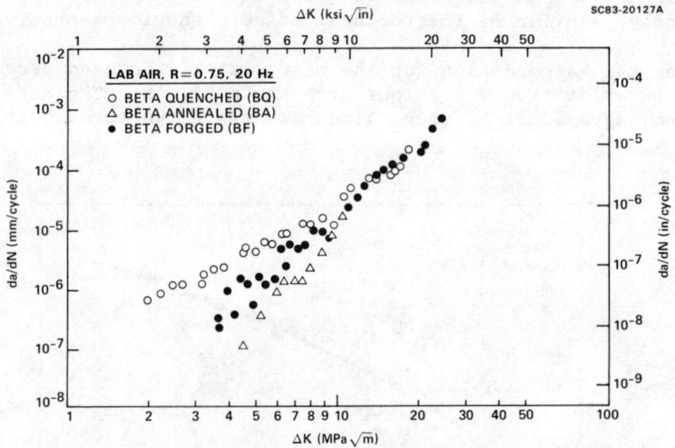

Figure 7 - FCP of Ti-6Al-4V forging as a function of microstructure, laboratory air, R = 0.75, 20 Hz.

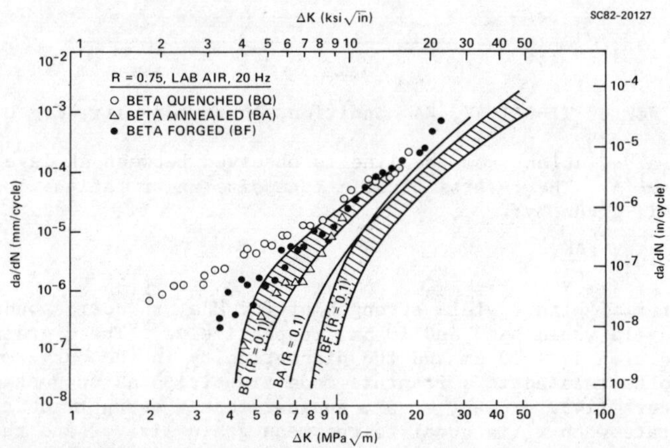

Figure 8 - FCP of Ti-6Al-4V forging as a function of microstructure and R ratio, laboratory air, 20 Hz.

Discussion

Effect of Microstructure at Constant Load Ratio

The effect of microstructure of plate Ti-6Al-4V cycled at constant load ratio will be discussed in terms of fatigue crack propagation behavior and predictive methods.

Fatigue Crack Propagation. The microstructures developed for the FCP study of plate Ti-6Al-4V were similar to those used in a previous study on pancake forgings of Ti-6Al-4V (8). Although data for the plate material in this program and pancake forgings in the previous program cannot be compared directly, similar microstructural effects should be observed.

Data for the RA condition for the plate material are compared to data for forged material from a previous program in Fig. 9. The data are seen to be in good agreement. A trend line has been drawn through the data

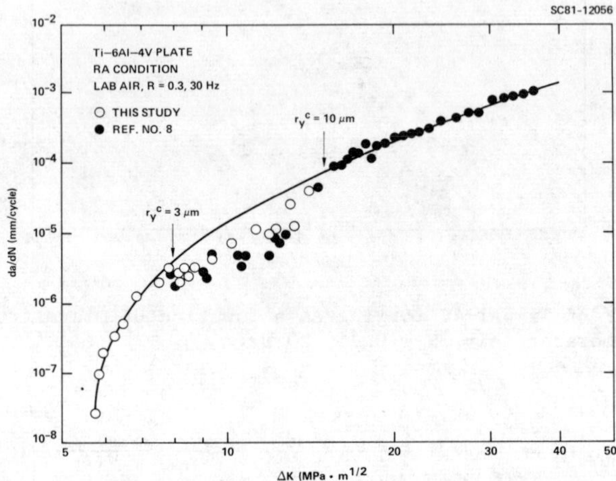

Figure 9 - FCP of Ti-6Al-4V, RA condition, laboratory air, R = 0.3, 30 Hz.

points and a deviation from the line is observed between ΔK level of ~ 8 and 15 MPa•m$^{1/2}$. The reverse plastic zone size under fatigue loading conditions is given by:

$$r_y^c = 0.033 \left(\frac{\Delta K}{\sigma_y}\right)^2 \qquad (1)$$

and for material with a yield strength of 862 MPa, r_y^c corresponding to these ΔK levels would be 3 and 10 μm, respectively. The α grain size of the RA condition is ~ 10 μm and the discontinuity in the curve might be expected to be related to a fracture mode transition as suggested by Yoder and co-workers (14). Yoder, et al, suggest that a sharp transition occurs in growth rate when r_y^c is equal to the mean grain size, $\bar{\ell}$ and that the sharpness of the transition decreases with decreased clustering of the grain size distribution about the mean value (15). In this study and work in a previous program (8), the transition has been observed to occur over a range of ΔK. The transition is characterized by a fracture mode transition from cyclic cleavage to striation formation, with mixed mode propagation occurring over a range of ΔK levels and crack growth rates. Such a fracture mode transition is illustrated in Fig. 10. Transgranular cyclic cleavage of the α-phase is the principal fracture mode at low growth rates where the plastic zone is wholly contained withing the α grains, Fig. 10a. The mixed mode fracture, at higher ΔK levels, can be seen in Fig. 10b and 10c where $r_y^c < \bar{\ell}$. For ΔK levels where $r_y^c > \bar{\ell}$, Fig. 10d, the absence of cyclic cleavage is noted. Most of the propagation at these higher ΔK levels is by striation formation although the striations are not resolved at this magnification.

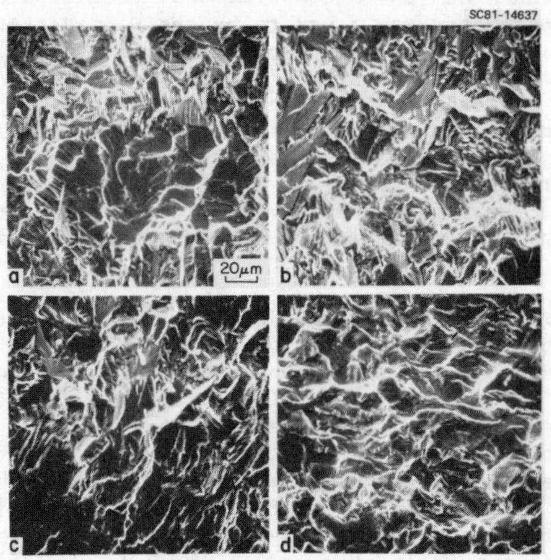

Figure 10 - SEM of Ti-6Al-4V FCP specimen, RA plate, laboratory air, R = 0.3, 30 Hz.
a) ΔK = 5.5 $MPa \cdot m^{1/2}$, da/dN = 8.5 x 10^{-9} mm/cycle.
b) ΔK = 8.3 $MPa \cdot m^{1/2}$, da/dN = 2.7 x 10^{-6} mm/cycle.
c) ΔK = 9.4 $MPa \cdot m^{1/2}$, da/dN = 4.5 x 10^{-6} mm/cycle.
d) ΔK = 18.7 $MPa \cdot m^{1/2}$, da/dN = 1.8 x 10^{-4} mm/cycle.

The ranking of the three plate microstructures in laboratory air at near-threshold growth rates is the same as the ranking from other studies (1,8), in that the STOA condition exhibits the fastest growth rate while the BA condition exhibits the slowest growth rate with the RA condition intermediate to the two. The lower growth rate of the RA microstructural condition with respect to BA microstructural condition at higher ΔK levels is associated with the dip in the curve for the RA material. Such behavior probably results from the combined effect of the grain size and texture on FCP. The material used in this study exhibited some texture as opposed to the weakly and uniformly pancake forgings used in the previous study. Beta annealed microstructures are generally found to exhibit the slowest growth rate of the three microstructural conditions at load ratios of 0.1 to 0.3.

Prediction of Fatigue Crack Growth Rate from LCF Properties Excellent results were obtained in predicting near-threshold fatigue crack growth behavior of all three microstructural conditions using the model of Chakrabortty (11) as modified by Heikkenen (12,13). In all three cases, the microstructural parameter which resulted in the predicted curve being in excellent agreement with experimental data, could be rationalized in terms of a microstructural constituent which was important to the slip and cracking behavior of that microstructure. For the recrystallization annealed (RA) microstructure, the microstructural parameter was the alpha-phase, mean free path which was essentially two alpha grain diameters in this material, Fig. 3a. In a material of this microstructural condition, which has significant texture, two adjacent alpha grains could be considered to act as a single slip unit and therefore be the controlling microstructural constituent.

In the beta annealed (BA) microstructural condition, the controlling microstructural parameter was found to be the Widmanstatten-alpha colony size as indicated in Fig. 4a. In a considerable amount of the literature concerning beta microstructures, the colony size is found to be a controlling feature for slip and cracking behavior. The alpha platelets in a single colony tend to act as a single slip unit especially in the absence of a significant amount of beta phase between the alpha platelets as shown by Hack and Leverant (16). Thus, it is not surprising that excellent agreement between the predictive curve and the experimental data is seen to correlate with the Widmanstatten-alpha colony size in this microstructural condition.

In the higher-strength, STOA condition, the controlling microstructural feature appears not to be the alpha grains but the transformed beta matrix. In Fig. 5a, the mean linear intercept between alpha phase particles is 14 μm. The predicted curve using this microstructural dimension showed excellent agreement with the experimental data. The use of the mean linear intercept through the transformed beta phase appears reasonable in terms of the relative strength and cracking resistance of the alpha phase and the transformed beta matrix. In the high-strength STOA condition, a considerable portion of the strength is derived from the fine transformed matrix, and this matrix is generally observed to crack in a flat, lower-energy mode when compared to the crack propagation through the alpha particles. Therefore, it seems reasonable that the overall growth rate is controlled by cracking of the matrix as opposed to cracking of the primary alpha particles.

In summary, excellent predictive capability of the model has been demonstrated. Selection of the controlling microstructural constituent requires careful analysis in that fatigue crack propagation behavior of titanium alloys is a complex interaction of the imposed plastic zone at the crack tip and the microstructural constituents on several scales, and can be affected strongly by the crystallographic texture that exists in the material. The controlling microstructural feature may not be the most obvious one, as illustrated by the case of the STOA condition previously described.

Effect of Microstructure and Load Ratio

In the beta processed microstructures studied in this part of the program, microstructure had a significant effect on fatigue crack propagation rates at constant load ratios of 0.1 and 0.75, Figs. 6 and 7.

The data at R = 0.1 for both the beta-forged (BF) and beta-quenched (BQ) microstructures exhibited considerable scatter. Crack propagation in the BF condition follows a transcolony path and local crack growth rates with changes in colony orientation. A combination of transcolony propagation and secondary cracking gives this microstructure the best propagation resistance of the beta processed forging material at R = 0.1.

The beta-annealed (BA) microstructure is less crack propagation resistant, probably as a result of the smaller colony size in the BA condition compared to the BF condition. Yoder and Eylon (17) have shown that crack propagation rate for these microstructures is inversely proportional to colony size for specimens with a specimen thickness (B) more than 100 times the colony size. It should be noted that the BA condition is slightly stronger and less ductile than the BF condition; this could contribute to the faster growth rates for beta-annealed microstructures. At R = 0.1, the beta-quenched microstructure exhibits the fastest growth rates. This behavior probably results from a combination of the higher strength of this condition and the absence of discrete colonies.

Microstructure also has a significant effect on growth rate at R = 0.75. The growth rates are fastest for the BQ condition. The effect of microstructure for the BA and BF conditions is reversed from that at R = 0.1. The slower growth rate of the beta-annealed material probably derives from secondary cracking along prior beta grain boundaries associated with the continuous alpha at these boundaries.

Load ratio has a small effect on BA, an intermediate effect on BF and larger effect on BQ (Fig. 8) microstructures. Low power optical examination of the fracture surfaces revealed that the beta-quenched fracture path is essentially transgranular at both load ratios. The fracture path in the beta-annealed microstructure is quite rough at both load ratios, suggesting that the path is transcolony. For the beta-forged microstructure at R = 0.1, a transcolony fracture path leads to a very rough fracture surface and low growth rate. The higher R specimen had a much flatter surface, suggesting a transgranular path in which the crack propagated across several colonies without changing direction. Thus, it would appear that high load ratios eliminate the crack path interaction with colonies in the beta-forged material.

Applicability of Threshold ΔK to Design

Fatigue crack propagation curves for a range of R values are characterized, assuming bilinear behavior, by four components. The four components are: 1) slope of the upper portion of the curve, 2) the slope of the lower portion of the curve, 3) the transition point between the two slopes and 4) the threshold ΔK value. Two crack growth life calculations are evaluated for design purposes: 1) the durability life and 2) the damage tolerance life. The durability life assumes an unflawed structure; "intrinsic manufacturing flaws" are assumed small enough that crack nucleation and near-threshold growth is controlling. Lower slope and threshold ΔK values are predominantly used with this analysis. Damage tolerant design assumes a pre-existing flaw which grows at upper slope rates. Depending on the level of the spectrum stresses and the crack size, the damage tolerance flaw can invoke the lower slope growth rates. In particular, this occurs at low spectrum stress levels when the crack is at the initial damage tolerance size (12.7 mm). Conversely, the durability life calculations can use part of the upper slope at higher spectrum stress levels.

Durability life and damage tolerance life are each calculated and the component tested for durability lifetimes or lifetimes required by contract. Upon completion of the lifetime test, a slot of damage tolerance criteria dimensions is electro-discharge machined in the component and the component is tested for the required damage tolerance lifetimes.

Microstructures that exhibit higher threshold ΔK values, such as the RA and BA plate material (Fig. 2), would be expected to have longer durability lifetimes. The data for the beta processed forgings, Fig. 8, suggest that, in the case of beta processed microstructures, the BA microstructure should provide a compromise microstructure for components designed to durability life and subjected to a significant number of high R cycles. In summary, application of durability life calculations requires knowledge of near-threshold fatigue crack propagation and threshold ΔK at several load ratios.

Another factor germane to design involving fatigue crack propagation, which has not been considered here, is the effect of periodic overloads on propagation behavior. Overload effects are well understood and embodied in damage tolerance design methodology (7), but are much less well understood for the case of near-threshold crack growth.

Conclusions

1. For the plate material tested at R = 0.3, the following rankings were observed for near-threshold growth: 1) STOA - fastest growth rate and smallest ΔK_{th}, 2) RA - intermediate growth rate and ΔK_{th}, and 3) BA - slowest growth rate and largest ΔK_{th}. These rankings are consistent with previous results for both forgings and plate Ti-6Al-4V.

2. Fatigue crack propagation predictions using the model of Chakrabortty and Starke as modified by Heikkenen were in excellent agreement with the experimental data. The controlling microstructural features were: 1) the mean linear intercept distance through the alpha-phase particles in the RA condition, 2) the Widmanstatten alpha colony size in the BA condition, and 3) the mean linear intercept distance through the beta phase in the STOA condition.

3. The ranking of beta-processed microstructures is dependent on load ratio. The beta-forged (BF) condition was most crack propagation resistant at R = 0.1, while the beta-annealed (BA) condition was best at R = 0.75. The beta-quenched (BQ) microstructure exhibited the fastest growth rates of the three microstructures at both load ratios. Of the three beta-processed microstructures, the BA condition appears to be most suitable for components subjected to loading histories containing a significant amount of high R-factor loading.

Acknowledgements

The authors wish to thank M. Calabrese, Dr. S.B. Chakrabortty, Dr. M.R. Mitchell, A.R. Murphy, L.F. Neverez, R.A. Spurling, P.Q. Sauers and J.D. Young for their cooperation and assistance during this program. The program was supported in part by Rockwell Independent Research and Development funds and in part by the Air Force Office of Scientific Research under Contract No. F49620-80-C-0030.

References

1. J. C. Chesnutt, C. G. Rhodes, and J. C. Williams, "Relationship Between Mechanical Properties, Microstructures, and Fracture Topography in $\alpha + \beta$ Titanium Alloys," pp. 99-138 in Fractography-Microscopic Cracking Processes, ASTM STP 600, 1976.

2. H. W. Rosenberg, J. C. Chesnutt, and H. Margolin, "Fracture Properties of Titanium Alloys," pp. 213-252 in Application of Fracture Mechanics for Selection of Metallic Structural Materials, American Society for Metals, 1982.

3. S. B. Chakrabortty, "Cyclic Ductility, Cyclic Strength and Fatigue Crack Propagation in Metals and Alloys," Technical Report 78-2, ONR Contract N00014-75-C-0349, May 16, 1978.

4. S. B. Chakrabortty and E. A. Starke, Jr., "Fatigue Crack Growth Behavior of Metastable Beta Titanium-Vanadium Alloys," presented at AIME Annual Meeting, New Orleans, February 19, 1979.

5. D. Eylon, J. A. Hall, C. M. Pierce, and D. L. Ruckle, "Microstructure and Mechanical Properties Relationships in the Ti-11 Alloy at Room and Elevated Temperatures," Metallurgical Transactions A, 7A (12), 1976, pp. 1817-1826.

6. "Military Standard, Aircraft Structural Integrity Program, Airplane Requirements," MIL-STE-1530a, December, 1975.

7. J. B. Chang, R. M. Hiyama, and M. Szamossi, "Improved Methods for Predicting Spectrum Loading Effects - Volume I - Technical Summary," AFWAL-TR-81-3092, Vol. I, Wright-Patterson AFB, OH, Nov. 1981.

8. J. C. Chesnutt, A. W. Thompson, and J. C. Williams, "Influence of Metallurgical Factors on the Fatigue Crack Growth Rate in Alpha-Beta Titanium Alloys," AFML-TR-78-68, Air Force Materials Laboratory, Wright-Patterson AFB, OH, May 1978.

9. R. J. Bucci, "Development of a Proposed ASTM Standard Test Method for Near-Threshold Fatigue Crack Growth Rate Measurement," pp. 5-28 in Fatigue Crack Growth Measurement and Data Analysis, ASTM STP 738, 1981.

10. R. O. Ritchie, "Near-Threshold Fatigue-Crack Propagation in Steels," International Metals Reviews, 24 (5 and 6), 1976, pp. 205-230.

11. Saghana B. Chakrabortty, "A Model Relating Low Cycle Fatigue Properties and Microstructure to Fatigue Crack Propagation Rates," Fatigue of Engineering Materials and Structures, 2 (3), 1979, pp. 331-344.

12. H. C. Heikkenen, "A Study of the Fatigue Behavior of an Al-6Zn-2Mg-0.1Zr Alloys," M.S. Thesis, Georgia Institute of Technology, November 1981.

13. H. C. Heikkenen, E. A. Starke, Jr., and S. B. Chakrabortty, "Modification of the Chakrabortty Model for Calculating Fatigue Crack Growth Rates (FCGR's)," Scripta Metallurgica, 16 (5), 1982, pp. 571-574.

14. G. R. Yoder, L. A. Cooley, and T. W. Crooker, "50-Fold Differences in Region II Fatigue Crack Propagation Resistance of Titanium Alloys: A Grain-Size Effect," Journal of Engineering Materials and Technology, 101 (1), 1979, pp. 86-90.

15. G. R. Yoder, L. A. Cooley and T. W. Crooker, "Quantatitive Analysis of Microstructural Effects on Fatigue Crack Growth in Widmanstatten Ti-6Al-4V and Ti-8Al-1Mo-1V," Engineering Fracture Mechanics, 11, 1979, pp. 805-816.

16. J. E. Hack and G. R. Leverant, "The Influence of Microstructure on the Susceptibility of Titanium Alloys to Internal Hydrogen Embrittlement," Metallurgical Transactions A, 13A (10), 1982, pp. 1729-1738.

17. G. R. Yoder and D. Eylon, "On the Effect of Colony Size on Fatigue Crack Growth in Widmanstatten Structure $\alpha + \beta$ Titanium Alloys," Metallurgical Transactions A, 10A (11), 1979, pp. 1808-1810.

ENVIRONMENTAL AND FREQUENCY EFFECTS ON NEAR-THRESHOLD FATIGUE CRACK PROPAGATION IN A STRUCTURAL STEEL

A. Bignonnet* D. Loison* R. Namdar-Irani*,
B. Bouchet** J.H. Kwon** J. Petit**

* IRSID
185 r. Président Roosevelt
78105 ST GERMAIN EN LAYE
FRANCE

** ENSMA
rue Guillaume VII
87034 POITIERS
FRANCE

The influence of environment and test frequency on near-threshold fatigue crack growth was investigated in a structural steel. Crack closure and crack surface oxide thickness measurements, were performed on specimens tested in air, nitrogen and vacuum. Threshold and near-threshold growth rates were found to depend upon R ratio, environment, and test frequency. The analysis of results incorporated three effects :

- Water vapor adsorption and subsequent diffusion of hydrogen, which tends, to accelerate the crack growth,

- Inhibition of water vapor adsorption, due to the occurence of oxidation,

- Frequency dependence of oxide thickening which results in higher closure level and hence lower growth rates and higher threshold values at higher frequencies.

Introduction

Over the past years, numerous research studies have been undertaken to understand the behaviour of fatigue cracks in the near-threshold region. Fatigue threshold studies of engineering materials are more and more relevant to ensure the reliability of high technology structures. Although many influencing factors on near-threshold behaviour have been identified, there still exist many discrepancies between results, and reliable data are scarce. It is now recognized that threshold values may be influenced by three main sources : microstructural parameters, crack tip deformation (mechanical parameters) and environmental effects. Although the two former have been the object of many studies (see (1)), little information is available on the influence of environment. However, recent studies have shown the influence of moist environments on the near-threshold behaviour of high and low strength steels (2, 3, 4), and aluminium alloys (5, 6, 7). These observations have been analysed in terms of crack closure, involving thickening of oxides observed on cracked surfaces. In addition, a frequency effect in ambient air environment reported by several authors (4, 8, 9) indicates the importance of adsorption and diffusion of damaging species at the crack-tip.

To further investigate the interaction of environmental and frequency effects, tests were carried out in moist air, vacuum and nitrogen with traces of oxygen and water vapor. The quenched and tempered structural steel studied here is a steel used in the new generation of offshore structures where reliable threshold data are of importance.

Experimental Procedures

The E460 structural steel, of composition shown in table I, was quenched after hot rolling and tempered at 690°C for 20 min. This treatment gives a fine tempered martensitic microstructure. The material was supplied in 30 mm thick plates. The mechanical properties are listed in table II.

Table I : Chemical composition of E460 steel (wt %)

C	Mn	Si	P	S	Ni	Nb
0.17	1.27	0.34	0.022	0.001	0.022	0.018

Table II : Mechanical properties of E460 steel

Yield Strength Monotonic	Yield Strength Cyclic	U.T.S.	Elong.	Reduct. in area	Charpy V Energy - 40°C
(MPa)	(MPa)	(MPa)	%	%	J
460	330	600	21	70	90

Fatigue crack propagation experiments were performed on an electro servohydraulic machine fitted with a chamber for low pressure or controlled atmosphere tests. The specimens used were of CT type with two sets of dimensions : CT-W = 80 mm - B = 28 mm and CT-W = 75 mm - B = 10 mm. Tests were performed in moist air (30% to 50% relative humidity), in nitrogen with 50 ppm oxygen and water vapor and in vacuum (10^{-3} Pa). All the experiments were carried out at ambient temperature. Threshold values were determined by the decreasing ΔK method. The decreasing parameters were $\frac{dK}{da} = -1$ MPa$\sqrt{m}$/mm and $\frac{1}{\Delta K}\frac{dK}{da} \sim 0.2$ mm^{-1}. Crack closure was assessed using compliance. Accurate measurement was obtained using a differentiation technique as described in a previous paper (9).

Crack surface oxides were characterized (nature and thickness) using a Secondary Ion Mass Spectrometer (SIMS). Full details of this technique are given in references (9, 10).

Experimental Results

Influence of environment and stress ratio on threshold

The results described in this section were obtained on CT specimens with W = 75 mm and B = 10 mm. The test frequency was 35 Hz. Crack growth and threshold tests were carried out under vacuum (10^{-3} Pa), nitrogen (50 ppm ($O_2 + H_2O$)) and laboratory air (30 to 50% relative humidity). For each environment, results were obtained for two R ratio : 0.1 and 0.7. The influence of environment on fatigue crack growth for R = 0.1 is given in figure 1, and for R = 0.7, in figure 2 with reference to the R = 0.1 results.

For the tests with R = 0.1 (figure 1), it is seen that at higher ΔK values crack growth rates in vacuum and in nitrogen are very similar ($\frac{da}{dN} > 10^{-8}$m/c); although, the rates in nitrogen remain slightly higher. The crack growth behaviour under nitrogen exhibits a plateau at a rate near 10^{-8}m/c similar to the one observed with aluminium alloys (6, 11). Below this plateau, crack growth rates are much higher than in vacuum and the threshold is reached at a relatively low ΔK value. The air results at R = 0.1 show higher fatigue crack propagation rates than those in vacuum, although below 10^{-9} m/c the crack growth rate sharply decreases and surprisingly reaches a threshold value higher than in vacuum. Compared to the nitrogen data, the air results show higher crack growth rates above the plateau and lower ones below it.

In figure 2 it is seen that in vacuum, crack propagation rates above 10^{-9} m/c do not depend upon the R ratio. At low ΔK, the knee observed at R = 0.1 is not so pronounced for R = 0.7 and the threshold value is lower. Contrary to the results with low R ratio, fatigue crack propagation rates at R = 0.7 are similar in nitrogen and in air. The crack growth rates in air are slightly lower especially in the mid-ΔK range, but threshold levels are identical. For both environments crack growth rate is higher than in vacuum and threshold values are lower.

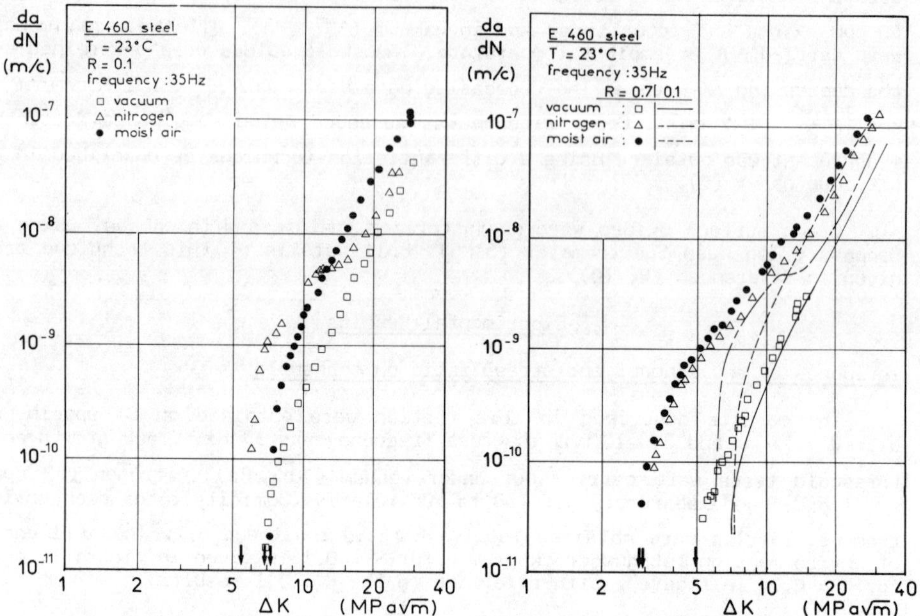

Figure 1 - Near threshold fatigue crack growth for E460 steel at R = 0.1 in vacuum, nitrogen and air.

Figure 2 - Near threshold fatigue crack growth for E460 steel at R = 0.7 in vacuum, nitrogen and air.

Influence of frequency and specimen thickness on threshold

Results obtained from tests on CT specimens with W = 75 mm and B = 28 mm conducted at 35 Hz and 7 Hz in air with R = 0.1 are compared with vacuum data in figure 3. In air the effect of frequency becomes pronounced below 10^{-8} m/c. This is in agreement with previous results on the same type of steel (9, 12). We should note that for low ΔK values, fatigue crack propagation is markedly higher for the test at low frequency. Results under vacuum emphasize that this frequency effect, in the domain investigated, is due to environmental phenomena. Near the threshold, the growth rates change gradually with ΔK in vacuum, while the near-threshold region in air is marked by sharp changes in growth rates. At 35 Hz the threshold value in air is much higher than the threshold in vacuum whereas at 7 Hz it is slightly lower.

Results obtained in nitrogen are compared in figure 4 with vacuum results. In this figure it is seen that in nitrogen fatigue crack propagation at rates above 10^{-8} m/c does not depend upon the test frequency and is close to that obtained in vacuum. For crack growth below 10^{-9} m/c the frequency effect is absent and crack growth rates are much higher than in vacuum. Around 10^{-8} to 10^{-9} m/c a transition is observed between these two regimes of growth. This transition is sentitive to the test frequency and the plateau observed at 35 Hz is very attenuated at 7 Hz.

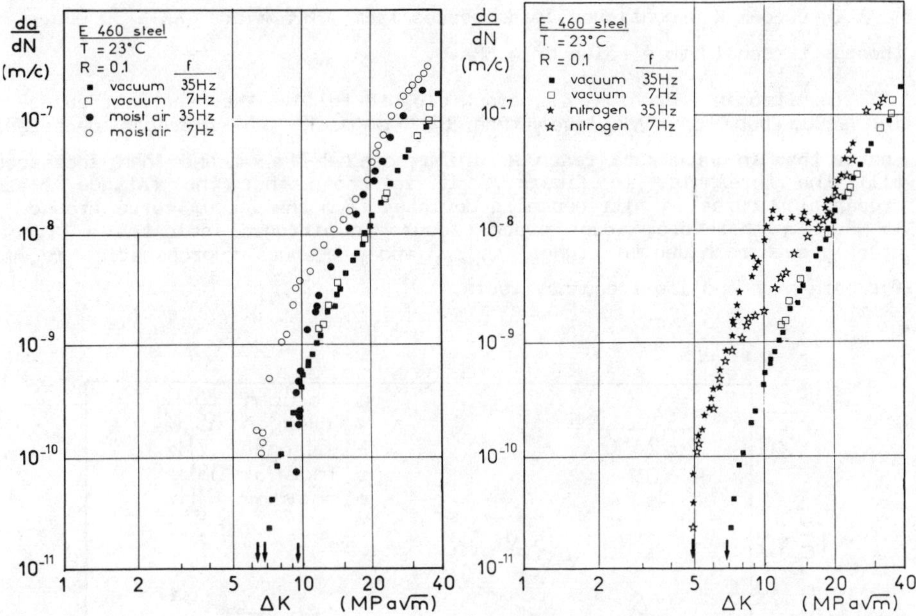

Figure 3 - Effect of frequency on near-threshold fatigue crack propagation in air and vacuum for E460 steel.

Figure 4 - Effect of frequency on near-threshold fatigue crack propagation in nitrogen and vacuum for E460 steel.

Threshold values (ΔK_{th}) defined at 10^{-11} m/c, are given on table III.

Table III : Threshold data for E460 steel in the different loading and environmental conditions investigated

	Environment	Air (30 - 50% RH)		Nitrogen (50 ppm ($H_2O + O_2$))		Vacuum (10^{-3} Pa)
	Frequency	7 Hz	35 Hz	7 Hz	35 Hz	35 Hz
ΔK_{TH} (MPa $\sqrt{m}$)	R = 0.1, B = 28 mm	6.4	9.2	5	5	6.8
	R = 0.1, B = 10 mm		7.1		5	6.8
	R = 0.7, B = 10 mm		2.9		3	5.3

Crack closure. Crack closure was assessed using an improved compliance technique (9). The dependence of the Opening Stress Intensity Factor (hereafter K_{op}) on ΔK is shown in figure 5.

In vacuum K_{op} continuously decreases from 7 MPa$\sqrt{m}$ at ΔK = 30 MPa$\sqrt{m}$ towards threshold to a value of 4 MPa$\sqrt{m}$.

In nitrogen, for high ΔK, crack closure follows the behaviour observed in vacuum but for ΔK lower than 15 MPa $\sqrt{m}$, K_{op} in nitrogen decreases faster than in vacuum to reach a minimum of 2.8 MPa $\sqrt{m}$ and then increases till the threshold. In figure 4 it is shown that the fatigue crack propagation curve in nitrogen also deviates from the vacuum curve around 15 MPa$\sqrt{m}$; therefore, lower crack closure in nitrogen indicates a higher crack growth rate due to higher ΔK_{eff}, and a change of propagation regime for both high and low frequency tests.

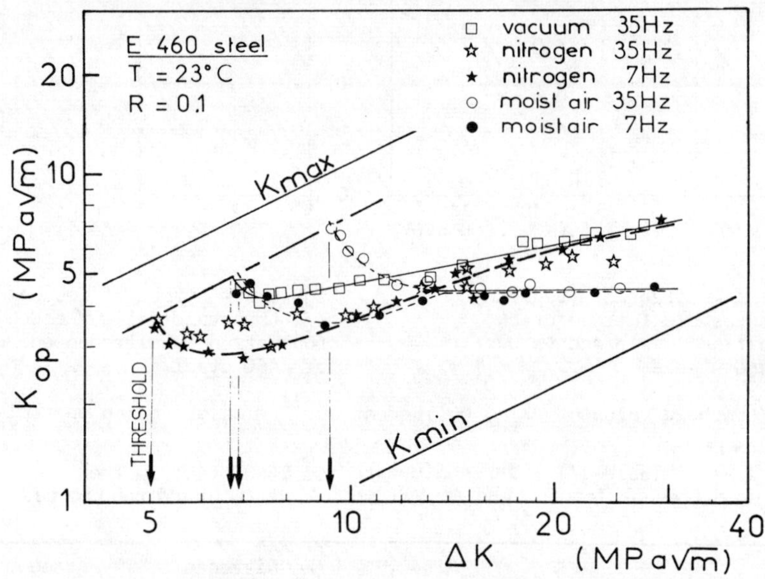

Figure 5 - Opening Stress Intensity Factor (K_{op}) versus ΔK at R = 0.1 in vacuum, nitrogen and air for E460 steel.

In air K_{op} remains roughly constant for high ΔK values (K_{op} = 4.5 MPa$\sqrt{m}$). and there is less crack closure than in vacuum. This is still in agreement with the higher growth rates observed in air as shown in figure 3. Once again below ΔK = 15 MPa$\sqrt{m}$ (da/dN ~ 10^{-8} m/c) a change in closure behaviour is found. For low frequency tests (7 Hz) K_{op} follows the trend of the results obtained in nitrogen before increasing towards threshold at 4.5 MPa $\sqrt{m}$. For higher frequency test (35 Hz) K_{op} does not follow this trend but begins to increase when ΔK equals 15 MPa $\sqrt{m}$ up to the threshold. From figures 3 and 4 one also can see that crack propagation in air below 15 Mpa$\sqrt{m}$ at 7 Hz tends to follow the results obtained in nitrogen, while at 35 Hz the propagation rate markedly decreases to a relatively high threshold value.

Crack surface analysis. In the low crack growth region, oxidation debris was seen on the fracture surface for propagation in air and to a lesser extent in nitrogen. One can see that this oxide is thicker at 35 Hz than at 7 Hz as previously observed (9, 12). The oxide was characterized using SIMS as described elsewhere (9, 10). The SIMS analysis shows two layers in this oxide : Fe_3O_4 at the metal-oxide interface, covered by a thinner Fe_2O_3 layer. The oxide thickness was determined by Argon ion sputtering on 500 μm^2 areas. The sputter rate was calibrated with a bulk Fe_3O_4 specimen and the interface criterion was chosen to be a 90% decrease of intensity between the maximum intensity and the steady state level. As the Pilling Bedworth ratio of Fe_2O_3 and Fe_3O_4 is close to 2 (1 vol. Fe gives 2 vol. Oxide) and assuming only thickness direction growth, half of the oxide produced on the cracked faces is in excess inside the crack. In figure 6 the excess oxide thickness is plotted versus ΔK. These measurements confirm that the thicker the oxide, the lower is the crack propagation rate and the higher is the threshold value. One should note that the traces of H_2O and O_2 in the nitrogen environment are sufficient to produce oxide in the threshold region. It is also noticeable that the increase in crack closure towards threshold (see figure 5) corresponds to the oxide thickening on the crack face.

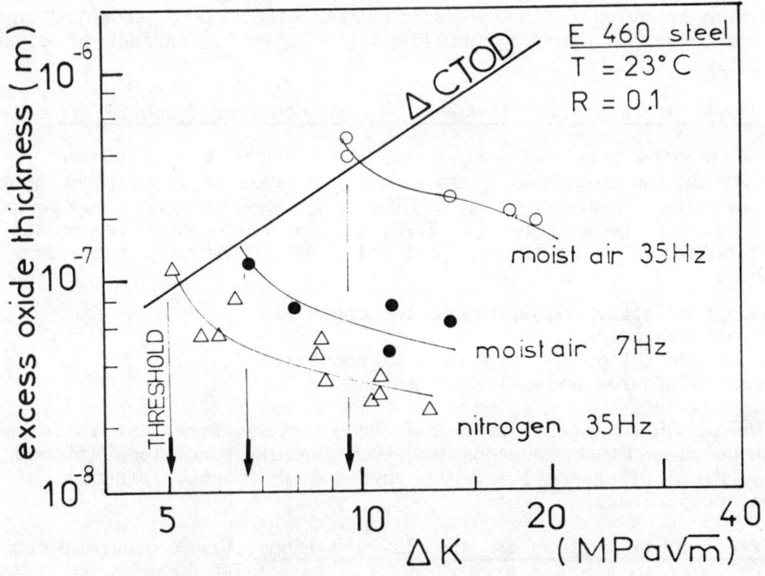

Figure 6 - Measurements of crack face oxide thickness using SIMS, and comparison with ΔCTOD in the near-threshold region in E460 steel.

The cyclic crack tip opening displacement is also plotted in figure 6 as $\Delta CTOD = 0.5 \frac{\Delta K^2}{2E\sigma_y}$. Though oxide thickness measurements are not very accurate, it is worthwhile to note that when the oxide is thick enough to wedge the crack closed, the crack stops growing and the threshold is reached. These experimental results support the oxide induced closure model proposed by Suresh et al (3) and Stewart (2) and the concept of threshold level determined by an oxide wedge effect. Nevertheless this model does not take into account the frequency effects.

Environmentally assisted fatigue crack propagation analysis

Reference crack propagation in vacuum

The fatigue crack growth data established in vacuum at high stress ratio (R = 0.7) can be described by a linear variation of log da/dN with log ΔK (the "Paris law") with an exponent m ~ 3.5. This is consistent with a damage accumulation process without an environmental effect.

The curve established at low stress ratio (R = 0.1) shows a gradual increase of the slope m towards threshold (figure 1). This behaviour can be essentially analysed in terms of crack closure. As shown in figure 5, the ratio $U = \Delta K_{eff} / \Delta K$ progressively decreases for decreasing ΔK indicating an increasing effect of closure. In addition one could expect an effect of a mode II displacement component (13-15) such that the relative mode II/mode I amplitude increases towards threshold. The energy dissipated in mode II being without effect on crack propagation, the effective energy spent per cycle in mode I therefore decreases approaching threshold. This crack propagation behaviour without environmental effect is represented in figure 7 by the curve (1) (high R ratio) and the curve (2) (effect of closure at low R ratio).

Crack growth behaviour in nitrogen (50 ppm (H_2O + O_2)) and in air

The results obtained show a large influence of environment on crack growth and on the threshold in ambient air as well as in nitrogen containing traces of water vapor and oxygen. This is in agreement with numerous observations for different alloys (2, 3, 6, 11) for which an environmental effect is attributed to water vapor. From Wei (16) a four step mechanism can be proposed :

- transport of active molecules to the crack tip
- physical adsorption
- chemical adsorption and molecular dissociation
- hydrogen diffusion and embrittlement.

The different steps do not occur in every case and their relative importance or activity depends upon the loading conditions (stress intensity, R ratio, frequency) and the environment (partial pressure of active species, temperature).

Effect of adsorption in the mid-rate range. Crack growth rates higher than 10^{-8} m/c, in air and at a given ΔK, have been shown to be independent of R ratio (figures 1 and 2), test frequency and specimen thickness (figure 3). Such a result suggests an influence of environment which is independent of time. In ambient air at atmospheric pressure, the partial pressure of water vapor is far higher than the critical value (5) for which the rate of surface coverage equals the rate of surface generation where saturation is reached (at least in the range of frequency investigated). In this case the environmentally-assisted process is not controlled by the transport of active species but is a surface reaction controlled process (5, 16).

The predominant surface effect on fresh cracked surfaces at the crack tip has been shown to be adsorption of water vapor (16, 17, 18, 19). As pointed out by Westwood (20), adsorbed surface active species induce a decrease in fatigue resistance as a consequence of reduction in surface free energy which is called the "Rehbinder effect" (21).

If the partial pressure of water vapour at the crack tip is sufficiently high to form a monolayer of adsorbed molecules, there is no substantial effect of transport on the environmental influence. The results obtained show that this is the case in ambient air at atmospheric pressure.

On the contrary, crack growth in nitrogen (with 50 ppm (H_2O + O_2)) in the mid-rate range appears to be dependent upon R ratio (figures 1 and 2) and frequency (figure 4). With the presence in the surrounding environment of less than 50 ppm of water vapour, an effect of time is detected. This indicates an influence of water vapour molecule transport to the crack tip ; one can assume that the formation of a monolayer of adsorbed molecules is not completely achieved at each cycle under these conditions.

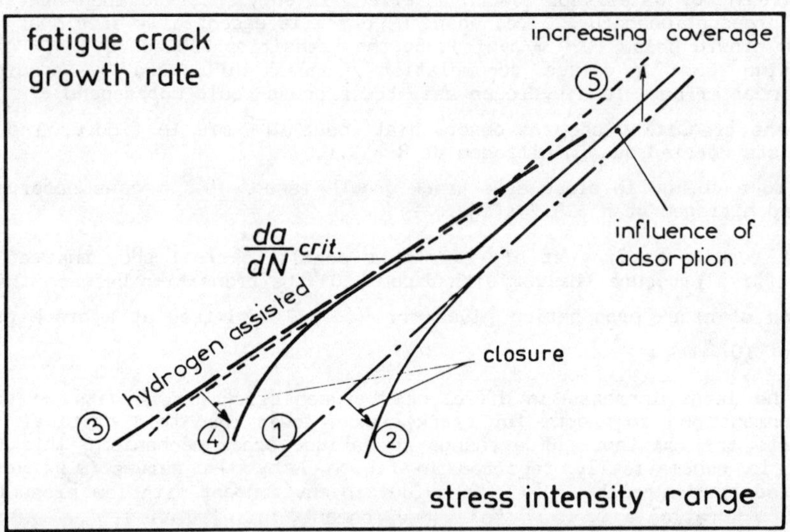

Figure 7 - Influence of adsorption and hydrogen embrittlement on fatigue crack growth characteristics.

1. Purely mechanistic, mode I crack propagation behaviour (in vacuum)
2. Effect of closure and mode II on the near-threshold, without environmental effect.
3. Hydrogen-assisted crack propagation behaviour
4. Influence of adsorption and H diffusion at low R ratio
5. Influence of adsorption and H diffusion at high R ratio

This analysis of the influence of environment in the mid-rate range is consistent with those proposed by Petit (11) for aluminium alloys and Wei and coworkers (5, 22) for aluminium alloys and steels. We believe that the results obtained on the steel studied can be analysed only on the basis of adsorption effects for this crack growth rate range. This behaviour is schematically represented in figures 7 and 8 by the upper part of the curves (4) and (5) .

Hydrogen-assisted near-threshold crack growth. Chemical adsorption and molecular dissociation lead to the formation of ions H^+ which pick up electrons to form atomic hydrogen. So in addition, one can expect a hydrogen embrittlement effect. As shown by Achter (23) and Bowles (24), the occurrence of such an effect requires diffusion of hydrogen into the process zone at the crack tip. These authors (23-24) estimated a critical growth rate, depending upon test frequency, R ratio and partial pressure of the damaging species, above which hydrogen embrittlement cannot occur.

The onset of a hydrogen-assisted crack growth process would be associated with a change in the slope of the log da/dN vs log ΔK curves for crack growth rates lower than a critical rate. In environments with a low content of water vapor, such an effect is only observed when non-steady crack propagation occurs, i.e. when the crack is essentially stationary (11) (crack growth below 10^{-8} m/cycle). So the transition between a crack growth resulting from a damage accumulation process influenced by adsorption (Rehbinder effect) to a hydrogen-assisted process could correspond :

- to the transition step as observed at about 10^{-8} m/c in figures 1 and 4, for tests carried out in nitrogen at R = 0.1,

- or to a change in slope at a crack growth rate $> 10^{-8}$ m/c as observed in air and nitrogen at R = 0.7 (figure 2).

Recent results by Gray et al (25) for a pearlitic steel show that at very low partial pressure (helium with 3 ppm H_2O) the transition between the two regimes of crack propagation gives for R = 0.7 a plateau at a crack growth rate of 10^{-8} m/c.

The large increase in difference between crack growth data in active environment and in vacuum for crack growth lower than this critical value supports the existence of a change in fatigue crack mechanism. This behaviour is schematically reported in figure 7 by the curves (3) (hydrogen assisted crack growth), (4) (behaviour in environment with low pressure of H_2O, low R ratio) and (5) (moist environment, high R ratio).

The difference in this transition step configuration as observed in figure 4 for tests in nitrogen performed at 7 and 35 Hz is difficult to analyse. However it can be suggested that the occurence at 35 Hz of an acceleration in hydrogen production could be related to a larger influence of mode II opening at a higher frequency. Such an assumed process could also lead to an enhancement of oxide thickening observed in air at 35 Hz as compared to the test at 7 Hz (see figure 6).

Crack surface oxidation. Crack closure leads to successive contacts between the opposite fracture surfaces and results in fretting corrosion (10). So the thickness of the oxide formed depends upon the parameters controlling fretting which are essentially : the oxygen supply, the amplitude of slip, the relative velocity of the sliding surfaces, the contact pressure (load), the number of cycles and the temperature (26).

The influence of these parameters on fretting corrosion experiments can be transposed to the fretting corrosion on fracture surfaces.

The influence of oxygen content on the oxide thickness (see figure 6) is consistent with earlier fretting corrosion results (26, 27) which indicate that environments with low oxygen and H_2O contents produce less oxide. High relative humidity increases lubrication and leads to less fretting damage and oxidation (27).

For very large numbers of fretting cycles, which is the situation likely to occur in the near-threshold, the rate of removal of material is directly proportional to the number of cycles (26). This is also consistent with the large oxide thickening at the threshold even at low oxygen contents (see figure 6 and references (4) (18)).

An increase in the contact pressure (load) or slip amplitude causes higher fretting and removing of material (26, 27, 28). This is consistent with the reported observations (2,3,9) that crack surface oxide deposits are thicker for low R ratio which results in larger contact pressure and local slip amplitude.

Far fewer results exist concerning the influence of frequency on oxide production by fretting. Most of the literature results are on fretting fatigue and show that decreasing the frequency has a detrimental effect, but expressed in terms of fatigue life or fretting damage. In fact these results, expressed in such terms, do not represent the production of oxide. Measurements of the coefficient of friction, by Endo et al (29), during the early stages of fretting show that the coefficient rises more rapidly and to a higher value at the lower frequencies. The lower coefficient of friction at high frequencies is attributed to the greater temperature rise and therefore greater oxidation rate. Waterhouse (24) proposes that the thicker oxide film acts as a more efficient lubricant. The rise of temperature with frequency suggested by Endo (29) was earlier noted by Feng and Uhlig (27) who report that the maximum temperature rise reached by test specimens increases appreciably with frequency. Considerable energy is dissipated in the fretting contact and it has been suggested that there are local temperatures rises which may amount to several hundred degrees C (30). Very careful measurements by Alyab'ev et al (31) show at 25 Hz a temperature rise between 60°C and 270°C depending on pressure and slip amplitude. A local rise in temperature due to the fretting is not unreasonable when one considers that it is a sliding process involving frictional forces. This would explain why debris produced by fretting on steel in ordinary air is $\alpha\ Fe_2O_3$ (24), a high temperature form of ferric oxide produce in the range 300 to 600°C. We should note that several authors found also Fe_3O_4, another high temperature oxide, on fretted mild steel (see 28) and on cracked surfaces of a structural steel tested in fatigue crack propagation experiments at 65 Hz (9).

Influence of frequency. From the present results and previous results reported in the literature one can reasonably state that the effect of frequency on the near-threshold crack growth in air is largely due to oxide build-up on the cracked surfaces. Oxide production is enhance by the effect of frequency on fretting as seen above, or by a possible larger importance of mode II opening at higher frequency which would also enhance fretting oxidation due to a larger slip amplitude. Thick oxide layers enhance the crack closure by a wedge effect and so the crack growth rates decrease as the effective stress intensity range decreases. Near-threshold growth and threshold values are controlled by the thickness of the oxide in excess inside the crack (see figure 6).

Fretting process could further induce two opposite effects :

- additional hydrogen production by a reaction at the oxide-metal interface along the cracked surfaces, supplied in water vapour by diffusion through the oxide after adsorption which is enhanced by fretting. This could be observed when oxide thickeness is small (i.e. low content of oxygen in the environment or low test frequency) ;

- trapping of water vapour by oxide deposits, reducing drastically the access of water vapour to the crack tip. The transport of water vapor could be limited also by oxide wedging in the near-threshold, which would result in a lower rms free path of water molecules.

In the near-threshold, oxide thickening becomes important even in environments with low oxygen pressure. The threshold level for tests conducted at R = 0.1 is controlled by crack wedging even in nitrogen. This behaviour is reported in figure 8, curve (6).

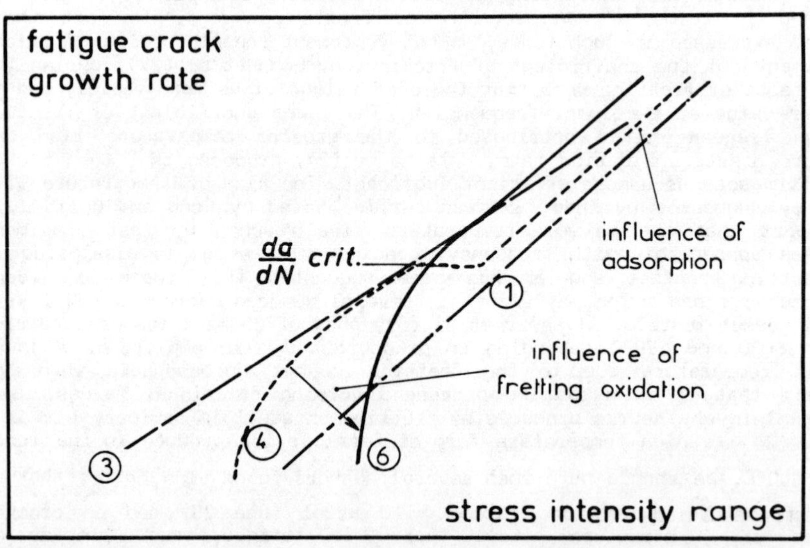

Figure 8 - Influence of oxide induced closure on fatigue crack growth characteristics.

(1) Purely mechanistic, mode I crack propagation behaviour (in vacuum)

(2) Effect of closure and mode II on the near-threshold, without environmental effect

(3) Hydrogen-assisted crack propagation behaviour

(4) Influence of adsorption and H diffusion at low R ratio

(6) Crack growth behaviour at low R ratio in moist environment

Thickness effect. Experimental results show a thickness effect on crack growth rates and threshold in air and to a lesser extent in nitrogen, but such an effect does not occur in vacuum (see figures 1 - 4 and table III). This suggests that the observed effect is due to a difference in the environmental activity at the crack tip which would depend upon specimen thickness. This could be a consequence of a difference in stress triaxiality and hence in shear displacements at the crack tip which could enhance fretting. This assessment is supported by the thicker oxide observed in the center of the specimens.

Conclusions

Fatigue crack growth behaviour in a structural steel has been shown to be strongly dependent on environmental conditions. Figures 7 and 8 summarize the effects of environment as follows :

- an increase in growth rates in the mid-rate range due to reduction in surface energy induced by adsorption, independent of R ratio and test frequency in air where a saturation level is reached, but dependent upon both these parameters in environment with a low water vapor content ;

- a hydrogen assisted cracking for rates lower than a critical value depending upon R ratio and test frequency or for rates where non-steady crack growth occurs (typically below 10^{-8} m/c). This results in marked acceleration of crack growth and decrease of threshold values ;

- enhancement of crack closure near threshold from oxide thickening, leading to crack wedging at a threshold level which depends upon R ratio, partial pressure of H_2O and O_2 and test frequency.

The influence of test frequency is largely related to its effect on oxide production by fretting corrosion. The observed effect of the specimen thickness could be a consequence of higher triaxiality in thick specimen which results in larger amplitude of mode II displacements and enhanced oxide production.

References

(1) R.O. Ritchie, "Near-threshold fatigue-crack propagation in steels", Int. Metals Reviews, 20 (5 & 6) (1979) pp. 205-230.

(2) A.T. Stewart, "The Influence of Environment and Stress Ratio on Fatigue Crack Growth at Near-threshold stress Intensities in Low Alloy Steels", Eng. Fract. Mech., 13 (1981), pp. 463-478.

(3) S. Suresh, G.F. Zamiski and R.O. Ritchie, "Oxide-Induced crack closure : An Explanation for Near-threshold Corrosion Fatigue Crack Growth Behavior", Met. Trans., 12 A (1981), pp. 1435-1443.

(4) A. Bignonnet, D. Loison, R. Namdar-Irani, J.H. Kwon, B. Bouchet and J. Petit, "Influence de la fréquence d'essai sur le seuil de propagation des fissures de fatigue dans un acier de construction". IRSID Report RE 1006, 1983, IRSID, St Germain en Laye, France.

(5) R.P. Wei and G.W. Simmons, "Recent progress in understanding environment assisted fatigue-crack growth", Int. Journ. of Fracture, 17 (2) (1981), pp. 235-247.

(6) J. Petit and A. Zeghoul, "Gaseous Environmental Effect on Threshold Level in High Strength Aluminium Alloys", pp. 563-580 in Fatigue Thresholds, J. Bäcklund, A.F. Blom and C.J. Beevers, Ed. ; EMAS Ltd, Warley U.K., 1982.

(7) A.K. Vasudevan and S. Suresh, "Influence of corrosion deposits on near-threshold fatigue crack growth behavior in 2xxx and 7xxx series aluminium alloys", Met. Trans., 13A (1982), pp. 2271-2280.

(8) H.G. Noack and K. Seifert, "Frequency Effects on Crack-Propagation", pp. 365-374 in Analytical and Experimental Fracture Mechanics, G.C. Sih and M. Mirabile, Ed. ; SIJTHOFF & NOORDHOFF, The Netherlands, 1981.

(9) A. Bignonnet, R. Namdar-Irani and M. Truchon, "Fatigue crack growth rate in air of a quenched and tempered steel", pp. 417-425 in Fracture and the Role of Microstructure, K.L. Maurer, F.E. Matzer, Ed. ; EMAS Ltd, Warley, U.K., 1982.

(10) D. Benoit, R. Namdar-Irani and R. Tixier, "Oxidation of fatigue fracture surfaces at Low Crack Growth Rates", Mat. Sci. Eng., 45 (1980), pp. 1-7.

(11) J. Petit, "Some aspects of Near-threshold Crack Growth : Microstructural and Environmental effects". "Concepts of Fatigue crack growth threshold", D.L. Davidson and S. SURESH, Ed., The Metallurgical Society of Aime, Warrendale, PA (1983), this volume.

(12) A. Bignonnet, R. Namdar-Irani and M. Truchon, "The influence of test frequency on fatigue crack growth in air, and crack surface oxide formation", Scripta Met., 16 -19 2), pp. 795-798.

(13) D.L. Davidson, "Incorporating threshold and environmental effects into the damage accumulation model for fatigue crack growth", Fat. Eng. Mat. and Struct., 3 (3) (1980), pp. 229-236.

(14) J. Lankford and D.L. Davidson, "Wear due to Mode II Opening of Mode I Fatigue Crack in an Aluminium alloy", Met. Trans., 14A (1983), pp. 1227-1230.

(15) D.L. Davidson and J. Lankford, "The effect of water vapor on fatigue crack tip stress and strain range distribution and the energy required for crack propagation in low carbon steel", Int. Journ. of Fracture, 17, (1981), pp. 257-275.

(16) R.P. Wei, pp. 816-840 in Fatigue Mechanisms, J.T. Fong Ed., ASTM-STP 675 (1979).

(17) T. Broom and A. Nicholson, "Atmospheric Corrosion-Fatigue of Age Hardened Aluminium Alloy", J. Institute of Metals, 89 (1960-1961), p. 183.

(18) B. Bouchet, J. de Fouquet, M. Aiguillon, "Influence de l'environnement sur les faciès de rupture par fatigue d'éprouvettes monocristallines et polycristallines d'alliages Al-Cu 4%", Acta Met., 23, (1975), pp. 1325-1336.

(19) M.R. Achter, "Effect on environment on fatigue cracks", pp. 181-204 in Fatigue Crack Propagation, ASTM-STP 415 (1966).

(20) A.R.C. Westwood, "Adsorption-sensitive flow and fracture of solids, Proceedings of "Physico-chimie de l'état solide : applications aux métaux et à leurs composés", Paris (France), 19-23 Septembre 1983. In press - Elsevier publ.

(21) P.A. Rehbinder et E.D. Scukin, "Les phénomènes de surface dans la déformation et la fracture des solides", in Proceedings "Séminaire de Mécanique des Surfaces", Paris, (1971), ISMCM ed Paris.

(22) True-Hwa Shih and R.P. Wei, "The effects of load ratio on environmentally-assisted fatigue crack growth", Eng. Fract. Mech., 18, (4), (1983), pp. 827-83.

(23) M.R. Achter, "The adsorption model for environmental effects in fatigue crack propagation", Scripta Met., 2, (1978), pp. 525-528.

(24) C.Q. Bowles, "The role of environment, frequency and wave shape during fatigue crack growth in aluminium alloys", Report LR-270 Delft Univ. of Technol., The Netherlands (1978).

(25) G.T. Gray III, J.C. Williams and A.W. Thompson, "Roughness induced crack closure : an explanation for microstructurally sensitive fatigue crack growth", Met. Trans., 14A, (1983), pp. 421-433.

(26) R.B. Waterhouse, "Fretting corrosion", pp. 106-127, Pergamon Press (1972).

(27) I. Ming Feng and Herbert H. Uhlig, "Fretting corrosion of mild steel in air and in nitrogen", J. Applied Mech., 21, (1954), pp. 395-400.

(28) H.H. Uhlig, W.D. Tierney and A. Mc Clellan, "Test equipment for evaluating fretting corrosion", p. 71, ASTM-STP 144 (1953).

(29) K. Endo, H. Goto and T. Nakamura, "Effects of cycle frequency on fretting fatigue life of carbon steel", Bulletin of JSME, 12, (54), (1969), pp. 1300-1308.

(30) R.B. Waterhouse, "Theories of fretting processes", pp. 203-219 in Fretting Fatigue, R.B. Watherhouse. Applied Science Publishers LTD-London (1981).

(31) A. Ya Alyab'ev, Yu. A. Kazimirchik and V.P. Onoprienko, Fiz. Khim. Mekh. Mat., 6, (1970), pp. 12-16.

NEAR-THRESHOLD FATIGUE CRACK PROPAGATION OF HY80 AND HY130 STEELS*

J. L. Horng and M. E. Fine

Department of Materials Science and Engineering and
Materials Research Center
Northwestern University
Evanston, Illinois 60201
USA

The role of microstructure in near-threshold fatigue crack propagation was studied in pressure vessel steels HY80 and HY130. Three different quenched and tempered structures of HY130 were produced keeping the strength constant. A small increase of threshold stress intensity range, ΔK_{th}, from increasing the tempering temperature correlates with increase in iron carbide size, replacement of Fe_3C by V_4C_3, and decrease in dislocation density. In HY80, dual phase structures with the ferrite surrounded by martensite were developed. As compared with the standard quenched and tempered heat treatment, the dual phase structures give improved resistance to near-threshold fatigue crack propagation. The rate was found to decrease with the combined width of the martensite and ferrite patches. SEM fractography, closure stress and surface roughness measurements were made. Roughness induced crack closure is shown to be an important factor but $\Delta K_{eff,th}$ did vary with microstructure.

* Supported by ONR Contract #N00014-78-C-0565

Introduction

At low values of stress intensity range near the threshold for fatigue crack propagation, ΔK_{th}, $(da/dN)|_{\Delta K}$ is known to be a function of microstructure, environment and load ratio. The role of microstructure (e.g., grain size) on ΔK_{th} is still controversial and crack closure plays a role. Crack closure is generally thought to derive from three sources (1-3): (1) plastic stretching of the metal at the crack tip, (2) wedging open of the crack by oxide deposit, and (3) non-mating rough fracture surfaces due to shear displacements. The purpose of the present work was to study the effect of microstructure on the near-threshold fatigue crack propagation rate and the role and origin of near-threshold crack closure. Such an understanding is needed before this region of fatigue crack growth can be theoretically modelled. Pressure vessel steels, HY80 and HY130, were chosen for this study not only because of their technical importance but because of a wide variety of microstructures may be produced through varying heat treatment. In HY130 three quenched and tempered structures were produced all having the same strength but much different structures taking advantage of its secondary hardening characteristics. HY80 was heat treated to have dual martensite-ferrite phase structures. McEvily and coworkers (4,5) have shown that such structures give very high values of ΔK_{th} in 1018 and 2.25Cr-1Mo steels. Specimen geometry was also studied since preliminary measurements indicated it affected the near threshold crack propagation behavior in some metals.

Experimental Procedures

The HY80 and HY130 steels, compositions shown in Table I, were donated

Table I. Compositions of HY80 and HY130 Steels

	Chemical Composition (wt.%)						
Alloy	C	Si	Ni	Cr	Mo	V	Mn
HY80	0.18	0.27	3.00	1.58	0.50	0.006	0.32
HY130	0.10	0.23	5.33	0.49	0.57	0.064	0.35

by the U. S. Steel Research Laboratory in the form of hot rolled one-half and one inch thick plates respectively. In HY130 the three quenched and tempered heat treatments (Q.T.(A), Q.T.(B), Q.T.(C)) shown in Table II were chosen to give essentially the same hardness even though the tempering temperatures varied from 400 to 610°C. This results from secondary hardness due to V_4C_3 precipitation. For HY80 steel, the two different dual phase microstructures (IAA and IAB) shown in Table II were developed for comparison with the microstructure given by the standard heat treatment (ST) for this steel.

The microstructures resulting from the dual phase IAA and IAB heat treatments are shown in Figs. 1(a) and 1(b) respectively. From TEM thin foil observation (not shown here), the average widths of the martensite regions are 1.1 μm and 0.7 μm, respectively, for the IAA and IAB structures. While the average width of the ferrite regions is 0.7 μm and 0.5 μm, both structures contain interconnected martensite regions inside the prior austenite grains. The prior austenite grain sizes for the dual phase IAA and IAB structures are 35 μm and 12 μm respectively, corresponding to the

Table II. Heat Treatments and Hardness Values in HY80 and HY130

Material	Structure	Heat Treatment Procedures	Hardness R_c
HY130	Q.T. A	815° C, 1 hr, water quenched, heated to 400° C, 10 hrs, water quenched	33
	Q.T. B	815° C, 1 hr, water quenched, heated to 550° C, 5 hrs, water quenched	33
	Q.T. C	815° C, 1 hr, water quenched, heated to 610° C, 1 hr, water quenched	33
HY80	S.T.	1000° C, 3 hrs, furnace cooled to 25° C, heated to 900° C, 1 hr, water quenched, heated to 700° C, 1 hr, water quenched	20
	IAA	1000° C, 3 hrs, furnace cooled to 25° C, heated to 750° C, 2 days, water quenched	38
	IAB	1000° C, 3 hrs, furnace cooled to 25° C, heated to 900° C, 1 hr, water quenched, heated to 700° C, 1 day, water quenched, heated to 750° C, 2 days, water quenched	35
	IAAT	IAA treatment + 650° C, 1 hr, air cooled	20

austenitizing temperatures of 1000 and 900° C (Table II). With the standard treatment, the width of the martensite plates before tempering was approximately 0.4 μm.

Fatigue tests were conducted at room temperature on a closed loop electrohydraulic MTS system of 100 KN capacity. All crack propagation tests were run in a dry argon atmosphere at a load ratio R = 0.05 and frequency of 30 Hz. The crack length was measured with a 40 times magnification traveling microscope. Panel center and single edge notch specimens (designated CN and SEN respectively), 100 x 30 x 3 mm, were used. The stress intensity amplitude range, ΔK, is calculated by the secant equation for CN specimens (6) and the Pook's equation for SEN specimens with a rigid grip system (7). Crack propagation rate da/dN, versus ΔK was measured by the load shedding technique, and ΔK_{th} was defined as the ΔK value below which no crack advance is detected in 2×10^6 cycles, i.e., da/dN less than 1Å/cycle. After ΔK_{th} was determined, the load was increased less than 5% each step and da/dN was measured. The da/dN versus decreasing ΔK and increasing ΔK curves were consistent. At least two specimens of each structure were tested, with good agreement between identical specimens. Each data point shown in the figure of da/dN versus ΔK represents one individual measurement near the threshold region, but the average of three consecutive measurements in the higher ΔK region.

The crack closure stresses were measured using small foil strain gages cemented to the specimens (8). The local strain and applied nominal stress were recorded and the closure stress was determined from the deviation of the linear compliance relationship of the unloading portion of the curve.

A Talysurf 4 profilometer was used to characterize the fatigue fracture surfaces as a function of ΔK. Surface roughness scans were taken with a 1.3 μm radius stylus in the specimen thickness direction to represent a constant ΔK. The root mean square of the recorded roughness profile at intervals of 25 μm was taken as the measure of surface roughness.

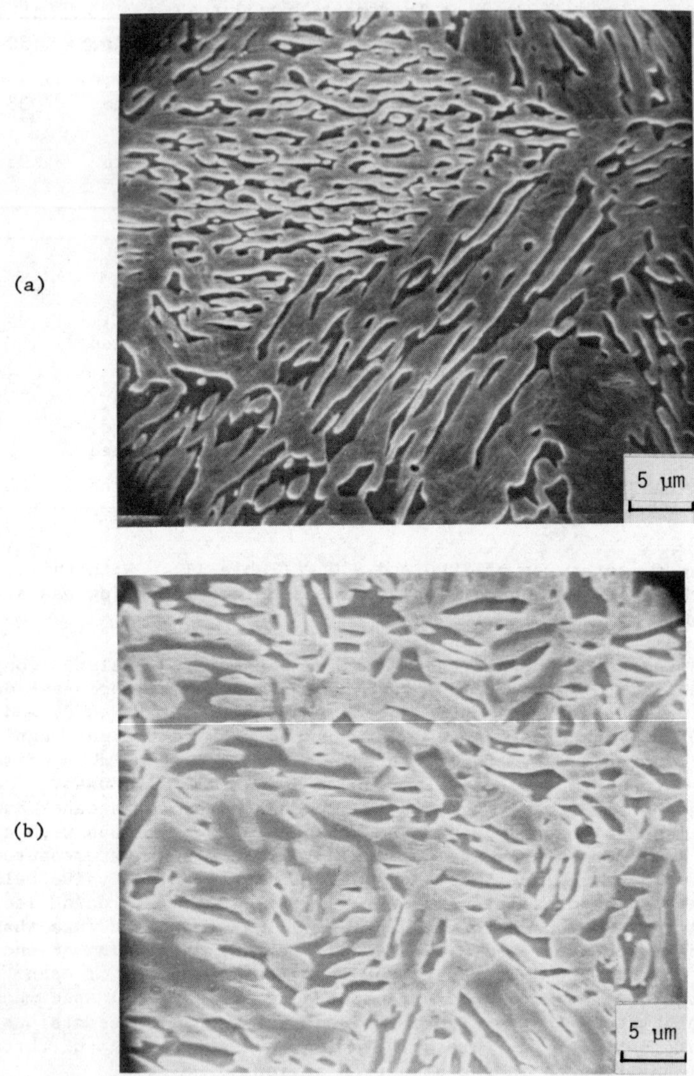

Figure 1 - Electron micrographs of HY80 dual phase steels.
(a) SEM of the dual phase IAA microstructure
(b) SEM of the dual phase IAB microstructure

Experimental Results

The crack propagation rates near threshold in HY130 with the different heat treatments A, B, and C (Table II) are shown in Fig. 2. The highest

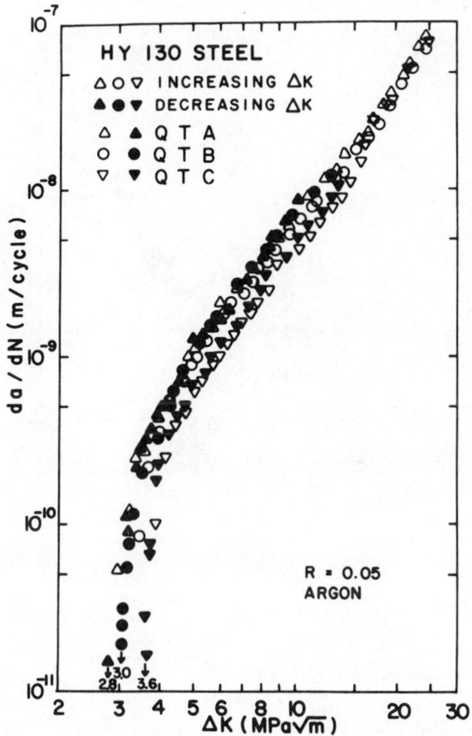

Figure 2 - Variation of fatigue crack propagation rates versus ΔK in HY130 steel with various tempering treatments.

tempering temperature, 1 hr at 610° C, gave the lowest da/dN and highest ΔK_{th}, 3.6 MPa√m. At higher ΔK's, all three heat treatments converged to one curve. As expected, microstructure plays a more important role in the near-threshold region than at higher ΔK's. Because there was so little difference among the 3 sets of data, this line of research was discontinued.

The near-threshold fatigue crack propagation rates of HY80 steel with the ST, IAA, and IAB heat treatments are plotted versus ΔK in Fig. 3. Compared to the standard heat treatment, where ΔK_{th} is 3.1 MPa√m, the dual phase heat treatments improve the near threshold fatigue crack propagation resistance. The value of ΔK_{th} is 4.3 MPa√m for the IAB structure and 5.4 MPa√m for the IAA structure. The IAA dual phase structure was additionally tempered at 650° C for 1 hr (IAAT) to give the same hardness as the ST heat treatment. A ΔK_{th} value of 4.4 MPa√m was obtained.

With respect to the closure stresses, the closure stress intensities (K_{cl}) versus ΔK are shown in Fig. 4(a). While the closure stress intensity

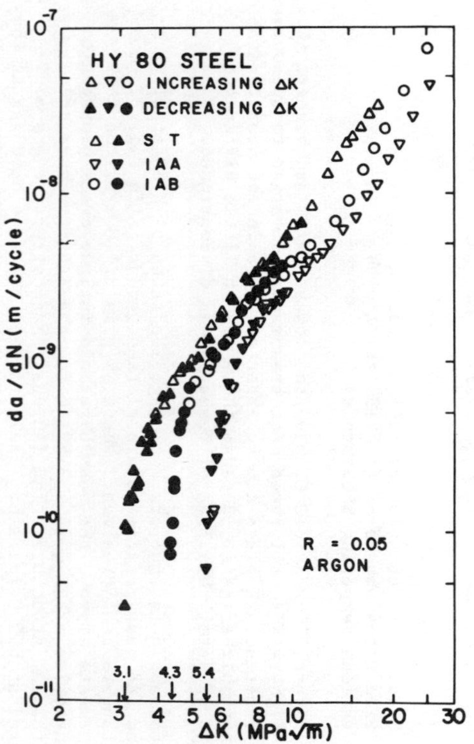

Figure 3 - Comparison of fatigue crack propagation rates versus ΔK of various heat treatments in HY80 steel.

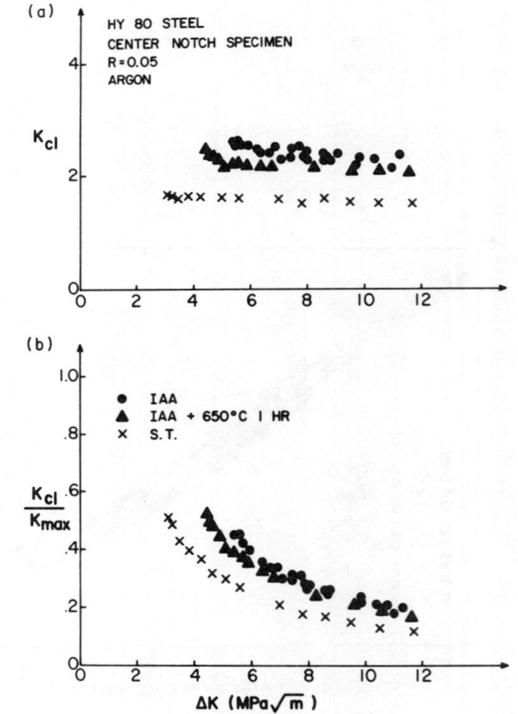

Figure 4 - The crack closure behavior of HY80 steel.

(a) Crack closure stress intensity K_{cl} vs. ΔK

(b) K_{cl}/K_{max} vs. ΔK

is basically constant for the higher ΔK region, it increases when ΔK_{th} is approached. The dual phase IAA structure which gave the highest ΔK_{th} also gives the highest K_{cl} values. When K_{cl} is normalized by K_{max}, as shown in Fig. 4(b), the importance of the closure effect in the near-threshold region becomes more obvious. The closure stress intensity near threshold is about 45 to 50% of K_{max} but it is only 10 to 20% of K_{max} at a ΔK of 12 MPa$\sqrt{m}$. Tempering after the dual phase IAA treatment (IAAT) reduced ΔK_{th}, but ΔK_{th} and K_{cl} were still higher than for the ST heat treatment.

Figure 5 shows the fatigue crack propagation rates in HY80 versus ΔK

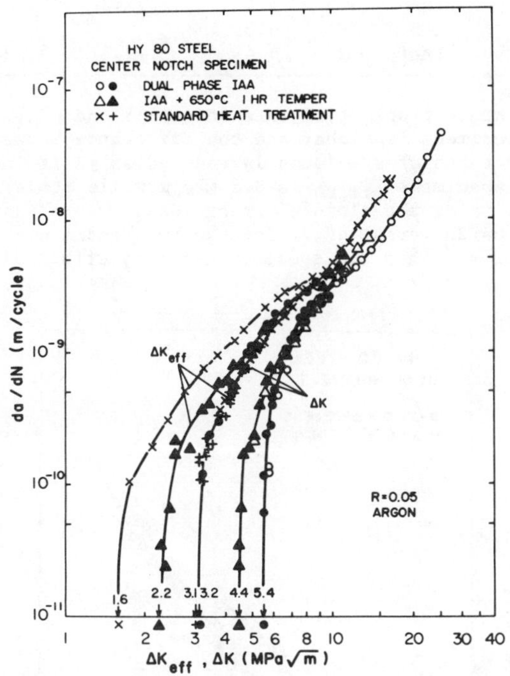

Figure 5 - Near-threshold fatigue crack propagation rate vs. ΔK and ΔK_{eff} in HY80 steel with various heat treatments.

and ΔK_{eff} where ΔK_{eff} is $K_{max} - K_{cl}$. In each HY80 steel structure, the convergence of the ΔK and ΔK_{eff} curves as ΔK is increased shows again that the crack closure effect is not significant just beyond the near-threshold region of ΔK. At low ΔK's the difference among ΔK_{eff} curves for the three structures is smaller than the difference among the ΔK curves. However, the ΔK_{eff} curves do not converge to one curve at low ΔK's. There is a microstructural effect on ΔK_{eff} near threshold. The results of ΔK_{th}, $K_{cl,th}$ and $\Delta K_{eff,th}$ in HY80 and HY130 are summarized in Table III.

The fatigue crack propagation tests covered up to this point were run using center notch specimens. A series of near-threshold fatigue crack propagation tests were done with both center notch (CN) and single edge notch (SEN) specimens of HY80-IAA structure. Figure 6 shows the near-

Table III. ΔK_{th}, $K_{cl,th}$ and $\Delta K_{eff,th}$ in HY130 and HY80 Steels-Center Notch Specimens

Material	Heat Treatment	ΔK_{th} (MPa$\sqrt{m}$)	$K_{cl,th}$ (MPa$\sqrt{m}$)	$\Delta K_{eff,th}$ (MPa$\sqrt{m}$)
HY130	Q.T. A	2.8	-	-
	Q.T. B	3.0	-	-
	Q.T. C	3.6	-	-
HY80	ST	3.1	1.7	1.6
	IAA	5.4	2.6	3.1
	IAB	4.3	-	-
	IAAT	4.4	2.4	2.2

threshold fatigue crack propagation rate versus ΔK and ΔK_{eff}. Obviously, ΔK_{th} of the SEN specimens is higher and the difference between the da/dN curves for the two types of specimens is reduced as ΔK is increased. Further, for the SEN specimens $K_{cl,th}$ is 3.8 MPa$\sqrt{m}$ while it is 2.6 MPa$\sqrt{m}$ for CN specimens. Utilizing the closure stress measurements shown in Fig. 6, the two curves of da/dN versus ΔK_{eff} for the two specimen geometries converge to one curve. Thus the specimen geometry effect disappears when

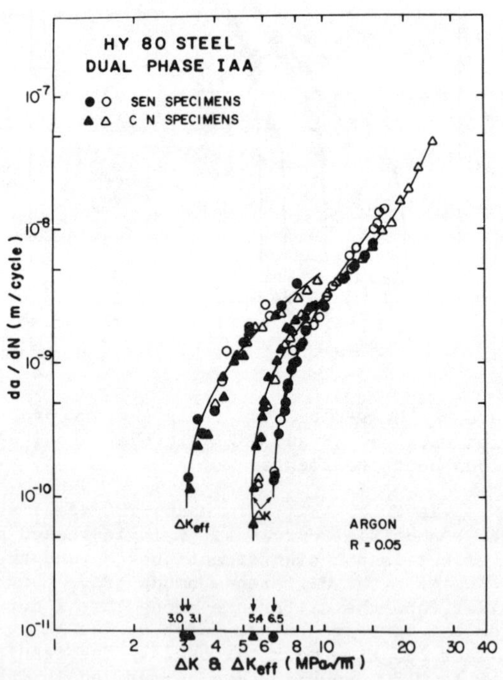

Figure 6 - Comparison of the fatigue crack propagation rates vs. ΔK and ΔK_{eff} in HY80 steel having dual phase IAA structure with different specimen geometries.

ΔK_{eff} is plotted rather than ΔK. Similar results were also found in 1018 steel with a dual phase structure and IAAT HY80 specimens (9).

The surface roughnesses of the HY80 IAA and ST structures were measured as previously described. The results are plotted in Fig. 7 along

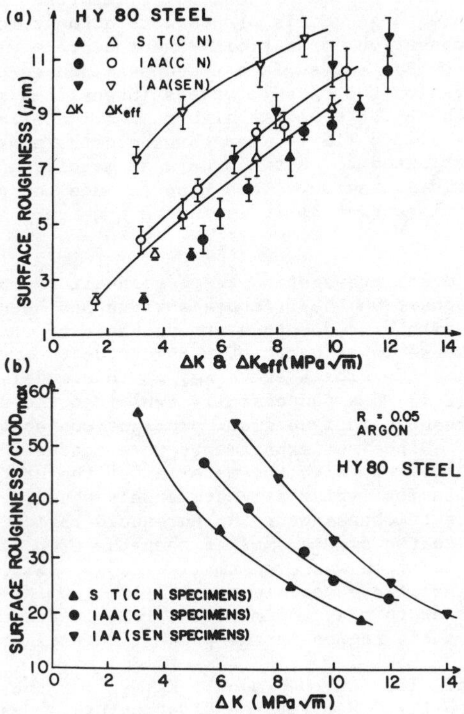

Figure 7 - Comparison of the surface roughness behavior in HY80 steel

(a) Surface roughness versus ΔK and ΔK_{eff}
(b) Surface roughness/$CTOD_{max}$ versus ΔK

with the ratio of surface roughness to $CTOD_{max}$ calculated using the equation $CTOD_{max} = 0.49K^2_{max}/E\sigma_y$. At any value of ΔK (Fig. 7(b)), this ratio is higher for the dual phase IAA structure than for the ST structure. The surface roughness monotonically increases as ΔK is increased which is consistent with the results reported by other researchers (10,11). Near the threshold region the roughness of the dual phase IAA heat treatment is higher than for ST heat treatment. As compared with the CN specimens, the roughness in the IAA heat treatment with the SEN specimens is always higher at the same ΔK. When surface roughness is plotted versus ΔK_{eff} rather than ΔK, the effect of specimen geometry on surface roughness becomes more apparent.

Discussion

Besides improving the fatigue properties of HY80 and HY130 the objectives of the present work were to hopefully advance our understanding of near-threshold fatigue crack propagation and the roles of microstructure and specimen geometry. The present experiments clearly show that crack closure in each structure becomes increasingly important as ΔK_{th} is approached. There are three models proposed to account for crack closure in fatigue crack propagation: plasticity-induced crack closure (1), oxide-induced crack closure (2), and surface roughness-induced crack closure (3). Near the threshold region the closure stress intensities in our results are always higher than those for somewhat higher ΔK's. Similar evidence was found in nickel-base alloys (12). Near threshold the plastic strain around the crack tip was estimated to be two orders of magnitude smaller than for the low end of the mid ΔK region. Thus Elber's model of plasticity-induced crack closure is not important as it is in the high part of the mid ΔK region.

During fatigue crack propagation testing in air or when water vapor is present, an oxide deposit on the fracture surface has been observed to increase in thickness with time due to a stress-assisted fretting mechanism. When the oxide thickness is of comparable magnitude to a value CTOD larger than $CTOD_{min}$ the crack tip closes above K_{min}. This model, first proposed by Ritchie and Suresh (2,13) has successfully explained the effect of environment on the near-threshold fatigue crack propagation behavior in a number of materials (2,13-15). Since our experiments were conducted in dry argon, interaction of H_2O and/or O_2 with the material at the crack tip was minimized. Recent results for medium strength steels which are similar to HY80 showed that the oxide thickness near the threshold region in the environment of air was one to two orders smaller than the CTOD at threshold (16). The oxide thickness in argon was even smaller (16). Similar results were obtained in fully pearlitic eutectoid steels (17). Oxide-induced crack closure does not appear to play an important role in the crack closure behavior near the threshold region in the present investigation.

As shown in Table III (CN specimens), $K_{cl,th}$ for the IAA structure is 2.6 $MPa\sqrt{m}$ compared to 1.7 $MPa\sqrt{m}$ for the ST structure. Crack surface roughness measurements show that near threshold the fracture surface of the IAA structure is two times rougher than for the ST structure. Since the measured crack surface roughness depends on the yield strength and applied loading condition, it suggests that the relative crack surface roughness, that is the measured crack surface roughness divided by $CTOD_{max}$, might be a better parameter than the measured roughness to correlate with the near-threshold fatigue crack propagation behavior. The relative crack surface roughness for the IAA structure is always higher than for the ST structure at the same ΔK or ΔK_{eff}. The good correlation of closure stress intensity with the relative crack surface roughness strongly suggests that in the present experiments, with other sources of crack closure essentially absent, the crack closure may be almost exclusively due to surface roughness induced crack closure (Fig. 7(b)). The origin of this roughness is due to shear displacements coupled with a normal fracture mode.

The present results also show that ΔK_{th} and the near-threshold fatigue propagation rate are functions of specimen geometry. The higher ΔK_{th} for SEN specimens compared to CN specimens correlates with higher $K_{cl,th}$. The higher $K_{cl,th}$ values in turn correlate with greater crack surface roughness. Near threshold the crack surface roughness at a given ΔK is larger for the SEN specimens than for CN specimens as illustrated in Fig. 7(a). This correlation is further evidence that the near-threshold crack closure effects is due to crack surface roughness.

The origin of the effect of specimen geometry on fatigue crack surface roughness is of great interest. Near ΔK_{th} the fatigue crack propagation is by a mixed shear-tensile mode, i.e., Mode I plus Mode II crack propagation occurs with shear facets seen on the fracture surfaces (18-20). Recently Suresh and Ritchie (3) have proposed a geometrical model of roughness-induced crack closure for the near-threshold fatigue crack propagation. The crack becomes wedge-closed along crack faces and stops advancing based on the assumption that near the threshold region a significant amount of Mode II displacements exists during cyclic loading when the applied stress intensity is above K_{cl}. Their equation to quantify the proportion of Mode II displacements (μ_{II}) to Mode I displacements (μ_I), x, is expressed by

$$x = \frac{(K_{cl}/K_{max})^2}{1-(K_{cl}/K_{max})^2} \frac{w}{2h} \qquad (1)$$

where w and h are the average wavelength and average height of the asperities respectively. Taking h equal to two times the measured surface roughness and w equal to the prior austenite grain size as an approximation, x was estimated for the different specimen geometries. The normalized Mode II displacement, $x_{II} \equiv x/(1+x)$, is found to correspond to the fraction of shear mode fracture near threshold (21). The relationships of x and x_{II} versus ΔK are shown in Fig. 8(a) and Fig. 8(b) respectively. It is found that the SEN specimens give higher amounts of Mode II displacements, i.e., x_{II} at a given ΔK is larger for SEN specimens. The structures giving higher ΔK_{th} values also have higher Mode II displacements and x_{II} which is in good agreement with the SEM fracture surface observations (9). Considering the effect of specimen geometry the question then becomes why is the ratio of Mode II to Mode I fatigue fracture greater for specimens with asymmetrical notches. The bending moment at the notch is suggested to play a role. If one corrects for crack closure to compute $\Delta K_{eff,th}$, then the specimen geometry effect disappears within experimental error. Combining the results of closure stress and crack surface roughness measurements gives further evidence that surface roughness induced crack closure is the origin of the crack closure effect in these experiments.

The effect of microstructure on ΔK_{th} can then be divided into two parts. The first part is due to the crack closure effect, i.e., the effect of microstructure on $K_{cl,\,th}$ keeping specimen geometry constant and the second part is due to the effect of microstructure on $\Delta K_{eff,th}$. The crack surface roughness and crack closure stress are expected to increase with the characteristic microstructural size parameter such as grain size. The effect of grain size on ΔK_{th} in iron is almost all due to $K_{cl,th}$ (20). In the present case, the values of $K_{cl,th}$ are almost the same for the IAA and IAAT structures presumably because the size of the ferrite patches are not changed during tempering. The higher $K_{cl,th}$ for HY80-IAAT compared to HY80-ST is attributed to the relatively large carbide free ferrite patches in the former.

The present results clearly show that $\Delta K_{eff,th}$ varies with microstructure in the same alloy. Table III shows that $\Delta K_{eff,th}$ is 3.1 MPa$\sqrt{m}$ for IAA and 1.6 MPa$\sqrt{m}$ for ST. It has been suggested that $\Delta K_{eff,th}$ is determined by the stress (σ_s) necessary to operate a dislocation source located a distance s from the crack tip. Accordingly, for low R values (20)

$$\Delta K_{eff,th} = \alpha\sigma_s\sqrt{2\pi s} \qquad (2)$$

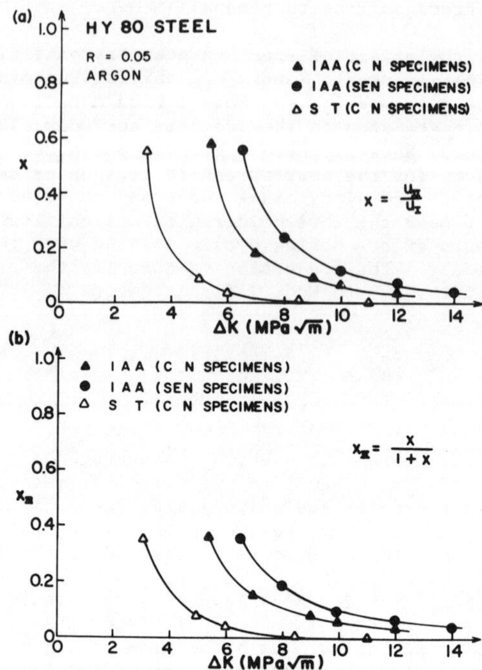

Figure 8. Comparison of Mode II displacements in HY80 steel.

(a) x versus ΔK

(b) x_{II} versus ΔK

where α is a factor which takes account of the fact that plasticity reduces the stress in the plastic zone from that calculated from fracture mechanics. From eq.(2), σ_s, which is related to yield strength, and s, which is related to the characteristic microstructural size parameter, are both larger after the IAA treatment than the standard heat treatment. The significant amount of isolated ferrite patches in the dual phase structures may alleviate the decrease of α which normally occurs with increase in σ_y. Tempering the IAA structure to give the IAAT structure reduces σ_s and $\Delta K_{eff,th}$. Comparing IAAT and ST structures, the strength levels are the same, thus α and σ_s are the same to a first approximation; however, $\Delta K_{eff,th}$ is higher for the IAAT structure. The microstructural size parameter is larger in the IAAT compared to the ST structure giving a larger value of s. Thus eq.(2) may be a basis for qualitatively understanding the effect of microstructural parameters on $\Delta K_{eff,th}$.

Since crack closure was not measured, it is not possible to discuss the variation of ΔK_{th} in HY130 with tempering temperature keeping hardness constant as definitively. However, increasing the tempering temperature is expected to increase the microstructural size parameter which could increase both crack surface roughness as well as s in eq.(2). The fracture surfaces were compared using SEM and 400° C tempered structure had the smoother fracture surface. It is thus likely that most of the increase in ΔK_{th} with tempering keeping strength constant is due to variation in crack closure.

Since much of the effect of microstructure on the near-threshold fatigue crack propagation is due to crack closure arising from crack surface roughness which in turn results from mixed Mode I-Mode II crack propagation, the ratio of near threshold Mode I-Mode II fatigue crack propagation is a function of specimen geometry. Thus near threshold fatigue crack propagation data must be used with extreme caution by the design engineer.

Conclusions

1. As compared with the standard quenched and tempered structure in HY80, the dual phase structures did improve the near-threshold fatigue crack propagation resistance as well as yield strength.

2. The measured crack surface roughness correlated very well with the closure stress intensity. Surface roughness was concluded to be responsible for crack closure in the present experiments. The effects of microstructure and specimen geometry on ΔK_{th} were explained on this basis.

3. ΔK_{eff} is a more fundamental parameter than ΔK for describing near threshold fatigue crack propagation behavior. $\Delta K_{eff,th}$ is independent of specimen geometry.

4. $\Delta K_{eff,th}$ is a function of microstructure and appears to be related to the material's strength and a microstructural size parameter.

Acknowledgements

This work was supported by the Office of Naval Research through Contract N00014-78-C-0565. The use of the facilities of Northwestern University's Materials Research Center, sponsored under the NSF-MRL Grant DMR79-23573 is acknowledged. The authors are grateful to Dr. L. F. Porter of the U. S. Steel Research Laboratory for providing the HY80 and HY130 steels used in this research.

References

1. W. Elber, "The Significance of Fatigue Crack Closure", pp. 230-242 in Damage Tolerance in Aircraft Structure, ASTM STP 486, Amer. Soc. Test. Mat., 1971.

2. S. Suresh, D. M. Parks and R. O. Ritchie, "Crack Tip Oxide Formation and Its Influence on Fatigue Threshold", p. 391 in Fatigue Threshold, Proc. 1st Int. Conf., Stockholm, J. Bäcklund, A. Blom, and C. J. Beevers, eds., EMAS Publ. Ltd., Warley, U.K., Vol. 1, 1982.

3. S. Suresh and R. O. Ritchie, "A Geometric Model for Fatigue Crack Closure Induced by Fracture Surface Roughness", Metallurgical Transactions, 13A (1982) pp. 1627-1631.

4. H. Suzuki and A. J. McEvily, "Microstructural Effects on Fatigue Crack Growth in a Low Carbon Steel", Metallurgical Transactions, 10A (1979) pp. 475-481.

5. K. Minakawa, Y. Matsuo and A. J. McEvily, "The Influence of a Duplex Microstructure in Steels on Fatigue Crack Growth in the Near-Threshold Region", Metallurgical Transactions, 13A (1982) pp. 439-445.

6. C. E. Feddersen, Discussion, p. 77, in Plane Strain Crack Toughness Testing of High Strength Metallic Materials, ASTM STP 410, Amer. Soc. Test. Mat., 1967.

7. L. P. Pook, "The Effect of Friction on Pin Jointed Single Edge Notch Fracture Toughness Test Specimens", International Journal of Fracture Mechanics, 4 (1968) pp. 295-297.

8. D. Gan and J. Weertman, "Crack Closure and Crack Propagation Rates in 7050 Aluminum", Engineering Fracture Mechanics, 15 (1981) pp. 87-106.

9. J. L. Horng, "Fatigue Crack Propagation of HY80, HY130 and 1018 Steels", Ph.D. Thesis, Northwestern University, Evanston, IL, 1983.

10. P. K. Liaw, A. Saxena, V. P. Swaminathan and T. T. Shih, "Effects of Load Ratio and Temperature on the Near-Threshold Fatigue Crack Propagation Behavior in a Cr-Mo-V Steel", Metallurgical Transactions, 14A (1983) pp. 1631-1640.

11. D. H. Park and M. E. Fine, "Near-Threshold Fatigue Crack Propagation in Fe and Al-3% Mg", paper presented at TMS-AIME Intl. Symposium: Concepts of Fatigue Crack Growth Threshold, Philadelphia, PA, October 1983.

12. M. Clavel and A. Pineau, "Fatigue Behavior of Two Nickel-base Alloys I: Experimental Results on Low Cycle Fatigue, Fatigue Crack Propagation and Substructures", Materials Science and Engineering, 55 (1982) pp. 157-171.

13. R. O. Ritchie, "Environmental Effects on Near-Threshold Fatigue Crack Propagation in Steels: A Re-Assessment", p. 503 in Fatigue Threshold, Proc. lst. Int. Conf., Stockholm, J. Bäcklund, A. Blom and C. J. Beevers, eds., EMAS Publ. Ltd., Warley, U.K., Vol. 1, 1982.

14. P. K. Liaw, T. R. Leax, R. W. Williams and M. G. Peck, "Influence of Oxide-Induced Crack Closure on Near-Threshold Fatigue Crack Growth Behavior", Acta Metallurgica, 30 (1982) pp. 2071-2078.

15. R. O. Ritchie, S. Suresh and C. M. Moss, "Near-Threshold Fatigue Crack Growth in $2\frac{1}{4}$Cr-1Mo Pressure Vessel Steel in Air and Hydrogen", Journal of Engineering Materials Technology, 102 (1980) pp. 293-299.

16. P. K. Liaw, S. J. Hudak, Jr. and J. K. Donald, "Influence of Gaseous Environments on Rates of Near-Threshold Fatigue Crack Propagation in Ni-Cr-Mo-V Steel", Metallurgical Transactions, 13A (1982) pp. 1633-1645.

17. G. T. Gray, III, J. C. William and A. W. Thompson, "Roughness-Induced Crack Closure: An Explanation for Microstructurally Sensitive Fatigue Crack Growth", Metallurgical Transactions, 14A (1983) pp. 421-433.

18. A. Otsuka, K. Mori and T. Miyata, "The Condition of Fatigue Crack Growth in Mixed Mode Condition", Engineering Fracture Mechanics, 7 (1975) pp. 429-439.

19. K. Minakawa and A. J. McEvily, "On Crack Closure in the Near-Threshold Region", Scripta Metallurgica, 15 (1981) pp. 633-636.

20. G. M. Lin and M. E. Fine, "Effect of Grain Size and Cold Work on the Near-Threshold Fatigue Crack Propagation and Crack Closure in Iron", Scripta Metallurgica, 16 (1982) pp. 1249-1254.

21. J. L. Horng and M. E. Fine, "Near-Threshold Mixed Mode I-II Fatigue Crack Propagation: Parallel Superposition Model", to appear in Scripta Metallurgica.

THE EFFECT OF MICROSTRUCTURE ON FATIGUE CRACK PATH

AND CRACK PROPAGATION RATE

G. T. Gray III*, A. W. Thompson**, and J. C. Williams**

*Formerly with Carnegie-Mellon University, is now a Visiting Scientist at the Technische Universität Hamburg-Harburg, Harburger Schloßstraße 20, 2100 Hamburg 90, West Germany

**Carnegie-Mellon University
Department of Metallurgical Engineering
and Materials Science
Pittsburgh, PA 15213 USA

The exact role played by microstructure in fatigue crack growth (FCG), even the existence of a microstructural role, has been the subject of considerable study during recent years. Studies at Carnegie-Mellon and elsewhere have now established that microstructure can have a considerable effect on FCG rate in a wide range of alloys and microstructures. Numerous factors have been identified as being important; these include local plasticity, slip character and slip reversibility. On a different size scale it has been shown that increasing fracture surface roughness tends to correlate with lower FCG rates and this has been related to variations in the ease and extent of crack closure. This paper examines the aforementioned factors and discusses their contributions to or role in FCG. The interaction between microscopic mechanism and roughness is also discussed and the various causes of changes in roughness (slip character, slip length, slip reversibility, microstructural scale) are examined.

Introduction

While widespread evidence (1,2) has shown that microstructure can have a pronounced influence on fatigue crack growth (FCG), particularly near the threshold, the exact micromechanisms responsible remain poorly understood. This quandary is further complicated by observations that the manner in which FCG actually occurs is a complex summation of several interrelated processes (1-3). Research on a variety of microstructures and alloys has concluded that: i) slip character/slip length, ii)local plasticity, iii) crack branching/secondary cracking and iv) crack-path tortuosity mechanisms may be important to understanding the influence of microstructure on FCG. These mechanisms have been utilized by numerous investigators to explain microstructural influences such as grain size, microstructural scale, alloying/interstitial additions, and yield strength on fatigue crack propagation (1-4).

Recently an additional mechanism, that of roughness-induced crack closure, has been suggested to contribute to or explain the influence of microstructure on FCG at low stress ratios (1). This process is thought to result when the topographic size-scale or height of the fracture surface roughness is comparable with the crack tip opening displacement and where Mode II displacements exist (a situation found at near-threshold levels at low load ratios) (5). According to this proposal, microstructures that possess rougher fracture topographies should exhibit a reducted effective ΔK or ΔK_{eff} value at the crack tip thereby shifting ΔK to higher values. Increasing the stress ratio, defined as R=(minimum load/maximum load), allows the crack to remain open ($\Delta K_{eff} \rightarrow \Delta K$) throughout the load cycle and crack closure is reduced. Studies on the influence of grain size on the FCG response at R=0.1 in pearlitic steels (1) and a nickel-base superalloy, 718 (6) found that FCG rates decreased with increased macro-fracture surface roughness. Additionally, data for microstructures exhibiting large variations in FCG rates at low R, with corresponding fracture surface roughnesses, were seen at high R values to collapse onto a single curve (1,7). Roughness-induced crack closure consequently appears to be a second-order mechanical phenomenon integrally linked to the micro-fracture mechanisms determining a microstructure's fracture topography.

The object of this paper is to provide an overview of the micromechanisms of FCG and their influence on fracture topography and crack propagation rate. These mechanisms will be discussed individually and then collectively examined in light of the growing evidence supporting the importance of roughness-induced crack closure.

Slip Character/Slip Length

The process of fatigue-crack propagation, whether by the "plastic blunting" model or another mechanism, is the consequence of slip at the crack tip (8). Accordingly, factors which influence slip behavior such as i) slip distribution, and ii) slip length may be important in governing crack paths in different materials. Experimental observations have linked these factors to changes in the stacking fault energy (SFE), alloying elements/interstitial content, and grain size or microstructural unit size (9-16). Among the factors determining how slip is distributed in a microstructure are: i) the propensity of dislocations within a material to be confined to distinct planes, regardless of the applied stress which can produce planar (as opposed to wavy) slip; ii) the tendency of that slip, whether planar or wavy, to be homogeneous (fine slip) or inhomogeneous (coarse slip); iii) the interaction of slip with particles, particularly inhomogeneous slip concentration due to particle shearing; and iv) a preferred crack path; inhomogeneous slip may occur in particle-free zones, for example.

Crack propagation rates have been observed to decrease in alloys with low SFE and resultant planar slip (9-11). Ishii and Weertman (10) attributed this behavior, in single crystals of Cu and Cu-Al, to the ease of cross slip on planes at the crack tip. This effect is shown in Figure 1. In their model, if cross slip cannot occur, or is impaired, the rate of crack growth is reduced. Research on Cu-Al polycrystalline alloys (12) has similarly shown a reduction in ΔK_{eff}, expressed as a reduction in threshold ΔK, ΔK_{th}, with increasing SFE over a range of grain sizes (Fig. 2).

The slip character and slip length of a material may determine whether crack propagation occurs through the matrix, along grain boundaries, or is concentrated along slip bands (13). Inhomogeneous or planar slip in a material can result in pile-ups against grain boundaries, and if the peak stress is sufficient the boundaries will fail by a brittle process (13). Cracks formed in this manner could then propagate either transgranularly or along the grain boundaries. If the inhomogeneity in the slip distribution of an alloy becomes more pronounced, with a high dislocation density in the slip bands, cracks will extend only along the slip bands (13).

While increased slip planarity may result in grain boundary or slip band cracking, slip planarity is thought to improve the chance of slip reversibility thereby reducing per cycle crack tip damage resulting in lower crack growth rates (4,14-17). Bathias and Pelloux (16) demonstrated the importance of slip character and slip reversibility on FCG rate of austenitic and martensitic steels. The difference in degree of slip reversibility involved in the process of crack tip advancement between an inhomogeneous and homogeneous material is schematically illustrated in Figure 3. More extensive slip reversibility resulting from more inhomogeneous slip and an increased slip length in the large grained microstructures is thought to be the reason for the observed reduction in propagation rates with increased grain size. Lindigkeit et al. (4) utilized this argument to explain the effect of grain size on the FCG response of Ti-8.6Al and Al-5.77Zn-2.5Mg-1.5Cu (Fig. 4). While this qualitative concept is only applicable in the low growth regime, where the plastic zone is small compared to the grain size*, it offers a plausible link between FCG response and the slip-dependent crystallographic fracture mode observed in inhomogeneously deforming materials.

If the ideas of slip character and slip length are taken collectively, alloys possessing both planar slip and a large grain size should exhibit greatly reduced FCG rates. This should be evident, for example, as a shift in threshold ΔK, which in turn implies a shift in ΔK_{eff}. The experimental results of Higo et al. (12), as shown in Figure 2, reveals such a trend.

Local Plasticity

The flow stresses within the cyclic plastic zone generated ahead of a growing crack are known to be influenced by the cyclic work hardening behavior and cyclic yield strength (18). Depending on whether a microstructure hardens or softens during cyclic loading, the rate at which dislocation damage accumulates at the crack tip may vary. The area of damage ahead of the crack tip, i.e. the plastic zone, is dependent on the cyclic yield strength (19). Combined, these two parameters are thought to govern the portion of

*When the plastic zone size, according to a fracture mechanics calculation, is small relative to the grain, uncertainty is appropriate as to the actual physical extent of dislocation motion, especially when deformation is inhomogeneous.

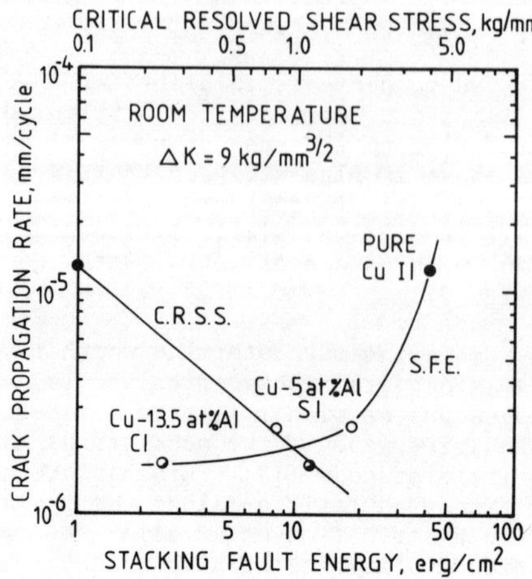

FIGURE 1 - Fatigue crack propagation rate as a function of SFE and of critical resolved shear stress (C.R.S.S.) for single crystals with approximately the same orientation (10).

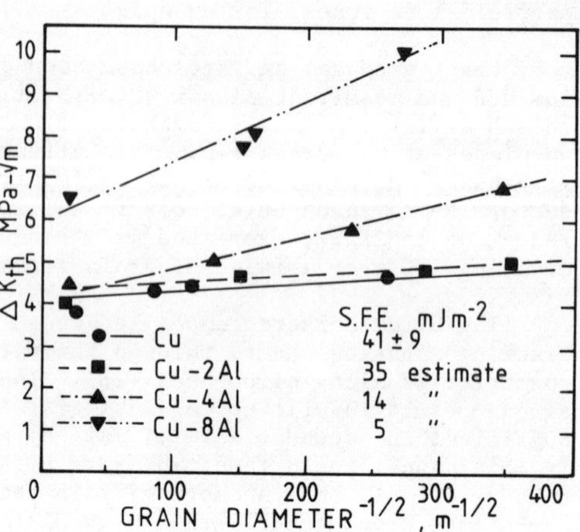

FIGURE 2 - The variation of ΔK_{th}, for tests at R = 0.5, with grain size and SFE (12).

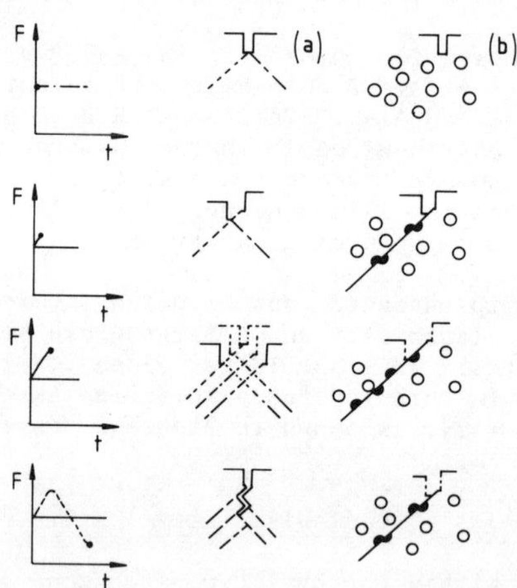

FIGURE 3 - Schematic representation of crack propagation during one cycle. (a) Microstructure with tendency toward homogeneous deformation: several slip planes and two slip systems operating. (b) Microstructure with tendency toward inhomogeneous deformation: slip in one plane and one system allows partial reversion of precipitates as the primary crack progresses. From Hornbogen and Zum Gahr (17).

the microstructure stressed at any given ΔK level and to influence the local dislocation response.

Numerous investigators have observed that increasing yield strength results in accelerated near-threshold FCG in a variety of alloys (2). Figure 5 illustrates this trend in a 300-M steel, tested in moist air, for two microstructures which possess the same monotonic yield strength of 1497 MPa (18). The variation in cyclic yield behavior of the Q and T microstructure, tested at R=0.05, was utilized to explain the large observed differences in the threshold ΔK. Cyclic softening in the quenched and tempered steels was attributed to a rearrangement of the dislocation substructure and a reduction in dislocation density with cyclic loading (18). Cyclic hardening was attributed to dynamic strain aging, which is characteristic of untempered and lightly tempered steels having high dislocation densities (18). While this research indicates how the cyclic work hardening behavior of a material may influence FCG, this rationale is considered to provide little insight into many materials that are cyclically stable (3).

The size of the reversed plastic zone, compared to the microstructural unit size, has additionally been suggested to determine whether the mode of FCG is "microstructurally sensitive" or insensitive (20). Robinson and Beevers (21) attributed the influence of grain size on FCG in α-titanium to grain-orientation control of the fracture path. This path dependence was suggested to occur when the extent of reversed plasticity at the crack tip was of the order of, or less than, the grain size. While the relative extent of the reversed plastic zone may be capable of activating a transition in fracture mode, rationalizing the influence of grain size, in many materials, with such a process seems doubtful in view of the fact that at near-threshold growth levels the reversed plastic zone is small compared to the grain size.

This correlation also appears suspect in view of the fact that in that study the influence of grain size diminished with increasing R. In light of recent evidence (1,6), which has linked grain size to roughness-induced crack closure, this model may need to be reassessed.

Crack Branching/Secondary Cracking

The rate with which a crack propagates through a material, if only microstructural effects are considered, can be viewed as dependent on the processes that dissipate the energy provided to the specimen plastic zone by the cyclic stresses. Crack branching and secondary cracking are presumed to lead to reduced crack growth due to increasing the total crack length while decreasing ΔK at each new crack tip when the crack branches. Assuming a reasonably constant energy requirement per unit length of a given fracture mode, additional energy is necessary to create the new surface area in a branched or secondary crack (3).

Once a branch occurs, the branching may be accompanied by a temporary decreasing growth rate effect, a crack arresting effect, or a constant velocity effect (22). For a branched or forked crack, the local Mode I and Mode II stress intensity factors, "k_1" and "k_2", respectively, can be expressed as functions of the stress intensities for the main crack "K_1" and "K_2", such that (23):

$$k_1 = \kappa_{11}(\alpha)K_1 + \kappa_{12}(\alpha)K_2$$

$$k_2 = \kappa_{21}(\alpha)K_1 + \kappa_{22}(\alpha)K_2$$

where "α" is the angle denoting the extent of crack branching. If we assume that each branch can grow only in the direction of K_2=0 (22), and that branching occurs at α=45°, the stress intensity factors k_1 and k_2 are found to be 0.8K_1 and 0.3K_1, respectively. The effective stress intensity at the tip of the branched crack is: k = 0.81K_1. Crack branching may therefore result in a reduction in the effective stress intensity at the crack tip of 19%. If the main crack forks, assuming each fork being of equal length, the driving force at the crack tip may be reduced by as much as 35% of K_1 (22). When branches are small, however, the applicability of conventional fracture mechanics may be questioned. Additional work on this topic would be welcome.

Based on these effects, microstructures exhibiting extensive secondary cracking, crack branching, or both, should have reduced crack growth rates compared to microstructures that exhibit little or no branching. The influence of microstructure on the FCG rate in Ti-6Al-4V (20,24) and IMI 685 (25) has been rationalized partly on this basis.

Crack Path Tortuosity

When the crack path is easiest along some path in the microstructure which is not continuous but is irregular with respect to the average crack growth direction, the crack travels a tortuous path, going "further" through the microstructure than the linear average fatigue crack path would suggest (24). Accordingly, microstructures which divert the crack from the plane normal to the stress axis, or provide a complex crack path, may exhibit reduced FCG rates. This crack path deviation is suggested to result when the microstructure has a preferred crack path or one in which crack propagation is impaired by boundaries (26).

A qualitative explanation of the influence of grain size on ΔK_{eff} has been based on a consideration of the "zig-zag" deviations of the fatigue crack (27,28). The original rationale of Masounave and Bailon (27) stated that a crystallographically caused "zig-zag" in the crack path, which increases with grain size, would reduce crack growth and therefore ΔK_{eff} because of an increased crack length traveled per unit crack advance. Priddle (28), while studying near-threshold da/dN in 316 stainless steel, similarly suggested that the influence of grain size on FCG rate was due to i) a reduction in K_1 with crack deviation from the plane of maximum tensile stress, and also ii) an increase in real crack surface energy compared to that obtained from macroscopic crack length and specimen width considerations. Research on a variety of titanium alloys (24-26) has similarly shown that microstructures having superior FCG resistance exhibit a great deal of crack growth at small angles with respect to the tensile axis and crack branching. Eylon and Bania (26) additionally observed a reduction in FCG rate in Ti-11, with a colony microstructure, to result from cracks halted at the colony boundaries and forced to reinitiate through a cycle consuming process into the next colony.

Roughness-Induced Crack Closure

Several investigators (1,7,29,30) have suggested that roughness-induced

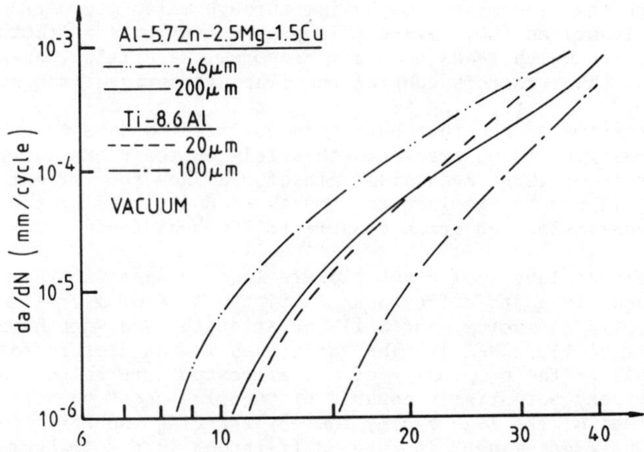

FIGURE 4 - Rate of FCG, da/dN, as a function of ΔK at R = 0.1 for the aluminum alloy, and at R = 0.2 for the titanium alloy, for two different grain sizes (4).

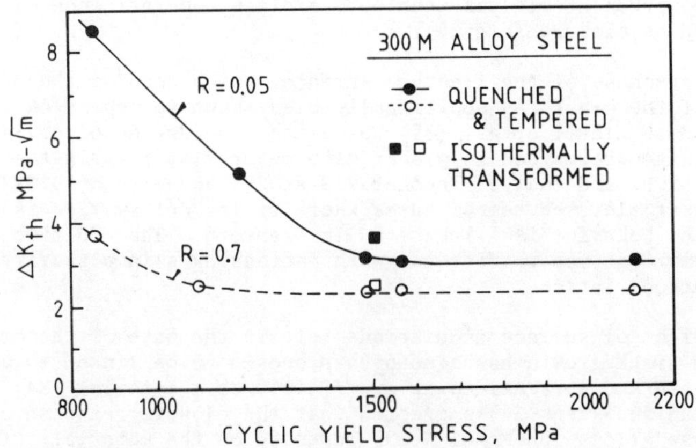

FIGURE 5 - Influence of cyclic yield strength, measured by 0.2 pct offset flow stress, on ΔK_{th} at R values of 0.05 and 0.7, in a 300-M steel tested in moist air. From Ritchie (18).

closure may be the predominant mechanism through which microstructure manifests its influence on FCG. Since this process reflects a fracture surface asperity mismatch which leads to crack closure, the critical question therefore posed is: What factors control the fracture surface topography of a microstructure? The preceding sections of this paper have reviewed some of the micromechanisms of FCG that have been utilized to date to explain influences of microstructure on crack growth solely on their own merit. Assessment of the role of these mechanisms, insofar as how they affect fracture topography, will now be examined as a means to understanding the contributions of roughness-induced crack closure to FCG behavior.

The different levels of crack closure load between varying microstructures have been linked to differences in i) the size of microstructural components, ii) slip character, and iii) the strength, and work hardening behavior of a material (1,7,29-34). The variations in the closure load in turn are thought to reflect the propensity of the microstructure to generate a fracture surface topography potentially capable of "wedging open" the crack and preventing reversal of the load cycle, thereby changing the effective ΔK driving the crack. The exact manner in which differences in the fracture roughness leads to varying amounts of crack closure is difficult to assess. A schematic illustration of how increasing the crack path tortuosity, macroscopic fracture surface roughness, or both, may result from an increase in grain size is shown in Figure 6.

Crack closure, due to Mode II loading causing a fracture surface mismatch, is thought to increase with the magnitude of the fracture roughness, causing reduced FCG rates. Such a relationship between crack topography and crack closure has been utilized to explain the decrease in FCG with increasing grain size in pearlitic steels (1), α-titanium (19), α/β-titanium (31), mild steel (25), and stainless steel (26). While several investigators have shown that roughness-induced closure appears to be the predominant mechanism responsible for the influence of grain size on FCG, a study by Nakai et al. (29) on low-carbon steels was unable to explain the influence of grain size solely based on closure (Fig. 7).

The unevenness of the fracture surface, which denotes the propensity to deviation of the crack has additionally been shown to depend on slip character (7,32,33). Clavel et al. (32) noted that the degree of deviation of the crack path was more marked in planar slip materials; this is the case of Waspaloy at 298K and that of Inconel 718 at 298 and also at 923K as compared to Waspaloy at elevated temperatures where cyclic deformation is more homogeneous. The behavior is illustrated in Figure 8. The FCG rate for both materials was observed to decrease with increasing slip planarity and corresponding crack deviation.

The height of surface undulations left in the wake of the crack tip during ductile crack growth has also been proposed to be linked to the cyclic yield strength and work hardening coefficient of a material (34). Purashothaman and Tien (34) suggested that the closure response of several aluminum-base alloys and Monel K-500 depended on the material's fracture ductility. Increasing fracture ductility was observed to correspond to increased crack closure and reduced FCG rates. Contrary to these findings, research on Ti-6242, conducted at 510°C, exhibited exactly the reverse correlation with increasing ductility accompanying increases in the FCG rate (7). Resolution of these differing views encompasses one of the stumbling blocks to an independent roughness closure mechanism; that being how to separate the influences of roughness from plasticity-induced crack closure. While increased ductility may increase the potential reduction in FCG due to roughness-induced closure, increased ductility may also reflect a less stabilized

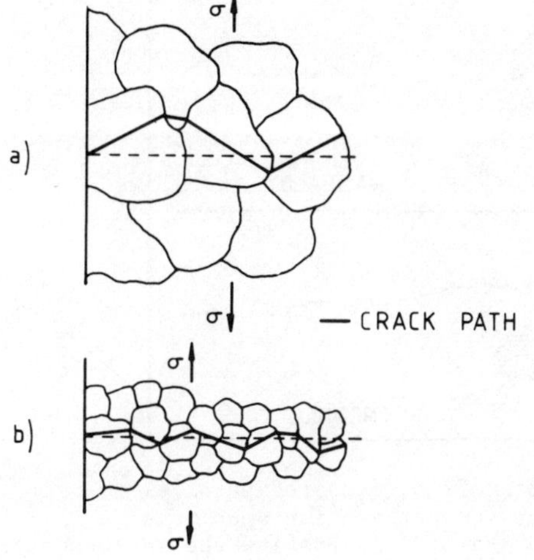

FIGURE 6 - Schematic representation of the fatigue crack path in (a) a coarse-grained, and (b) a fine-grained microstructure (27).

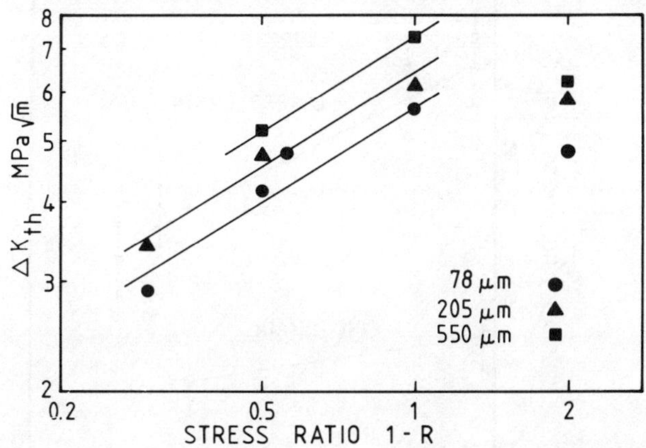

FIGURE 7 - Threshold ΔK as a function of stress ratio, expressed as (1-R), and grain size in a low-carbon steel (29).

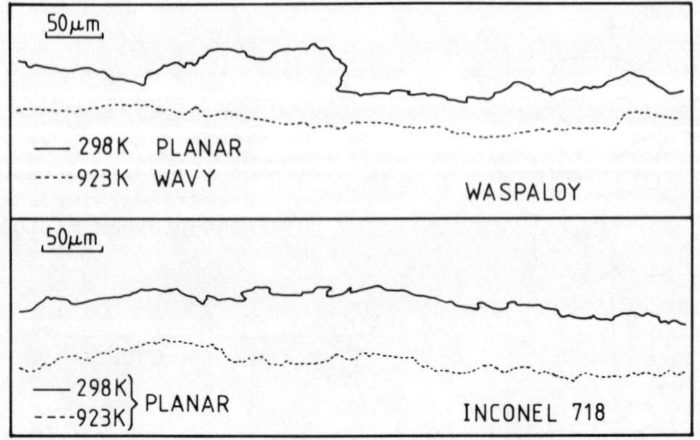

FIGURE 8 - Fracture surface profile, measured parallel to crack growth direction, for compact tension specimens fatigued at ΔK of about 30 $MPa\sqrt{m}$, illustrating the influence of slip character on crack path tortuosity. From Clavel, et al. (32).

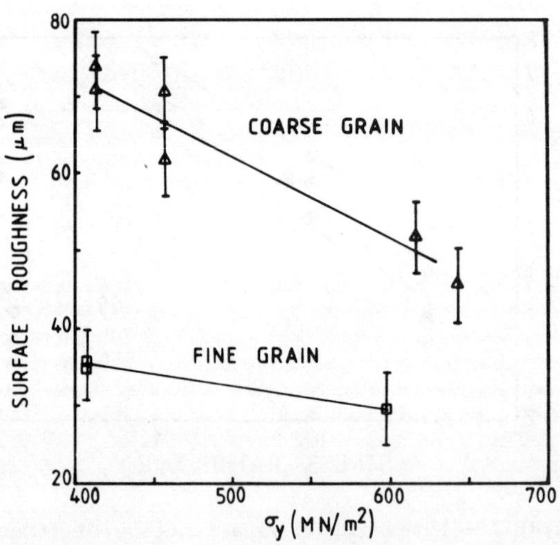

FIGURE 9 - Dependence of fracture surface roughness on yield strength, for measurements of the surface near the threshold, of pearlitic steel with either fine (25 μm) or coarse (130 μm) grain size. From Gray, et al. (1).

plastic zone, i.e. cyclic softening, ahead of the crack tip leading to a potential opposite reduction in plasticity-induced closure increasing the local FCG rate. There also are likely to be unresolved microplasticity as well as mechanics issues in connecting closure to monotonic ductility properties.

The cyclic hardening response of a material has also been suggested by Budiansky and Hutchinson (35) to be important to closure behavior. According to their model, cyclic hardening should lead to increased closure and cyclic softening to reduced closure. The dependence of yield strength in pearlitic steels, where increasing yield strength corresponds to a refinement in the pearlitic interlamellar spacing, may also reflect a combined effect of cyclic hardening and roughness-induced closure on near-threshold FCG. Coarse-lamellar microstructures were observed to possess a ΔK_{th} approximately 1.5 MPa $\sqrt{m}$ higher than the fine lamellar microstructures and an increased fracture roughness (Fig. 9). The cyclic deformation behavior in eutectoid pearlitic steels has been shown to depend strongly on the interlamellar spacing, with cyclic softening observed in fine pearlite and cyclic hardening observed in coarse pearlite. The observed decrease in ΔK_{th} with decreasing lamellar spacing may therefore include contributions of both plasticity and roughness-induced crack closure.

Summary and Conclusions

In this paper, an attempt has been made to review the existing information on the micromechanisms by which microstructure influences FCG. Within a broad class of metals and alloys, FCG rate can be reduced by a microstructure which: i) delays the accumulation of dislocation damage ahead of the crack tip, and/or ii) provides a multiplicity of crack paths, and/or iii) produces substantial out of plane deviation of the fatigue crack leading to "non-closure" of the crack. These processes have been linked to changes in the SFE, slip length, yield strength, cyclic hardening behavior, crack branching, and roughness-induced crack closure behavior of a material. FCG rates have been observed to be reduced in materials or microstructures possessing:

i) a low SFE, planar slip, or inhomogeneous slip tendency which inhibits cross-slip, favoring slip reversibility and crack path deviation leading to crack closure,

ii) increased slip length, increasing slip reversibility and more out-of-plane crack growth linked to roughness-induced crack closure,

iii) extensive crack branching, secondary cracking, or both, reducing the ΔK_{th} driving the crack,

iv) an irregular crack path leading to increased crack length and out-of-plane crack growth,

v) a reduced yield strength and cyclic hardening which may increase crack closure.

The mechanism of roughness-induced crack closure appears particularly helpful in understanding the influence of microstructure and stress ratio on FCG rate, especially near-threshold. This process, while resulting from a mechanical "wedging" of the crack tip, reflects the fracture surface topography which is a manifestation of the underlying microstructure. Modification of a materials structure and properties therefore remains the important variable which may reduce or enhance the contribution of crack closure to

FCG behavior.

Continuing research is needed to quantify and model the micromechanisms of FCG to understand more fully the interrelationship between intrinsic material properties and crack closure leading to crack propagation.

Acknowledgments

We appreciate helpful discussions with several of our colleagues, notably R. O. Ritchie, S. Suresh, G. Luetjering and A. Gysler. Some of the preparation of this manuscript was supported by the Association of American Railroads and by The Center for the Joining of Materials at Carnegie-Mellon University.

References

1. G. T. Gray, III, J. C. Williams, and A. W. Thompson, Metall. Trans. A, 1983, 14A, pp. 421-433.

2. R. O. Ritchie, Int. Metall. Rev., 1979, vol. 24, pp. 205-230.

3. J. C. Chesnutt, A. W. Thompson, and J. C. Williams, Influence of Metallurgical Factors on the Fatigue Crack Growth Rate in Alpha-Beta Titanium Alloys, AFML TR-78-68, Air Force Materials Laboratory, Wright-Patterson AFB, OH, 1978.

4. J. Lindigkeit, G. Terlinde, A. Gysler, and G. Luetjering, Acta Metall., 1979, vol. 27, pp. 1717-1726.

5. R. O. Ritchie and S. Suresh, Metall. Trans. A, 1982, 13A, pp. 937-940.

6. J. L. Yuen, Ph.D. Dissertation, 1982, Stanford University, Stanford, CA.

7. J. E. Allison, Ph.D. Dissertation, 1983, Carnegie-Mellon University, Pittsburgh, PA.

8. C. Laird, Fatigue and Microstructure, 1978 ASM Materials Science Seminar, pp. 149-203, ASM, Metals Park, OH, 1979.

9. A. J. McEvily and R. C. Boettner, Acta Metall., 1963, vol. 2, pp. 725-43.

10. H. Ishii and J. Weertman, Metall. Trans., 1971, 2, pp. 3441-52.

11. G. A. Miller, D. H. Avery, and W. A. Backofen, Trans. TMS-AIME, 1966, 236, pp. 1667-73.

12. Y. Hiso, A. C. Picard, and J. F. Knott, Metal Sci., 1981, vol. 15, pp. 233-240.

13. G. Luetjering, Slip Distribution and Mechanical Properties of Metallic Materials, 1976, ESA-TT-245, European Space Agency.

14. H. M. Kim, H. G. Paris, and J. C. Williams, Titanium '80, Science and Technology, H. Kimura and O. Izumi, eds., TMS-AIME, 1981, pp. 1825-35.

15. J. C. Williams and G. Luetjering, Titanium '80, Science and Technology, H. Kimura and O. Izumi, eds., TMS-AIME, 1981, p. 671.

16. C. Bathias and R. M. Pelloux, Metall. Trans. A, 1973, 4, pp. 1265-73.

17. E. Hornbogen and Z.H. Zum Gahr, Acta Metall., 1976, vol. 24, pp. 581-92.

18. R. O. Ritchie, J. Eng. Mat. Tech., 1977, vol. 99, pp. 195-204.

19. J. R. Rice, Fatigue Crack Propagation, ASTM STP 415, Amer. Soc. Testing Materl, 1967, pp. 247-309.

20. G. R. Yoder, L. A. Cooley, and T. W. Crooker, Metall. Trans. A, 1977, 8A, pp 1737-43.

21. J. L. Robinson and C. J. Beevers, Metal. Sci., 1973, vol. 7, pp. 153-59.

22. H. Kitagawa, R. Yuuki, and T. Ohira, Eng. Frac. Mech., 1975, vol. 7, pp. 515-29.

23. B. A. Bilby, G. E. Cardew, and I. C. Howard, Fracture 1977, D. M. Taplin, ed., University of Waterloo Press, 1977, pp. 197-200.

24. A. W. Thompson, J. C. Williams, J. D. Frandsen, and J. C. Chesnutt, Titanium and Titanium Alloys, J. C. Williams and A. F. Belov, eds., Plenum Press, New York, 1982, pp. 691-704.

25. D. Eylon, Metall. Trans. A, 1979, 10A, pp. 311-17.

26. D. Eylon and P. J. Bania, Metall. Trans. A, 1978, 9A, pp. 1273-79.

27. J. Masounave and J. P. Bailon, Scripta Met., 1976, vol. 10, pp. 165-70.

28. E. K. Priddle, Scripta Met., 1978, vol. 12, pp. 49-56.

29. Y. Nakai, K. Tanaka, and T. Nakanishi, Eng. Frac. Mech., 1981, vol. 15, pp. 291-302.

30. M. D. Halliday and C. J. Beevers, J. Testing and Eval., 1981, vol. 9, pp. 195-201.

31. M. D. Halliday and C. J. Beevers, Int. J. Fract., 1979, vol. 15, p. R27.

32. M. Clavel, C. Levaillant, and A. Pineau, Creep-Fatigue Environmental Interactions, R. M. Pelloux and N. S. Stoloff, eds., AIME, Warrendale, PA, 1980, pp. 24-45.

33. R. J. Asaro, L. Hermann, and J. M. Baik, Metall. Trans. A, 12A, pp. 1133-35.

34. S. Purushothaman and J. K. Tien, Strength of Metals and Alloys, vol. 2, Pergamon Press, New York, 1979, pp. 1267-71.

35. B. Budiansky and J. W. Hutchinson, Trans. ASME, 1978, vol. 45, pp. 267-76.

36. H. Sunwoo, M. E. Fine, M. Meshii, and D. H. Stone, Metall. Trans. A, 1982, 13A, pp. 2035-47.

ORIGIN OF CRACK CLOSURE IN THE NEAR-THRESHOLD FATIGUE CRACK PROPAGATION OF Fe AND Al-3% Mg

D. H. Park and M. E. Fine

Department of Materials Science and Engineering and
Materials Research Center
Northwestern University
Evanston, Illinois 60201
USA

Near threshold fatigue crack propagation rates were determined in annealed iron at 298° K in dry argon and Al-3% Mg at 298° K in dry argon and 77° K in liquid nitrogen for load ratios of 0.05 and 0.5. The crack closure levels and fracture surface roughness were also measured. Fractographs were examined to give the areal fractions of shear mode and normal mode fracture. While ΔK_{th} is a function of load ratio, ΔK_{eff}^{th} is not at 77° K or 298° K. Both ΔK_{th} and ΔK_{eff}^{th} at 77° K are higher than the corresponding values of 298° K. The effect of crack closure on ΔK (ΔK_{cl}) increases as ΔK_{th} is approached and also as the load ratio is decreased. While the mechanism for crack closure at low load ratio appears to be surface roughness coupled with shear mode displacements, at high load ratios plasticity induced crack closure also appears to play a role due to the high mean stress. The Suresh and Ritchie equation for mixed fracture mode crack closure was modified to introduce the second contribution.

Introduction

Since crack closure is known to play an important role in the near-threshold stress intensity region (region I) of fatigue crack propagation (1-12), it is desirable to quantify the crack closure stress intensity and its relation to fracture surface morphology, load ratio and temperature in order to give a better understanding of the phenomenon. At high values of ΔK, crack closure is considered to be mainly due to plastic stretching at the crack tip (13-15); however, at low values of ΔK and low load ratios (R), two sources of crack closure are considered to be important: (1) oxide wedging of the crack tip (2,8,10,16) and (2) an irregular crack surface coupled with shear mode displacements (1,3-7,9). In an inert environment only the latter is operative at low load ratios but at high load ratios, as shown by the present research, plasticity induced crack closure may play a role due to the high mean stress level. The geometrical model presented by Suresh and Ritchie (7) for crack closure relates the extent of crack closure to the irregular crack surface and the crack tip displacements due to shear mode deformation. Quantitative analyses of the fracture surface are necessary to examine this model and the role that irregular crack surface and shear mode displacements play in region I of the crack propagation rate (da/dN) vs. stress intensity range (ΔK) curve. The major effect of increase in the near-threshold fatigue crack propagation rate with increase in load ratio ($R = \sigma_{min}/\sigma_{max}$) is well known; however, most previous studies have been conducted in an aggressive environment (2-5,8-11,19-24) such as laboratory air. Although some previous studies were done in an inert environment, the results are controversial. Some investigators (17,21,24) observed the absence of a load ratio effect on the stress intensity range threshold for fatigue crack propagation (ΔK_{th}) in vacuum, but others, Stewart in vacuum (10), Liaw in liquid helium and nitrogen (16) and Gray in purified helium (5) found that ΔK_{th} is affected by load ratio even though the effect is reduced. The effect of crack closure was not well quantified in any of these studies.

In the present study most of the research was conducted in an inert environment, dry argon or liquid nitrogen, although some tests were done in laboratory air for comparison. Two materials were selected for study, vacuum melted iron and Al-3% Mg alloy. Studies of the latter alloy were done in liquid nitrogen partly to determine the effect of temperature on the region I fatigue crack propagation behavior and partly to provide an inert environment since purified gases and the best vacuum available contain some residual impurities. Reduction in thermal activation due to lowering the temperature is expected to completely remove oxidation as a factor in tests done in liquid nitrogen thereby simplifying analysis of the data.

Once ΔK_{th} has been corrected for the closure stress intensity, it is of interest to determine the influence of dislocation dynamics on ΔK^{th}_{eff}, i.e., $(K_{max}-K_{cl})$. The effect of temperature on ΔK^{th}_{eff} is thus of considerable interest.

Experimental Matters

The materials examined in this study were vacuum melted commercially pure iron (99.8+% purity*) donated by the Inland Steel Research Laboratory

* Chemical composition in wt. pct.: 0.008C, 0.01 Mn, 0.03S, 0.023Si, 0.010(O), 0.03N, 0.008P, 0.013Al and balance Fe.

and high purity aluminum-3% magnesium alloy** donated by ALCOA Technical Center in the form of 6 mm thick, 150 mm wide and 11 mm thick, 300 mm wide hot rolled plates, respectively.

The iron samples were machined and homogenized at 750° C for 2 hrs in 1×10^{-6} torr vacuum and furnace cooled. The Al-3% Mg alloy samples were solutionized at 430° C for 5 hrs, oil quenched, reheated to 250° C for 10 hrs to dissolve any precipitates which may have occurred during quenching and finally oil quenched again before machining. The mechanical properties are shown in Table I. Using a 0.2 mm copper wire tool on an "ELOX" electric

Table I. Grain Size and Mechanical Properties of Vacuum Melted Iron and Al-3% Mg

	Temp.	σ_y (MPa)	σ_y' (MPa)	Grain Size (μm)
Fe	298° K	128	204	80
Al-3% Mg	298° K	52	86	170
	77° K	85	-	170

σ_y and σ_y' are 0.2% offset monotonic and cyclic yield stress, respectively.

discharge machine, 3 mm long single edge notches were introduced into panel specimens for the near-threshold fatigue crack propagation studies. The specimens were 100 mm long, 40 mm wide and 3, 4, or 6 mm thick after heat treatment and polishing with a series of grits down to 1 μm diamond paste.

The fatigue crack growth tests were conducted under load control on a closed-loop servohydraulic MTS machine in dry argon or liquid nitrogen at R-values of 0.05 or 0.5. A sinusoidal waveform was used with 30 Hz frequency at 298° K and 20 Hz frequency at 77° K. The 298° K measurements were done in dry argon; the specimen was completely immersed in liquid nitrogen during the 77° K tests. The stress intensity range ΔK at the crack tip corresponding to crack length is given by (25)

$$\Delta K = \Delta\sigma(\pi a)^{\frac{1}{2}} [20-13(a/w)-7(a/w)^2]^{-\frac{1}{2}} \tag{1}$$

with $a/w \leqq 0.6$.

Threshold stress intensity levels were approached by the load-shedding technique (24). The test was started at a ΔK somewhat above ΔK_{th} to initiate the crack at the notch and then ΔK was progressingly decreased in steps 3% or less of the previous load to approach ΔK_{th}. Specifically, ΔK_{th} was determined as no observable crack advance after 5×10^6 cycles. After ΔK_{th} was determined, the load amplitude was slightly increased and data were taken until stage II was nearly reached. Crack length was monitored with a 40 times magnification traveling microscope. The tests were done at least twice for each set of experimental conditions. The data sets presented are the combined sets of the individual tests.

** Chemical composition in wt. pct.: 3.01Mg, 0.006Si, 0.002Fe, 0.003Mn, 0.003Cu, 0.003Zr, 0.003Ti and balance Al.

Crack closure was determined using the foil strain-gage method (26). After ΔK_{th} was reached, 0.2 mm wide and 0.25 mm long strain gages (Micro-Measurement MA-06-008CL-120 for iron and Micro-Measurement MA-13-008CL-120 for Al-3% Mg) were mounted at selected locations near the expected crack path. The crack closure load was determined from the change in the slope of the load vs. strain curve during unloading recorded on an X-Y recorder at a frequency of 0.1 Hz. The distance between the center of the strain gage and crack tip was within the range 0.5 to 1.2 mm.

After the tests, the specimens were broken open and scanning electron microscopy and profilometry were used to characterize the fatigue fracture surfaces. To measure the fraction of the propagation by shear modes, a series of fractographs across the specimen thickness corresponding to desired values of ΔK at and near the threshold stress intensity range were prepared and the areal fractions of shear facets were determined by point counting. The fracture surfaces were examined with a Federal Surfanalyzer System 2000 profilometry using a 2.54 μm radius stylus. The accuracy was thus $\pm$ 2.5 μm. The fracture surfaces were scanned in the thickness direction three times over the same region corresponding to a constant value of ΔK. The average roughness $\overline{H}$ is given by

$$\overline{H} = \frac{\sum_{i=1}^{n} 2H_i}{n} \tag{2}$$

with H_i being the amplitude of each individual surface irregularity and $\frac{n}{2}$ the number of irregularities along the specimen thickness direction. Figure 1(a) is a schematic diagram of a fracture surface. Small local irregularities and a longer range fluctuation in surface height were noted. The latter is denoted "waviness" and has been subtracted out to give Fig. 1(b) which shows only the local irregularities. The average height of the local irregularities $\overline{H}_\ell$ is defined as the local roughness.

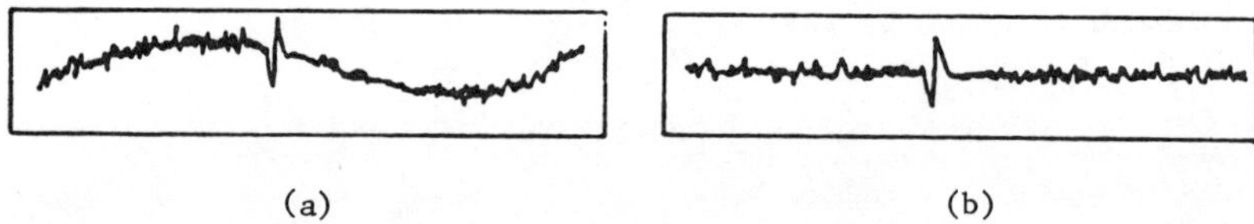

(a) (b)

Figure 1 - Schematic drawing of fracture surface along specimen thickness direction: (a) short and long wavelength fluctuation are noted, the latter is denoted waviness; (b) waviness subtracted out showing only local irregularities.

Experimental Results

Fatigue Crack Propagation

The effects of load ratio on the near threshold fatigue crack propagation rate of iron and Al-3% Mg alloy in dry argon at 298° K are shown in Figs. 2 and 3 as conventional double log plots of da/dN vs. ΔK. The data for the Al-3% Mg alloy at 77° K are given in Fig. 4. In these data, the near-threshold fatigue crack propagation rates are observed to increase with increasing R value at a given ΔK as previously observed by many others (2,

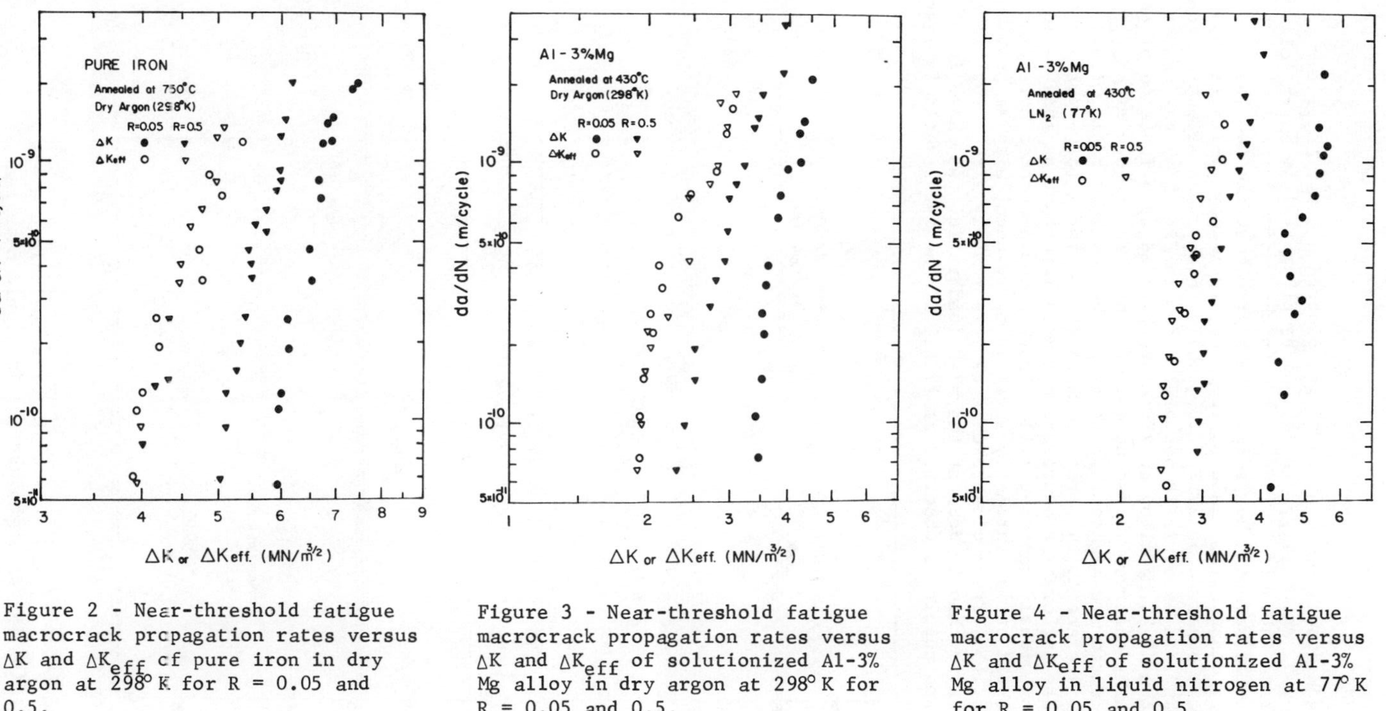

Figure 2 - Near-threshold fatigue macrocrack propagation rates versus ΔK and ΔK_{eff} of pure iron in dry argon at 298° K for R = 0.05 and 0.5.

Figure 3 - Near-threshold fatigue macrocrack propagation rates versus ΔK and ΔK_{eff} of solutionized Al-3% Mg alloy in dry argon at 298° K for R = 0.05 and 0.5.

Figure 4 - Near-threshold fatigue macrocrack propagation rates versus ΔK and ΔK_{eff} of solutionized Al-3% Mg alloy in liquid nitrogen at 77° K for R = 0.05 and 0.5.

3,5,8-11,17-24).

When the crack growth rates at 298° K are plotted against ΔK_{eff} for R = 0.05 and R = 0.5, where ΔK_{eff}, as defined by Elber (13) is $K_{max} - K_{cl}$, K_{cl} being the stress intensity corresponding to the measured crack closure level, the difference between the fatigue crack growth at the two load ratios essentially disappears, again as reported previously (2-3,9,11,18). The same result was observed at 77° K, i.e., while there is an effect of load ratio on ΔK_{th}, ΔK_{eff}^{th} is essentially not affected. In Fig. 5(a) the values of $\Delta K_{eff}/\Delta K$ for the Al-3% Mg alloy at 298° K are plotted as a function of ΔK. Similar results were observed in vacuum melted iron. The values of $\Delta K_{eff}/\Delta K$ at the high load ratio are higher than those at the low load ratio, i.e., as the load ratio is decreased, the crack remains closed for a large portion of loading cycle, that is, crack closure becomes more important as load ratio is decreased as previously reported (11). Crack closure becomes more important as the threshold stress intensity level is approached, also as previously reported (1,2,4-6,11). Crack closure changes more rapidly with change in ΔK at the low load ratio than it does at the high load ratio. As shown in Fig. 5(b), the same result is observed at 77° K. The near-threshold fatigue

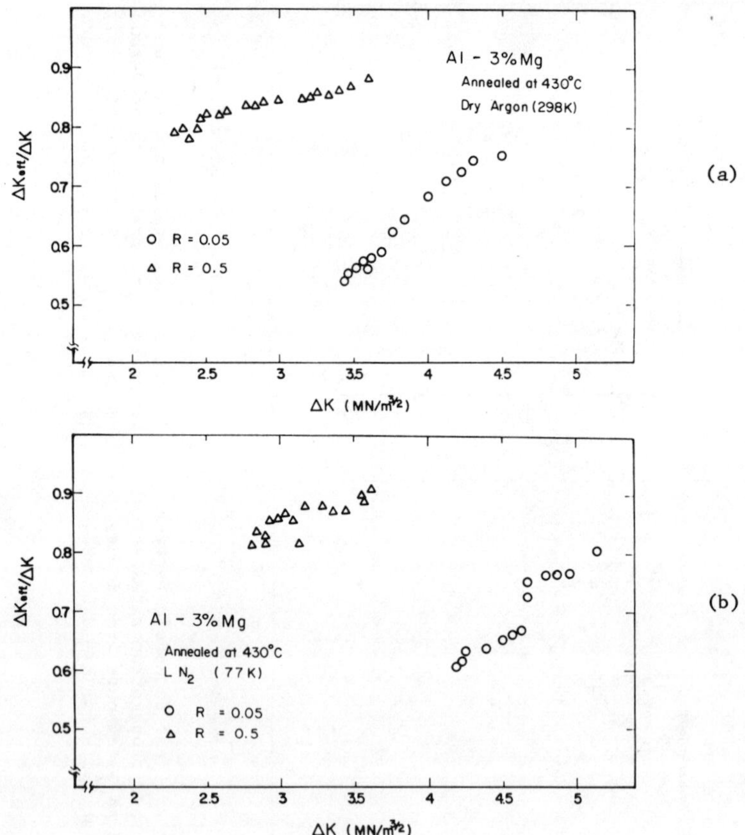

Figure 5 - $\Delta K_{eff}/\Delta K$ versus ΔK for solutionized Al-3% Mg alloy: (a) in dry argon and (b) in liquid nitrogen.

crack growth rate data for the Al-3% Mg alloy at 298° K and 77° K are compared in Figs. 6 and 7. In Fig. 6 the near threshold fatigue crack growth rates

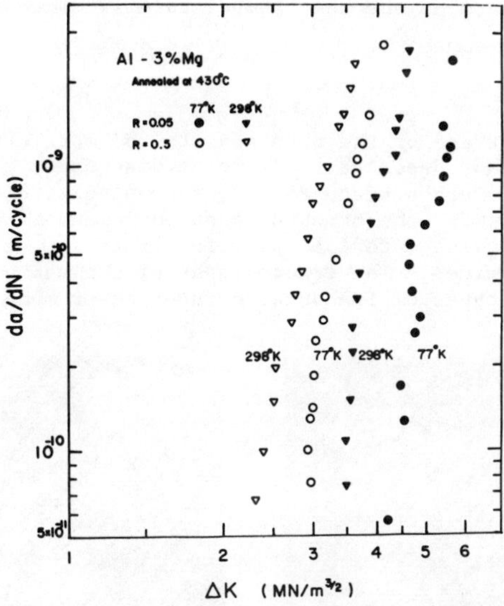

Figure 6 - Comparison of near-threshold fatigue macrocrack propagation rates versus ΔK at 298° K and 77° K in solutionized Al-3% Mg alloy.

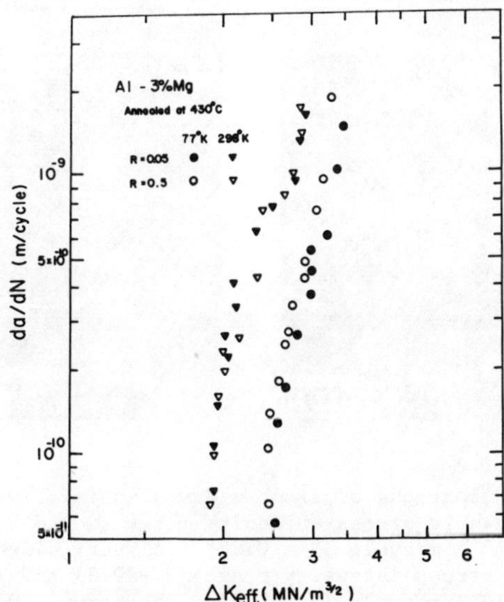

Figure 7 - Comparison of near-threshold fatigue macrocrack propagation rates versus ΔK_{eff} at 298° K and 77° K in the solutionized Al-3% Mg alloy.

are shown to decrease for both load ratios as the test temperature is decreased from 298° K to 77° K. The same trend was found in other alloy systems (16,28-32). The data is replotted in Fig. 7 in terms of ΔK_{eff}. Clearly ΔK^{th}_{eff} increases as the test temperature is decreased.

Fracture Morphology

The fracture surface of the specimens tested were examined by SEM and profilometry as already described. As to be discussed, the results strongly suggest that the fracture surface roughness coupled with shear mode displacements are responsible for the crack closure phenomena in these specimens at low R values but in these materials plastic stretching at the crack tip comes in at high R values. The fractographs of the vacuum melted iron, Fig. 8, exhibit two types of fracture surface: shear mode facets which

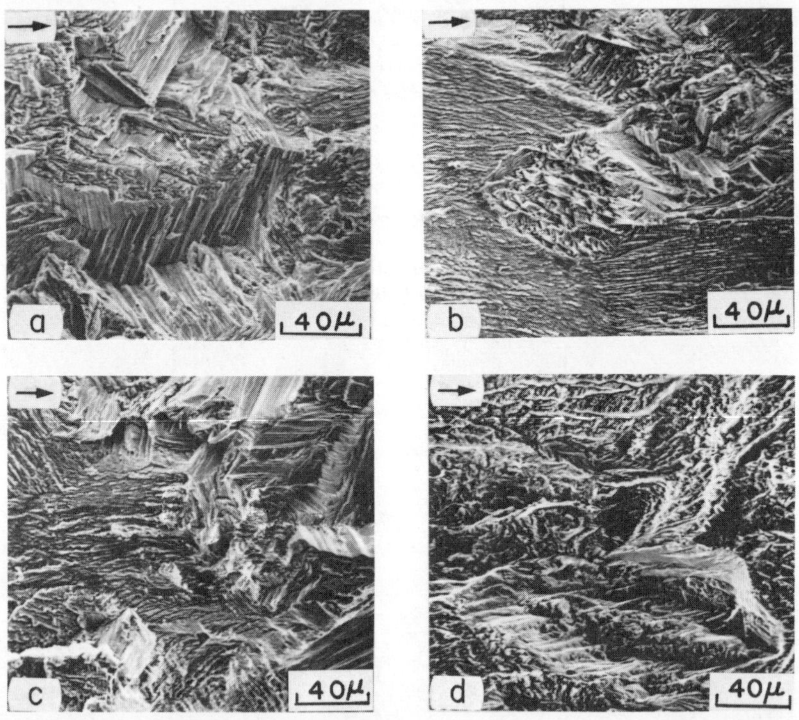

Figure 8 - Fractographs of iron in dry argon: (a) very slightly above the threshold stress intensity range (R = 0.05), (b) at a da/dN of 2×10^{-9} m/cycle (R = 0.05), (c) very slightly above the threshold stress intensity range (R = 0.5) and (d) at a da/dN of 2×10^{-9} m/cycle (R = 0.5). The arrow indicates the direction of crack propagation.

provide strong evidence for the occurrence of shear mode cracking processes and regions characteristic of transgranular tensile mode cracking. Shear mode fractographic features were previously observed by Otsuka (27) and other researchers (1,4,9,20). At R = 0.05 the proportion of shear mode fracture is much decreased as ΔK is increased from threshold to the higher growth rate region (compare Figs. 8(a) and 8(b). Similar results were also observed for the Al-3% Mg alloy at 298° K (Fig. 9). As the load ratio is

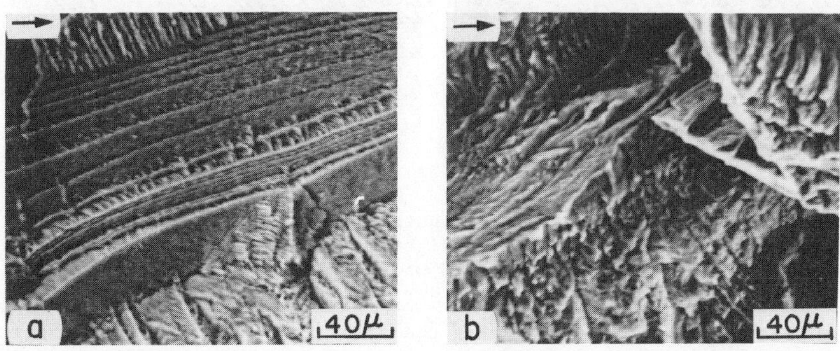

Figure 9 - Fractographs of solutionized Al-3% Mg alloy in dry argon: (a) very slightly above the threshold stress intensity range, (b) at a da/dN of 2×10^{-9} m/cycle at R = 0.05.

increased, the proportion of shear mode facets decreases as shown in Figs. 8(c) and 8(d). The increase in fraction of shear mode facets as ΔK_{th} is approached and as load ratio is lowered is quantified in Figs. 10(a) and 10(b) which are plots of the areal fractions of shear mode facets as a function of load ratio and (ΔK-ΔK_{th}) in iron and Al-3% Mg alloy, respectively. The fatigue fracture surface appearance at 77° K (Fig. 11) is much smoother and there is no big change as ΔK is increased. It was not possible to separate the surface into mode I and mode II fracture surfaces, but it is suspected shear mode fracture is occurring since crack closure is observed

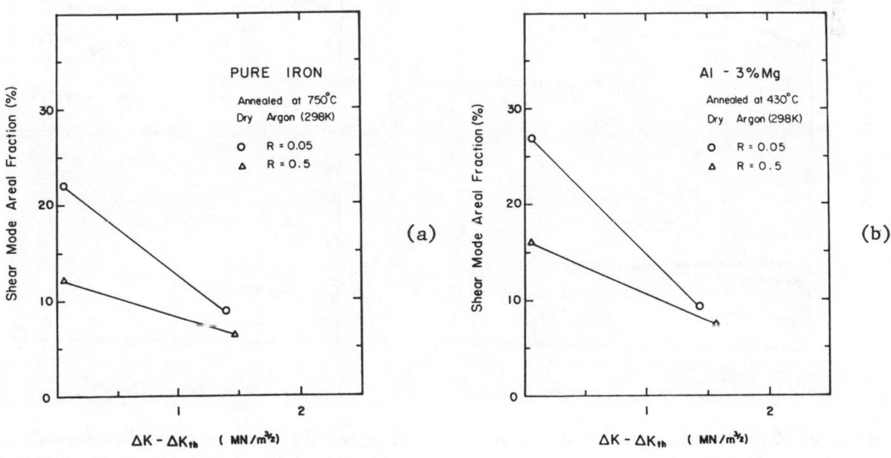

Figure 10 - Areal fraction of shear mode fracture as a function of ΔK-ΔK_{th} in dry argon for both R of 0.05 and 0.5: (a) iron and (b) solutionized Al-3% Mg.

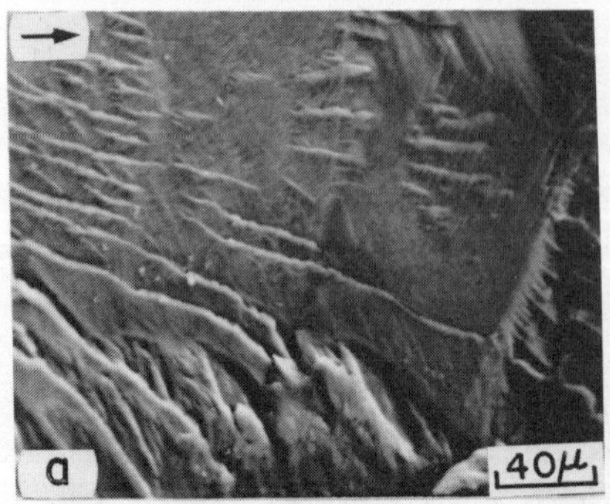

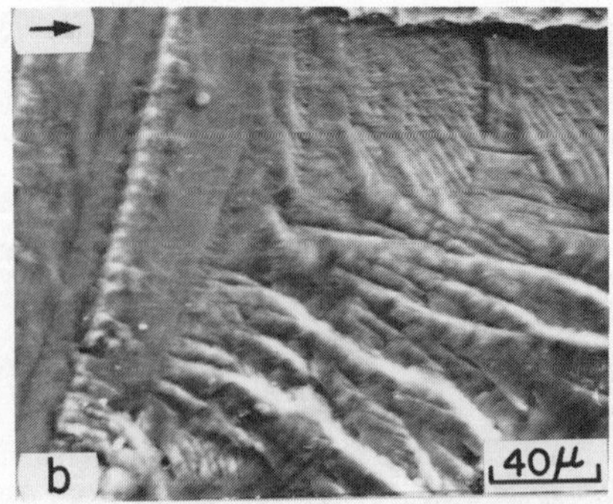

Figure 11 - Fractographs of solutionized Al-3% Mg alloy in liquid nitrogen for R = 0.05: (a) very slightly above the threshold stress intensity range and (b) at a da/dN of 2×10^{-9} m/cycle.

and the fracture surface doesn't have the usual appearance of the transgranular tensile mode. The values of the average fracture surface roughness $\overline{H}$ and the average local roughness near threshold and at the higher growth rate region at the two load ratios are shown in Fig. 12 as a function of $(\Delta K-\Delta K_{th})$ for iron. The average surface roughness increases somewhat with ΔK near-threshold and also with load ratio. Similar results were reported in other studies (3,5,6). The average local surface roughness which has a value within the range 5 to 7 μm is almost independent of load ratio and ΔK. In Fig. 13 the average surface roughness values for the Al alloy at 298° K and

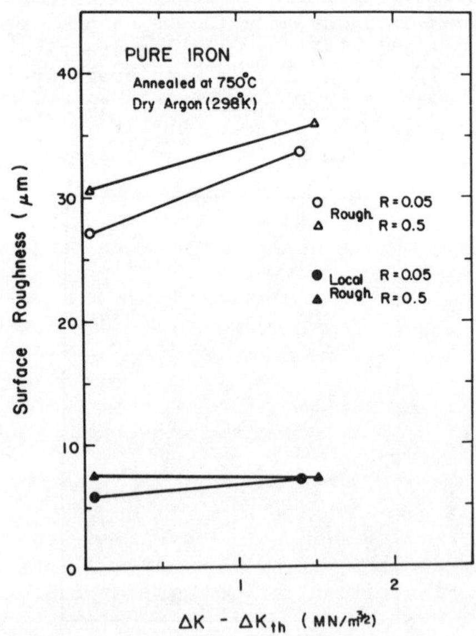

Figure 12 - Roughness $\overline{H}$ and local roughness $\overline{H}_{\ell}$ (waviness subtracted out) versus $\Delta K-\Delta K_{th}$ in iron for both R = 0.05 and 0.5.

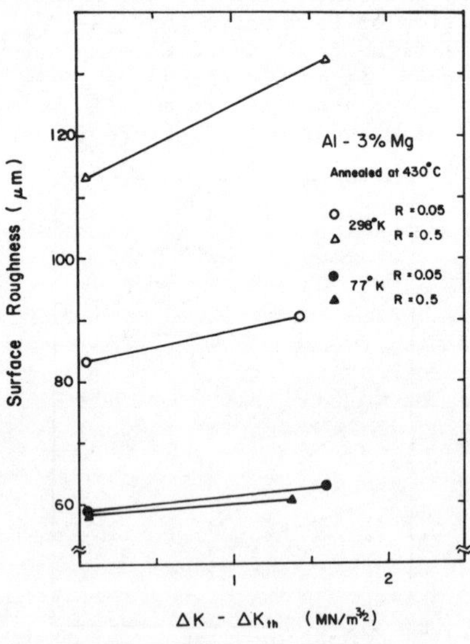

Figure 13 - Roughness $\overline{H}$ versus $\Delta K-\Delta K_{th}$ in solutionized Al-3% Mg alloy at 77° K and 298° K for both R = 0.05 and 0.5.

77° K are presented. The average surface roughness decreases with decrease in temperature in the near-threshold region. The variations of load ratio and ΔK have little influence on the fracture surface roughness at 77° K.

For purposes of comparison, ΔK_{th} and ΔK^{th}_{eff} together with the related information concerning crack closure stress intensities are given in Table II.

Table II. Near-Threshold Fatigue Crack Propagation Properties

	R $\frac{\sigma_{min}}{\sigma_{max}}$	Temp (° K)	ΔK_{th} ($MN/m^{3/2}$)	ΔK^{th}_{eff} ($MN/m^{3/2}$)	Surface Rough. at ΔK_{th} (μm)	Shear Mode Areal Fraction at ΔK_{th} (%)	$\sqrt{\frac{\delta_{cl(1)}}{\delta_{max}}}$	$\sqrt{\frac{\delta_{cl(2)}}{\delta_{max}}}$
Fe	0.05	298	5.9	3.8	28	22	0.41	0
	0.5		5.0	3.9	31	13	0.32	0.53
Al-3% Mg	0.05	298	3.4	1.9	83	27	0.52	0
	0.5		2.3	1.8	114	16	0.45	0.40
	0.05	77	4.2	2.5	58	-	-	-
	0.5		2.8	2.4	58	-	-	-

Discussion

Elber (13) initially observed crack closure in the mid-ΔK stage II region of the fatigue crack growth curve; subsequently, many investigators (1-12) reported crack closure in the near-threshold, stage I region of the fatigue crack propagation curve where the relative closure stress increases as ΔK is reduced at low load ratios. There has been considerable research to try to understand the role and origin of crack closure during near-threshold fatigue crack growth. As summarized by Suresh et. al. (7), it is considered that oxide wedging of the crack tip (2,8,10,16) and crack surface roughness coupled with shear mode displacements (1,3-7,9) are responsible for the low ΔK behavior while plasticity induced crack closure is responsible for the high ΔK behavior (13-15). In their model the effect of increase in load ratio and the attendant increase in mean stress level were not considered. As discussed subsequently at high R values even though ΔK is near ΔK_{th}, plasticity induced crack closure should be considered.

Since the present tests were done in dry argon or liquid nitrogen, oxide wedging is not considered to be important. In this investigation on iron and Al-3% Mg alloy (Figs. 2-4), a load ratio effect on ΔK_{th} was observed but not on ΔK^{th}_{eff} at either 77° K or 298° K. These results extend to non-aggressive environments the conclusion of Schmidt et al. (19) that the effect of load ratio on the near-threshold growth rate is due to crack closure. Most previous studies (2-5,8-11,19-24) of the effect of load ratio on ΔK_{th} were carried out in a corrosive environment such as moist air and thus are more difficult to interpret than the present results. Thus the effect of load ratio on ΔK_{th} disappears if K_{cl} is subtracted from K_{max} to give ΔK_{eff} even if oxide bridging does not play a role. The effect of load ratio on ΔK_{th} and the near-threshold crack growth will next be discussed in light of the fracture surface morphology results.

According to Suresh et al. (7), the stress intensity level at closure may be evaluated from

$$\frac{K_{c1}}{K_{max}} = \sqrt{\frac{\delta_{c1}}{\delta_{max}}} = \sqrt{\frac{2\gamma x}{1+2\gamma x}} \tag{3}$$

where the δ's are the crack opening displacements at crack closure and at maximum load, $\gamma = \overline{H}/W$, $\underline{x} = U_{II}/U_{I}$ with U_{I} and U_{II} being the mode I and mode II displacements, $\overline{H}$ the average roughness and W the average roughness wavelength. The latter is comparable with grain size in this study. In the present study it was assumed that the magnitude of the shear mode displacements are proportional to the shear mode areal fraction of the fracture surface at a given value of ΔK. In Figs. 14(a) and 14(b) the K_{c1}/K_{max} data

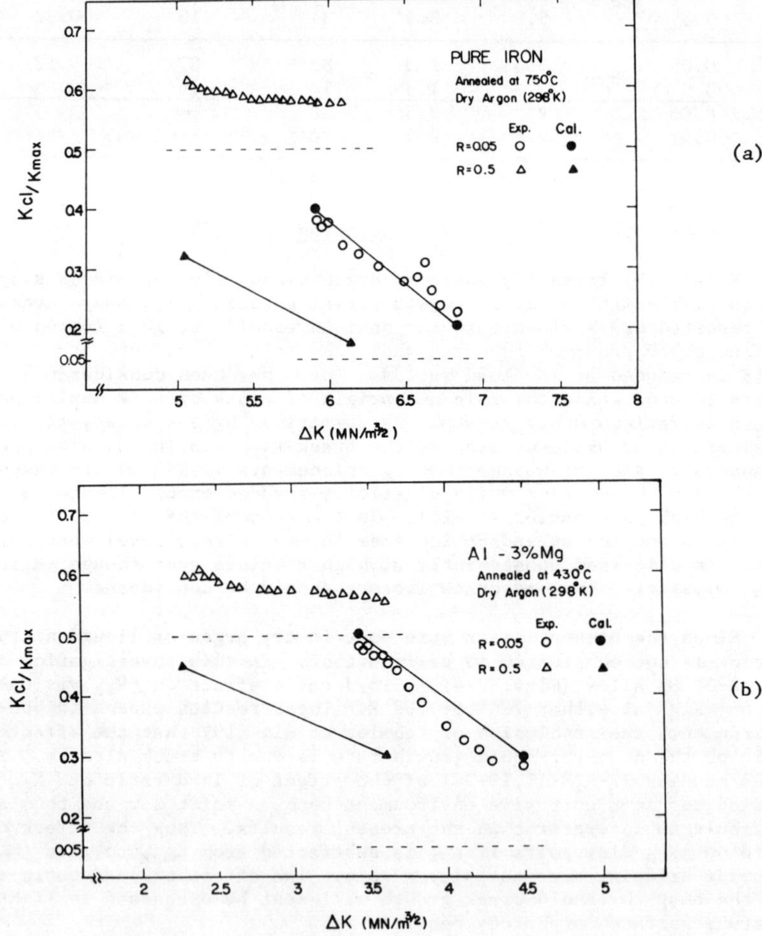

Figure 14 - K_{c1}/K_{max} in dry argon versus ΔK from experimental results and from eq.(3): (a) iron and (b) solutionized Al-3% Mg alloy. Results for both R = 0.05 and 0.5 are plotted.

which were measured using foil strain gages and obtained by eq.(3) in iron and the Al alloy are compared. It is noted that K_{cl}/K_{max} may be larger but cannot be smaller than the R value since the limiting value of K_{cl} is K_{min}. While K_{cl}/K_{max} increases with increase in R, as shown in Fig. 14, $\Delta K_{cl}/\Delta K$ decreases, that is the effect of closure on ΔK decreases as R is increased and ΔK_{eff} approaches ΔK. The experimental results are in good agreement with eq.(3) at the low load ratio (R = 0.05). However, at the high load ratio (R = 0.5) the calculated K_{cl}/K_{max} using eq.(3) is much smaller than R which is not possible. Thus at large values of R the crack closure level is not explained fully by surface roughness coupled with shear mode displacements. Another crack closure mechanism is needed. Equation (3) is modified as

$$\frac{K_{cl}}{K_{max}} = \sqrt{\frac{\delta_{cl(1)} + \delta_{cl(2)}}{\delta_{max}}} \tag{4}$$

with $\delta_{cl(1)}$ and $\delta_{cl(2)}$ being respectively the magnitude of the crack closure due to surface roughness coupled with shear displacements and due to another crack closure mechanism which was assumed to be plasticity induced crack closure. The ratio of effective stress intensity range to stress intensity range then becomes

$$\frac{\Delta K_{eff}}{\Delta K} = \frac{K_{max} - K_{cl}}{K_{max}(1-R)} = \frac{1}{1-R}\left\{\frac{\sqrt{\delta_{max}} - \sqrt{\delta_{cl(1)} + \delta_{cl(2)}}}{\sqrt{\delta_{max}}}\right\} . \tag{5}$$

Combining eqs.(4) and (5) gives

$$\sqrt{\frac{\delta_{cl(2)}}{\delta_{max}}} = \left\{\left[1 - (1-R)\frac{\Delta K_{eff}}{\Delta K}\right]^2 - \frac{\delta_{cl(1)}}{\delta_{max}}\right\}^{\frac{1}{2}} . \tag{6}$$

The values of $(\delta_{cl(2)}/\delta_{max})^{\frac{1}{2}}$ from eq.(6) are plotted in Fig. 15. They are essentially zero near the threshold stress intensity range at low load ratios in both materials, that is, there is no contribution from plasticity induced crack closure. Thus at low load ratios the crack closure phenomenon is fully explained by coupled roughness and shear. At R = 0.5, $(\delta_{cl(2)}/\delta_{max})^{\frac{1}{2}}$ has a much higher value which increases as ΔK is increased. The increase of the mean stress level as R is increased is suggested to cause plastic extension of the crack tip. The low yield stress of the alloys studied (Table I) no doubt increases this effect.

At liquid nitrogen temperature the same mechanisms as at room temperature are available to explain the observed crack closure. The surface roughness resulting from testing at 77°K is not as large as that resulting at 298°K (Fig. 13). The fracture morphology (Fig. 11) shows a rather smooth surface which does not have the appearance of ordinary transgranular tensile mode fracture. Considerable shear mode cracking has likely occurred and increase in x in eq.(3) is thought to have compensated for a decrease in γ. The temperature dependence of near-threshold fatigue crack growth rate is reported for both load ratios in Fig. 6. Even when ΔK^{th}_{eff} is considered, a temperature effect exists, i.e., as the temperature increases, ΔK^{th}_{eff} decreases. This is considered to be an intrinsic effect.

In two phase iron base alloys (29-30,32), it has been suggested that thermal stresses are responsible for the effect of temperature on the near-

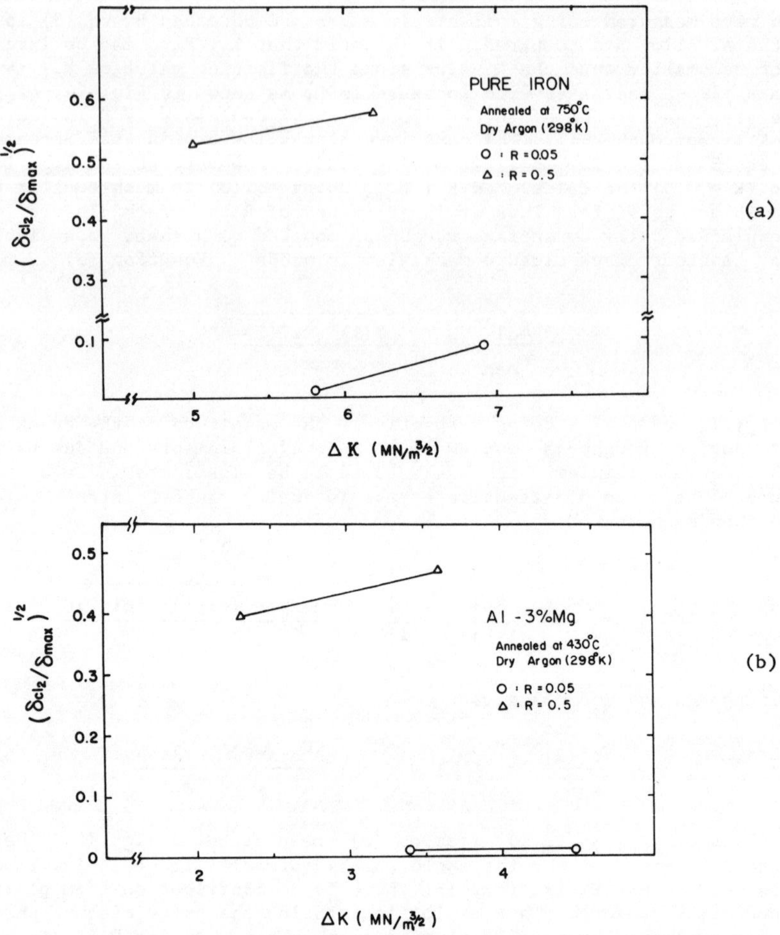

Figure 15 - Calculated $\sqrt{\delta_{c1(2)}/\delta_{max}}$ from eq.(6): (a) iron and (b) solutionized Al-3% Mg alloy. Results from both R = 0.05 and 0.5 are plotted.

threshold fatigue crack propagation rate. In this single phase Al-3% Mg alloy, the effect of temperature on ΔK_{th} and ΔK^{th}_{eff} may not be attributed to thermal stresses. It was previously suggested by Fine (33) that ΔK is determined by the stress necessary to activate a dislocation source near the crack tip. Two kinds of dislocation sources are available to assist crack propagation, namely, Frank-Reed sources and double cross-slip. With the absence of precipitates and inclusions in this high purity alloy, the latter is expected to be the major dislocation source. Since cross-slip is a thermally activated process, as temperature is reduced cross-slip becomes more difficult. It is suggested that the increase in ΔK^{th}_{eff} on cooling is due to the increase in difficulty for cross-slip.

Summary and Conclusions

1) While ΔK_{th} depends on load ratio at both 77 and 298° K in the alloys studied, ΔK^{th}_{eff} does not. The effect of load ratio on ΔK_{th} is caused by crack closure.

2) The effect of crack closure on ΔK, that is $K_{cl}-K_{min} = \Delta K_{cl}$ or $\Delta K_{cl}/\Delta K$, increases as ΔK_{th} is approached and also as the load ratio is decreased.

3) The mechanism for crack closure at low load ratios in an inert environment is surface roughness coupled with shear mode displacements; however, at high load ratios plasticity induced crack closure also plays a role due to the high mean stress.

4) Fractographs of the fracture surface show two types of morphology: (a) shear mode facets which provide strong evidence for the occurrence of shear mode cracking processes and (b) transgranular tensile crack mode regions. The areal fraction of shear mode fracture increases as ΔK_{th} is approached and as the load ratio is reduced.

5) At 298° K the fracture surface roughness increases with ΔK near-threshold and also with load ratio. The fracture surface roughness near-threshold decreases with decrease in temperature. Variation of load ratio or ΔK has little influence on the fracture surface roughness at 77° K.

6) Both ΔK_{th} and ΔK^{th}_{eff} at 77° K are larger than those at 298° K. The temperature effect on ΔK^{th}_{eff} is explained by an intrinsic effect such as the increase in difficulty for cross-slip as the temperature is lowered.

Acknowledgements

The author (D. H. Park) would like to thank the Korean Government for their financial support during his stay at Northwestern University. This research was partially supported under the NSF-MRL program through the Materials Research Center of Northwestern University (Grant No. DMR79-23573). The authors appreciate the assistance of the Inland Steel Research Laboratory and the ALCOA Technical Center for providing the iron and Al-3% Mg alloy. The fracture surface roughness analyses were made possible by the generous help of Drs. D. Ro and H. Sunwoo, Imperial Clevite Inc. Technology Center, Cleveland, Ohio who made their surface analysis facility available for this research.

References

1. G. M. Lin and M. E. Fine, "Effect of Grain Size and Cold Work on the Near Threshold Fatigue Crack Propagation Rate and Crack Closure in Iron", Scripta Metallurgica, 16 (1982) pp. 1249-1254.

2. P. K. Liaw, S. J. Hudak, Jr. and J. K. Donald, "Influence of Gaseous Environments on Rates of Near-Threshold Fatigue Crack Propagation in NiCrMoV Steel", Metallurgical Transactions, 13A (1982) pp. 1633-1645.

3. P. K. Liaw, A. Saxena, V. P. Swaminathan and T. T. Shih, "Effects of Load Ratio and Temperature on the Near-Threshold Fatigue Crack Propagation Behavior in a CrMoV Steel", Metallurgical Transactions, 14A (1983) pp. 1631-1640.

4. K. Minakawa, Y. Matsuo and A. J. McEvily, "The Influence of a Duplex Microstructure in Steels on Fatigue Crack Growth in the Near-Threshold Region", Metallurgical Transactions, 13A (1982) pp. 439-445.

5. G. T. Gray, III, J. C. Williams and A. W. Thompson, "Roughness-Induced Crack Closure: An Explanation for Microstructurally Sensitive Fatigue Crack Growth", Metallurgical Transactions, 14A (1983) pp. 421-433.

6. J. L. Horng, "Fatigue Crack Propagation of HY80, HY130 and 1018 Steels", Ph.D. Dissertation, Northwestern University, Evanston, IL (1983).

7. S. Suresh and R. O. Ritchie, "A Geometric Model for Fatigue Crack Closure Induced by Fracture Surface Roughness", Metallurgical Transactions, 13A (1982) pp. 1627-1631.

8. S. Suresh, G. H. Zamiski and R. O. Ritchie, "Oxide-Induced Crack Closure: An Explanation for Near-Threshold Corrosion Fatigue Crack Growth Behavior", Metallurgical Transactions, 12A (1981) pp. 1435-1443.

9. Y. Nakai, K. Tanaka and T. Nakanishi, "The Effects of Stress Ratio and Grain Size on Near-Threshold Fatigue Crack Propagation in Low-Carbon Steel", Engineering Fracture Mechanics, 15 (1981) pp. 291-302.

10. A. T. Stewart, "The Influence of Environment and Stress Ratio on Fatigue Crack Growth at Near-Threshold Stress Intensities in Low-Alloy Steels", Engineering Fracture Mechanics, 13 (1980) pp. 463-478.

11. J. A. Vazquez, A. Morrone and H. Ernst, "Experimental Results on Fatigue Crack Closure for Two Aluminum Alloys", Engineering Fracture Mechanics, 12 (1979) pp. 231-240.

12. Private communication, K. A. Esaklul, University of Minnesota, Dept. of Chemical Engineering and Materials Science, Minneapolis, MN, Dec. 1982.

13. W. Elber, "The Significance of Fatigue Crack Closure", ASTM STP 486 (1971) pp. 230-242.

14. C. L. Ho, O. Buck and H. L. Marcus, "Application of Strip Model to Crack Tip Resistance and Crack Closure Phenomena", ASTM STP 536 (1973) pp. 5-21.

15. B. Budiansky and J. W. Hutchinson, "Analysis of Closure in Fatigue Crack Growth", J. Appl. Mech., 45 (1978) pp. 267-276.

16. Private communication, P. K. Liaw, W. A. Logsdon and M. H. Attaar, Westinghouse R&D Center, Pittsburgh, PA, 1983.

17. R. J. Cooke, P. E. Irving, G. S. Booth and C. J. Beevers, "The Slow Fatigue Crack Growth and Threshold Behavior of a Medium Carbon Alloy Steel in Air and Vacuum", Engineering Fracture Mechanics, 7 (1975) pp. 69-77.

18. M. Jolles, "Constraint Effects on the Prediction of Fatigue Life of Surface Flows", Journal of Engineering Materials and Technology, 105 (1982) pp. 215-218.

19. R. A. Schmidt and P. C. Paris, "Threshold for Fatigue Crack Propagation and the Effects of Load Ratio and Frequency", ASTM STP 536 (1973) pp. 79-94.

20. D. L. Davidson, "Incorporation Threshold and Environmental Effects into the Damage Accumulation Model for Fatigue Crack Growth", Fatigue of Engineering Materials and Structures, 3 (1981) pp. 229-236.

21. B. R. Kirby and C. J. Beevers, "Slow Fatigue Crack Growth and Threshold Behavior in Air and Vacuum of Commercial Aluminum Alloys", Fatigue of Engineering Materials and Structures, 1 (1979) pp. 203-215.

22. J. Masounave and J. P. Bailon, "The Dependence of the Threshold Stress Intensity Factor on the Cyclic Stress Ratio in Fatigue of Ferritic-Pearlitic Steels", Scripta Metallurgica, 9 (1975) pp. 723-730.

23. R. J. Cooke and C. J. Beevers, "The Effect of Load Ratio on the Threshold Stresses for Fatigue Crack Growth in Medium Carbon Steels", Engineering Fracture Mechanics, 5 (1973) pp. 1061-1071.

24. P. E. Irving and A. Kurzfeld, "Measurements of Intergranular Failure Produced during Fatigue Crack Growth in Quenched and Tempered Steels", Metal Science (Nov. 1978) pp. 495-502.

25. L. P. Pook, "The Effect of Friction on Pin Jointed Single Edge Notch Fracture Toughness Test Specimens", International Journal of Fracture Mechanics, 4 (1968) pp. 295-297.

26. D. Gan and J. Weertman, "Crack Closure and Crack Propagation Rates in 7050 Aluminum", Engineering Fracture Mechanics, 18 (1981) pp. 87-106.

27. A. Otsuka, K. Mori and T. Miyata, "The Condition of Fatigue Crack Growth in Mixed Mode Condition", Engineering Fracture Mechanics, 7 (1975) pp. 429-439.

28. L. H. Burck and J. Weertman, "Fatigue Crack Propagation in Iron and Fe-Mo Solid Solution Alloys (77 to 296 K)", Metallurgical Transactions, 7A (1976) pp. 257-264.

29. W. W. Gerberich and N. R. Moody, "A Review of Fatigue Fracture Topology Effects on Threshold and Growth Mechanisms", ASTM STP 675 (1979) pp. 292-341.

30. E. Tschegg and S. Stanzl, "Fatigue Crack Propagation and Threshold in B.C.C. and F.C.C. Metals at 77 and 293 K", Acta Metallurgica, 29 (1981) pp. 33-40.

31. J. McKittrick, P. K. Liaw, S. I. Kwun and M. E. Fine, "Threshold for Fatigue Macrocrack Propagation in Some Aluminum Alloys at 77 and 300 K", Metallurgical Transactions, 12A (1981) pp. 1535-1539.

32. W. Yu and W. W. Gerberich, "On the Controlling Parameters for Fatigue Crack Threshold at Low Homologous Temperatures", Scripta Metallurgica, 17 (1983) pp. 105-110.

33. M. E. Fine, "Fatigue Resistance of Metals", Metallurgical Transactions, 11A (1980) pp. 365-379.

THE EFFECTS OF GRAIN SIZE AND STRESS RATIO

ON FATIGUE CRACK GROWTH IN 7091 ALUMINUM ALLOY

P. E. Bretz, J. I. Petit, and A. K. Vasudevan

Aluminum Company of America
Alcoa Laboratories
Alloy Technology Division
Alcoa Center, PA 15069
USA

Increasing grain size by thermomechanical processing can significantly increase near-threshold fatigue crack growth (FCG) resistance. At least three mechanisms have been proposed to explain this increase: a more tortuous crack path, crack closure, and crack deviation. In this study, wrought P/M aluminum alloy 7091 in the as-extruded (fine grained) and two cold rolled + recrystallized (coarse grained) conditions was tested to determine which mechanisms control FCG response. Fatigue tests were conducted at stress ratios, R, of 0.1 and 0.8 to maximize and minimize, respectively, the effect of crack closure. At the higher stress ratio, FCG rates increased for each grain structure in response to decreased closure levels. A factor of two increase in the threshold for FCG (ΔK_{th}) with grain size was measured at both R ratios, clearly indicating that crack closure alone is not responsible for grain size effects on fatigue behavior at a particular R ratio. Crack tip deviation is suggested as an important mechanism by which near-threshold FCG rates are reduced as grain size increases.

Introduction

A significant body of literature is available in which the effect of grain size on fatigue crack growth (FCG) resistance is examined. Yoder and co-workers have studied this effect in Ti alloys (1), steels (2), and Al alloys (3). Other published information includes work on steels (4, 5), Ti (6), a Ni-based superalloy (7), and Al alloys (6, 8, 9). In all cases, an increase in grain size results in decreased FCG rates, primarily in the near-threshold (ΔK_{th}) region. An extensive list of additional references is included in (5). Various explanations for the grain size effect at ΔK_{th} are available in these references. These explanations are outlined below and addressed more completely in the Discussion. Recently, several authors (10-12) have examined and attempted to model the relationship between grain size and ΔK_{th}. The changes in fracture surface roughness and fatigue crack closure levels with grain size are believed to influence strongly this relationship (11, 12).

While it has been shown that increasing the grain size of wrought aluminum alloys improves FCG resistance in the near-ΔK_{th} region, the underlying mechanism responsible for this improvement is not known. The purpose of this investigation is to examine the details of FCG in P/M alloys of various grain structures and determine which of the following mechanisms actually control FCG resistance.

A. A More Tortuous Crack Path. The fracture surface of a coarse-grained alloy is much rougher than that of a fine-grained material. Therefore, the actual path which the crack travels is longer in the large-grained alloy, and the macroscopically-measured FCG rates are lower.

B. Crack Closure. Because fracture surface roughness increases with grain size, there is a greater tendency for contact between mating surfaces during fatigue loading in a large-grained alloy. This contact (closure) reduces the effective crack tip driving force (ΔK_{eff}) and lowers FCG rates.

C. Crack Deviation. Increased fracture surface roughness indicates that the crack is deflected away from a straight line path to a greater degree in a large-grained alloy. This crack deflection (sometimes referred to as branching) also can reduce ΔK_{eff} and lower FCG rates.

Materials and Procedures

The starting material for this program was a piece of 7091-F extruded bar, 12.7 mm x 158.7 mm (0.5" x 6.25") in cross-section. One segment of the bar was given a solution heat treatment of a 1-1/2 hour soak at 488°C (910°F), a cold water quench, and a 1-1/2 to 2% stretch. This material retained the unrecrystallized, as-extruded structure and was designated "A-E". The remainder of the bar was processed to produce samples with different recrystallized grain structures, starting with the following annealing treatment.

Soak 2-1/2 hours at 413°C (775°F)
Furnace cool to 204°C (400°F)
Reheat to 232°C (450°F)
Soak 4 hours at 232°C (450°F)
Air cool to room temperature

One segment of the annealed bar was cold rolled to 8.9 mm (0.35") thickness for a total of 30% cold reduction and designated 30% C.R. The other segment was cold rolled to 6.4 mm (0.25") thickness for a total of 50%

cold reduction (50% C.R.). The cold rolled pieces were solution heat treated, quenched, and stretched along with the A-E bar.

Artificial aging treatments were chosen in an effort to bring both the recrystallized and unrecrystallized materials to the same overaged strength level. All materials were aged for 24 hours at 121°C (250°F). The A-E material was second step aged for 8 hours at 163°C (325°F), while the two CR alloys were aged for 2 hours at 163°C. These overaged tempers are both designated -T7X.

The microstructures of these materials were characterized by optical metallography. Tensile tests were run on 3.2 mm (0.125") diameter specimens machined in both longitudinal (L) and long-transverse (LT) orientations. Fracture toughness was measured in the corresponding L-T and T-L directions, using the slow bend Charpy indicator test.

FCG tests using constant amplitude loading were conducted on all three materials at a frequency of 25 Hz in high humidity (> 90% relative humidity) room temperature air at stress ratios, R*, of 0.1 and 0.8. All tests were run in the L-T orientation (crack growth perpendicular to the extrusion/rolling direction) using 76 x 79 x 6 mm (3 x 3.12 x .25 in.) compact specimens.

These ratios were selected to maximize and minimize, respectively, the influence of closure on FCG behavior. A double cantilever clip gauge was placed on the specimen to measure crack opening displacement (COD). The load (P) versus COD signal was recorded periodically to measure the applied load above which the crack is fully open (P_{op}).

Two sets of FCG tests were run, each including all three materials tested at both R ratios. One group of tests was performed by Del Research (Hellertown, PA) and covered growth rates from 2×10^{-11} to 1×10^{-8} m/cycle (1×10^{-9} to 5×10^{-7} in./cycle). These near-threshold tests were entirely computer controlled, using a constant rate of decrease in ΔK. The second test series was conducted at the Alcoa Technical Center under manual control by load shedding down to 2×10^{-11} m/cycle and then holding the load range constant to allow crack growth up to the highest rates measurable, approximately 5×10^{-6} m/cycle (2×10^{-4} in./cycle).

All fractography was performed on an SEM, using a LeMont analyzer to measure surface roughness at specified points on the fracture surface, using 30 lines/frame at 35 times magnification. The scan profile line was perpendicular to the crack growth direction for these measurements. This analysis system uses both the secondary electron detector and a 4-element backscattered electron detector to measure surface topography. Its operation is described in references 13 and 14.

Results

Microstructure

Three-dimensional optical micrographs of the materials are presented in Figures 1-3. The A-E material (Figure 1) has a very fine, uniform grain/subgrain structure typical of an extruded P/M alloy; the longitudinal intercept length for this structure is 2-8 μm (Table I). The structures of the

---

$*R = \dfrac{\text{minimum stress}}{\text{maximum stress}}$

other two materials clearly are much coarser, with the 30% C.R. alloy having the largest grain size of the three.

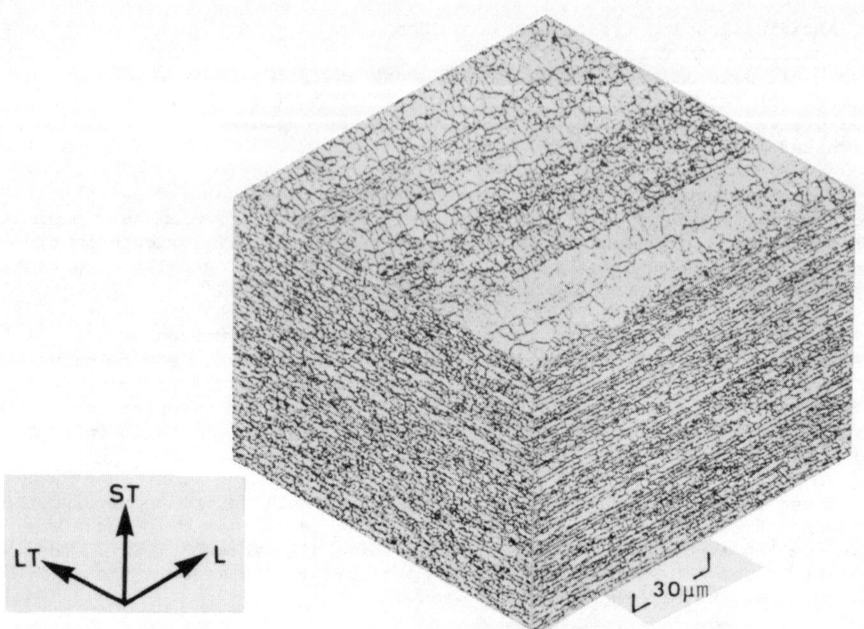

Figure 1 - 3-D optical micrograph of A-E 7091-T7X.

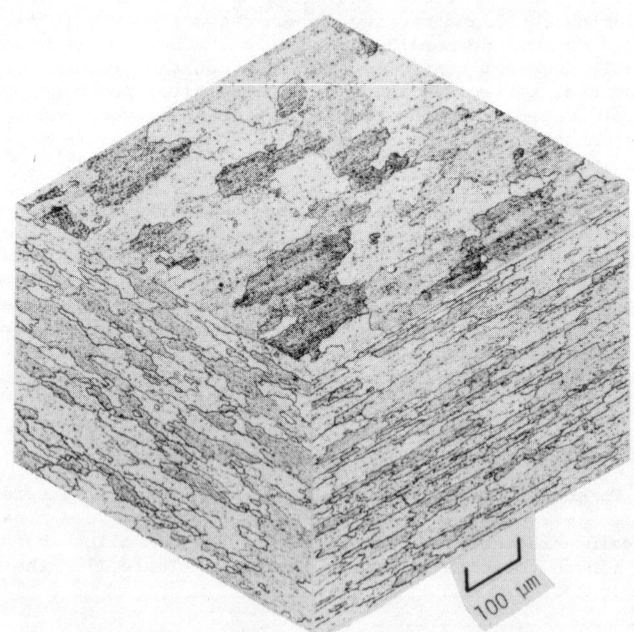

Figure 2 - 3-D optical micrograph of 50% C.R. 7091-T7X.

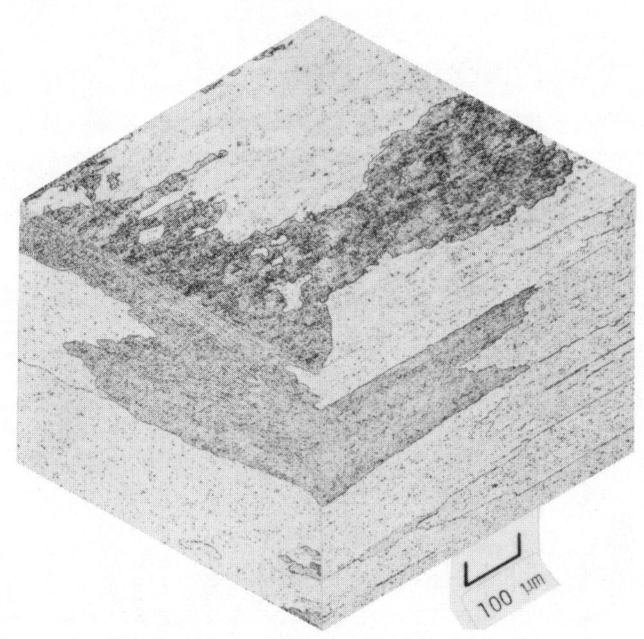

Figure 3 - 3-D optical micrograph of 30% C.R. 7091-T7X.

Table I. Grain Boundary Intercept Length Measurements

	In Microns		
	L	LT	ST
As-Extruded (A-E)	2-8	2-8	--
50% C.R.	80	70	16
30% C.R.	450	450	60

Mechanical Properties

The results of tensile and toughness indicator tests are listed in Table II. The strengths of the two C.R. alloys are nearly equal and are 28-35 MPa (4-5 ksi) lower than for the as-extruded material. This small difference in strength should not influence FCG behavior under constant amplitude loading significantly. The toughness of each C.R. material is about 8-9 MPa$\sqrt{m}$ (7-8 ksi$\sqrt{in.}$) lower than that of the as-extruded alloy.

Table II. Mechanical Properties of Recrystallized 7091-T7X

	Orientation	Yield Strength, MPa (ksi)	Tensile Strength, MPa (ksi)	El., %		Slow Bend Charpy Fracture Toughness, MPa$\sqrt{in.}$ (ksi$\sqrt{in.}$)
As-Extruded (A-E)	L	531 (77.0)	566 (82.1)	12.2	L-T	* *
	LT	519 (75.3)	552 (80.1)	14.2	T-L	43.6 (39.7)
50% C.R.	L	509 (73.8)	539 (78.2)	12.2	L-T	34.9 (31.8)
	LT	485 (70.3)	529 (76.7)	14.2	T-L	35.5 (32.3)
30% C.R.	L	502 (72.7)	534 (77.4)	12.2	L-T	35.8 (32.6)
	LT	468 (67.8)	512 (74.3)	12.2	T-L	34.1 (31.0)

All values are averages of duplicate tests.
*Insufficient material to run these tests.

Fatigue Crack Growth Behavior

Figures 4 and 5 compare the FCG behavior of the three materials at R = 0.1 and 0.8, respectively. The substantial increase in low-ΔK FCG resistance with increasing grain size is apparent at both R ratios. ΔK_{th} for the 30% C.R. material is a factor of about 2 higher than that for the A-E structure in both cases (Table III), and ΔK_{th} for 50% C.R. also is relatively high.

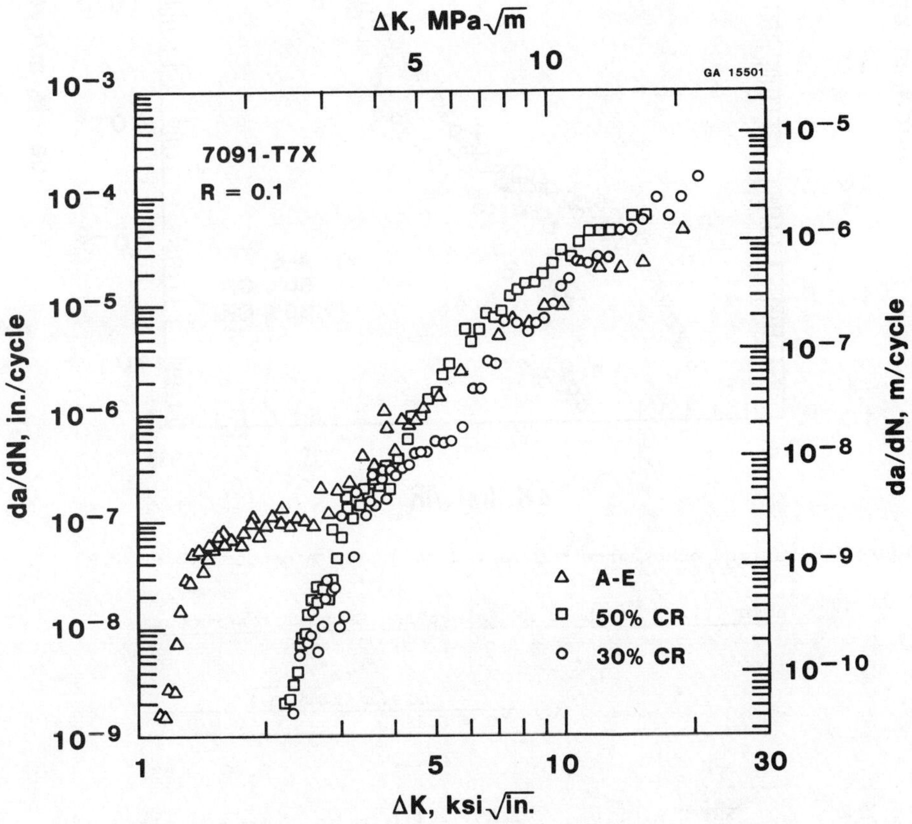

Figure 4 - Fatigue crack growth data for the three microstructures at R = 0.1.

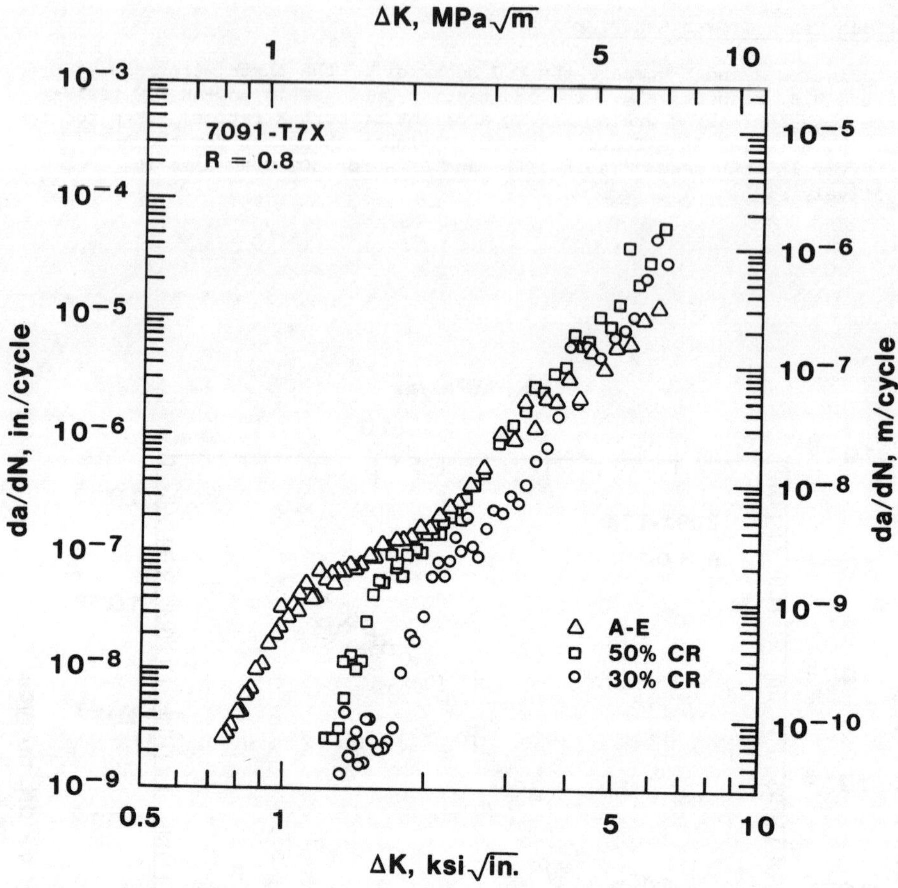

Figure 5 - Fatigue crack growth data for the three microstructures at R = 0.8.

Table III. Threshold ΔK Values for Each Microstructure

	ΔK_{th}*, MPa√m (ksi√in.)	
	R = 0.1	R = 0.8
A-E	1.24 (1.13)	0.76 (0.69)
50% C.R.	2.38 (2.17)	1.22 (1.11)
30% C.R.	2.52 (2.29)	1.45 (1.32)

*ΔK_{th} defined as the ΔK value for a growth rate of 2.5×10^{-11} m/cycle (1×10^{-9} in./cycle).

To assess the influence of closure on these results, closure loads were monitored for all tests conducted at both Del Research and ATC. Figures 6a and 6b show these closure data in the form of P_{op}/P_{max} versus ΔK for the largest and smallest grain sizes tested at R = 0.1, combining data from both test laboratories taken during the K-decreasing and K-increasing portions of the FCG tests. It is clear for the 30% C.R. material (Figure 6a) that closure loads can rise to high levels as the stress intensity approaches ΔK_{th}. As ΔK increases, on the other hand, the closure load decreases toward 10% of P_{max}, which is the minimum applied load for these tests at R = 0.1. (For this reason, no closure data are plotted for values less than 10%.) Comparison of the data shows that closure loads at all values of ΔK rise with increasing grain size, suggesting that the grain size effect on low-ΔK FCG behavior can be due to closure. In contrast, no crack closure loads greater than the minimum applied load were measured for any material at R = 0.8. The significance of this lack of closure is considered further in the Discussion.

An additional observation of note is that FCG rates in the intermediate growth rate region (approximately 2.5×10^{-9} to 1×10^{-6} m/cycle) do not vary substantially among the three materials at either R ratio. Thus, larger grain size does not increase FCG rates in the region frequently of prime interest to airframe manufacturers for use in current alloy selection/ life prediction methodologies. However, crack growth rates at very high ΔK levels should be faster for the large-grained materials, reflecting the decrease in toughness (Table II).

Fractography

Figure 7 shows the fracture surface appearance at ΔK_{th} of all three materials for FCG tests at R = 0.1. The relatively flat fracture topography of the A-E material is quite different from the rough, coarse surface of the 30% C.R. specimen. For the intermediate grain size material (50% C.R.), the fracture topography also is rough but on a somewhat finer scale than the 30% C.R. specimen. Higher magnification fractographs (Figure 8) show a transgranular crystallographic fracture mechanism for the recrystallized alloys, while the actual fracture path for the A-E material is not clear. In fact, the FCG mechanism at lower ΔK levels in fine-grained Al P/M alloys has never been identified adequately. Nevertheless, the difference in surface roughness among these three materials is apparent. The differences among these fracture surfaces are very similar for specimens tested at R = 0.8.

The fracture surfaces of A-E and 30% C.R. material at the same intermediate growth rate of 2.5×10^{-7} m/cycle (which corresponds to a ΔK of about 10.4 $MPa\sqrt{m}$ for both specimens) are shown in Figure 9. Notice that the fracture surface of the A-E material still is quite fine. For the 30% C.R. material, the fracture mechanism remains predominantly crystallographic and the surface still is much rougher than that of the A-E alloy.

The LeMont surface analyzer was used to quantify the differences in fracture surface roughness which clearly are evident in Figures 7-9. The roughness determined by this instrument is the arithmetic average of the distance of all points along a fracture surface from a mean line drawn through the profile (Figure 10a). The data are plotted in Figure 10b as average roughness versus K_{max}, a format selected to compare data from different R ratio tests at the same maximum crack tip plastic zone size, r_y, which is proportional to K^2. If the size of crystallographic facets on the fracture surface is a function of the extent of crack tip plasticity, then surface roughness for R = 0.1 and 0.8 specimens should be equivalent at the same K_{max}. The data in Figure 10b do suggest that surface roughness is

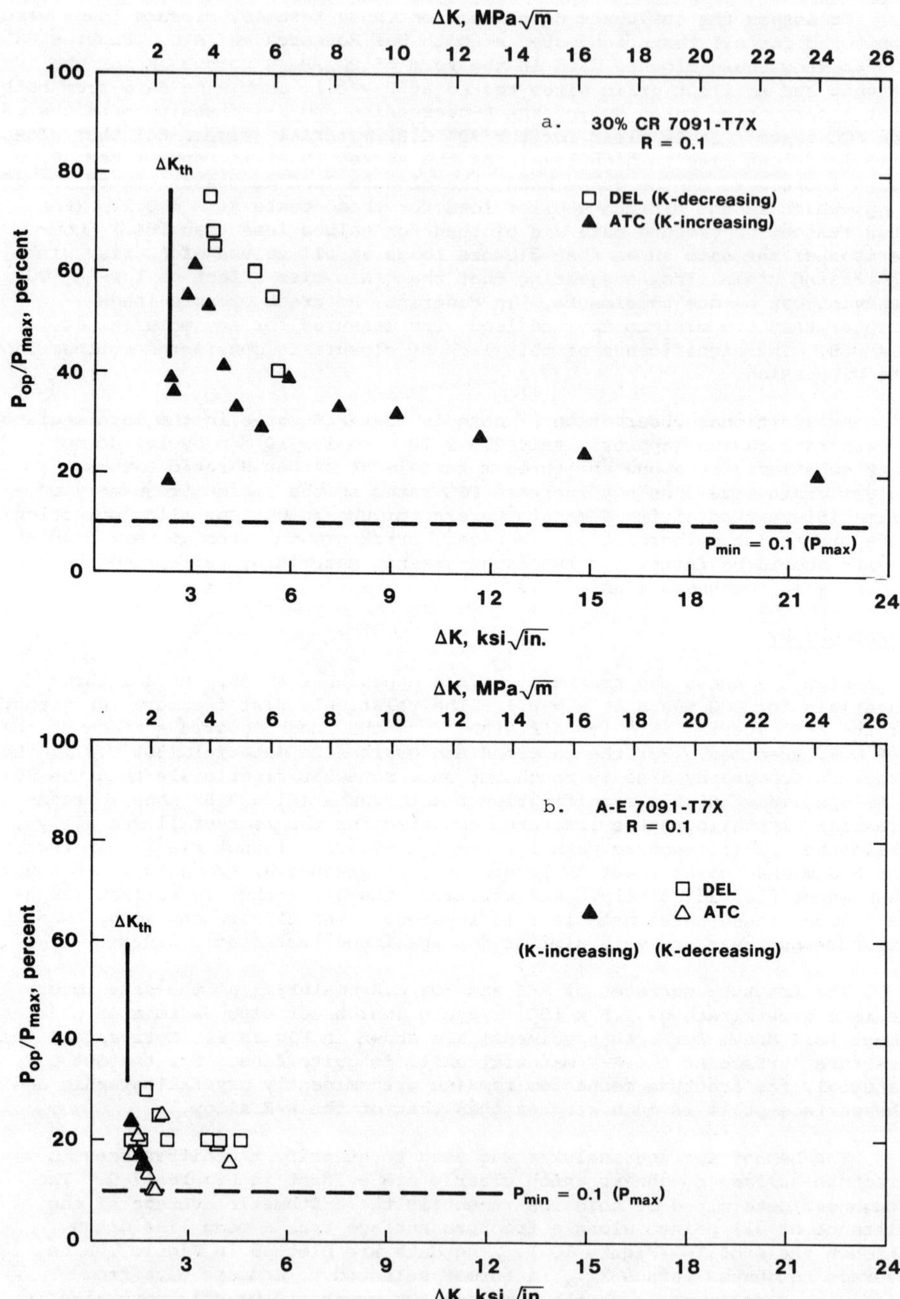

Figure 6 - Crack closure measurements (P_{op}/P_{max} versus ΔK) for both K-decreasing and K-increasing tests conducted at Alcoa Technical Center (ATC) and Del Research (DEL).

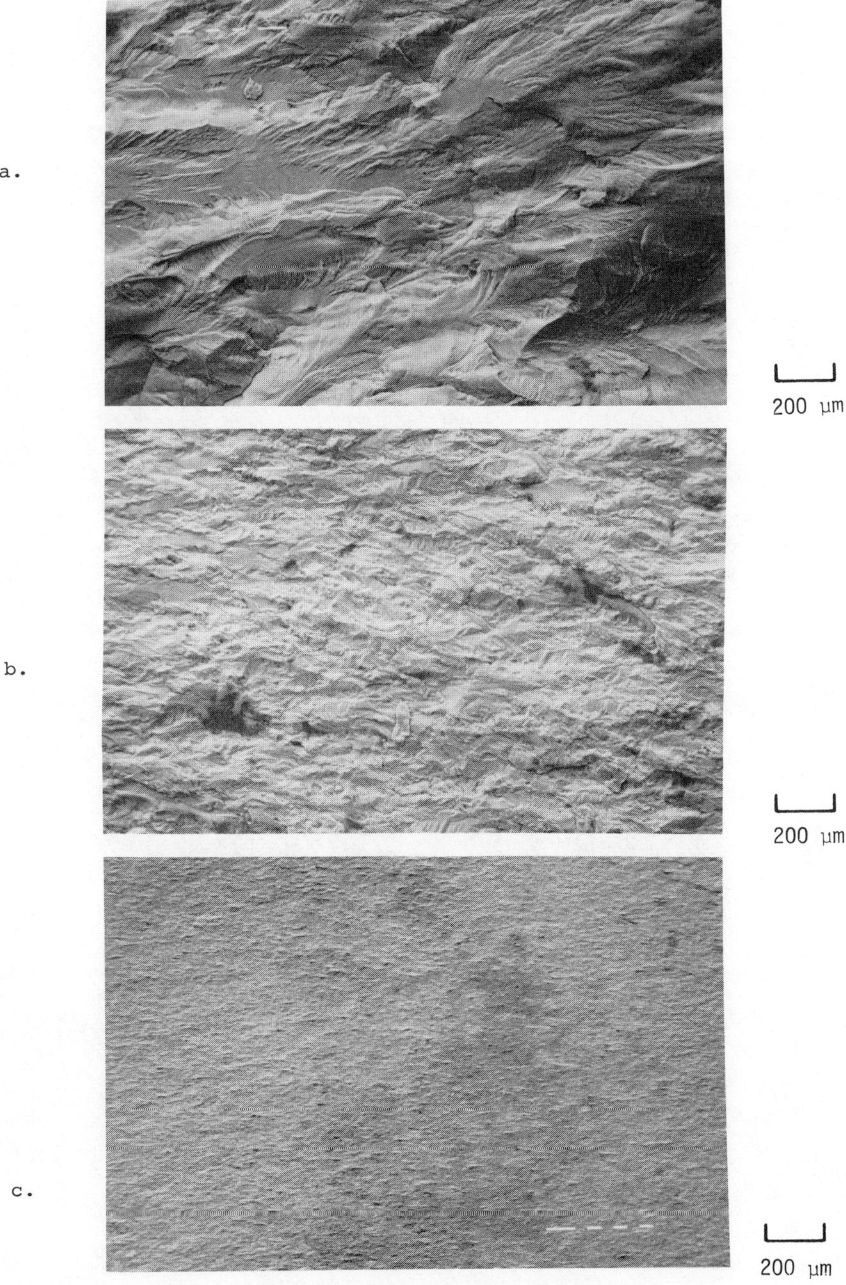

Figure 7 - Fracture surface at threshold, R = 0.1.

a. 30% C.R., ΔK_{th} = 2.52 MPa$\sqrt{m}$.
b. 50% C.R., ΔK_{th} = 2.38 MPa$\sqrt{m}$.
c. A-E, ΔK_{th} = 1.24 MPa$\sqrt{m}$.

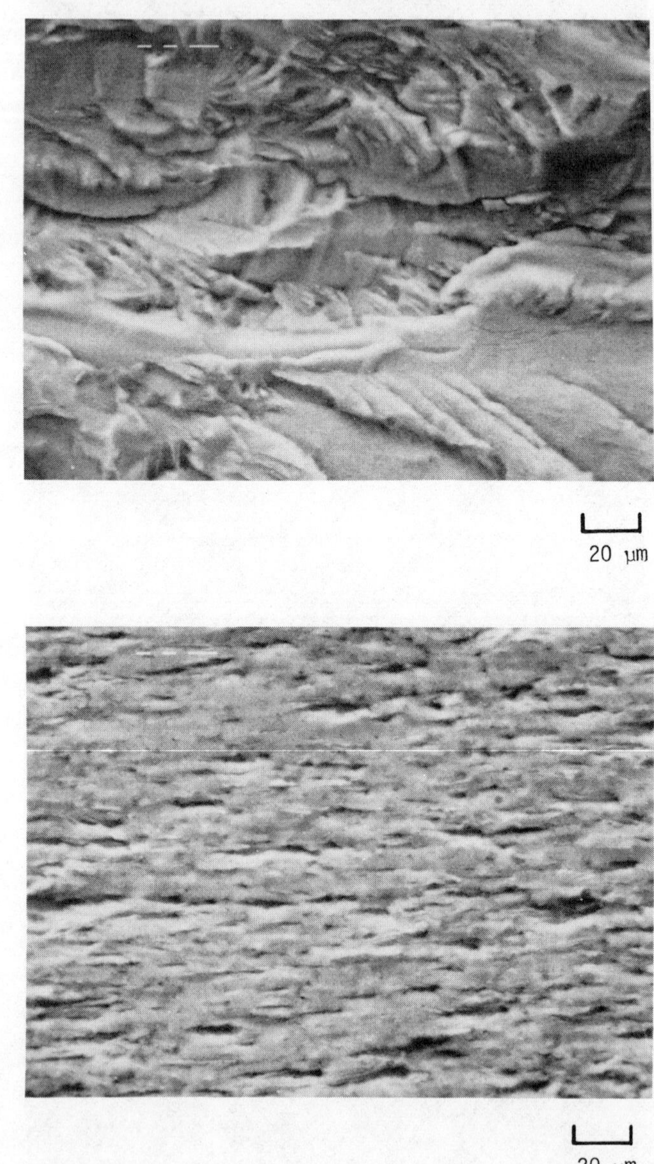

Figure 8 - Higher magnification of near-threshold fracture surfaces in Figure 7 at R = 0.1.

a. 50% C.R., crystallographic fracture (refer to Figure 7b).
b. A-E, smooth fracture appearance (refer to Figure 7c).

a.

200 μm

b.

Figure 9 - Fracture surfaces at $\Delta K = 16.4\ MPa\sqrt{m}$, R = 0.1.

a. 30% C.R.
b. A-E.

independent of R ratio for each material. Interestingly, the degree of this macroscopic roughness (at x35) does not appear to be a function of K_{max} (except perhaps for the 50% C.R. material). Consistent with visual observations, the 30% C.R. samples show the roughest fracture surface and the A-E material the smoothest surfaces over the entire K_{max} range.

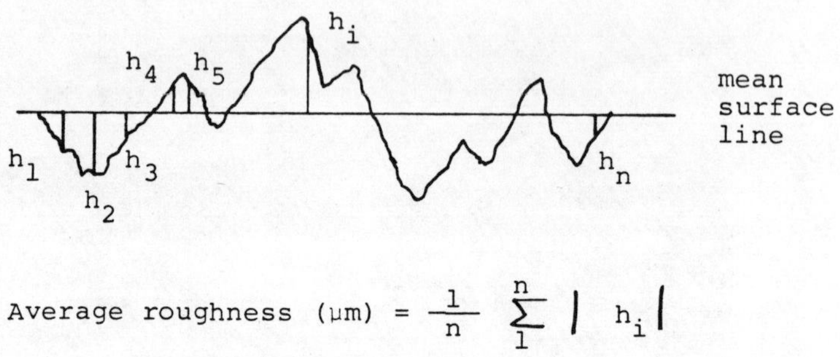

$$\text{Average roughness } (\mu m) = \frac{1}{n} \sum_{1}^{n} | h_i |$$

Figure 10a - Schematic diagram illustrating calculation of surface roughness.

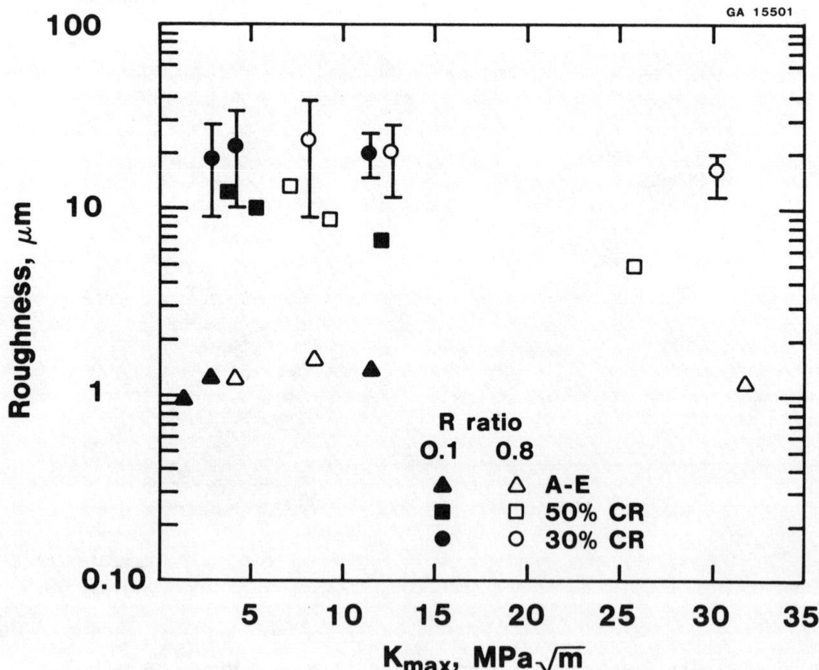

Figure 10b - Surface roughness measurements versus K_{max} for all three microstructures. Error bars = ±2 standard deviations.

Discussion

The two-fold increase in ΔK_{th} with grain size at R = 0.1 agrees with previous studies of grain structure/FCG behavior relationships. In contrast, the similar increase in ΔK_{th} at R = 0.8 is unexpected for two reasons. The popular arguments based on crack closure would not predict a grain size effect at R = 0.8 in the absence of closure, a topic to be dealt with shortly. Furthermore, the very recent work by Gray, et al., (5) in a 0.8% C steel showed no grain size effect on low-ΔK fatigue behavior at R = 0.7 despite a pronounced effect at R = 0.1. Because the results of the present study are somewhat surprising, it is instructive to examine the data in relation to the three mechanisms listed in Section II.

A More Tortuous Crack Path. This mechanism originally was proposed by Masounave and Bailon (15), who reasoned that the rougher fracture surface in a coarse-grained alloy reflects a longer actual crack path. Since, however, the FCG rate measurement implicitly assumes a linear crack path, the macroscopic growth rate appears to decrease with increasing grain size (i.e., more deviation from the linear path).

One can argue that the angle of deviation ought to approach ±45° about the nominal crack plane, since this is the direction of maximum shear stresses. In this case, a simple calculation shows that the new crack path is longer by a factor of 1.4 (= 1/cos 45°); therefore, macroscopic crack growth rates should be 71% (= 1/1.4) of the straight line da/dN. Although fractography does show that the A-E fatigue surfaces are much smoother, near-ΔK_{th} FCG rates for 50% C.R. and 30% C.R. materials are as much as x100 lower. Clearly, the crack tortuosity argument alone cannot account for such differences.

A somewhat related hypothesis has been offered by Yoder and co-workers (1-3), who suggest that near-ΔK_{th} behavior occurs when the crack tip plastic zone size (PZS) decreases to an appropriate microstructural dimension, in this case presumably the grain size ($\bar{d}$). The point at which PZS = $\bar{d}$ is assumed to correspond to a knee in the FCG rate data, which ought to be the one at about 2 x 10^{-9} m/cycle in Figures 4 and 5. Yoder has proposed an equation to predict the ΔK value at which this point occurs (ΔK_T), which is of the form:

$$\Delta K_T = 5.5\ \sigma_{YS}\sqrt{\bar{d}}\ . \tag{1}$$

Using the data in Tables I and II, this equation predicts that increasing the grain size should change the value of ΔK_{th} from 6.5 MPa$\sqrt{m}$ for the A-E material to 58 MPa$\sqrt{m}$ for 30% C.R.; this does not compare favorably with experimental data.

Crack Closure. The contact between mating fatigue fracture surfaces which occurs even at positive loads is assumed to reduce the ΔK level at the crack tip (ΔK_{eff}) and decrease crack growth rates. This contact is believed to occur because of fracture surface mismatch, which results from a Mode II (shearing) displacement at the crack tip and is accentuated with increasing grain size (11, 12). We used measurements of the load at which contact stops (P_{op}) during the fatigue tests to estimate ΔK_{eff}. The following equation can be derived to relate ΔK_{eff} to measured P_{op} values and the applied loading conditions ($\Delta K_{applied}$, P_{max}, and R).

$$\Delta K_{eff} = \Delta K_{applied} \frac{(1 - P_{op}/P_{max})}{1 - R} \text{ for } P_{op} > P_{min}. \tag{2}$$

The closure data in Figure 6 were used to shift the FCG results according to equation 2; the adjusted data are compared in Figure 11. As expected, the low closure levels in A-E material cause only a minor shift in the FCG data, whereas a large shift is calculated for the 30% C.R. specimens as a result of the very high closure loads. Based on comparisons such as that shown in Figure 11, in which the FCG rate data for material of various grain sizes more or less overlap when plotted versus ΔK_{eff}, others have argued that the grain size effect must be due to closure. Recently, this closure has been termed "roughness-induced" (12) to specify its origin and to distinguish it from "oxide-induced" closure (16), a separate subject to be discussed below.

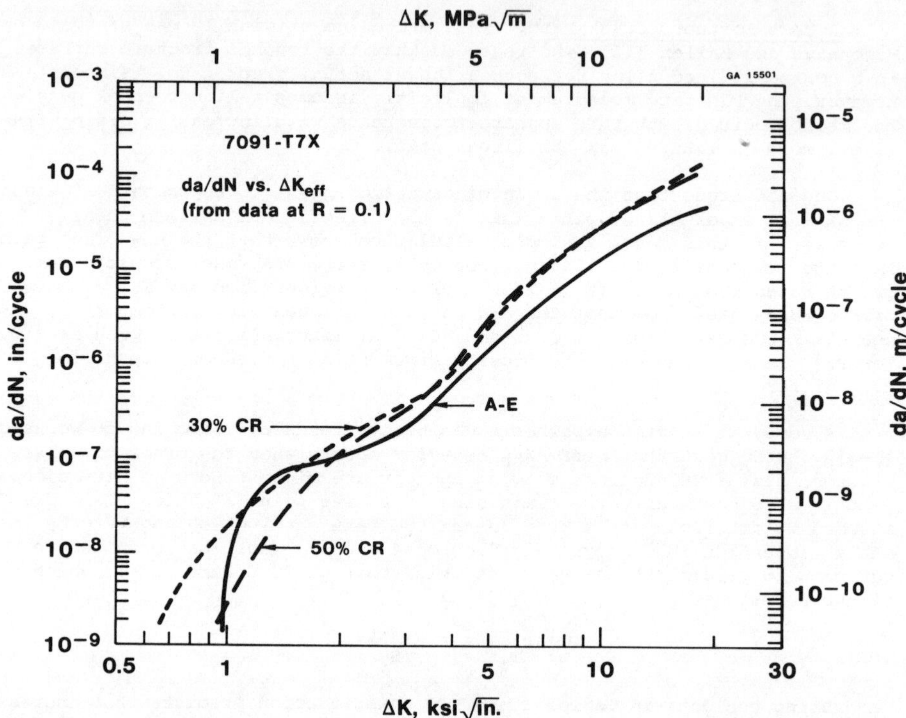

Figure 11 - FCG rates plotted against ΔK_{eff} showing general overlap of trend lines.

Further evidence suggesting a link between closure and grain size effects comes from the work of Gray, et al. (5), previously mentioned. These authors argued that the lack of a grain size effect on FCG behavior at R = 0.7 in a steel was due to the absence of roughness-induced closure at this high stress ratio. In reference 5, surface roughness measurements were made, the results of which are quite similar to those in Figure 10b. In both studies, the degree of roughness appears to be independent of both R ratio and K (or ΔK).

Despite similarities between the current study and that in reference 5 and the closure-corrected results in Figure 11, the FCG data in Figure 5 show unquestionably a two-fold increase in ΔK_{th} with grain size at R = 0.8. It is important to reiterate that no closure loads greater than P_{min} were measured in any of the three materials at R = 0.8, even at ΔK_{th}. Thus, the common statement that "near-threshold FCG behavior is a closure phenomenon" cannot be generally true.

Another type of crack closure, that due to the presence of oxide debris on the fracture surface, already was mentioned briefly. While oxide-induced closure is believed to influence strongly near-ΔK_{th} behavior in steels (16), its effect on crack growth in Al alloys seems to be less important (17). In the present study, it certainly is unlikely that any surface debris could affect fatigue behavior at R = 0.8, especially since no closure was measured.

Although crack closure may occur due to microstructural changes (grain size) or to the presence of oxide debris, its effect on fatigue behavior is mechanical in the sense that the degree of closure is dependent on stress ratio. In contrast, the mechanism responsible for the increase in ΔK_{th} at R = 0.8 appears to be microstructural in origin. If the increase in ΔK_{th} at R = 0.1 is due to the combined effects of these two mechanisms, how important is crack closure vis-a-vis any intrinsic microstructural influence on ΔK_{th}? Figure 12 shows ΔK_{th} plotted as a function of log (grain size), using the average intercept length in the LT direction (parallel to the crack growth direction) as an indicator of grain size. The dashed line through R = 0.8 data suggests the intrinsic microstructural effect (as yet undefined) of grain size on ΔK_{th}. If we translate this line to pass through the R = 0.1 point for A-E material, then we can estimate the relative amounts of closure and microstructurally-related contributions at larger grain sizes. (Recall that the amount of closure is small for the A-E material.) For the 30% C.R. material, this analysis suggests that these two contributions toward increasing ΔK_{th} are nearly equal, as indicated by the shaded regions in Figure 12.

Crack Deflection. The increasing degree of fracture surface roughness which accompanies an increase in grain size implies a greater deviation of the crack from a straight line path through the microstructure. Suresh (18, 19) recently considered the effect of this deviation on the magnitude of ΔK at the crack tip for cases of both simple and multiple deflection (i.e., branching). Depending upon the angle of deviation and the degree of complexity of the branch (e.g., the number of secondary branches), ΔK_{eff} can reduce crack growth rates dramatically in the near-threshold region where the slope of da/dN vs. ΔK is quite high. In his most recent treatment of crack deflection, Suresh (19) has considered not only the reduction in ΔK_{eff} but also the increased crack length (i.e., tortuosity) which accompanies greater surface roughness. In combination, these two effects can indeed reduce FCG rates by several orders of magnitude in the near-threshold regime.

While increasing fracture surface roughness does indicate increased crack deviation, it does not provide any direct evidence regarding crack branching (i.e., multiple crack deflection). Metallographic profiles of the fracture surfaces, on the other hand, clearly show that branching occurs in the 30% C.R. material but not in the A-E material (compare Figures 13a and d, respectively). Furthermore, this extensive branching occurs near ΔK_{th} (Figure 13a) but not at higher ΔK levels (Figures 13b and c) in the 30% C.R. sample. These metallographic results suggest that the greater influence of grain size on near-threshold FCG behavior (as compared to that at higher ΔK levels) may be attributed to extensive crack branching in the absence of crack closure effects.

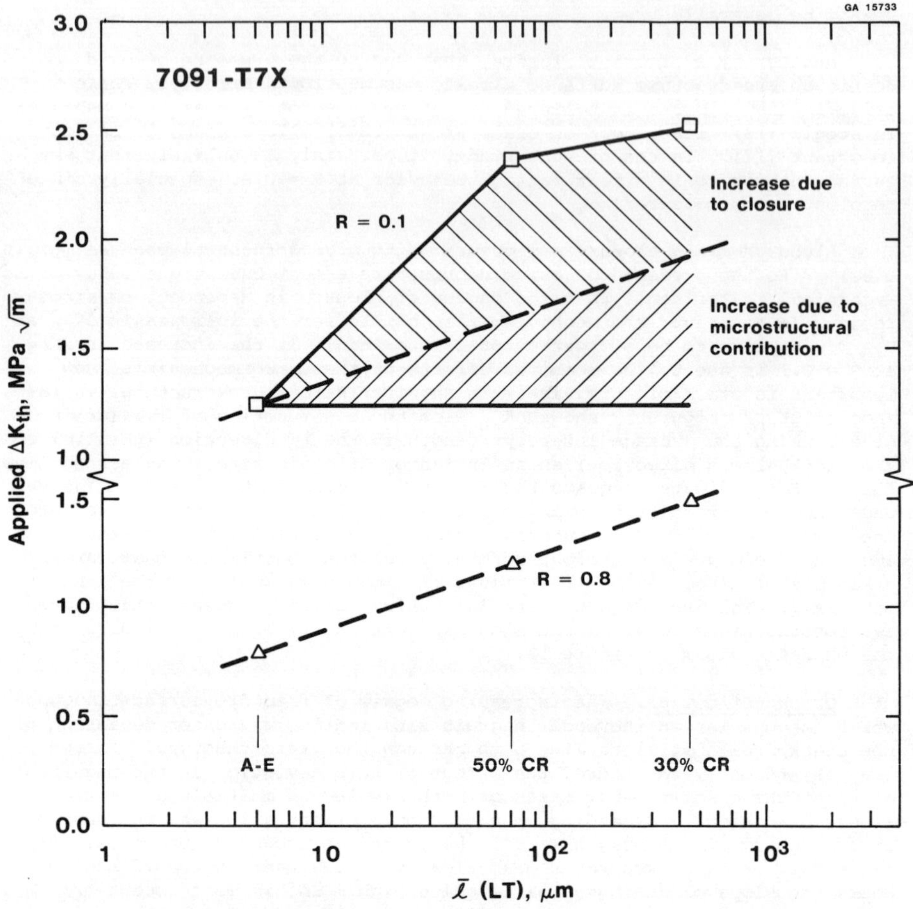

Figure 12 - Effect of grain size (l) on ΔK_{th} at R = 0.1 and 0.8. Note approximately equal contributions of closure and microstructure (crack branching) to ΔK_{th} increase in 30% C.R. 7091-T7X.

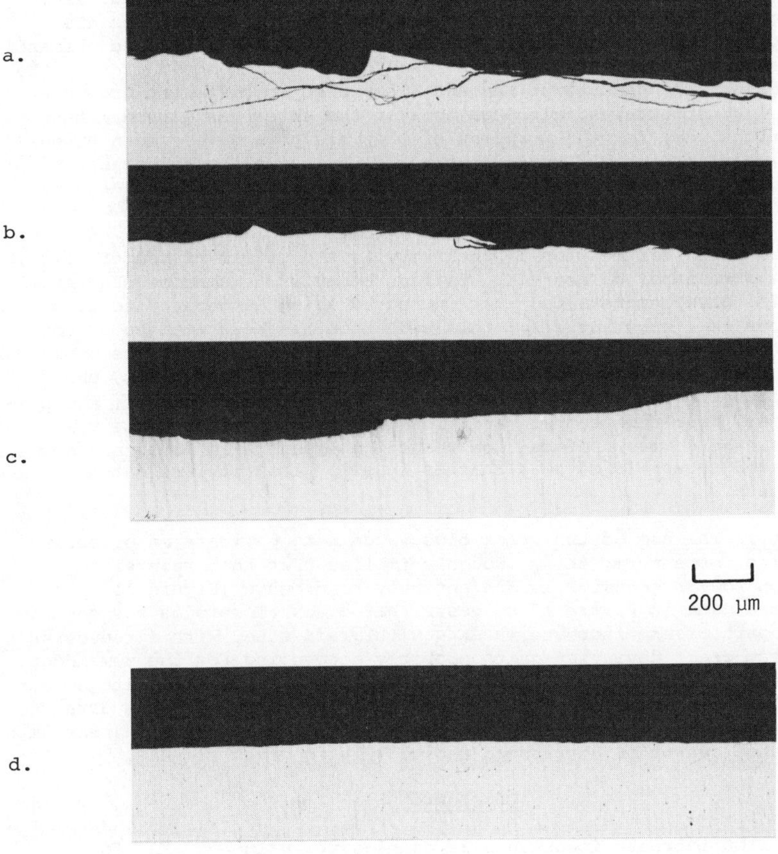

Figure 13 - Metallographic profiles of fatigue cracks in 30% C.R. (a-c) and A-E (d) 7091-T7X for tests at R = 0.8. Polishing plane near center of specimen thickness.

a. 30% C.R. at threshold; ΔK_{th} = 1.4 MPa$\sqrt{m}$ (1.3 ksi$\sqrt{in.}$).

b. 30% C.R.; ΔK = 2.5 MPa$\sqrt{m}$ (2.3 ksi$\sqrt{in.}$).

c. 30% C.R.; ΔK = 6.0 MPa$\sqrt{m}$ (5.5 ksi$\sqrt{in.}$).

d. A-E at threshold; ΔK_{th} = 0.76 MPa$\sqrt{m}$ (0.69 ksi$\sqrt{in.}$).

To this point, the three materials have been discussed as though they differ only in grain size. The situation actually is more complex, since the A-E material is unrecrystallized while the two C.R. structures are fully recrystallized. Thus, grain boundaries in the A-E alloy are primarily low-angle subgrain boundaries, in contrast to high-angle boundaries in the C.R. materials. The texture of the A-E alloy also is different from that of the two C.R. structures, although none of the three has a random texture. Nevertheless, for the purposes of this study we have concentrated on grain size differences. Further work will be necessary to establish grain boundary and texture effects independently, since neither has been reported for aluminum alloys.

A final point not mentioned previously is the effect of temper (i.e., precipitate structure) on near-ΔK_{th} fatigue behavior. Numerous studies (e.g., 20-23) have demonstrated that increased aging (underaged to overaged) reduces ΔK_{th} substantially; this reduction has been attributed to the loss in slip planarity with aging. Since the A-E material was overaged for 8 hours as opposed to 2 hours for the two C.R. materials, one might expect that some of the difference in ΔK_{th} is due to temper. We presume, however, that this temper variation does not influence these results to a significant degree. Deformation should be essentially homogeneous in all three overaged materials, so that changes in slip planarity are not a relevant issue.

Summary. The use of the crack closure data to estimate an effective crack driving force term, ΔK_{eff}, roughly implies that the grain size influence on low-ΔK behavior is due entirely to closure (Figure 11). In contrast, the data in Figure 12 suggests that crack closure is responsible for only a part of the increase in ΔK_{th} with grain size, with a reduction in ΔK_{eff} from crack deviation quite probably accounting for the remainder. Without measuring the degree of crack deviation and calculating its effect on ΔK levels, this argument remains essentially speculative. The data clearly do demonstrate, however, that the effect of grain size on near-ΔK_{th} FCG resistance cannot be attributed exclusively to crack closure.

Conclusions

A. A x100 increase in grain size doubles the value of ΔK_{th} for 7091-T7X at stress ratios of both 0.1 and 0.8. FCG resistance in the coarse-grained alloy is better than in the unrecrystallized extrusion for all growth rates below about 2.5×10^{-9} m/cycle (1×10^{-7} in./cycle). For growth rates between 2.5×10^{-9} and 1×10^{-6} m/cycle (1×10^{-7} to 4×10^{-5} in./cycle), FCG resistance in the coarse-grained materials is essentially equivalent to that in the fine-grained A-E structure.

B. Because ΔK_{th} doubles even at R = 0.8, the mechanical effect of fatigue crack closure cannot be exclusively responsible for the influence of grain size on near-threshold fatigue behavior at lower R ratios. Rather, closure appears to be responsible for roughly one-half of the observed ΔK_{th} increase at R = 0.1 when grain size increases by 100 times.

C. The results of this study, in concert with theoretical calculations by Suresh (18, 19), suggest that the apparent increase in crack deflection with grain structure coarsening can account for the fatigue threshold increase at R = 0.8. Deflection also may be responsible for a portion of the ΔK_{th} increase at R = 0.1.

References

1. G. R. Yoder, L. A. Cooley, and T. W. Crooker in Titanium '80 (Proc. 4th International Conference on Titanium, Kyoto), 3, H. Kimura and O. Izumi, Eds., TMS-AIME, 1981, 1865.

2. Ibid in Fracture Mechanics (Proc. 14th National Symposium on Fracture Mechanics, Los Angeles), J. C. Lewis and G. Sines, Eds., ASTM STP XXX, 1983, in Press.

3. Ibid, Scripta Met., 16, 1982, 1021.

4. G. M. Lin and M. E. Fine, Scripta Met., 16, 1982, 1249.

5. G. T. Gray, III, J. C. Williams, and A. W. Thompson, Met. Trans., 14A, 1983, 421.

6. J. Lindigkeit, G. Terlinde, A. Gysler, and G. Lutjering, Acta Met., 27, 1979, 1717.

7. J. E. King, Metal Science, 16, 1982, 345.

8. V. W. C. Kuo and E. A. Starke, Jr., Met. Trans., 14A, 1983, 435.

9. J. I. Petit and P. E. Bretz, High-Strength P/M Aluminum Alloys, M. J. Koczak and G. J. Hildeman, Eds., TMS-AIME, 1982, 147.

10. Y. Mutoh and V. M. Radhakrishnan, J. Engr. Mat. Tech., 103, 1981, 229.

11. K. Minakawa and A. J. McEvily, Scripta Met., 15, 1981, 633.

12. S. Suresh and R. O. Ritchie, Met. Trans., 13A, 1982, 1627.

13. J. Lebiedzik and E. W. White in Scanning Electron Microscopy/75 (Part I), IIT Research Institute, 1975, 181.

14. J. Lebiedzik, Scanning, 2, 1979, 230.

15. J. Masounave and J. P. Bailon, Scripta Met., 10, 1976, 165.

16. S. Suresh, G. F. Zamiski, and R. O. Ritchie, Met. Trans., 12A, 1981, 1435.

17. A. K. Vasudevan and S. Suresh, Met. Trans., 13A, 1982, 2271.

18. S. Suresh, Engr. Frac. Mech., 18, 1983, 577.

19. S. Suresh, Met. Trans., 14A, 1983, 2375.

20. E. Hornbogen and K. H. ZumGahr, Acta Met., 24, 1976, 581.

21. B. R. Kirby and C. J. Beevers, Fat. Engr. Mat. Str., 1, 1979, 203.

22. J. F. Knott and A. C. Pickard, Metal Science, 1979, 399.

23. P. E. Bretz, A. K. Vasudevan, R. J. Bucci, and R. C. Malcolm, Final Contract Report, Naval Air Systems Command Contract N00019-79-C-0258, 1981 October 09.

INFLUENCE OF TEMPERATURE ON THE NEAR-THRESHOLD FATIGUE CRACK GROWTH BEHAVIOR OF A NICKEL BASE SUPERALLOY

J. L. Yuen* and P. Roy
General Electric Company
Advanced Reactor Systems Department
Sunnyvale, California 94086

SUMMARY

The effect of temperature on the near-threshold fatigue crack growth behavior of a nickel base precipitation hardenable superalloy, Alloy 718,was studied in air. The tests were performed from room temperature up to 649°C. Over the range of growth rates from 10^{-8} to 10^{-4} mm/cycle, both accelerating and decelerating crack growth mechanisms were observed. At intermediate growth rates (rates above 10^{-5} mm/cycle), accelerating mechanisms dominated and at near-threshold growth rates (rates below 10^{-5} mm/cycle), decelerating mechanisms were found to be operating. The room temperature ΔK_{th} value was higher than the elevated temperature ΔK_{th} values. In considering only the elevated temperature ΔK_{th} values, the values increased with increasing temperature. These results are explained with a mechanism that involves the accumulation of oxidation products within the crack enhancing crack closure. At room temperature, the fracture appearance consisted of fatigue striations in the intermediate growth regime and became crystallographic in the near threshold growth regime. The fracture morphology at elevated temperatures was similar. At near-threshold growth rates, the fracture surface consisted of smooth transgranular regions about the size of a grain diameter whereas fatigue striations were observed at intermediate growth rates.

*Presently with Standard Oil Company of California
Engineering Department, Richmond, CA 94804

INTRODUCTION

For many years, a large effort has been devoted towards understanding fatigue crack growth behavior at rates above 10^{-5} mm/cycle. Recently, near-threshold fatigue crack growth rates (below 10^{-5} mm/cycle) are needed to assess the lifetime of components that are subjected to high cycle low amplitude loads. In the liquid metal fast breeder reactor, there are components that are subjected to high cycle thermal fatigue which arise from the mixing of different temperature streams of sodium. The fluctuating temperatures produce thermal stresses on the surfaces of these components (1,2). Alloy 718 is a candidate material for these components in which high cycle fatigue resistance is required. Alloy 718 is a precipitation hardening nickel base superalloy that has good elevated temperature fatigue strength. The purpose of this study is to determine the effect of temperature on the fatigue crack growth behavior of Alloy 718.

The environmental effects on fatigue crack growth can result in an acceleration or deceleration in growth rates (3). The accelerating and decelerating mechanisms act competitively with the dominating mechanism being dependent upon synergistic effects from the environment and the mechanical fatigue processes at the crack tip. Accelerating mechanisms may involve the absorption of hydrogen (3,4,5) or oxygen (6,7) into the lattice embrittling the metal ahead of the crack tip. Decelerating mechanisms may involve crack tip blunting (8,9), crack branching (10,11), or oxide-induced crack closure (12,13). The latter will be explored in more detail in this paper in relation to the increased oxidation kinetics at elevated temperatures.

MATERIALS

A 19 mm thick plate of Alloy 718 was received in the annealed condition. The chemical composition is given in Table I. Compact tension specimens 12.7 mm thick were fabricated in the T-L orientation. After final machining, the specimens were heated to 954°C for one hour and air cooled. They were precipitation hardened by aging at 718°C for 8 hours, furnace cooled to 621°C, and held at 621°C for a total aging time of 18 hours. Finally the specimens were air cooled to room temperature. This heat treatment produced an average grain diameter of 22 μm. The mechanical properties of the material are listed in Table II.

Table I. Alloy 718 Chemical Composition

Element	Percent by Weight
Ni	51.22
Cr	17.97
Fe	19.07
Cb+Ta	5.07
Ti	0.99
Al	0.57
Mo	2.88
Co	<0.05
Ta	<0.01
Mn	0.28
Si	0.12
C	0.05
Cu	<0.01
B	0.003
P	<0.005
S	<0.002

Table II. Mechanical Properties of Alloy 718 in Transverse Orientation

Test Temperature °C	0.2% Offset Yield MPa	Ultimate MPa	Elongation,% in 25.4 mm	Reduction in Area, %
24	1092	1355	23.4	40.1
24	1054	1345	23.8	42.7
538	906	1156	30.2	40.7
538	934	1174	36.4	44.6

EXPERIMENTAL PROCEDURE

The fatigue crack growth tests were performed with a 30 kN servohydraulic test system that was interfaced with a computer for test control and data acquisition. A 100 Hz sinusoidal waveshape was used at load ratios ($R = K_{min}/K_{max}$) of 0.1 and 0.5. The tests were performed in air with a relative humidity of 50-60%. The test temperatures were 24°, 427°, 538°, and 649°C. The specimens were heated with a 10 kW induction heating unit that operated at 50 kHz. The crack lengths were measured with a traveling microscope which had a resolution of 0.1 mm and with an elastic compliance method that is described elsewhere (14).

The fatigue crack growth rates were measured with a load shedding and load increasing procedure. The load shedding procedure consisted of a repeated sequence of load shedding and subsequent crack growth. For each load, a minimum of 0.5 mm of growth was obtained prior to reducing the load which was no more than 10%. This sequence was repeated until a growth rate of 10^{-7} mm/cycle was reached and was defined as an operational definition of ΔK_{th}. Upon establishing ΔK_{th}, the test was resumed under load increasing conditions. Crack growth rates were calculated by dividing the crack growth increment (nominally 0.5 mm) by the number of elapsed cycles. The guidelines given in ASTM Standard E 647-78T were also satisfied.

The oxide thicknesses on the fracture surface were measured using secondary ion mass spectrometry (SIMS). A 7 μm diameter beam of ^{18}O ions (13.5 keV impact energy) was rastered over an area 36 x 36 μm. The current density was 0.2 mA/cm^2. Ion mass 68, CrO^+, was monitored as a function of evolution time and collected only when the beam was in the central 25% of the rastered area. The sputter rate was 12.0 nm per minute, based on an AISI Type 316 stainless steel standard. The oxide thickness was defined as the evolution time required for the CrO^+ intensity to decay by 90% of its initial value.

RESULTS AND DISCUSSION

The room temperature fatigue crack growth rates (da/dN) as a function of the alternating stress intensity factor (ΔK) are shown in Figure 1. The room temperature threshold ΔK value was 7.9 MPa$\sqrt{m}$. The crack growth curve through the data shows an unusual behavior just above ΔK_{th} in which it is concave upward.

Scanning electron microscope (SEM) examination of the fracture surface revealed facets that had a strong crystallographic appearance at near

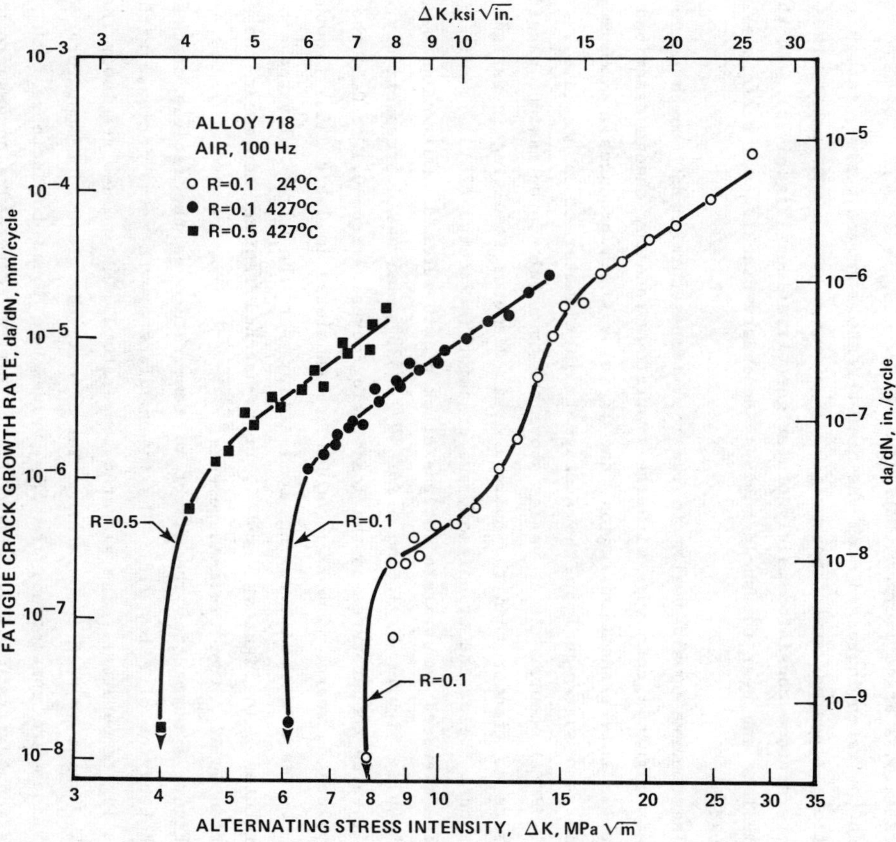

Figure 1 - Fatigue crack growth behavior of Alloy 718 in air at 24° and 427°C.

threshold growth rates (see Figure 2a). Many of the facets are about the size of a grain diameter. The facets appeared to have resulted from decohesion along active slip bands which have been suggested to be the $\{111\}$ (15,16). The decohesion along intense slip bands can be seen in Figure 2b in which the face of the compact tension specimen (free surface) is shown along with the adjoining fracture surface. Decohesion occurred in adjacent slip bands that were parallel to the crack which was oriented 45° to the vertical tensile axis, a Stage I crack (17).

As the rates increased from the near threshold-growth regime, fatigue striations appeared more frequently. At intermediate growth rates, the fracture surface consisted mainly of fatigue striations (Figure 2c). In the intermediate and near threshold regimes, secondary cracking was occasionally observed.

The concave upward region in the room temperature crack growth curve shown in Figure 1 coincides with the fracture morphology change described above. In this transition region, the ΔK value which produces a maximum plastic zone size equal to the average grain diameter of 22 μm (i.e., $\Delta K = 11\ MPa\sqrt{m}$) occurs. At ΔK values above the transition, the maximum plastic zone size is greater than the average grain diameter resulting in fatigue striations. Below the transition, the cracking is faceted (26).

The elevated temperature fatigue crack growth rates as a function of ΔK are shown in Figures 1, 3, and 4 for 427°, 538°, and 649°C, respectively. For all three temperatures as the load ratio increases from 0.1 to 0.5, the ΔK_{th} values decrease. The ΔK_{th} values are tabulated in Table III. Growth rates increase with increasing load ratios for the three elevated temperatures. In Figures 3 and 4, cross-hatched regions represent the variation in ΔK_{th} from two measurements.

Figure 5 summarizes the effect of temperature on the fatigue crack growth behavior of Alloy 718. The intermediate growth rates increase with increasing temperature. The elevated temperature ΔK_{th} values are lower than the ΔK_{th} value at room temperature. Yet if only the elevated temperature ΔK_{th} values are considered, the ΔK_{th} values are found to increase with increasing temperature. As a result, the higher temperature curves cross over the lower temperature curves.

There is little difference in fracture appearance from the three elevated temperature tests. Load ratio does not significantly affect the fracture appearance. In the near threshold growth regime at 427°, 538°, and 649°C, the fracture surface consists of smooth transgranular regions about the size of a grain diameter that changed orientation from grain to grain as seen in Figures 6a, 6b, and 6c, respectively. In a few of the smooth

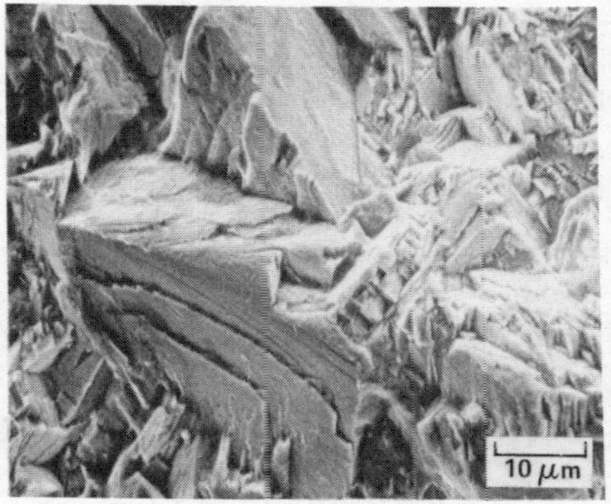

a) Faceted fracture at near threshold growth rates, $\Delta K = 3 MPa\sqrt{m}$.

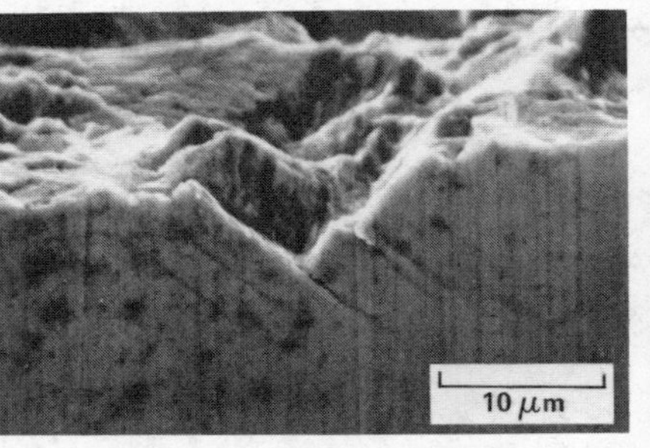
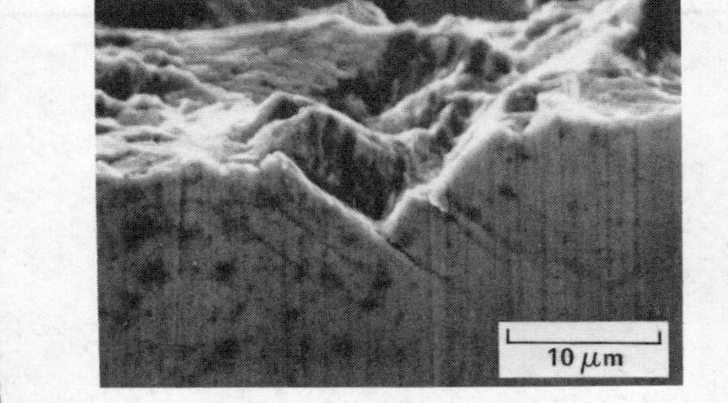

b) Decohesion along active slip bands at near threshold growth rates as seen on the face of the compact tension specimen, $\Delta K = 13 MPa\sqrt{m}$.

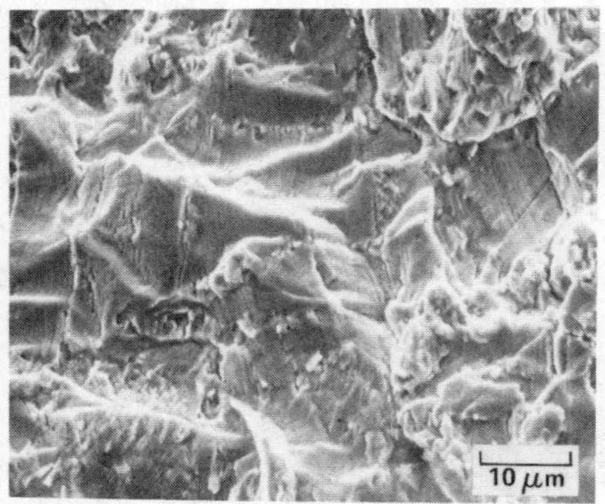

c) Fatigue striations at intermediate growth rates, $\Delta K = 28 MPa\sqrt{m}$

83-501-02

Figure 2 - Fatigue fracture morphology at 24°C and R = 0.1.

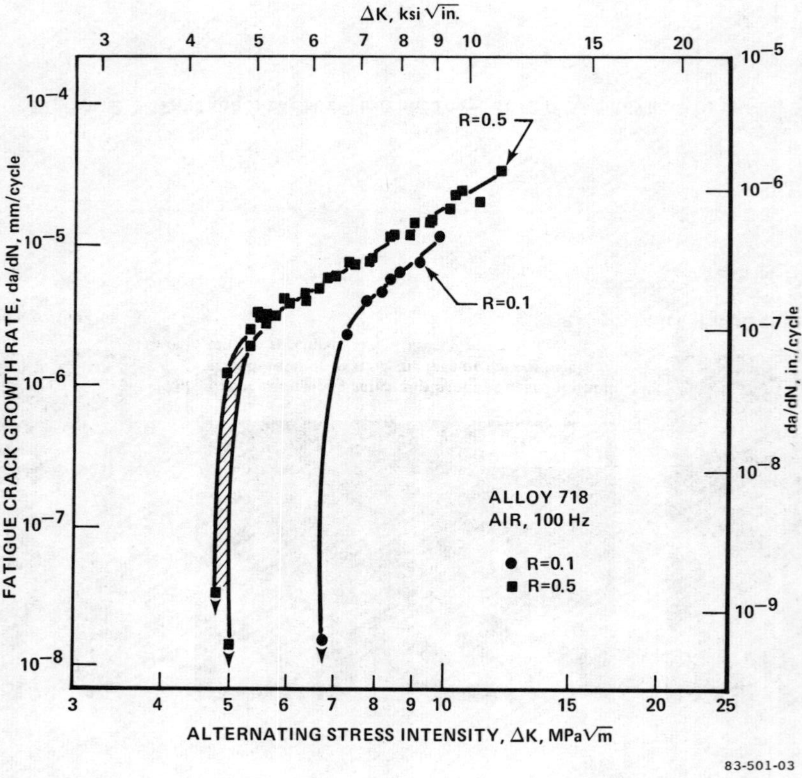

Figure 3 - Variation of fatigue crack growth behavior with load ratio of Alloy 718 in air at 538°C.

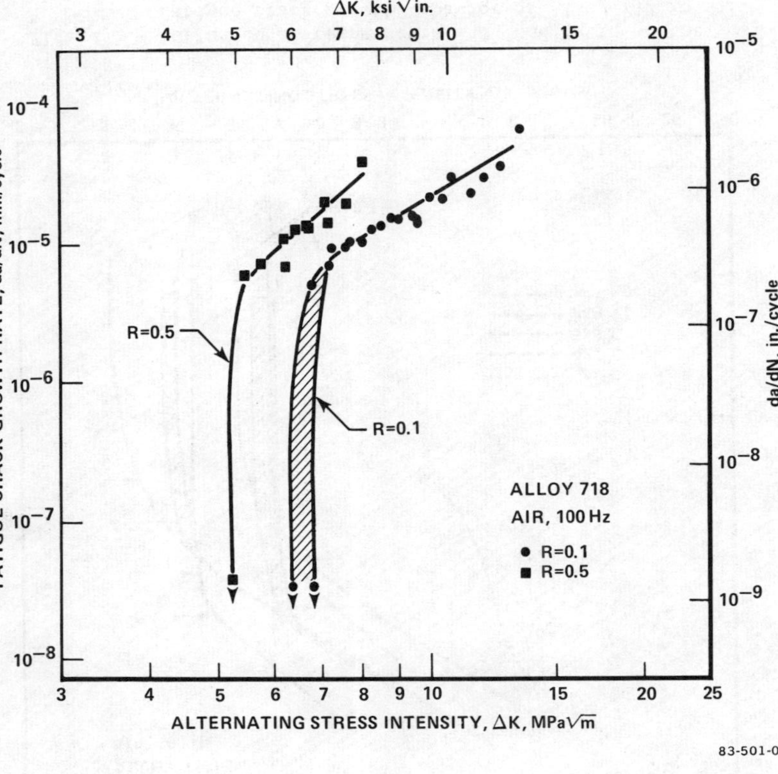

Figure 4. - Variation of fatigue crack growth behavior with load ratio of Alloy 718 in air at 649°C.

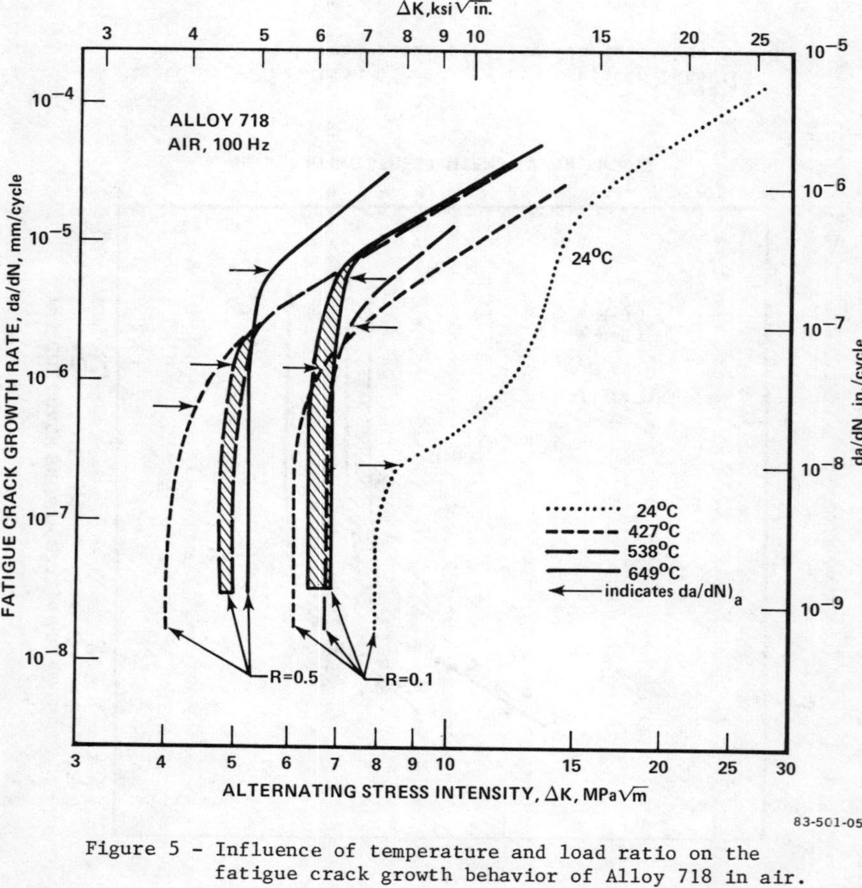

Figure 5 - Influence of temperature and load ratio on the fatigue crack growth behavior of Alloy 718 in air.

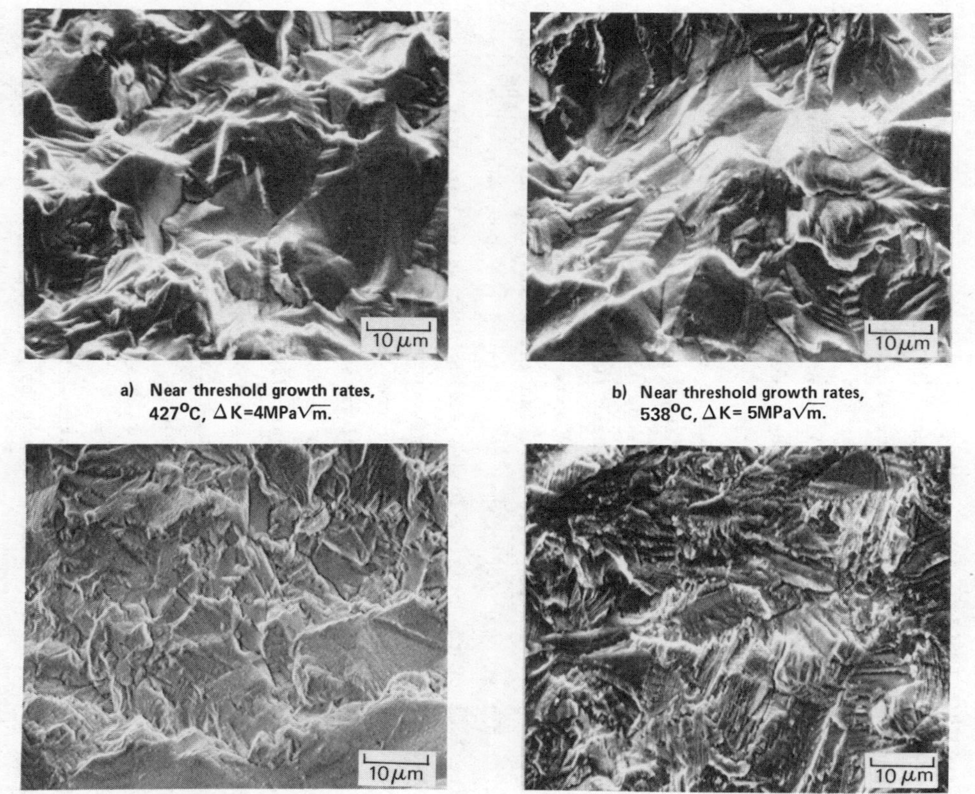

a) Near threshold growth rates, 427°C, $\Delta K = 4 MPa\sqrt{m}$.

b) Near threshold growth rates, 538°C, $\Delta K = 5 MPa\sqrt{m}$.

c) Near threshold growth rates, 649°C, $\Delta K = 6 MPa\sqrt{m}$.

d) Intermediate growth rates, 538°C, $\Delta K = 9 MPa\sqrt{m}$.

Figure 6 - Elevated temperature fatigue fracture morphology at R = 0.5.

regions, parallel markings were observed which were too coarse to be fatigue striations. Mills and James (15) observed similar features in their study of Alloy 718 and suggested that they were slip offsets. At intermediate growth rates, fatigue striations were observed. Figure 6d shows an example of fatigue striations observed at elevated temperatures in the specimen tested at 538°C with R = 0.5. As seen in Figure 6 in both crack growth regimes, secondary cracks were occasionally observed in fracture surfaces from the elevated temperature tests.

The near threshold fatigue crack growth behavior at room and elevated temperatures can be explained with a decelerating crack growth mechanism. This involves the accumulation of oxide within the crack which enhances crack closure [oxide-induced crack closure (12)]. At near threshold growth rates at room temperature and at low load ratios, the oxide deposits are thickened by repeated fretting in which the oxide repeatedly ruptures, reforms, and compacts between the fracture surfaces due to the contact of crack surfaces caused by residual stresses left in the wake of the crack (12,13,18-20). Surface abrasion is further promoted by Mode II shear which has been suggested to be the dominant displacement at near threshold growth rates (21-24).

The oxide thicknesses are shown in Figure 7, as measured by SIMS. The oxide thicknesses were plotted such that the crack length where ΔK_{th} was measured for each test, was designated as origin of the abscissa. For crack lengths to the left of the origin, load shedding procedures were used; to the right of the origin, load increasing procedures were used. The oxide thicknesses at 24°C gradually increase as the crack length increases towards the origin. In other words, the oxide thicknesses increase with decreasing ΔK. The maximum oxide thickness is found at the crack length where ΔK_{th} was measured, i.e., origin of abscissa in Figure 7. This oxide thickness of 0.18 μm corresponds to the maximum crack tip opening displacement (CTOD) listed in Table III. This oxide is an order of magnitude thicker than the oxide thickness formed on the fracture surfaces at higher growth rates. With increasing crack length from the origin (i.e., increasing ΔK values), the oxide thicknesses decrease.

The data indicate that the maximum oxide thickness formed at 24°C and at ΔK_{th} is sufficient to enhance crack closure by wedging the crack closed for the entire load cycle, thus raising the crack closure stress intensity, K_{cl}. The effective ΔK, i.e., $K_{max}-K_{cl}$, is reduced to a negligible value. At this point, a threshold ΔK value is measured.

At elevated temperatures, oxidation kinetics play an important role in controlling the oxide thickness and can be used to rationalize the increase in ΔK_{th} with increasing temperature. As the load is reduced, the maximum

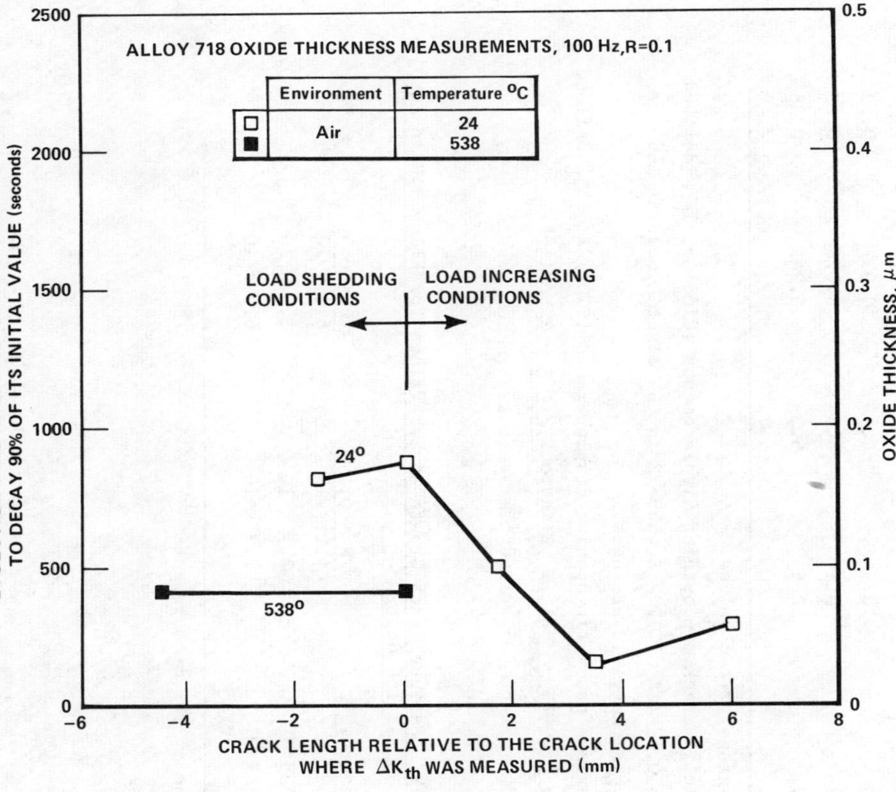

Figure 7. - Oxide thickness as a function of crack length relative to the crack location where ΔK_{th} was measured. Measurements were made with secondary ion mass spectrometry.

Table III. Fatigue Crack Growth Thresholds for Alloy 718 in Air at 100 Hertz

Temperature °C	Load Ratio, R	ΔK_{th} $MPa\sqrt{m}$	Cyclic CTOD(b) μm	Maximum CTOD(c) μm
24	0.1	7.9	0.072	0.18
427	0.1	6.1	0.053	0.13
427	0.5	4.0	0.023	0.19
538	0.1	6.8	0.072	0.17
538	0.5	4.9 (a)	0.037	0.30
649	0.1	6.6 (a)	0.076	0.19
649	0.5	5.3	0.049	0.38

(a) Average of two values.

(b) Cyclic crack tip opening displacements (CTOD) at ΔK_{th} defined as 0.49 $(\Delta K_{th})^2/2\sigma_{ys}E$ where E is the elastic modulus and σ_{ys} is the yield strength.

(c) Maximum crack tip opening displacements (CTOD) at ΔK_{th} defined as 0.49 $(K_{th,max})^2/\sigma_{ys}E$ where $K_{th,max} = \Delta K_{th}/(1-R)$.

Table IV. Fatigue Crack Growth Arrest Values for Alloy 718 in Air at 100 Hertz

Temperature, °C	R = 0.1	R = 0.5
427	1.2×10^{-6}	6.1×10^{-7}
538	2.4×10^{-6}	1.6×10^{-6}*
649	7.3×10^{-6}*	6.0×10^{-6}

*Average of two measurements

CTOD will decrease correspondingly and eventually will approach the thickness of the oxide layer that forms on the fracture surface. At this load, the oxide can wedge the crack closed, thereby reducing the effective ΔK to a negligible value and arrest the crack growth (i.e., establish a threshold ΔK). With increasing temperature, a thicker oxide is produced for a given period of time. This thicker oxide can wedge the crack closed at a higher ΔK, thus causing the ΔK_{th} values to increase with increasing temperature.

The crack arrest rates, $da/dN)_a$, shown in Figure 5 with horizontal arrows increase with temperature for both load ratios of 0.1 and 0.5 (see Table IV). These crack arrest rates can be used to determine the rate controlling mechanism for crack arrest. As the rate of environmental interaction reaches the crack growth rate, the crack will arrest. But if the crack growth rate exceeds the kinetics of the environmental interaction processes, the crack continues to propagate (14). The apparent activation energy for the crack arrest phenomenon can be obtained from an Arrhenius plot of log $[da/dt)_a]$* versus 1/T as shown on Figure 8. The apparent activation energy was found to be 55.7 ± 21 kjoule/mole. If the rate controlling mechanism for the crack arrest phenomenon involves oxidation, then the apparent activation energy should be comparable to the apparent activation energy for oxidation. Oxidation data for the nickel base alloy Rene 80 are used for qualitative comparisons since there is limited oxidation information for Alloy 718. The current value is in the range of apparent activation energy for parabolic oxidation of Rene 80 which is 40.2 kjoule/mole (25). This suggests that the mechanism for crack arrest involves oxidation. Oxidation kinetics play an important role in determining the oxide thickness on the fracture surfaces. Therefore, the oxide should be uniform over the fracture surface, independent of the ΔK.

The SIMS measurements of the oxide on the fracture surface of a specimen tested at 538°C verify this hypothesis. As seen in Figure 7, the oxide thickness is uniform over the fracture surface. Since the calculated maximum CTOD at ΔK_{th} for 24°C and 538°C were nearly identical (see Table III), the maximum oxide thicknesses for 24°C would be expected to be similar to that of the 538°C. The reason for the 50% difference between the measured oxide thicknesses at 538°C and the maximum oxide thickness at 24°C is unclear.

The ΔK_{th} values found in this study are dependent upon the interactions between the environmental processes and the mechanical fatigue processes at the crack tip. Changes in the test variables, such as temperature, have been shown to alter the ΔK_{th} values. Oxidation kinetics were important in

---

*$da/dt)_a = da/dN)_a \times 100$ Hz

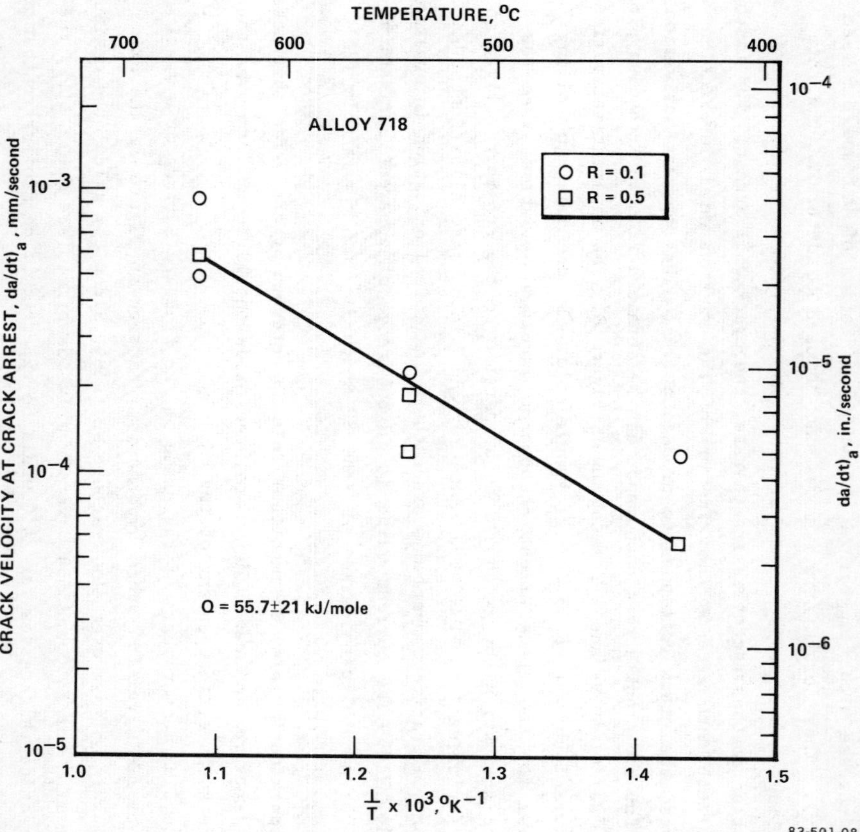

Figure 8 - Apparent activation energy for fatigue crack growth arrest in air of Alloy 718.

establishing the oxide thickness at elevated temperatures, whereas at room temperature, an oxide fretting mechanism was required to thicken the oxide in the crack. At elevated temperatures as the temperature was increased, a thicker oxide formed for a given period of time which enhanced crack closure at higher ΔK values, i.e., higher ΔK_{th} values. Altering the relationship between the environmental processes and mechanical fatigue processes will alter the ΔK_{th} values. This suggests that, for alloys in which the oxide-induced crack closure mechanism is applicable, changes in those variables that increase the environmental interaction processes such as decreasing the cyclic frequency or increasing the temperature will cause the threshold ΔK values to increase at low load ratios.

CONCLUSION

The room temperature ΔK_{th} value was higher than the elevated temperature ΔK_{th} values. But, if only the elevated temperature threshold ΔK_{th} values were considered, the values increased with increasing temperature. These results were explained with a decelerating crack growth mechanism that involved the accumulation of oxide within the crack enhancing crack closure. At room temperature, an oxide fretting mechanism was necessary to build up the oxide debris in the crack. Oxidation kinetics were important in determining the oxide thicknesses in the crack at elevated temperatures.

The fracture morphology at room temperature was highly faceted (crystallographic) at near threshold growth rates and changed to fatigue striations at intermediate growth rates. At elevated temperatures, the fracture appearances were quite similar. The fracture morphology at elevated temperatures consisted of smooth transgranular regions that changed orientation from grain to grain at near threshold growth rates while at intermediate growth rates, fatigue striations were observed.

ACKNOWLEDGEMENT

The support of this work by the General Electric Company is gratefully acknowledged.

REFERENCES

1. L. Gruter and W. Huget, Int. J. Pres. Ves. and Piping, 1982, Vol. 10, p. 335.

2. P. Marshall and C. R. Brinkman, Nuclear Energy, BNES, 1981, Vol. 20, No. 3, p. 257.

3. A. K. Vasudevan and S. Suresh, Metall. Trans., 1982, Vol. 13A, p. 2271.

4. R. P. Wei and G. W. Simmons, Int. J. of Fract., 1981, Vol. 17, No. 2, p. 235.

5. P. K. Kiaw, S. J. Hudak, and J. K. Donald, Metall. Trans., 1982, Vol. 13A, p. 1633.

6. J. D. Frandsen, N. E. Paton, and H. L. Marcus, Metall. Trans., 1974, Vol. 5, No. 7, p. 1655.

7. J. W. Swanson and H. L. Marcus, Metall. Trans., 1978, Vol. 9A, No. 2, p. 291.

8. J. C. Radon, C. M. Branco, and L. E. Culver, Int. J. of Fract., 1976, Vol. 12, p. 467.

9. F. P. Ford and P. W. Emigh, NACE Int. Corrosion Forum, March 22-26, 1982, Houston, Texas, Paper No. 248.

10. L. K. L. Tu and B. B. Seth, J. of Testing and Evaluation, 1978, Vol. 6, No. 1, p. 66.

11. W. Hoffelner and M. O. Speidel, Proceedings Corrosion and Mechanical Stress, V. Guttmann and M. Merz, eds., Petten, Netherlands, May 1980, p. 275.

12. R. O. Ritchie, S. Suresh, and C. M. Moss, J. of Eng. Mat. and Tech., 1980, Vol. 102, No. 3, p. 293.

13. S. Suresh, G. F. Zamiski, and R. O. Ritchie, Metall. Trans., 1981, Vol. 12A, p. 1435.

14. J. L. Yuen, Ph.D. Dissertation, 1982, Stanford University, Stanford, CA.

15. W. J. Mills and L. A. James, J. of Eng. Mat. and Tech., 1979, Vol. 101, No. 3, p. 205.

16. M. Clavel and A. Pineau, Metall. Trans., 1978, Vol. 9A, No. 4, p. 471.

17. C. Laird, Fatigue and Microstructure, ASM Materials Sci. Sem., St. Louis, Mo., October 14-15, 1978, ASM, 1979, p. 149.

18. P. K. Liaw, T. R. Leax, R. S. Williams, and M. G. Peak, Metall. Trans. A., 1982, Vol. 13A, p. 1607.

19. A. T. Stewart, Eng. Fract. Mech, 1980, Vol. 13, No. 3, p. 463.

20. D. Benoit, R. Namdar-Irani, and R. Tixier, Mat. Sci. and Eng., 1980, Vol. 45, No. 1, p. 1.

21. R. O. Ritchie and S. Suresh, Metall. Trans., 1982, Vol. 13A, p. 937.

22. R. J. Asaro, L. Hermann, and J. M. Baik, Metall. Trans. A., 1981, Vol. 12A, No. 6, p. 1133.

23. K. Minakawa and A. J. McEvily, Scripta Met., 1981, Vol. 15, No. 6, p. 633.

24. A. J. McEvily, Met. Sci., 1977, p. 274.

25. S. D. Antolovich, R. Baur, and S. Liu, Superalloy 1980, Fourth Int. Symposium on Superalloys, Sept. 21-25, 1980, J. K. Tien et al., eds., ASM, 1980, p. 605.

26. P. E. Irving and C. J. Beevers, Mat. Sci. and Eng., 1974, Vol. 14, p. 229.

INFLUENCE OF TEMPERATURE AND LOAD RATIO ON NEAR-THRESHOLD FATIGUE CRACK GROWTH BEHAVIOR OF CrMoV STEEL

P. K. Liaw and A. Saxena
Metallurgy Department
Westinghouse R&D Center
Pittsburgh, PA 15235

V. P. Swaminathan and T. T. Shih
Westinghouse Steam-Turbine Generator Division
Orlando, Florida

The near-threshold fatigue crack growth rate behavior in a CrMoV steel was characterized in the temperature range of 24 to 427°C and at load ratios from 0.1 to 0.8. It was observed that increasing temperature from 24 to 149°C increased the near-threshold crack propagation rates. However, a further increase in temperature from 260 to 427°C decreased the near-threshold crack growth rates. The near-threshold fatigue crack propagation rates increased with an increase in load ratio. The observed temperature and load ratio effects on near-threshold crack growth characteristics were rationalized in terms of oxide and surface roughness-induced crack closure phenomena.

Introduction

In order to accurately predict the fatigue lives of components which are subjected to low-amplitude, high-frequency loading, it is essential to consider the near-threshold fatigue crack growth rate behavior in structural materials. Therefore, this topic has received considerable attention among researchers[1-14]. Most of these investigations, however, have been conducted at near ambient temperatures, and relatively little research has been performed to examine the effects of temperature on the near-threshold crack growth characteristics.

This paper describes the influence of temperature (24 to 427°C) and load ratio (0.1 to 0.8) on the near-threshold fatigue crack growth behavior in a CrMoV steel. It is a continuation of previously reported work in a limited temperature regime of 24 to 260°C[14]. Additionally, a thorough characterization of the fracture surfaces including fractography, oxide thickness and fracture surface roughness measurements were conducted. The results of fracture surface characterization coupled with crack closure measurements were utilized to rationalize the effects of temperature and load ratio on the near-threshold fatigue crack growth behavior.

Experimental Procedures

Material and Specimen

The material investigated was a forged ASTM A470 Class 8 CrMoV steel. The chemical composition in weight percent was 0.32C, 0.78Mn, 0.28Si, 1.20Cr, 1.18Mo, 0.23V, 0.012P, 0.011S, 0.13Ni, 0.05Cu, 0.005Al, 0.010Sn, 0.008As and balance Fe. The material was austenitized at 950°C, air cooled and tempered at 680°C. The microstructure was bainitic with an average prior austenite grain size of approximately 35 μm. The mechanical properties are presented in Table I.

Compact type (CT) specimens with H/W = 0.6 (where H and W were half-height and width of the specimen, respectively) were used to develop fatigue crack growth rate data. The thickness of the specimen, B, was 12.7 mm, and the value of W was 50.8 mm. Prior to testing, the CT specimens were precracked according to ASTM standard E647-81.

Table I

Mechanical Properties

Temperature (°C)	0.2% Offset Yield Strength (MPa)	Ultimate Tensile Strength (MPa)	Elongation* (%)	Reduction In Area (%)
24	623	776	14	39
427	516	625	14	53

*Using 5 cm gage length.

Near-Threshold Fatigue Crack Propagation Testing

The fatigue crack growth experiments were conducted using a computerized electrohydraulic fatigue machine the details of which have been reported earlier[15-17]. This system utilized the compliance method for determining crack length. The test frequency was 100 Hz with a sinusoidal waveform and the load ratios ($R = P_{min}/P_{max}$ where P_{min} and P_{max} were the minimum and maximum applied loads, respectively) were 0.1, 0.3, 0.5 and 0.8.

The near-threshold crack propagation tests were performed at four temperatures of 24, 149, 260 and 427°C in laboratory air. The relative humidity at 24°C was approximately 40%. Heating tapes were wrapped around the specimens to provide the desired temperatures of 149 and 260°C. The 427°C tests were conducted in a test chamber which was enclosed in a three-zone furnace. A thermocouple attached on the specimen near the crack path was used to monitor the temperature.

Crack closure loads were determined by the unloading compliance technique which was described elsewhere[14,18]. The crack closure level was defined at the point where the load vs. displacement curve began to deviate from the unloading linear elastic line.

Fractography

The fracture surfaces were carefully investigated using scanning electron microscopy (SEM). In specimens tested at the highest temperature of 427°C, heavy oxide deposits were visible on the fracture surfaces. An alkaline solution, ENDOX 214 commercially produced by Enthone Inc., was used to electrolytically clean the oxides on the fracture surfaces prior to the SEM investigation. The details of the cleaning process have been documented previously[19].

Oxide Thickness and Surface Roughness Measurements

The thickness of the oxide deposits on the fracture surfaces was determined using Auger spectroscopy or electron microprobe analysis. At 24, 149 and 260°C, oxide thicknesses were measured by Auger spectroscopy because the thicknesses were small. The experimental details of Auger spectroscopy have been described previously[12-14].

In the 427°C fatigue crack growth specimens, oxide layers were fairly thick. Auger spectroscopy was first attempted to determine oxide thickness, but it was deemed impractical because of a large amount of time needed per specimen. A simpler technique was found to be much more effective, which is described below. In this technique, the fracture surfaces were first nickel-plated. The specimens were then sectioned along the crack growth direction and carefully polished. Microprobe analysis was employed to map oxygen K_a X-rays, as shown in Fig. 1. The bright band in Fig. 1 indicated the presence of the oxide layer. The thickness of the oxide deposit was determined as the average value of the width of the oxygen map.

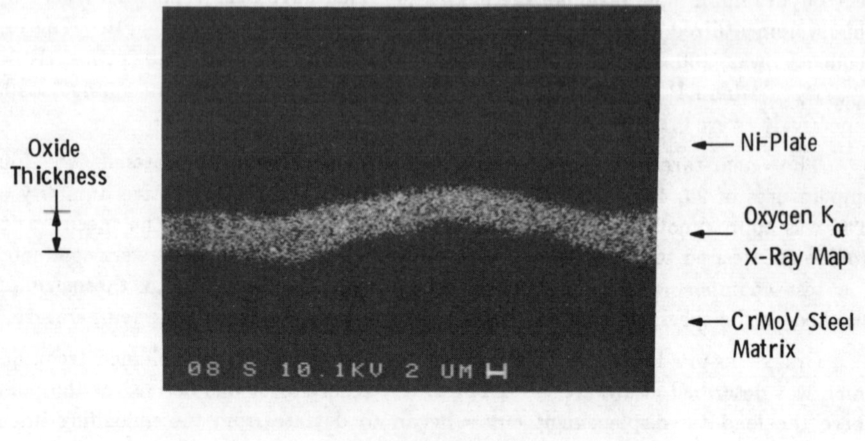

Fig. 1 - Oxygen mapping of nickel-plated fracture surface (using an electron microprobe)

The roughness of fracture surfaces was measured by a light-section microscope. The details of the measurements have been reported earlier[14]. Briefly, a narrow band of light in the microscope was projected on the fracture surface at an angle of 45 degrees and the image was observed on the other 45 degree plane. The image thus developed closely followed the profile of the fracture surface and enabled us to measure surface roughness. For determining quantitative roughness data corresponding to a given value of stress intensity range, ΔK, several photographs of roughness profiles were taken across specimen thickness. The value of roughness was defined as the average height of the roughness peaks on the various photographs.

Experimental Results

Effects of Load Ratio and Temperature

The influence of load ratio on near-threshold crack growth rates at 24, 260 and 427°C is presented in Figs. 2(a), (b) and (c), respectively. Increasing the load ratio from 0.1 to 0.8 generally increases the rates of crack propagation at all temperatures. The data at 149°C follow the same trend. Decreasing ΔK was found to increase the difference in the crack growth rates at various load ratios, as has been observed in several previous studies[1,2,8,10,12-14].

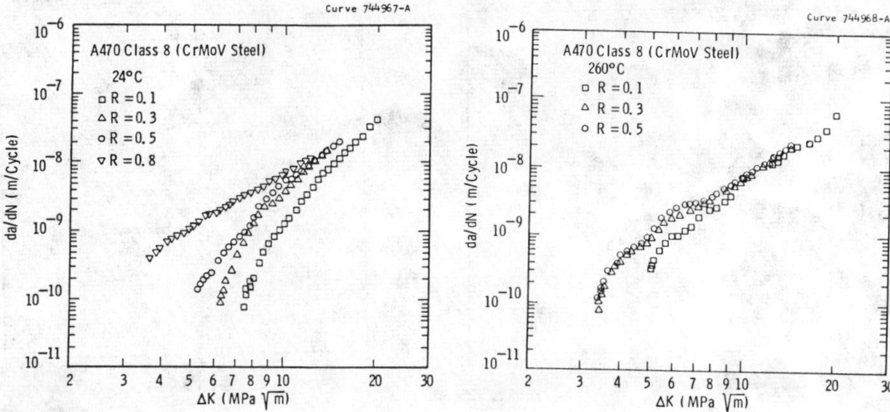

Fig. 2(a) - Effect of load ratio on fatigue crack growth rates at 24°C

Fig. 2(b) Effect of load ratio on fatigue growth rates at 260°C

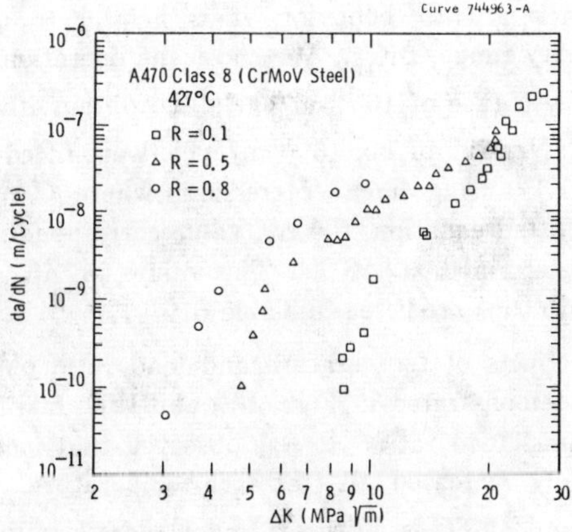

Fig. 2(c) - Effect of load ratio on fatigue crack growth rates at 427°C

The effect of temperature on crack propagation rates at R = 0.1 and 0.5 is exhibited in Figs. 3(a) and (b), respectively. The rates of crack propagation at 24°C are slower than those at 149 and 260°C. The crack growth rates at 149 and 260°C are essentially comparable. Surprisingly, increasing the temperature from 260 to 427°C significantly decreases the crack propagation rates in the near-threshold region both at R = 0.1 and at R = 0.5. It was also noted that the effect of temperature on near-threshold crack growth rates increased with decreasing ΔK. Comparing Figs. 3(a) and (b), it appears that increasing load ratio tends to decrease the influence of temperature on the crack growth rates.

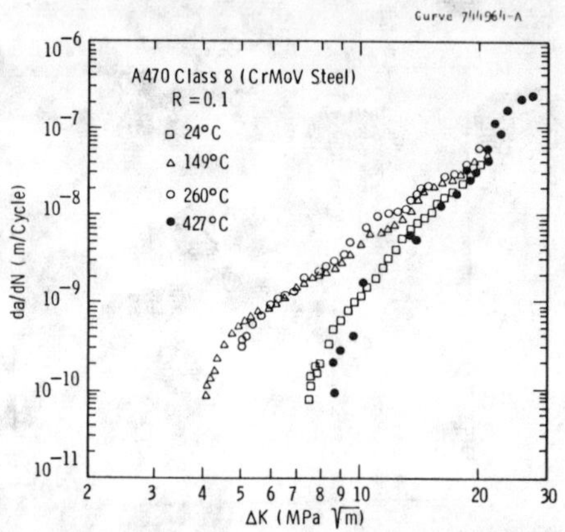

Fig. 3(a)- Effect of temperature on fatigue crack growth rates at R = 0.1

Fig. 3(b) - Effect of temperature on fatigue crack growth rates at R = 0.5

In order to quantify the effects of temperature and load ratio on the near-threshold fatigue crack growth behavior, it is helpful to define an operational threshold stress intensity range, ΔK_{th}. We chose the operational threshold to be the value of ΔK at a growth rate of 10^{-10} m/cycle. To obtain ΔK_{th}, the da/dN versus ΔK data below a da/dN value of 7 x 10^{-10} m/cycle were fitted to an equation of the type $da/dN = C(\Delta K)^m$ using linear regression, where C and m are the fitted constants. From the fitted equation, the ΔK value corresponding to a growth rate of 10^{-10} m/cycle was determined at ΔK_{th}. The values of ΔK_{th} obtained at various temperatures and load ratios are listed in Table II.

The combined effects of temperature and load ratio on near-threshold crack growth behavior are demonstrated in Figs. 4(a) and (b). In Fig. 4(a), the values of ΔK_{th} are plotted against load ratio. It was observed that increasing the load ratio from 0.1 to 0.8 generally decreased ΔK_{th} at each temperature. The influence of load ratio on ΔK_{th} at 24 and 427°C is, however, much more significant than that at 149 and 260°C. Moreover, increasing load ratio decreases the influence of temperature on ΔK_{th}. At a high load ratio of 0.8, the value of ΔK_{th} is essentially independent of test temperature.

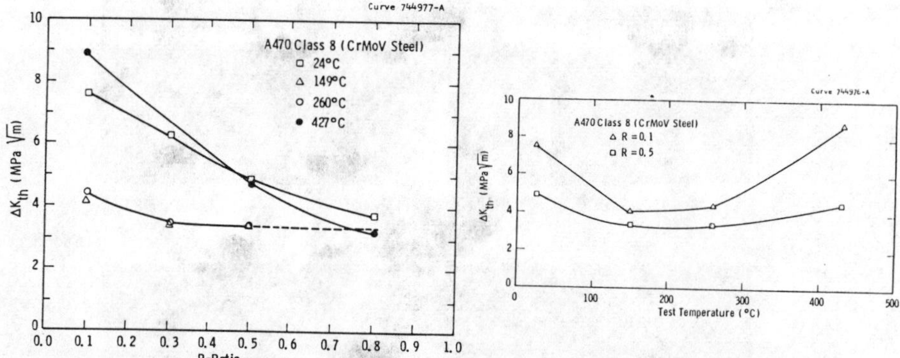

Fig. 4(a) - Effects of load ratio and temperature on ΔK_{th}

Fig. 4(b) - The relationship between ΔK_{th} and temperature

In Fig. 4(b), the value of ΔK_{th} is plotted versus temperature. At load ratios of 0.1 and 0.5, increasing the temperature from 24 to 149°C decreases ΔK_{th}. However, a further increase in temperature from 149 to 427°C increases ΔK_{th}. Thus, the value of ΔK_{th} in CrMoV steel goes through a minimum as the temperature varies between 24 and 427°C. Exactly the same trend was reported in two pressure vessel steels[20] and one nickel-base alloy[21]. It is also worth mentioning that in Fig. 4(b), the influence of temperature on ΔK_{th} at $R = 0.5$ appears to be less pronounced than that at $R = 0.1$.

Fractography

Figures 5(a)-(g) present the fracture modes at various temperatures. As shown in Figs. 5(a)-(d), at high ΔK levels, a transgranular fracture mode was observed at 24°C and as ΔK decreases, intergranular fracture begins to appear. The intergranularity reaches a maximum at some intermediate ΔK and then decreases as ΔK_{th} is approached. Near the threshold level, the fracture is a mixture of intergranular and transgranular modes at 24°C. Also at a fixed ΔK, the percentage of intergranular fracture generally increases with decreasing load ratio[14]. However, at 149, 260 and 427°C, the fracture mode is transgranular regardless of load ratios and ΔK values as shown in Figs. 5(c)-(g).

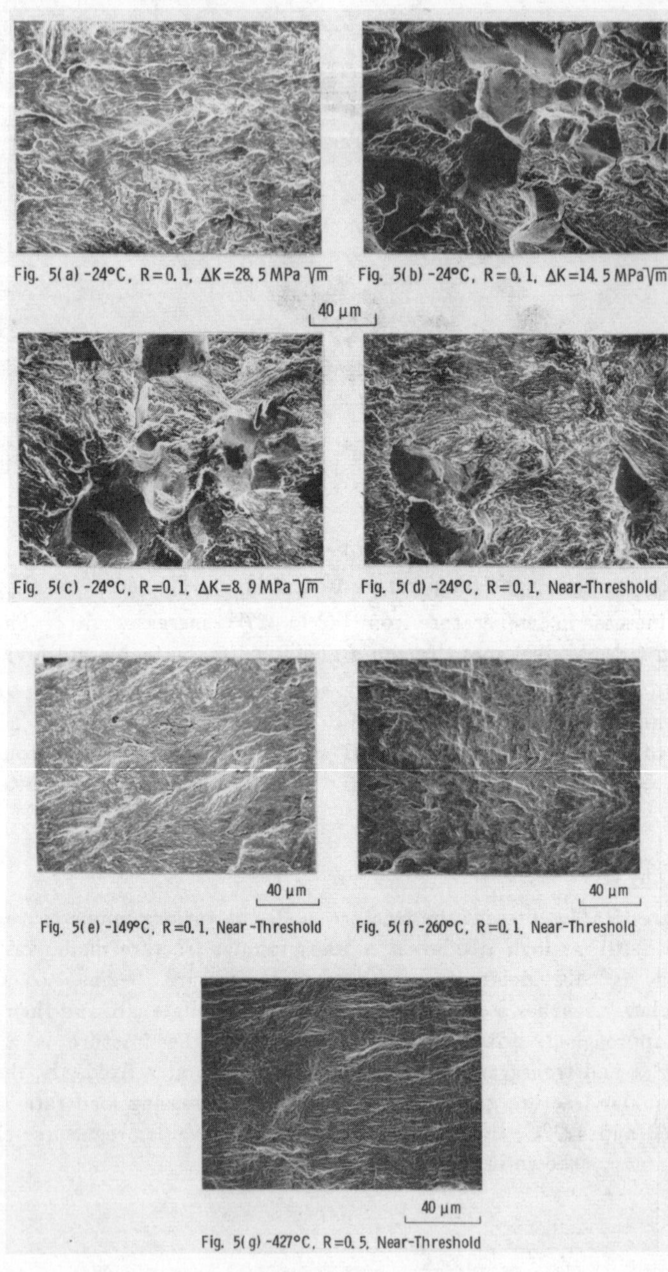

Fig. 5(a) -24°C, R=0.1, ΔK=28.5 MPa$\sqrt{m}$

Fig. 5(b) -24°C, R=0.1, ΔK=14.5 MPa$\sqrt{m}$

Fig. 5(c) -24°C, R=0.1, ΔK=8.9 MPa$\sqrt{m}$

Fig. 5(d) -24°C, R=0.1, Near-Threshold

Fig. 5(e) -149°C, R=0.1, Near-Threshold

Fig. 5(f) -260°C, R=0.1, Near-Threshold

Fig. 5(g) -427°C, R=0.5, Near-Threshold

Fig. 5 - Fracture Morphology

Crack Growth Rate Versus ΔK_{eff}

In Figs. 6(a), (b) and (c), the crack growth rates at various load ratios are plotted in terms of effective stress intensity range ($\Delta K_{eff} = K_{max} - K_{closure}$ where K_{max} and $K_{closure}$ are the maximum and the crack closure stress intensities) at 24, 260 and 427°C, respectively. It is interesting to note that the fatigue crack growth rate behavior at various load ratios is consolidated by using ΔK_{eff} at each temperature. Similar behavior was reported previously[12-14,22-26].

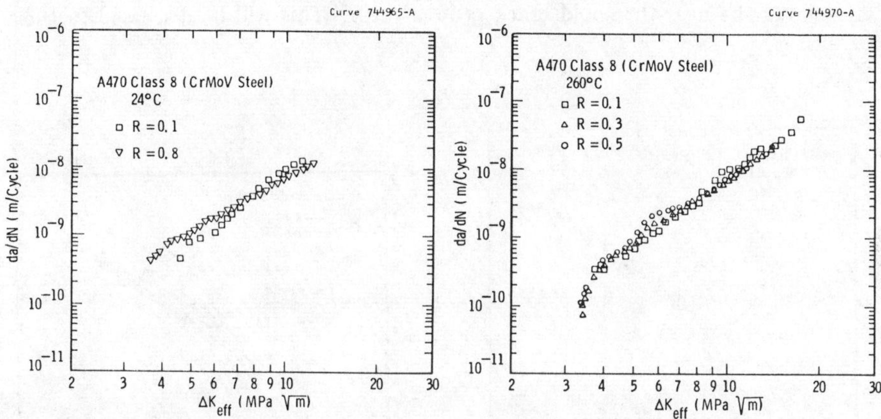

Fig. 6(a) - Plot of da/dN vs. ΔK_{eff} at 24°C

Fig. 6(b) - Plot of da/dN vs. ΔK_{eff} at 260°C

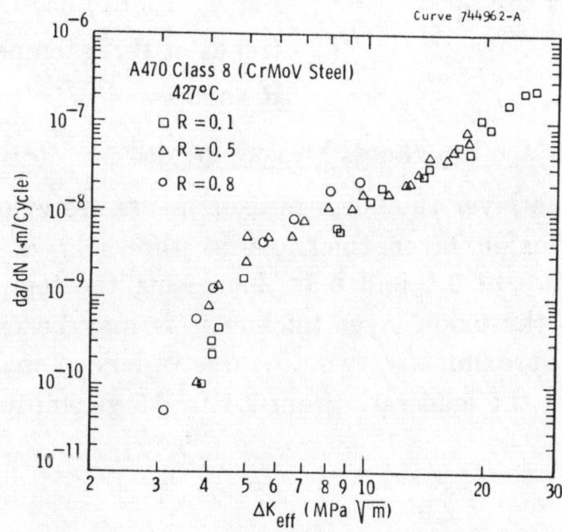

Fig. 6(c) - Plot of da/dN vs. ΔK_{eff} at 427°C

In Fig. 7, the crack propagation rates at three temperatures of 24, 260 and 427°C are presented as a function of ΔK_{eff} at a load ratio of 0.1. It was observed that the rather large effect of temperature on crack propagation rates [Fig. 3(a)] disappeared when the data were plotted with respect to ΔK_{eff}. The trend curves of da/dN vs. ΔK_{eff} at the various temperatures and load ratios investigated are plotted in Fig. 8. It is evident that all of the near-threshold crack growth rate results converge in a narrow band regardless of test temperatures and load ratios. Thus, the influence of load ratio and temperature on the rates of near-threshold crack propagation can be reconciled by ΔK_{eff}, which implies the dominance of crack closure in determining the near-threshold crack growth rates. This will be discussed further later.

Curve 744969-A
A470 Class 8 (CrMoV Steel)
R = 0.1
24°C
260°C
427°C
da/dN (m/Cycle)
ΔK_eff (MPa √m)

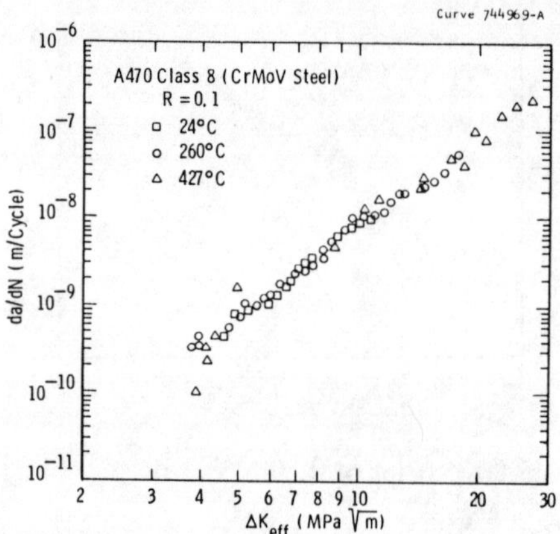

Fig. 7 - Plot of da/dN vs. ΔK_{eff} at three temperatures

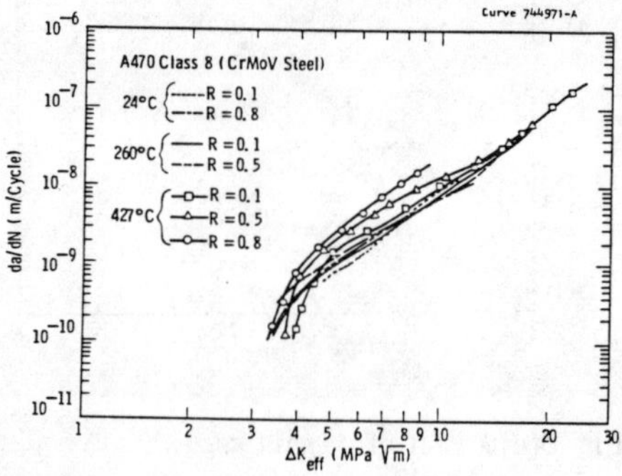

Curve 744971-A
A470 Class 8 (CrMoV Steel)
24°C
R = 0.1
R = 0.8
260°C
R = 0.1
R = 0.5
427°C
R = 0.1
R = 0.5
R = 0.8
da/dN (m/Cycle)
ΔK_eff (MPa √m)

Fig. 8 - Plot of da/dN vs. ΔK_{eff} trends at three temperatures and three R values

Oxide Thickness and Surface Roughness Measurements

The results of oxide layer thickness measurements are exhibited in Fig. 9 and Table II. In Fig. 9, oxide layer thickness at threshold is plotted versus test temperature at load ratios of 0.1 and 0.5. Increasing the temperature from 24 to 427°C greatly increases the oxide layer thickness, as may be expected. The oxide thickness at 427°C is approximately two to three orders of magnitude larger than that at 24°C. Increasing the load ratio from 0.1 to 0.5 generally decreases the oxide layer thickness.

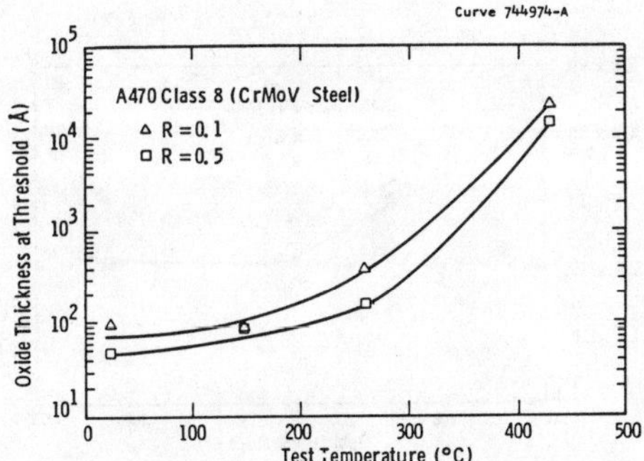

Fig. 9 - Effect of temperature on oxide thickness

TABLE II

VALUES OF THRESHOLD STRESS INTENSITY RANGE (ΔK_{th}), OXIDE THICKNESS, FRACTURE SURFACE ROUGHNESS AND CRACK TIP OPENING DISPLACEMENT AT THRESHOLD

Temp. (°C)	Load Ratio	ΔK_{th} (MPa$\sqrt{m}$)	Oxide Thickness (Å)	Surface Roughness (Å)	Cyclic Crack Tip Opening Displacement* (Å)	Maximum Crack Tip Opening Displacement** (Å)
24	0.1	7.6	96	39 x 10^4	1137	2807
24	0.3	6.3	25	43 x 10^4	769	3139
24	0.5	4.9	47	40 x 10^4	473	3784
24	0.8	3.6***	--	--	--	--
149	0.1	4.1	87	30 x 10^4	362	894
149	0.3	3.4	90	35 x 10^4	242	988
149	0.5	3.4	91	30 x 10^4	241	1928
260	0.1	4.4	400	19 x 10^4	418	1032
260	0.3	3.5	264	22 x 10^4	256	1045
260	0.5	3.4	160	22 x 10^4	249	1992
427	0.1	8.9	2.5 x 10^4	21 x 10^4	1715	4234
427	0.5	4.8	1.6 x 10^4	22 x 10^4	520	4160
427	0.8	3.2	0.7 x 10^4	22 x 10^4	240	12100

*Cyclic crack tip opening displacement (CTOD) = 0.49 $\frac{\Delta K_{th}^2}{2\sigma_y E}$ where σ_y is yield strength and E is Young's modulus[34].

**Maximum crack tip opening displacement (CTOD) = 0.49 $\frac{K_{th,max}^2}{\sigma_y E}$ where $K_{th,max}$ is the maximum stress intensity at threshold[34].

***The lowest ΔK value in Fig. 2(a).

The results of surface roughness measurements are presented in Fig. 10 and Table II. Increasing the temperature from 24 to 260°C decreases surface roughness at threshold for two load ratios of 0.1 and 0.5. Similar behavior was also observed in Ni-base alloys[27] and pressure vessel steels[28]. The surface roughness was found to be comparable at 260 and 427°C in Fig. 10. At a given temperature, surface roughness appears not to be affected by load ratio, as was previously found for pearlitic steels[29].

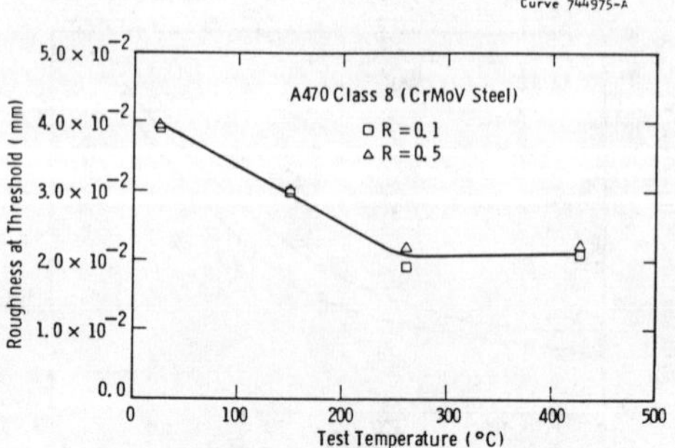

Fig. 10 - Effect of temperature on surface roughness

Discussion

The influence of temperature and load ratio on near-threshold fatigue crack growth behavior can be briefly summarized as follows. The values of ΔK_{th} decreases with an increase in temperature from 24 to 149°C, and then increases with further increasing temperature to 427°C. At all temperatures tested, increasing load ratio decreases the value of ΔK_{th}. The effects of temperature and load ratio on near-threshold crack propagation rates can be rationalized by crack closure. In the following discussion, the crack closure mechanisms governing near-threshold crack growth characteristics of CrMoV steel are examined in light of oxide thickness and surface roughness measurements.

At the threshold levels, crack tip opening displacements (CTOD) are small. Hence, oxide debris with a thickness comparable to CTOD or fracture surface roughness with a mis-match between the mating faces will cause early contact of fracture surfaces and raise the crack closure loads. Therefore, besides Elber's plasticity-induced crack closure[30], there are two other more important crack closure mechanisms; i.e., oxide and roughness-induced crack closure, which operate in the near-threshold region. A detailed discussion of these mechanisms of crack closure is documented elsewhere[2-5,11-14,27,31-32].

For the purpose of discussion, it is sufficient to recognize that the larger the oxide thickness or the rougher the fracture surface, the higher will be the crack closure level[2-6,12-14]. A higher crack closure level reduces the effective crack extension force, ΔK_{eff}, and, thus, decreases the crack growth rate. Near the threshold level, this behavior leads to an increased value of ΔK_{th}. It should be noted that the extent of oxide and roughness-induced crack closure is significant only at low load ratios ($\lesssim$ 0.5) since at high load ratios there exist high mean loads at the crack tip, which minimizes crack closure.

Effect of Temperature

In order to understand the influence of temperature on ΔK_{th}, it is important to consider the variations in surface roughness and oxide layer thickness as a function of temperature. Also, comparisons should be made between the oxide layer thickness, and the cyclic and the maximum levels of CTOD at threshold. The following two equations are used to estimate CTOD[34].

$$CTOD_{cyclic} = 0.49\ \Delta K_{th}^2/(2\sigma_y E)$$

$$CTOD_{max} = 0.49\ K_{th,max}^2/(\sigma_y E)$$

where $K_{th,max}$ is the maximum stress intensity at threshold. σ_y is yield strength and E is Young's modulus. The values of CTOD are listed in Table II for comparison with oxide thickness and surface roughness.

At 24 and 149°C, the oxide layers are considerably thin, ranging from 25 to 95A which are approximately one order of magnitude smaller than the cyclic and the maximum CTOD. Therefore, the oxide layer is not likely to cause significant wedging to affect the crack closure loads. In contrast, surface roughness is nearly four orders of magnitude larger than oxide thickness. Consequently, it is suggested that in the temperature range of 24 to 149°C, roughness-induced crack closure is dominant in influencing near-threshold crack growth behavior[14]. The quantitative relationship between surface roughness and crack closure loads, however, is not clear since roughness-induced crack closure is also related to mode II displacements[3,14,33] which are difficult to measure. Therefore, the surface roughness numbers shown in Table II can only be treated in a qualitative sense.

Above 149°C, oxide layer thickness increases rapidly, while surface roughness continues to decrease with increasing temperature to 260°C beyond which it remains constant [Figs. 9 and 10, and Table II]. It should be noted that at load ratios of 0.1 and 0.5, oxide thicknesses at 260°C are in the order of CTOD and those at 427°C are nearly one order of magnitude larger than CTOD. Thus, oxide-induced crack closure is expected to be important at 260 and 427°C.

Values of oxide thickness, surface roughness and ΔK_{th} are normalized relative to their respective data at 149°C so that they can be shown on the same plot. Figures 11 and 12 present these plots at R = 0.1 and 0.5, respectively.

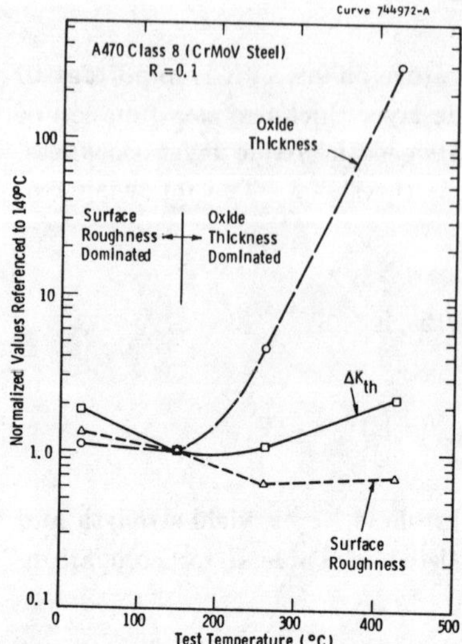

Fig. 11 - Variation in oxide layer thickness, surface roughness and ΔK_{th} as a function of temperature at R = 0.1. All values have been normalized to their respective values at 149°C

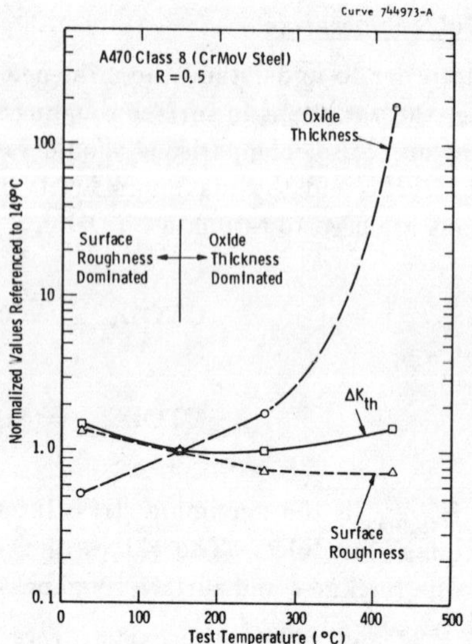

Fig. 12 - Variation in oxide layer thickness, surface roughness and ΔK_{th} as a function of temperature at R = 0.5. All values have been normalized to their respective values at 149°C

As mentioned before, we postulate that the crack closure level in the temperature range of 24 to 149°C is only dominated by surface roughness, Figs. 11 and 12. A decrease in roughness-induced crack closure with increasing the temperature from 24 to 149°C will be accompanied by an increase in the near-threshold crack growth rate and a decrease in ΔK_{th}, which agrees with the results. Thus, the decrease in ΔK_{th} from 24 to 149°C can be rationalized by roughness-induced crack closure[14]. Similar behavior was observed in Inconel 718[27]. It was stated earlier that at 24°C, the fracture mode was partially intergranular at threshold while it was completely transgranular at 149°C (Fig. 5). The presence of intergranular facets at 24°C may have caused the fracture surface to be rougher at 24°C than at 149°C[14].

An increase in temperature beyond 149°C results in extensive thermal oxidation and a larger oxide layer, which in turn promotes significant oxide-induced crack closure. As can be seen in Figs. 11 and 12, the increase in oxide thickness with temperature at threshold is very rapid above 149°C. We postulate that the increase in oxide layer thickness is sufficient to overwhelm the influence of slight decrease in surface roughness with increasing temperature from 149 to 260°C. This behavior results in a larger ΔK_{th} at 260°C than at 149°C. Above 260°C, surface roughness remains the same, but oxide layer thickness continues to increase sharply, thereby causing a greater extent of oxide-induced crack closure and resulting in a larger ΔK_{th}

value at 427°C than at 260°C.

Summarizing the above discussion, two temperature regimes can be identified in Figs 11 and 12. In the first temperature regime of 24 to 149°C, ΔK_{th} is a function of fracture surface roughness which can be highly microstructure-sensitive. In the second temperature regime of 149 to 427°C, ΔK_{th} is controlled by oxide thickness. Thus, oxide and/or roughness-induced crack closure mechanisms provide an explanation for the influence of temperature on the near-threshold crack growth behavior in CrMoV steel.

In Figs. 3(a), 3(b), 4(a), and 4(b), increasing load ratio tends to reduce the temperature effect on near-threshold crack propagation behavior. In pressure vessel steels, Paris, et al.[20] observed that at load ratio of 0.1 and in the temperature range of 24 to 343°C, the value of ΔK_{th} was a minimum at 149°C, and ΔK_{th} increased with decreasing temperature from 149 to 24°C or with increasing temperature from 149 to 343°C, which is consistent with the present results at load ratio of 0.1 (Fig. 4(b)). Nevertheless, at a high load ratio of 0.7, there is essentially no temperature effect on ΔK_{th}[20]. A similar trend was reported in Ni-base alloys[27]. Therefore, increasing load ratio decreases the temperature effect on near-threshold crack growth performance. This behavior is readily explained by the crack closure concept. The extent of oxide and roughness-induced crack closure decreases with increasing load ratio as reported earlier. At higher load ratios, the reduced extent of crack closure gives a decreased effect of temperature on the rates of near-threshold crack propagation.

Effect of Load Ratio

At temperatures of 24 and 149°C, surface roughness is insensitive to load ratio [Fig. 10 and Table II]. The degree of roughness-induced crack closure has been shown to be proportional to roughness and mode II displacement[3-5]. Increasing load ratio decreases the mode II displacement at threshold[33]. Moreover, increasing load ratio increases mean load at the crack tip, which reduces crack closure at higher load ratios, as mentioned before. Thus, it is expected that decreasing load ratio increases the extent of roughness-induced crack closure, decreases ΔK_{eff} and thus the crack growth rates, as compared to the results at higher load ratios. Figure 4(a) illustrates a greater effect of load ratio on ΔK_{th} at 149°C than at 24°C; this is in agreement with the decreased roughness-induced crack closure at 149°C (Fig. 10).

In Table II and Fig. 9, at 260 and 427°C oxide thickness increases with decreasing load ratio. Thus, at these two temperatures, decreasing load ratio increases oxide-induced crack closure and decreases ΔK_{eff}, thereby resulting in slower crack propagation rates and a larger ΔK_{th} at lower load ratios [Fig. 2(c) and Table II]. It should be noted that roughness-induced crack closure may also contribute to the load ratio effect at 260 and 427°C. Thus, oxide and/or roughness-induced crack closure provides a rationale to account for the effect of load ratio on near-threshold crack growth rates in CrMoV steel at the temperatures investigated.

Conclusions

1. In CrMoV steel, increasing the load ratio from 0.1 to 0.8 decreases the resistance to near-threshold fatigue crack propagation in the temperature regime of 24 to 427°C.

2. At load ratios of 0.1 and 0.5, increasing temperature from 24 to 149°C decreases the value of threshold stress intensity range, ΔK_{th}. However, a further increase in temperature from 149 to 427°C increases ΔK_{th}. The influence of temperature on ΔK_{th} tends to decrease with increasing load ratio.

3. At threshold, increasing temperature from 24 to 427°C generally decreases surface roughness but increases oxide thickness.

4. The phenomenon of crack closure can be used to rationalize the effects of temperature and load ratio on near-threshold crack growth behavior. The decreasing roughness-induced crack closure with increasing temperature from 24 to 149°C has been correlated with decreasing ΔK_{th}. The increasing oxide-induced crack closure with increasing temperature from 149 to 427°C has been correlated with increasing ΔK_{th}.

5. At 24°C, fracture morphology is a mixture of intergranular and transgranular fracture. The amount of intergranular fracture increases with increasing ΔK, reaches a maximum and then decreases. At 149, 260 and 427°C, the fracture mode is transgranular regardless of ΔK values and load ratios.

Acknowledgments

The authors wish to thank Dr. J. D. Landes and Dr. J. Schreurs for the review of the manuscript. We are grateful to D. Detar, A. Karanovich, R. W. Palmquist and P. M. Yuzawich for performing fracture surface characterization, and P. Barsotti, M. G. Peck, R. Brown and F. X. Gradich for conducting the fatigue crack growth experiments. The Mechanical Behavior Laboratory for crack growth testing is under the direction of R. E. Gainer. This work is suppported by Westinghouse Steam Turbine-Generator Division.

References

1. R. O. Ritchie, Int. Metals Rev., 1979, Nos. 5 and 6, p. 205.

2. S. Suresh, G. F. Zamiski and R. O. Ritchie, Met. Trans. A, 1981, Vol. 12A, p. 1435.

3. S. Suresh and R. O. Richie, Met. Trans. A, 1982, Vol. 13A, p. 1627.

4. K. Minakawa and A. J. McEvily, Scripta Met., 1981, Vol. 15, p. 633.

5. K. Minakawa, Y. Matsuo and A. J. McEvily, Met. Trans. A, 1982, Vol. 13A, p. 439.

6. G. T. Gray, J. C. Williams and A. W. Thompson, Met. Trans. A, 1983, Vol. 14A, p. 421.

7. J. Lankford and D. L. Davidson, Met. Trans. A, 1983, Vol. 14A, p. 1227.

8. W. W. Gerberich, W. Yu and K. A. Esaklul, "Fatigue Threshold Studies in Fe, Fe-Si and HSLA Steel - Part I: Effect of Strength and Surface Asperities on Closure", submitted to Met. Trans..

9. G. M. Lin and M. E. Fine, Scripta Met., 1982, Vol. 16, p. 1249.

10. R. J. Bucci, W. G. Clark and P. C. Paris, ASTM STP 513, 1972, p. 177.

11. N. Walker and C. J. Beevers, Fat. Eng. Mat. Struct., 1979, Vol. 1, p. 135.

12. P. K. Liaw, T. R. Leax, R. S. Williams and M. G. Peck, Met. Trans. A, 1982, Vol. 13A, p. 1607.

13. P. K. Liaw, T. R. Leax, R. S. Williams and M. G. Peck, Acta Met., 1982,Vol. 30, p. 2071.

14. P. K. Liaw, A. Saxena, V. P. Swaminathan and T. T. Shih, Met. Trans. A, 1983, Vol. 14A, p. 1631.

15. A. Saxena, S. J. Hudak, Jr., J. K. Donald and D. W. Schmidt, J. Test. Eval., 1978, Vol. 6, p. 167.

16. R. S. Williams, P. K. Liaw, M. G. Peck and T. R. Leax, Eng. Fract. Mech., 1983, Vol. 18, p. 953.

17. R. C. Brown and N. E. Dowling, ASTM STP 738, 1981, p. 58.

18. M. Kikukawa, M. Jona and K. Tanaka, Proceedings of the Second International Conf. on Mechanical Behavior of Materials, N. Promisel and V. Weiss, eds., Boston, MA, ASM, Metals Park, OH, 1976, p. 716.

19. P. M. Yuzawich and C. W. Hughes, Practical Metallography, 1978, Vol. 15, p. 184.

20. P. C. Paris, R. J. Bucci, E. T. Wessel, W. G. Clark and T. R. Mager, ASTM STP 513, 1972, p. 141.

21. J. L. Yuen, Ph.D. Thesis, Stanford University, Palo Alto, CA, 1982.

22. Y. Nakai, K. Tanaka and T. Nakaniski, Eng. Fract. Mech., 1981, Vol. 15, p. 291.

23. P. K. Liaw, S. J. Hudak, Jr. and J. K. Donald, Met. Trans. A, 1982, Vol. 13A, p. 1633.

24. R. A. Schmidt and P. C. Paris, ASTM STP 536, 1973, p. 79.

25. J. A. Vazquez, A. Morrone and H. Ernst, Eng. Fract. Mech., 1979, Vol. 12, p. 231.

26. J. A. Lewis, Fatigue Thresholds, Proceedings of First International Conf., Stockholm, J. Backlund, A. Blom and C. J. Beevers, eds., EMAS Publ. Ltd., Warley, U.K., 1982, p. 1127.

27. M. A. Hicks and J. E. King, Int. J. Fatigue, 1983, Vol. 5, p. 67.

28. P. K. Liaw and W. A. Logsdon, "The Influence of Load Ratio and Temperature on the Near-Threshold Fatigue Crack Growth Rate Behavior of Pressure Vessel Steels", to be published in J. Eng. Mat. and Tech..

29. G. T. Gray, Ph.D. Thesis, Carnegie-Mellon University, Pgh., PA, 1981.

30. W. Elber, ASTM STP 486, 1971, p. 230.

31. D. Benoit, R. Namdar-Tixier and R. Tixier, Mat. Sci. Eng., 1980, Vol. 45, p. 1.

32. P. K. Liaw, W. A. Logsdon and T. R. Leax, Acta Met., 1983, Vol. 31, p. 1581.

33. D. L. Davidson, Fatigue Eng. Mat. Struct., 1980, Vol. 3, p. 229.

34. D. M. Tracey, Engineering Materials and Technology, Trans. ASME, Series H, 1976, Vol. 98, p. 146.

Crack Closure

NEAR-THRESHOLD FATIGUE CRACK PROPAGATION: A PERSPECTIVE ON THE ROLE OF CRACK CLOSURE

S. Suresh[1] and R. O. Ritchie[2]

[1]Division of Engineering, Brown University, Providence, RI 02912, USA

[2]Materials and Molecular Research Division, Lawrence Berkeley Laboratory, and Department of Materials Science and Mineral Engineering University of California, Berkeley, CA 94720, USA

In recent years, mechanistic and continuum studies on fatigue crack propagation, particularly at near-threshold levels, have highlighted a dominant role of crack closure in influencing growth rate behavior. In this paper we review and model the various sources of closure induced by cyclic plasticity, corrosion deposits, irregular fracture morphologies, viscous fluids and metallurgical phase transformations. It is shown that many of the commonly observed effects of mechanical factors, such as load ratio, microstructural factors, such as strength and grain size, and certain environmental conditions can be traced to the extrinsic influence of closure in modyfying the effective driving force for crack extension. The implications of such closure mechanisms are discussed in the light of constant and variable amplitude fatigue behavior, the existence of a threshold stress intensity for no fatigue crack growth and the validity of such threshold concepts for the case of short fatigue cracks.

Introduction

The elastic-plastic models of McClintock (1), presented twenty years ago, postulated the existence of a propagation threshold for long fatigue cracks. The condition that fatigue cracks cease to propagate when the extent of cyclic plastic zone becomes comparable to some characteristic microstructural size-scale (ρ), was derived from failure criteria based on the attainment of a critical local strain or cumulative damage over the characteristic dimension ρ ahead of the crack tip (1). In 1966, Frost and co-workers (2) reported experimental evidence supporting the existence of a fatigue threshold when the value of the parameter ($\sigma_a{}^3 a$), where σ_a is the alternating stress and a the crack length, reached a critical value. With the advent of linear elastic fracture mechanics and its adoption to characterize the propagation of fatigue cracks, it soon became apparent, particularly from the work of Paris and co-workers (3), that the threshold for non-propagation of fatigue cracks can be denoted by a critical value of the stress intensity range ΔK_0, below which crack growth is not experimentally detectable.

The application of such fracture mechanics concepts to fatigue crack propagation, as originally postulated by Paris and co-workers (4), is based on the implicit assumption that the **nominal** value of crack driving force, taken for the case of linear elastic behavior as the range of stress intensity factor ΔK, uniquely and autonomously characterizes the crack tip stress and strain fields. Although this is clearly a simplification, since the analysis is based on monotonic loading of a stationary crack, using this linear elastic fracture mechanics (LEFM) approach the value of the alternating stress intensity ΔK can then be taken as a correlator for fatigue crack extension such that:

$$\frac{da}{dN} = C(\Delta K)^m \quad , \tag{1}$$

where (da/dN) is the crack propagation rate per cycle, and C and m are scaling constants (4). Experimental studies of fatigue crack propagation in engineering materials, however, have begun to reveal that the premise on which Eq. (1) is based may lead to overestimates of crack advance because of a phenomenon which has become known as crack closure.

Early observations of crack closure can be traced back to the work of Endo et al. (5) who suggested in 1969 that the slower crack growth rates of a low strength steel in water compared to air could be attributed to crack surface corrosion deposits in the former medium which prematurely closed the crack. Endo et al.'s results (5) were presented in terms of conventional stress-life approach and LEFM terminology was not used. The first mechanistic justification for crack closure in terms of fracture mechanics concepts was presented by Elber (6) in 1971. Based on compliance measurements of fatigue cracks at high ΔK levels, Elber (6) proposed that premature contact between the crack faces can occur even during the tensile portion of the fatigue cycle because of the permanent residual displacements, arising from prior plastic zones, left in the wake of a growing fatigue crack. Since the crack cannot propagate whilst it remains closed, the consequence of such plasticity-induced crack closure (6) is to modify the actual stress intensity factor range experienced by the crack tip from a nominal value of:

$$\Delta K = K_{max} - K_{min} \quad , \tag{2}$$

based on global measurements of geometry, applied loads and crack size to some near tip effective value of:

$$\Delta K_{eff} = K_{max} - K_{cl} \quad . \tag{3}$$

Here, K_{max} and K_{min} are the maximum and minimum stress intensity factors, respectively, during the fatigue cycle and K_{cl} is the stress intensity factor at which the two fracture surfaces first come into contact during the unloading portion of the fatigue cycle. It was postulated by Elber (6) that fatigue crack propagation rate is then governed by the effective value of the stress intensity factor range ΔK_{eff}, such that:

$$\frac{da}{dN} = C(\Delta K_{eff})^m = C(U\Delta K)^m \quad , \tag{4}$$

where U is an empirically-measured value defined as $\Delta K_{eff}/\Delta K$.

Since Elber's pioneering work, the role of crack closure in influencing fatigue crack growth has remained a topic of considerable research interest and practical significance, although it has not been until fairly recently that microscopic descriptions of closure have been developed. Despite the application of crack closure arguments to rationalize a number of fatigue characteristics in engineering materials, several independent experimental studies have revealed convincing evidence that the **plasticity-induced** crack closure phenomenon plays a role principally under **plane stress** loading conditions (e.g., refs. 7 and 8). Yet, there has long been considerable information in the literature indicating that marked crack closure can occur even in the near-threshold regime where predominantly **plane strain** conditions exist (e.g., ref. 9). Such apparently anomalous and conflicting experimental data led to the conclusion by Ritchie (10) that plasticity-induced crack closure must be questioned as an important mechanism for near-threshold fatigue crack growth. More recently, finite element studies by Newman (11) showed that plasticity-induced closure, modelled for plane strain conditions, was inadequate to explain the marked role of load ratio for fatigue crack growth at near-threshold levels. The reasons for this apparent paradox have only recently become evident from the observations that Elber's mechanism involving cyclic plasticity is not the sole source of closure. Numerous independent experimental studies have clearly identified several other sources of crack closure, which can become of major importance particularly at near-threshold growth rates. These include, closure arising from i) corrosion deposits formed within the crack (oxide-induced crack closure), ii) irregular crack surface morphologies (roughness-induced crack closure), iii) viscous media penetrated inside the crack (viscous fluid-induced crack closure) and iv) phase transformations due to crack-tip plasticity (transformation-induced crack closure), as illustrated schematically in Figure 1.

In this paper, we summarize, in as quantitative terms as possible, the various micro-mechanisms underlying different crack closure processes and assess the significance of such mechanisms to the propagation of near-threshold fatigue cracks. The implications of crack closure phenomena are also discussed in the light of various aspects of fatigue behavior involving short cracks, variable amplitude loading, as well as the question of the measurement and very existence of a fatigue threshold.

Description of Closure Mechanisms

Plasticity-Induced Crack Closure

Plasticity-induced closure (Fig. 1a) is considered to arise from the elastic constraint in the wake of the crack tip of material elements permanently stretched within prior plastic zones, leading to an interference between mating crack surfaces (6). Based on experimental compliance measurements from fatigue tests conducted at growth rates above 10^{-6} mm/cycle in 2024-T3 aluminum alloy, Elber (6) proposed the following **empirical** relationship between the ratio of the closure stress intensity to maximum stress intensity (K_{cl}/K_{max}) and load ratio, R:

$$\frac{K_{cl}}{K_{max}} = 0.5 + 0.1R + 0.4R^2 \quad . \tag{5}$$

It should be noted, however, that Eq. (5) pertains to crack growth at stress intensities much greater than ΔK_0, where deformation conditions approach

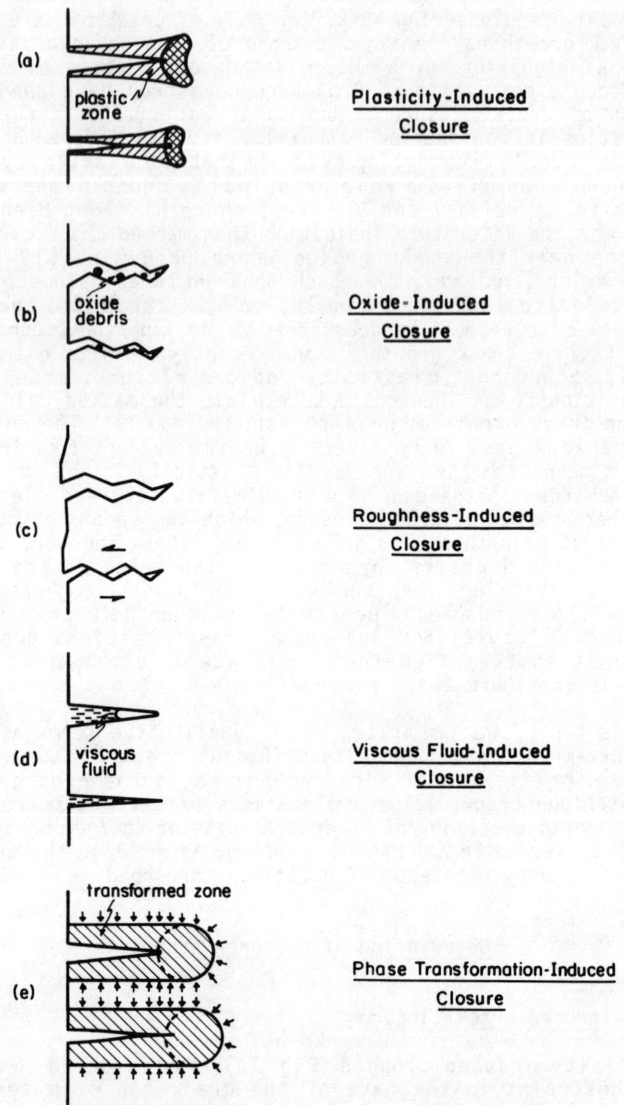

Figure 1 - Schematic illustration of the various mechanisms of fatigue crack closure.

that of plane stress, especially at the higher load ratio values. However, since the mechanism results from the constraint of surrounding elastic material on the plastic zone enclave surrounding the crack, such closure will become insignificant at very high stress intensity levels once the material becomes fully plastic. Based on extensive crack closure measurements in aluminum alloys, Schijve (12) modified Elber's empirical model to account for load ratio effects at the higher ΔK values by incorporating higher order terms in R such that:

$$\frac{K_{cl}}{K_{max}} = 0.45 + 0.2R + 0.25R^2 + 0.1R^3 \quad . \tag{6}$$

For such plane stress crack growth, physically-based analytical models have been developed by Budiansky and Hutchinson (13) and by Kanninen and Atkinson (14), whereas Newman (15,16) has developed numerical models involving finite-element analyses of crack tip deformation to represent plasticity-induced closure under both plane stress and plane strain conditions.

Despite the lack of convincing analytical formulations for plane strain closure, Schmidt and Paris (9) used crack closure concepts* to account for the dependence of ΔK_o on R (Fig. 2). The variation of threshold ΔK_o with R was found to show two distinct regimes when the following assumptions were considered: a) the closure stress intensity K_{cl} at the threshold is independent of R and b) a constant **effective** threshold stress intensity range ΔK_{th} is necessary to produce fatigue crack growth. As illustrated schematically in Figure 2, and by experimental results on 2 1/4Cr-1Mo steels in Figure 3, at low load ratios where $K_{min} \leq K_{cl}$, the maximum stress intensity at the threshold ($K_{o,max}$) appears constant, since:

$$K_{o,max} = K_{cl} + \Delta K_{th} \quad , \qquad K_{min} \leq K_{cl} \quad , \tag{7}$$

whereas at high R values where $K_{min} \geq K_{cl}$, the threshold stress intensity range (ΔK_o) appears constant, since:

$$\Delta K_o = \Delta K_{th} \quad , \qquad K_{min} \geq K_{cl} \quad . \tag{8}$$

Such results imply that there exists a critical value of the load ratio, R_{cr}, above which crack closure effects become insignificant. In Figure 2,

$$R = R_{cr} \text{ when } K_{min} = K_{cl}, \text{ such that } \Delta K_{th} = \Delta K_o \quad . \tag{9}$$

The nature of load ratio dependence of ΔK_o, as predicted by the Schmidt and Paris model (9), has been found consistent with the experimental results obtained for aluminum alloys (9) and steels (17,18), as shown in Figure 3. The model, however, does not account for the significant effects of environmental and microstructural factors on near-threshold crack propagation rates and on the value of ΔK_o, as indicated in Figure 3 by the difference in behavior between moist air and dry hydrogen. Moreover, there are existing experimental data, notably those by Lindley and Richards (7),

*The exact nature of the crack closure process was not defined (9), although it was presumably related to plasticity-induced closure.

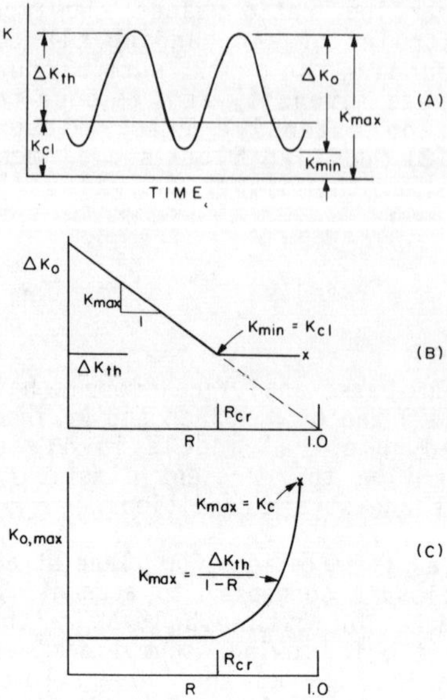

Figure 2 - Schematic representation of the model of Schmidt and Paris (9) for the dependence of fatigue thresholds on load ratio R, showing a) definition of the nominal and effective stress intensity ranges, ΔK_o and ΔK_{th}, respectively, b) the predicted variation in ΔK_o with R and c) the predicted variation of the (nominal) maximum stress intensity at threshold $K_{o,max}$ with R. K_{cl} and K_c are the stress intensities at closure and at final failure, respectively, R_{cr} is the critical load ratio above which closure effects are minimal.

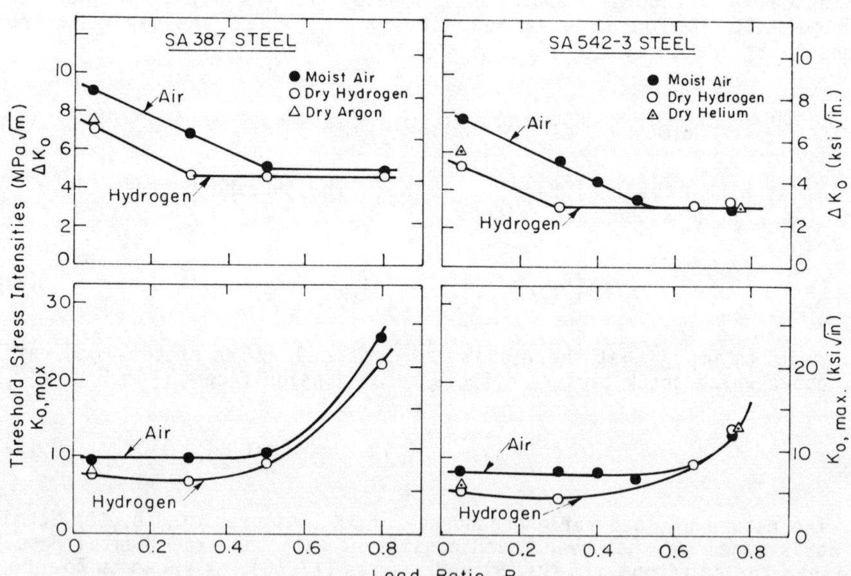

Figure 3 - Experimentally-measured variation of alternating and maximum threshold stress intensities (ΔK_o and $K_{o,max}$, respectively) with load ratio (R) in two 2¼Cr-1Mo steels, SA387 and SA542-3 (σ_y = 290 and 500 MPa, respectively). Tests in room temperature environments of moist air and dehumidified hydrogen gas. Sample points shown for dry inert gas atmospheres of argon and helium. Data from refs. 17 and 25.

which indicate that plasticity-induced closure plays a far less significant role in influencing fatigue crack propagation rates under plane strain conditions, such as those at near-threshold stress intensities. Thus, it is clear that the effects of various mechanical, environmental and microstructural variables on near-threshold crack growth rates cannot be explained solely on the basis of plasticity-induced crack closure and that other sources of crack closure need to be taken into consideration.

Oxide-Induced Crack Closure

It has long been realized that corrosion products formed within growing cracks (Fig. 4) can significantly influence their subsequent crack extension rates. Endo et al. (5) published early observations of slower crack growth in aqueous media than in gaseous environments and attributed the effect to the presence of oxide layers between the fracture surfaces, although their results were presented only in terms of crack length vs. number of cycles data. A number of independent studies, notably those by Paris et al. (19), Lindley (20), Tu and Seth (21), Skelton and Haigh (22), Kitagawa and co-workers (23), Stewart (24) and Suresh and Ritchie (17,25,26), have, since then, suggested the possibility of crack closure due to corrosion debris influencing near-threshold growth rates and threshold ΔK_0 values. Suresh, Ritchie and co-workers (25) reported the first experimental attempts to **quantify** the effects of corrosion deposits on fatigue crack propagation rates, especially in the near-threshold regime where the extent of oxide formation within the crack (estimated with the aid of scanning Auger spectroscopy) was found to be comparable to the scale of crack tip opening displacements.

The concept of oxide-induced crack closure (Fig. 1b) (17,24,25) is based on the phenomenon that at low load ratios, near-threshold growth rates are significantly reduced in moist environments (such as air or water), compared to dry environments (such as hydrogen or helium gas), due to the presence of corrosion deposits on crack faces. Moist atmospheres lead to the formation of oxide layers within the crack, which are thickened at low load ratios by "fretting oxidation," i.e., a continual breaking and reforming of the oxide scale behind the crack tip due to the repeated contact between the fracture surfaces arising from Mode II displacements and plasticity-induced closure (17,25,26). Such oxide debris, whose formation is limited in dry, moisture-free environments or at high load ratios (where there is no plasticity-induced closure), provides a mechanism for enhanced closure by resulting in an earlier contact between the fracture surfaces through an increased closure stress intensity K_{cl}. The formation of corrosion deposits and the process of oxide-induced crack closure are promoted by a) small crack tip opening displacements (CTOD), such as in the near-threshold regime, which are comparable to the thickness of the **excess** debris within the crack (25), b) highly oxidizing media (such as water), where substantial thermal oxidation is possible even in the absence of fretting (18), c) low load ratios, which facilitate repeated contact (and hence fretting) between the fracture surfaces through small CTOD values (17,24-26), d) rough fracture surfaces, which at low ΔK values, promote relative sliding and rubbing between mating crack faces (25) and e) lower strength materials, where the plasticity-induced closure mechanism (and hence fretting) is dominant and the base metal is soft enough to undergo considerable fretting damage (17,24-26).

The concept of oxide-induced crack closure has been successfully employed by a number of investigators to rationalize, either wholly or partially, the near-threshold corrosion fatigue characteristics in a wide

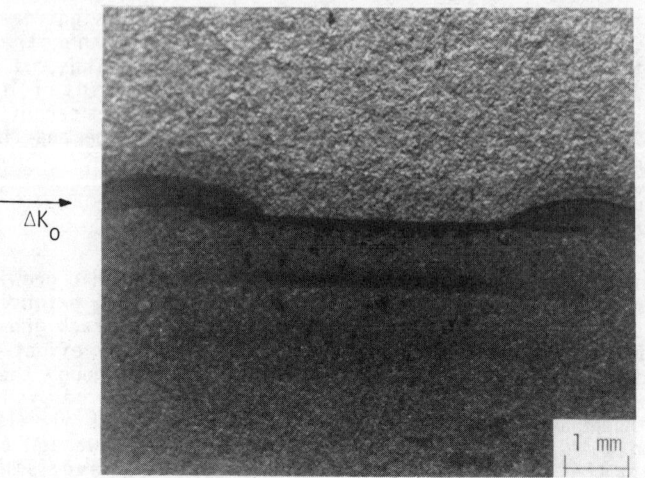

Figure 4 - Macrograph of a fatigue fracture surface in X-60 HSLA steel (σ_y = 414 MPa) showing accumulation of oxide debris near ΔK_o at the center of the test piece. Note how the crack, which is fully arrested at the center, is still trying to propagate at the surfaces where the accumulation of oxide is less. The crack is propagating from bottom to top. After ref. 18.

range of engineering materials comprising ferritic-pearlitic (26), ferritic-bainitic (17), fully bainitic (25) and fully martensitic (26,27) steels, as well as nickel-base superalloys (28,29), copper (30) and aluminum alloys (31). In steels, the influence of oxide-induced closure is found to be inversely related to strength level. In lower strength steels (yield strength $\lesssim$ 1000 MPa), closure mechanisms appear to dominate over conventional corrosion fatigue processes, such as hydrogen embrittlement or anodic dissolution, in influencing **near-threshold** environmental behavior (25,26), whereas in ultrahigh strength steels the reverse is found to be true (32). In contrast to such behavior for steels, the work of Vasudēvan and Suresh (31) has shown that the crack tip oxidation behavior of high strength aluminum alloys (e.g., 7075) at low ΔK levels is very sensitive to aging treatment, processing methods, and composition even for alloys which fall in a narrow range of yield strength values. This is presumably due to the dependence of oxidation on the extent of copper in solution as well as on the differences in the tenacity of the oxide layers produced in different aging conditions (31).

Order of magnitude estimates of the extent of oxide-induced crack closure have been reported by Suresh and Ritchie (26,33) by considering stress intensity solutions for a rigid wedge inside a linear elastic crack (Fig. 5). Assuming only a mechanical closure phenomenon arising from the presence of the excess oxide debris on the crack faces and ignoring plasticity and hysteresis effects, the closure stress intensity value K_{cl} at the crack tip can be obtained for plane strain conditions as (26,33):

$$K_{cl}\Big|_{x=0} = \frac{d_o E}{4\sqrt{\pi \ell}(1-\nu^2)} \,, \tag{10}$$

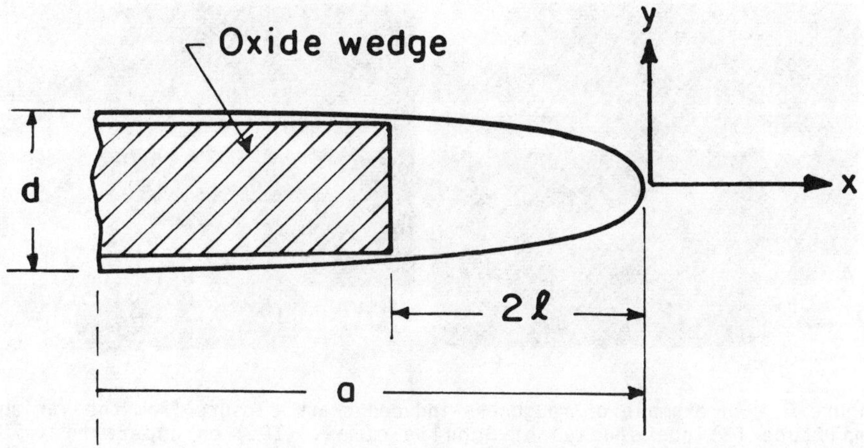

Figure 5 - Idealization of the role of crack face oxide debris in influencing fatigue crack advance by oxide-induced crack closure, in terms of a rigid wedge within a linear elastic crack. After ref. 33.

where d_o is the maximum excess oxide thickness, 2ℓ the location behind the tip corresponding to the thickest oxide formation and $E/(1 - \nu^2)$ the effective Young's modulus in plane strain. However, due to the assumptions pertaining to the rigidity of the oxide wedge, the non-uniform thickness of the oxide deposits along the crack length and along the crack front and the uncertainties in the estimates of d_o, ℓ and δ_{max}, Eq. (9) is realistically capable of providing only a crude description of the extent of oxide-induced closure (33).

Roughness-Induced Crack Closure

A third source crack closure can arise at near-threshold stress intensities due to the rough nature of the fracture surface (Fig. 1c) when crack tip opening displacements become comparable to the size-scale of the fracture surface asperities (Fig. 6). While it is widely acknowledged that crack propagation in the Paris regime (i.e., typically $10^{-6} \lesssim da/dN \lesssim 10^{-3}$ mm/cycle) occurs by a striation mechanism induced by concurrent or alternating crack tip shear (Stage II in Forsyth's terminology (34)), there is growing evidence that near-threshold crack advance in certain microstructures takes place primarily along a single active slip system (Stage I in Forsyth's terminology (34)) (35-37). Such single-shear Stage I growth, which occurs primarily where maximum plastic zone sizes at near-threshold growth rates are typically smaller than the significant microstructural unit, e.g., the grain size, results in crystallographic or generally faceted fracture features, an irregular surface morphology and **locally** mixed-mode crack growth (Fig. 7). While an ideally elastic crack unloads reversibly and does not undergo any premature closure, the irreversible nature of inelastic crack tip deformation and the possibility of slip-step oxidation in moist environments, in reality, lead to mismatch between the fracture surface asperities during the unloading portion of the fatigue cycle (38). The occurrence of such non-uniform crack tip opening and closure and the resulting Mode II crack opening displacements have been

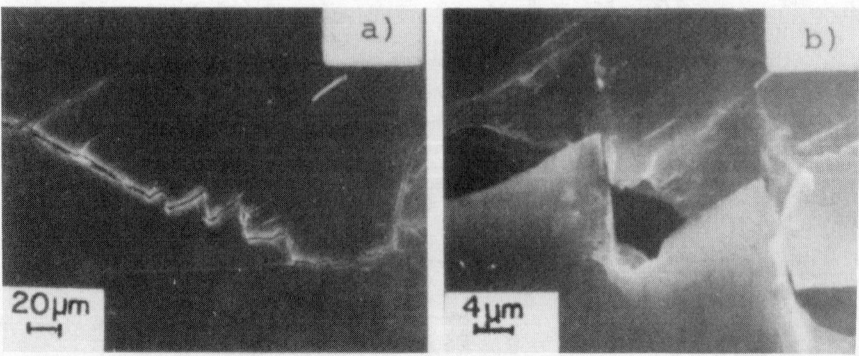

Figure 6 - An example of roughness-induced crack closure from the variable amplitude fatigue studies of Schulte et al. (105) on underaged X-7075 aluminum alloy tested in vacuum. Here the characteristic crystallographic mode of crack advance gives rise to markedly serrated crack paths, thereby promoting closure through fracture surface asperity contact.

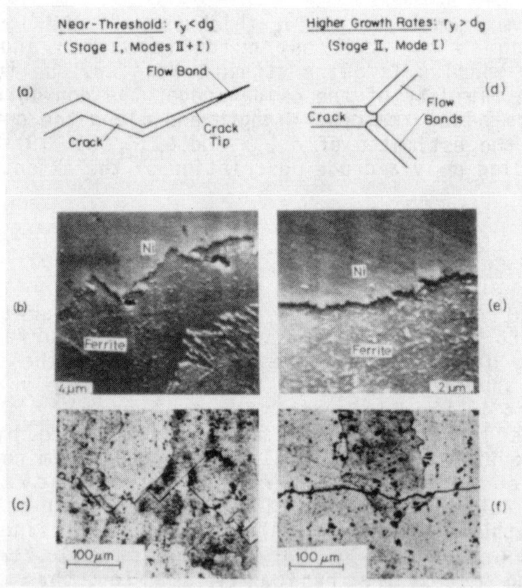

Figure 7 - Crack opening profiles and resulting crack path morphologies corresponding to a), b), c), near-threshold (Stage I) and d), e), f), higher growth rate (Stage II) fatigue crack propagation. b) and e) show sections through fatigue cracks in 1018 steel (after Minakawa and McEvily (35)) whereas c) and f) show sections in 7075-T6 aluminum alloy (after Louwaard (36)). r_y is the maximum plastic zone size, compared with the mean grain diameter d_y. After ref. 47.

experimentally verified by Davidson (39) using **in situ** observations of cyclic deformation in the scanning electron microscope. It was first suggested by Purushothaman and Tien (40) and by Walker and Beevers (41) that the interference between the mating fracture surface asperities behind the crack tip can lead to apparently higher (roughness-induced) crack closure loads and consequently, lower crack propagation rates. Minakawa and McEvily (35) have, subsequently, presented a qualitative model schematically describing the Mode II and Mode I displacements accompanying near-threshold crack advance (Fig. 8).

Roughness-induced crack closure is promoted by a) small plastic zone sizes at the crack tip (e.g., typically less than a grain diameter), which induce single shear mechanisms, b) small crack tip opening displacements, which are of a size scale comparable to the asperity height (such as in the near-threshold regime) (25), c) coarse-grained materials and microstructures with coherent and shearable precipitates capable of inducing coarse planar slip (43-45), d) crack deflection mechanisms (38), induced by stress-state, environment, microstructure or load excursions and the resulting non-linear crack profiles (46), e) inelastic crack tip deformation and oxidation of slip steps in moist environments, both of which can lead to mismatch between the fracture surfaces during the decreasing portion of the fatigue loading cycle (38) and f) low load ratios where the minimum crack tip opening displacements may be substantially smaller than the size of the surface asperities (18,25).

Estimates of (roughness-induced) closure stress intensities were first reported by Purushothaman and Tien (40) who denoted the fracture surface undulation by:

$$h = h_o \exp(\varepsilon_f) \quad , \tag{11}$$

where h_o and h are length dimensions describing the initial and final fracture surface roughness, and ε_f the true fracture ductility representative of the stress state ahead of the crack tip. By equating the change in asperity height $\alpha(h - h_o)$, where α is an empirical parameter of value less than unity, to the crack tip opening displacement, estimates of K_{cl} were derived (40). This model, however, does not incorporate the role of Mode II displacements, which are typical of near-threshold crack advance, in influencing roughness-induced closure.

In order to estimate the influence of Mode II displacements and fracture surface roughness on closure stress intensity levels, Suresh and Ritchie (47) developed a simple, two-dimensional geometric model. By assuming equal-sized asperities, the non-dimensional closure stress intensity at the point of first asperity contact was derived to be (47):

$$\left(\frac{K_{cl}}{K_{max}}\right) = \sqrt{\frac{2\gamma x}{1 + 2\gamma x}} \quad , \tag{12}$$

where γ is the non-dimensional surface roughness parameter taken as the ratio of the height, h, to the width, w, of the asperities and x the ratio of the Mode II to Mode I crack tip displacements (u_{II}/u_I). Although Eq. (12) must only be considered as a first order model for roughness-induced

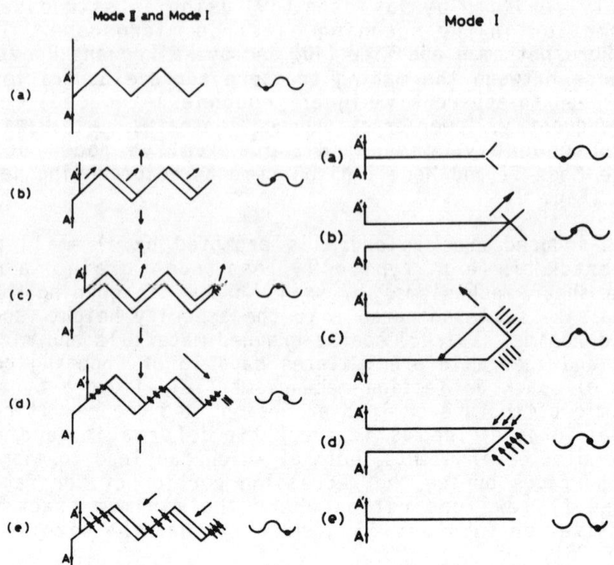

Figure 8 - Schematic illustration of the development of roughness-induced crack closure during "Stage I type" crack growth due to fracture surface asperity contact from Mode II and Mode I crack tip displacements, compared to the absence of such closure at higher strength intensities during "Stage II type" crack growth involving predominately Mode I displacements. After ref. 35.

crack closure, it does provide quantitative estimates for the extent of such closure in terms of the degree of fracture surface mismatch and the relative magnitude of the crack tip shear displacements. Such estimates for K_{cl}/K_{max} as a function of γ are shown in Fig. 9 (47), and indicate reasonably good agreement with experimental results derived from data (35,43) on a 1018 mild steel and a fully pearlitic rail steel.

Viscous Fluid-Induced Crack Closure

An additional source of closure can arise from the presence of a viscous fluid penetrated within a growing crack (Fig. 1d). Due to the approaching velocity of the crack walls, the presence of the fluid can give rise to a hydrodynamic wedging action counteracting the closing of the crack (Fig. 10). As experimentally demonstrated by Endo et al. (48,49) and more recently by Tzou et al. (50), this mechanism of viscous-fluid-induced crack closure becomes relevant for fatigue crack propagation in oil environments, where the rate of crack extension has been found to be sensitive to the kinematic viscosity (η) of oil. As shown in Figure 11, fatigue crack growth rates in chemically-inert oil environments at low load ratios tend to be in excess of those in moist air at near-threshold levels, due to a suppression of oxide-induced closure. However, such growth rates are slower than in moist air above $\sim 10^{-6}$ mm/cycle, due to a minimal effect of corrosion fatigue mechanisms (active path corrosion/hydrogen embrittlement), and are found to be faster in oils of higher viscosity (50) in the present case of a

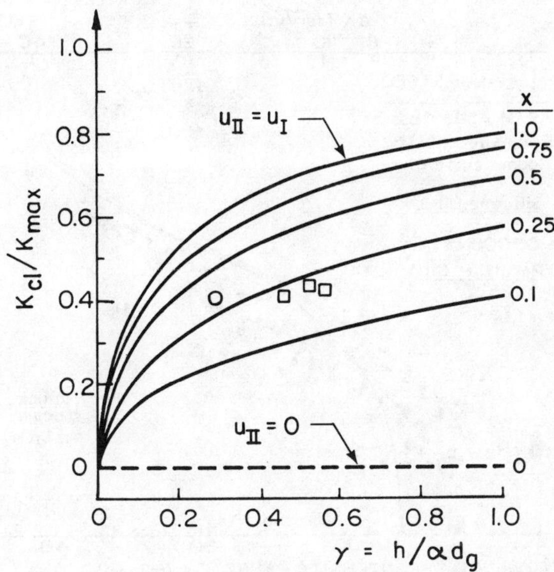

Figure 9 - Predicted variation from Eq. (12) of the extent of roughness-induced closure, in terms of K_{cl}/K_{max}, with fracture surface roughness factor γ as a function of x, the ratio of unloading Mode II and Mode I displacements (u_{II} and u_I, respectively). Experimental data points derived from results on 1018 steel after Minakawa and McEvily (35) and on fully pearlitic rail steel after Gray et al. (43). After ref. 47.

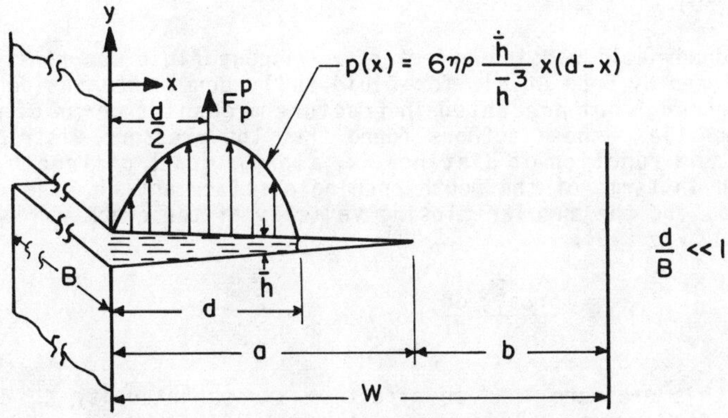

Figure 10 - Idealization of the distribution of pressure p(x) and resultant force (F_p) due to the presence of a viscous fluid inside a fatigue crack, of length (a), where the fluid has only partially penetrated the crack along a distance d. The resultant stress intensity arising from such internal oil pressure is a function of a, d, the average crack opening ($\bar{h}$), the closing velocity of the crack walls (v) and the absolute viscosity ($\eta\rho$). After ref. 51.

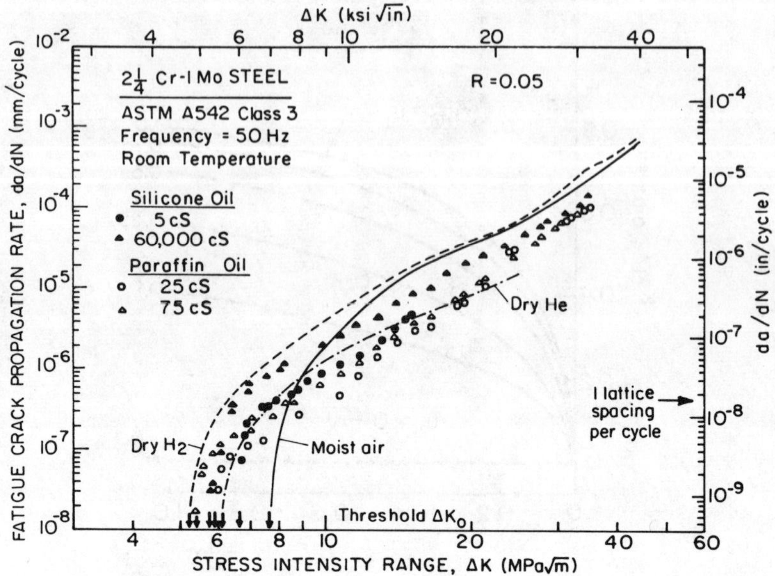

Figure 11 - Comparison of fatigue crack propagation behavior at R = 0.05 in a 2¼Cr-1Mo steel (σ_y = 500 MPa) tested at 50 Hz in high and low viscosity silicone and paraffin oils with behavior in gaseous environments of moist air, dry hydrogen and dry helium gases. After ref. 50.

bainitic steel tested at 50 Hz in silicone and paraffin oils. At high load ratios, however, the influence of viscosity on growth rate behavior largely disappears (50).

The hydrodynamic wedging effect of a viscous fluid was first quantitatively analyzed by Endo et al. for fluid fully penetrated inside the crack (48,49). Although not presented in fracture mechanics terms of resultant stress intensities, these authors found that the pressure distribution due to the oil as a function of distance, x, along a crack of length, a, could be presented in terms of the mouth opening displacement, h, the density of the fluid, ρ, and the angular closing velocity of the crack walls $(d\theta/dt) \approx (1/a)(dh/dt)$, viz (49):

$$p(x) = \frac{6\eta\rho a^2}{h^3}\ \frac{dh}{dt}\log(1 - \frac{x}{a}) \quad . \tag{13}$$

Converting this pressure into an effective stress intensity K_p due to the presence of the viscous fluid using standard linear elastic K_I solutions indicates that the magnitude of the fluid-induced closure should increase with increasing viscosity, yet the experimental data in Figure 11 show an opposite trend in that the highest growth rates are observed with the highest viscosity oils. This discrepancy can be explained by recognizing that the fluid pressure induced by the higher viscosity oils is reduced by the kinetic limitations of such oils in penetrating the crack. Accordingly, Tzou et al. (51) proposed an alternative quantitative description for viscous-fluid-induced crack closure based on the partial penetration of the oil inside the crack. Here the penetration distance d is computed as a function of time t from capillary flow analysis as:

$$d = \left[\frac{\gamma_L \cos \beta}{3\eta\rho} \int_0^t \bar{h} \, dt \right]^{\frac{1}{2}} , \qquad (14)$$

where γ_L is the surface tension, β the wetting angle and $\bar{h}$ the average crack opening, such that the subsequent pressure distribution within the crack becomes (51):

$$p(x) = \frac{6\eta\rho}{\bar{h}^3} \frac{d\bar{h}}{dt} x(d - x) , \qquad (15)$$

where $v = d\bar{h}/dt$ is the approaching velocity of the crack walls (Fig. 10). The extent of viscous-fluid-induced closure is then determined numerically by superposition of the resultant K_p values to the applied stress intensities from simultaneous solution of Eqs. (14) and (15). Based on the experimental data shown in Figure 11, this analysis indicated that in the lower viscosity oils (e.g., 5 to 75 cS), where the oil fully penetrated the crack, the extent of closure at R = 0.05 was characterized by K_p/K_{max} values of between 0.25 and 0.50 at ΔK levels between 10 and 20 $MPa\sqrt{m}$. In the higher viscosity oils (e.g., $\gtrsim$12500 cS), however, where only partial penetration occurred the extent of closure was somewhat less in that K_p/K_{max} values were estimated to lie between 0.22 and 0.43. Although the penetration rates of the fluids are much more rapid at high load ratios, no closure was predicted for R = 0.75 because of the much larger crack openings there.

Such closure is clearly promoted by low load ratios and is affected by such factors as oil viscosity and chemistry, test frequency and strength level/elastic modulus (through their influence on the extent of crack opening). However, it is difficult to simply indicate a trend for the latter variables since although the maximum extent of closure, i.e., K_p values for full oil penetration, is enhanced by increasing viscosity, increasing frequency and decreasing crack opening, such factors concurrently act to minimize the penetration of the oil. In view of the numerical estimates of steady state values of K_p under the equilibrium conditions of infinite time and full penetration (51), it would appear that closure induced by viscous fluids is a less potent mechanism than that generated by oxide debris or irregular crack paths.

Phase Transformation-Induced Crack Closure

In materials which undergo stress- or strain-induced phase transformations under cyclic loading, a further mechanism of fatigue crack closure can result from the so-called TRIP effect, i.e., arising from TRansformation Induced Plasticity (Fig. 1e). In analogous fashion to the increase in fracture toughness observed under monotonic loading in both metals (52), e.g., metastable austenitic stainless and high strength TRIP steels, and ceramics (53), e.g., tetragonal zirconia, due to deformation-induced martensitic transformations, the constraint of surrounding elastic material on regions which have transformed and thus undergone a positive volume change will place such regions under compression. Once the crack penetrates these transformed regions, generally corresponding to some process zone at the crack tip, they will act to close the crack and hence reduce the nominal stress intensity to some lower effective value at the tip.

Such transformation-induced closure will be promoted by metallurgical phase changes which show a large increase in volume (i.e., the martensitic transformation in steels is typically 4%), and by conditions which enhance the transformation, i.e., lower temperatures, higher strain rates, etc. However, analogous to the plasticity-induced mechanism, the closure effect arises from the constraint of surrounding elastic material on the transformed regions such that if the material is completely transformed prior to cracking, no closure effect will be felt. Furthermore, the transformation of regions remote from the tip will similarly not cause closure; the effect will only be experienced as the crack penetrates and is enveloped by the transformed region.

Rigorous analysis of the TRIP effect on crack closure has yet to be analyzed for deformation conditions involving deviatoric strains or cyclic inelasticity (54). However, McMeeking and Evans (55) and Budiansky et al. (56) have analyzed the effect under monotonic loading for purely stress-induced dilatant transformations in ceramics and shown the reduction in stress intensity due to closure to be given by:

$$K_I - K_{eff} = \frac{0.22}{(1-\nu)} V_f \, \varepsilon_T \, E\sqrt{r_t} \quad , \qquad (16)$$

where K_I and K_{eff} are the nominal (applied) and effective (near tip) stress intensities, E the elastic modulus, ε_T the transformation strain, V_f the volume change and r_t is the width of the transformed zone. It is worth noting that the above analyses indicate a resistance curve effect in that the full influence of transformation-induced closure is only felt when the crack has penetrated the transformation region to a distance comparable with approximately 5 times the width of the zone r_t (55). This highlights a common characteristic of all closure mechanisms in that closure acts in the wake of the crack tip. Thus, for small cracks, which by definition have a limited wake (i.e., in the present case, of a length less than $5r_t$) the extent of closure will be lessened and hence the effective driving force at the crack tip will be correspondingly larger.

Implications of Crack Closure

The implications of such fatigue crack closure mechanisms are now considered for the complex and often seemingly inconsistent nature of near-threshold fatigue crack growth. The understanding of the many sources of closure outlined above has enabled a much more consistent explanation to be obtained for the effects of many mechanical, microstructural and environmental factors which have been documented over the years to influence near-threshold growth rates and the value of the fatigue threshold. Although in most instances, direct experimental closure measurements have confirmed a primary role of crack closure, a common trend can be seen in that where closure is dominant, the effect of these factors, whether they involve strength, grain size or moist environments, etc., will be most pronounced at low load ratios with long cracks. In fact, when assessed at high load ratios or with small cracks, the effect may well be severely diminished, non-existent or even different since under these conditions closure mechanisms cannot be so effective. We now examine several of the major variables which affect thresholds and discuss the respective implications of closure.

Mechanical Variables

Load Ratio. Of the mechanical factors influencing fatigue crack growth, clearly one of the most important is that of mean stress, characterized in terms of the load or stress ratio as $R = K_{min}/K_{max}$. Numerous experimental investigations, reviewed in refs. 10 and 18, have shown that with increase in R crack growth rates for tension/tension loading are enhanced, although the effect generally predominates only at very high growth rates close to final instability (i.e., as K_{max} approaches K_{Ic}) and at near-threshold levels (i.e., as ΔK approaches ΔK_0) (10). The former effect of load ratio in accelerating growth rates close to instability is usually associated with the occurrence of "static" fracture modes, i.e., cleavage, intergranular and fibrous fracture, in addition to ductile striation growth (57). However, the near-threshold load ratio effect, shown in Figure 12 for SA542-3 steel tested in laboratory environment (18), occurs without dramatic changes in fractography and is attributed primarily to mechanisms of crack closure.

The analysis of Schmidt and Paris (9) reviewed in the section on plasticity-induced crack closure clearly shows the trend of closure effects on the threshold (Fig. 2) in that as the mean level is raised, the role of K_{cl} in modifying the effective ΔK is lessened. Studies on the role of load ratio in influencing ΔK_0 values in dehumidified gaseous hydrogen, as compared to moist air atmospheres, show behavior similar to that predicted by Schmidt and Paris (9). The critical load ratio R_{cr} above which ΔK_0 values are insensitive to R, however, is lower in the hydrogen environments (Fig. 3). As R_{cr} represents the load ratio above which closure effects are insignificant, such behavior can be attributed in this case to a much smaller effect of oxide-induced crack closure, consistent with the reduction in crack surface corrosion debris with increasing R. In hydrogen, which merely plays the role of a dry environment in low strength steels, oxide-induced closure is less significant (18). Such an interpretation of a lessening effect of closure with increasing R and in dry environments is consistent with numerous experimental closure measurements of K_{cl}. An example of such results, relevant to the threshold data in Figure 3, is presented in Figure 13, taken from an assessment of K_{cl} values at ΔK_0 in a bainitic steel using an ultrasonic technique (26). It is apparent that the opening of the crack is most delayed at R = 0.05 in moist air where both plasticity-induced and oxide-induced closure can occur.

The load ratio dependence of fatigue thresholds is often quite different to that depicted in Figure 2 in completely inert or chemically very aggressive environments. In either case the reduction in ΔK_0 with increase in R is far less marked compared to moist air (58,18); observations which are again consistent with oxide-induced closure concepts. In inert environments, such as high vacuum, the shallow dependence of ΔK_0 on R has been attributed to the fact that little or no corrosion deposits are formed at any load ratio such that, unlike moist air, the magnitude of oxide-induced closure does not vary with R (18). Similarly in aggressive environments, such as water (Fig. 14), fracture surface oxide deposits can form and thicken at all load ratios by natural thermally-activated mechanisms so that the extent of oxide-induced closure does not decay with increasing R (18).

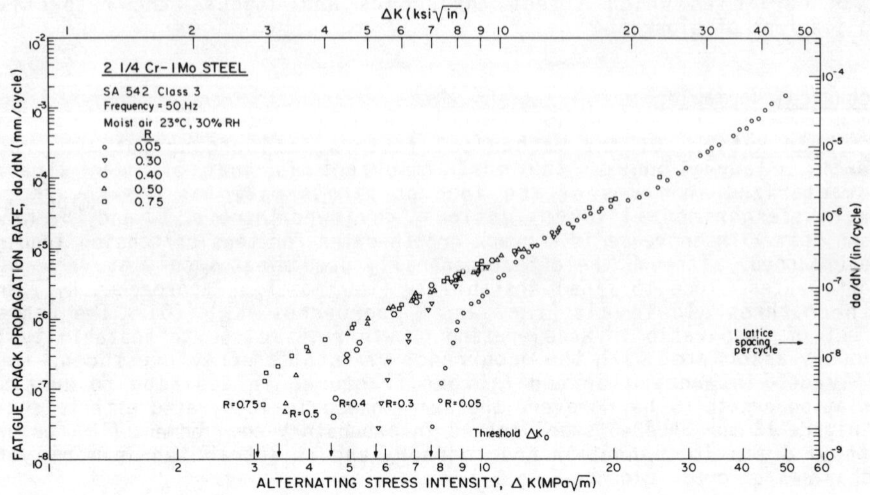

Figure 12 - Influence of load ratio R on near-threshold and higher growth rates in 2¼Cr-1Mo steel tested at 50 Hz in moist air. Whereas above $\sim 10^{-5}$ mm/cycle, behavior is essentially insensitive to R, below $\sim 10^{-6}$ mm/cycle, growth rates are increased and ΔK_0 values decreased with increasing positive load ratio from R = 0.05 to 0.75. After ref. 18.

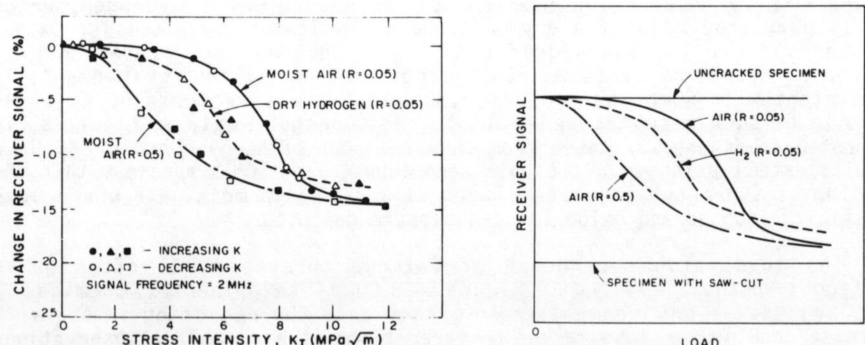

Figure 13 - Experimental measurements of crack closure using an ultrasonics technique indicating the variation in ultrasonic signal transmitted across a near-threshold crack during a fatigue cycle at ΔK_0. Data, after ref. 26, for a 2¼Cr-1Mo steel tested in moist air and dry hydrogen gas at R = 0.05 and 0.50.

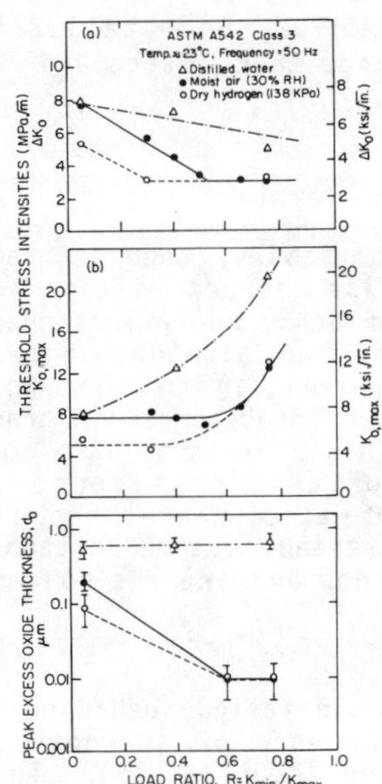

Figure 14 - Variation of fatigue thresholds ΔK_o and $K_{o,max}$ with the maximum excess oxide thickness d_o, as a function of load ratio R, for a bainitic 2¼Cr-1Mo steel (σ_y = 500 MPa) tested at 50 Hz in room temperature environments of moist air, dry gaseous hydrogen and distilled water. After ref. 18.

There are several other instances where the load ratio dependence of fatigue thresholds is decreased due to a reduced role of closure. In addition to behavior in inert or very aggressive environments (18) discussed above, the variation in ΔK_o with R is significantly less marked in ultrahigh strength steels (59), where oxide-induced closure effects are minimized (32), and in finer grained materials (60), where roughness-induced closure effects are minimized (37). Such behavior has been described in detail in ref. 18.

Behavior at negative load ratios is far less understood. However, the indications are that in certain materials near-threshold growth rates at R = -1 may marginally exceed corresponding growth rates at R = 0 (e.g., refs. 61,62). Based on limited experiments closure measurements (62), it would appear that such behavior is associated with a reduction in crack closure due to "smoothening" of asperity contacts behind the crack tip from compressive contact between crack surfaces.

Cyclic Frequency. Studies by Schmidt and Paris (9) in 2024-T3 aluminum alloy first revealed that the threshold stress intensity value is significantly influenced by the cyclic frequency, although no definitive trend could be observed for the variation of ΔK_o with frequency. More recently, Bignonnet et al. (63) found lower ΔK_o values and correspondingly less oxide debris during low load ratio tests in a quenched and tempered low alloy steel tested at a frequency of 7 Hz as compared to 65 Hz. The mechanism by which increasing frequency enhances the formation of crack surface corrosion deposits (and hence promoted oxide-induced crack closure) was not specified (63). The observations of Bignonnet et al. (63) are consistent with the results for 7475-T7 aluminum alloy (64) where the ΔK_o values and oxide thickness at ΔK_o are found to be higher at 200 Hz than at 25 Hz for tests conducted in 95% humid air. Studies over a wider range of frequencies (5 to 500 Hz) in SA542-3 steel, however, have failed to substantiate this effect above 50 Hz, although oxide thicknesses and ΔK_0 values were somewhat smaller at 5 Hz (65).

Microstructural Variables

It has long been known that microstructural variables, such as grain size, precipitate distribution and morphology, slip characteristics and duplex structures, can have a more significant influence on fatigue crack propagation behavior in the near-threshold regime than at higher growth rates (e.g., refs. 8-10,37,41,43,45-47,64). Furthermore, in most instances it has been shown that such microstructural effects on near-threshold crack growth can be significantly lessened for tests at high, as opposed to low, load ratios (10,37,43,45,47). The reasons for this prominence of microstructural-sensitive behavior for low load ratio, near-threshold conditions is again largely a consequence of closure and in some instances related mechanisms of crack deflection (38). We now examine the role of each of these metallurgical factors in turn:

Strength Level. Early studies of near-threshold fatigue behavior in ferrous alloys (59,67,68) indicated a large influence of strength in increasing ultra-low growth rates and decreasing ΔK_0 (Fig. 15). This effect, however, is less prominent at high load ratios (59), for small cracks (69) and for non-ferrous alloys (70).

The major role of yield strength in influencing near-threshold fatigue crack growth in ferrous alloys is consistent with a primary role of closure. Recent studies (32) in steels over a very wide range of strength levels (i.e., 290 to 1740 MPa) have confirmed that the extent of crack surface corrosion debris and hence oxide-induced closure is sharply decreased with increasing strength at R = 0.05 (Fig. 16). This was attributed to a reduced role of plasticity-induced closure (which promotes crack surface contact to enhance fretting oxidation) in higher strength steels and by the fact that the oxide products formed will be less potent in causing more fretting debris on the harder substrate (24,25,27,32). Furthermore, it has also been pointed out that higher strength steels are more prone to environmental effects (10) and that the microstructures typically utilized for highest strength levels, i.e., tempered martensites, are generally of considerably finer scale than for lower strength structures, such that with the resulting smoother fracture surfaces, roughness-induced closure may also be lessened (66).

The effect of strength level on near-threshold crack growth in non-ferrous alloys is somewhat more complex in that, although behavior is similar to steels in copper alloys (30), there is no marked trend of increasing or decreasing ΔK_0 values with change in strength level in precipitation hardened systems (70). In aluminum alloys, this has recently been interpreted in terms of aging treatment, processing methods, composition and slip mode, involving a mutual competition between resistance to corrosion fatigue mechanisms and oxide-induced closure (which are promoted in overaged structures) and roughness-induced closure (which are promoted in underaged structures) (31,44,45,76). Thus in view of the different hardening mechanisms responsible for a given yield strength in such alloys (e.g., shearing versus by-passing of precipitates) and the effect such slip characteristics can have on resultant closure mechanisms (e.g., 31,45), it is inappropriate to identify a direct correlation of yield strength and near-threshold behavior in non-ferrous systems.

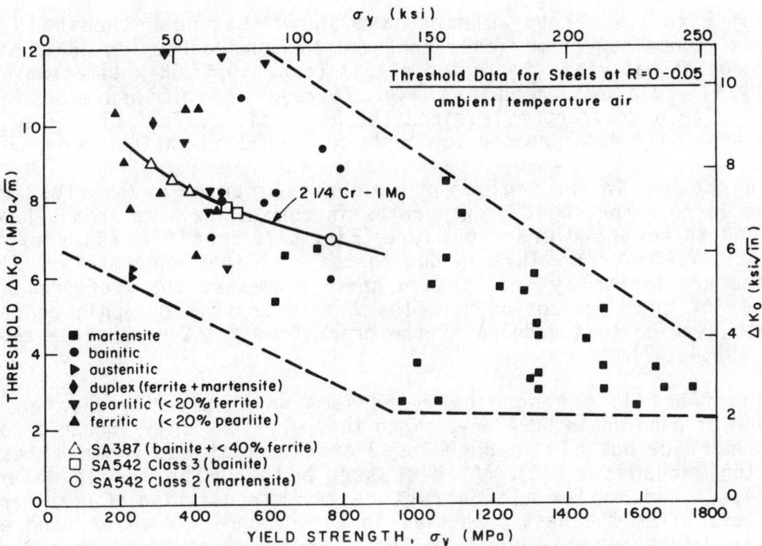

Figure 15 - Compilation from ref. 10 of fatigue threshold ΔK_0 data as a function of yield strength σ_y for steels, tested at low load ratios in room air.

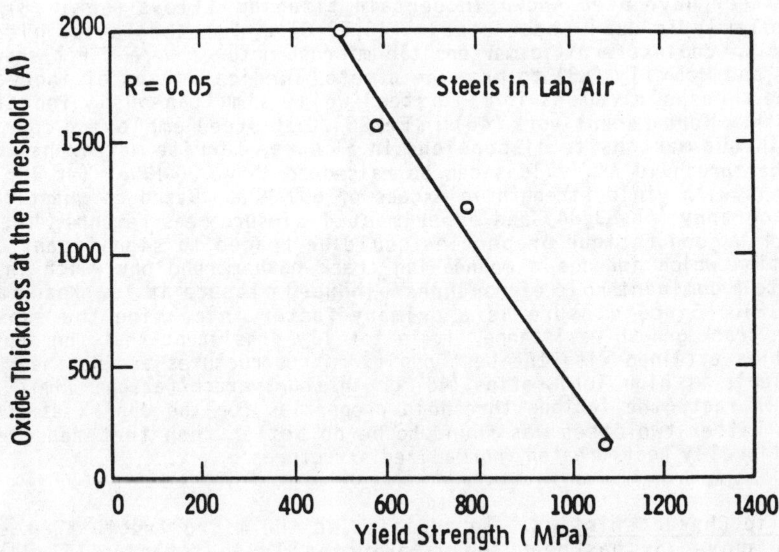

Figure 16 - Variation of maximum excess oxide thickness, measured on the fracture surface at the threshold ΔK_0 at R = 0.05, with yield strength σ_y for a range of steels tested at 50 Hz in moist air. After ref. 32.

Grain Size. Numerous authors have shown that near-threshold crack growth resistance can be markedly enhanced through coarsening grain size in a wide range of materials including steels (e.g., 10,43,68), titanium alloys (e.g., 71-74) and aluminum alloys (e.g., 75,76). Again closure mechanisms appear to play a dominant role since the beneficial effect of coarse grain sizes is generally much reduced for tests at high load ratios (e.g., 37,43).

As discussed in the section on roughness-induced crack closure, the reduction in near-threshold growth rates in coarse grain materials has been attributed to enhanced crack closure (Fig. 17), specifically roughness-induced (25,37,43,47,72). This is due to the fact that when maximum plastic zone sizes are typically less than a grain diameter, the preferred single shear mode of crack extension promotes a more crystallographic or faceted crack path leading to a wedging of the crack from fracture surface asperity contact (35,41,47).

This effect is enhanced by mechanisms which promote greater crack deflection at each grain boundary, since they will not only induce a rougher fracture surface but also reduce local crack tip stress intensities from deflection mechanisms (38). As discussed below, planar slip materials (44,45,74,77) and duplex microstructures (46), consisting of both hard and soft phases, offer the best potential in this regard. However, the effect is likely to be much reduced in inert environment where the lack of oxidation of exposed fracture surface at the crack tip will aid reversibility of slip there, thereby reducing the extent of closure (76).

Duplex Microstructures. Duplex microstructures seem to provide the most effective means to promote roughness-induced crack closure. Here dispersions of hard and soft phases can lead to significant crack deflection resulting in the formation of tortuous crack paths. Notable examples of this effect have been shown in certain titanium alloys (e.g., 78), and particularly in dual-phase steels (46,79-81). For the latter class of materials, duplex ferritic-martensitic microstructures were first shown by Suzuki and McEvily (83) to provide a metallurgical means of increasing fatigue threshold values in mild steel while simultaneously increasing strength. More recent work (46) in Fe/2Si/0.1C steel employing coarse or fine fibrous martensite dispersions in a coarse ferrite matrix has shown that the threshold ΔK_0 values can be raised to above 20 MPa$\sqrt{m}$ (at R = 0.05) in steels with yield strength in excess of 600 MPa. Based on quantitative metallography (46,83,84) and experimental closure measurements (46), such unusually good fatigue properties could be traced to significant crack deflection which induces a meandering crack path morphology which in turn leads to a dominant role of roughness-induced closure at low load ratios (Fig. 18). Since closure is a primary factor in causing the superior fatigue crack growth resistance, it is totally consistent that the superior properties attained with the best duplex microstructures are not nearly as appreciable at high load ratios (46) or in short crack/fatigue limit tests (82). In fact, the fatigue threshold properties for the duplex structures in the latter two cases was found to be no better than that measured in conventionally heat-treated, normalized structures.

Slip Characteristics. In addition to the microstructural factors listed above, it has been shown, principally in titanium (71-74) and aluminum (44,45,76,77) alloys that microstructures which promote coarse planar slip, as opposed to wavy slip, generally give rise to superior near-threshold fatigue crack growth resistance. Such microstructures are found in materials of low stacking fault energy or in dispersion hardened systems employing small coherent precipitates, and under near-threshold fatigue

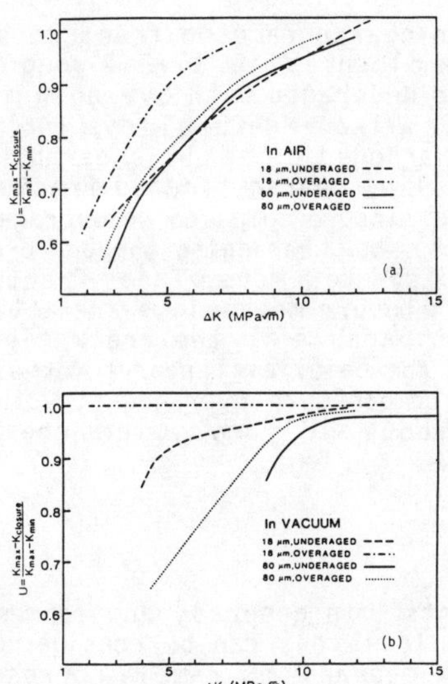

Figure 17 - Variation in experimentally measured crack closure, with ΔK for a 7475 aluminum alloy (σ_y = 445-505 MPa) tested in a) room air and b) vacuum. Closure data are compared between fine and coarser grained microstructures in the underaged and overaged conditions. After ref. 76.

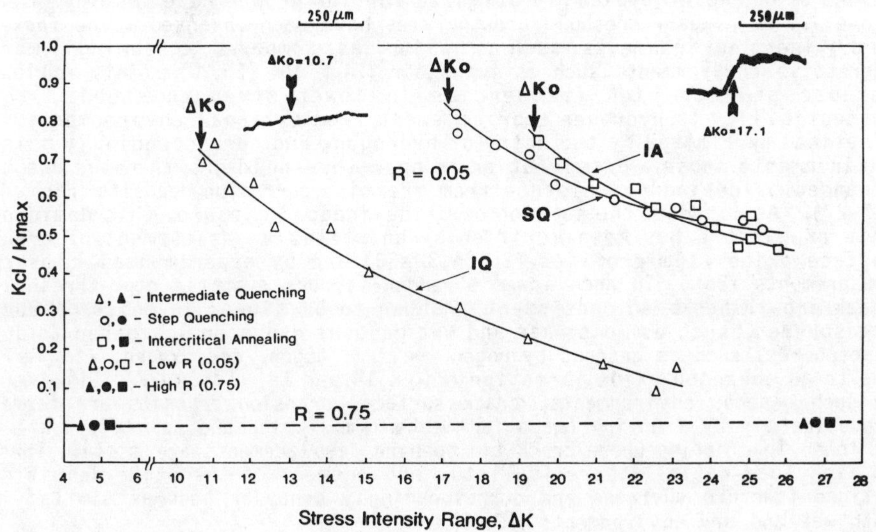

Figure 18 - Experimental closure data as a function of ΔK for dual phase Fe/2Si/0.1C steel ($\sigma_y \approx$ 600 MPa) tested at R = 0.05 and 0.75. Certain morphologies of the duplex ferritic/martensitic microstructures promote crack deflection causing significant (roughness-induced) closure, e.g., for the IA and SQ conditions, compared to more linear crack paths in the IQ condition. After ref. 46.

conditions give rise to coarse faceted fracture surfaces which enhance roughness-induced crack closure (Fig. 17). A good example of this effect can be seen by comparing underaged with overaged microstructures (at the same strength level) in Al-Zn-Mg-Cu alloys (45). Here the underaged structures, which are hardened by small shearable coherent precipitates, show extensive planar slip, coarse faceted fracture surfaces and hence greater roughness-induced closure compared to overaged structures, where the wavy slip characteristics, i.e., hardening through Orowan bypassing of large coherent precipitates, lead to a more planar fracture morphology. Since such roughness-induced closure mechanisms are active where the fracture surface roughness is comparable to the crack tip displacements, it is interesting to note that the beneficial near-threshold fatigue crack growth resistance of underaged compared to overaged structures is not retained at high propagation rates above $\sim 10^{-6}$ mm/cycle where the role of closure is severely diminished (45).

Environmental Variables*

Gaseous Environments. In general, environmentally-influenced crack extension at near-threshold levels can be considered in terms of a mutual competition of two basic mechanisms, namely corrosion fatigue processes, such as hydrogen embrittlement or active path corrosion which accelerate growth rates and the resultant closure mechanisms which decelerate growth rates. Where closure mechanisms dominate, such as at ultralow growth rates in lower strength, coarse-grained materials at low load ratios, environmentally-influenced fatigue behavior may be different than that conventionally reported for higher growth rates above typically 10^{-6} mm/cycle. For example, for lower strength steels tested at low load ratios at the frequencies typically utilized for low growth rate testing (i.e., above $\sim$20 Hz), near-threshold growth rates have been observed to be faster in dry inert environments, such as helium gas, compared to seemingly more aggressive environments such as moist air (Fig. 19) (25,26). This follows because at such high frequencies in lower strength steels, the susceptibility to hydrogen embrittlement from external environments is kinetically limited by the entry of hydrogen, and correspondingly moist environments cause a deceleration in near-threshold growth rates due to enhanced oxide-induced closure from fretting corrosion deposits (Fig. 14) (17-27). As noted in the section on oxide-induced closure, this dominant role of closure has been verified by quantitative assessment of crack surface oxide film profiles (25,26,30-33) and by experimental closure measurements (26). In such lower strength ferrous systems, near-threshold crack growth has been consistently shown to be slower in moist gaseous atmospheres, such as moist air and wet gaseous hydrogen, compared to dry atmospheres, such as gaseous hydrogen, helium, argon, dry oxygen, etc., all due to an enhanced oxide formation (Figs. 14 and 19) (17,24-27). However, in such gaseous environments, crack surface corrosion deposits are formed predominately by fretting oxidation mechanisms, which are mainly operative at lower load ratios where crack tip opening displacements are small. Thus, at high load ratios, there is little enhancement of the oxide debris on fatigue fracture surfaces and correspondingly behavior becomes similar in both wet and dry environments (17,25).

---

*In general, the precise influence of environment on crack growth is very specific to the particular material/environment system. Below we merely discuss the role of simple environments, e.g., air, water, hydrogen, etc., in affecting the behavior of engineering alloys.

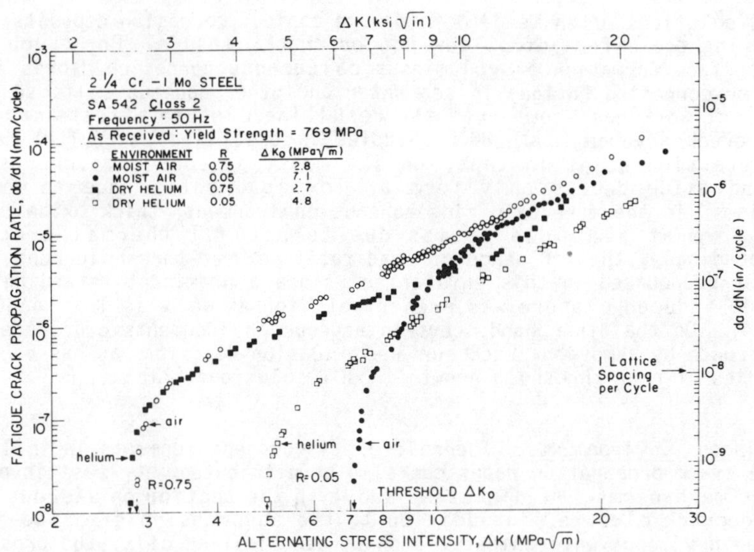

Figure 19 - Fatigue crack propagation behavior in a martensitic 2¼Cr-1Mo steel (σ_y = 769 MPa) tested at R = 0.05 and 0.75 in moist air and dehumidified gaseous helium, showing **faster** near-threshold crack growth rates in the inert gas at low load ratios only. After ref. 25.

This simple rationalization for near-threshold fatigue behavior in gaseous atmospheres in terms of oxide-induced closure becomes more complex in alloy/environment systems where conventional corrosion fatigue mechanisms, such as hydrogen embrittlement, assume greater importance. For example, the extrinsic role of external gaseous hydrogen in simply providing a dry environment which minimizes the oxide-induced closure effect in low strength steels at near-threshold levels (25) must be contrasted with the intrinsic role of hydrogen charging where accelerated near-threshold growth rates due to actual hydrogen embrittlement have been observed (83). However, the latter effect can be further complicated by the additional roughness-induced closure arising from the resulting intergranular fracture (83,84). Moreover, in ultrahigh strength steels (UTS $\gtrsim$ 1000 MPa) tested in moist air and dry hydrogen environments, the extent of corrosion debris and hence oxide-induced closure is extremely limited at all load ratios and accordingly behavior is dominated by hydrogen embrittlement processes even for external environments (27,32). Similarly for 7075 aluminum alloys tested in moist air environments (31,45,85), the enhanced oxide-induced crack closure, observed in overaged (T7) microstructures, does not necessarily lead to slower growth rates in this structure compared to under- or peak-aged conditions. This is because in the latter structures, resistance to hydrogen embrittlement/active path corrosion processes tends to be greater (31) and furthermore the planar slip characteristics of underaged structures can enhance additional roughness-induced closure (45).

Aqueous Environments. Behavior in aqueous environments, such as distilled water or sodium chloride solutions, is largely similar to that described for moist gaseous environments above with the exception that such aqueous solutions often tend to form more copious corrosion deposits thereby increasing the effect of oxide-induced crack closure. For example, the crack surface formation of voluminous **calcareous** corrosion debris is well known for corrosion fatigue in sea water and other aqueous salts solutions, and recent work has shown that the resulting closure effects can indeed retard crack advance (e.g., 86). Studies in distilled water (18) revealed an interesting point in that, unlike moist gaseous atmospheres where enhanced oxide debris only form at low load ratios due to fretting mechanisms, in the more oxidizing aqueous environment, thick oxide deposits could form at all load ratios due to natural thermal processes. Correspondingly, the influence of load ratio on near-threshold behavior was far less pronounced in this environment since a prominant retarding effect of oxide-induced closure was present at high as well as low load ratios (Fig. 14). On the other hand, certain aqueous environments can "accelerate" crack growth by removing crack surface oxidation deposits, as has been shown by studies of near-threshold growth in HCl solutions (23).

Viscous Environments. The role of viscous environments in influencing fatigue crack propagation rates can also be attributed, at least in part, to closure mechanisms. As discussed above in the section on viscous fluid-induced crack closure, in addition to the suppression of oxide-induced closure in viscous environments, such as dehumidified oils, the presence of a viscous fluid within the crack walls can induce crack closure effects from the hydrodynamic wedging effect of the oil (48-51). This results in an effect of viscosity on growth rates (Fig. 11), although as discussed above the precise trend in behavior with increasing viscosity depends upon a balance between the internal pressure generated by the oil wedge (which is enhanced with increasing η) and the kinetic ability of the oil to flow into the crack (which is decreased with increasing η). In general though behavior in inert viscous environments will resemble that in other inert atmospheres, such as dry gaseous helium, such that the suppression of oxide-induced closure at near-threshold levels will lead to faster growth rates compared to moist air, whereas the opposite effect will occur at higher growth rates due to the suppression of conventional corrosion fatigue mechanisms in the inert fluid (Fig. 11).

Temperature. The influence of temperature on specific mechanisms of closure has not been studied extensively to date, presumably due to the difficulty of isolating the individual effect on closure in experiments at different temperatures. Yuen and co-workers (29), however, have attributed the enhanced oxide formation, and hence oxide-induced closure, to the occurrence of premature thresholds during elevated temperature fatigue of nickel-base superalloys. Corresponding studies on the role of closure at low temperatures are complicated by the intrinsic effects of low temperature deformation characteristics on crack extension mechanisms (83). However, it is clear that oxide-induced closure is likely to be inhibited at low temperatures due to lower oxidation rates whereas phase transformation-induced closure may well be enhanced due to the greater driving force for the deformation-induced reaction.

Applications

The various crack closure phenomena, discussed thus far, not only play an important role in influencing the constant amplitude fatigue crack growth rates, but also have a marked effect on several applications such as the short crack problem, variable amplitude crack advance comprising overloads and underloads, the existence of a fatigue threshold and its uniqueness for a given material and testing procedure. We now examine each in turn.

Short Crack Behavior

Cracks are defined as being short i) when their length is small compared to relevant microstructural dimensions (a continuum mechanics limitation), ii) when their length is small compared to the scale of local plasticity (a linear elastic fracture mechanics limitation), and iii) when they are merely physically small (i.e., crack length $\lesssim$ 0.5-1.0 mm) (87). Recent studies have shown that all three types of short cracks grow substantially faster than (or at least as fast as) the equivalent long cracks subjected to the same nominal stress intensity factor range (e.g., refs. 88-94). In particular, the first two types of short flaws are known to propagate at high velocities even **below** the threshold for long cracks, ΔK_0 (88-90). The initially high growth rates of such short cracks are known to decelerate progressively (or even arrest completely, in some cases, to form so-called non-propagating cracks) before merging with the long crack data (88-90,92). Recent studies have suggested that part of the reason for the apparently faster growth of short cracks, even below the long crack threshold, stems from their limited **wake** and the consequent absence of appreciable crack closure (11,87,89,93). There is growing experimental evidence indicating that the progressive deceleration and/or arrest of cracks, which are of a length comparable to the scale of local plasticity or microstructure, and the apparently faster growth of physically-short flaws can be accounted for, at least partially, through arguments based on the development of crack closure with increasing crack length (11,38,87,89,93). Morris et al. (89) first showed that for short flaws ($a \sim 50$ to 500 μm) in titanium alloys, crack closure (presumably that induced by roughness) decreased with decreasing crack length. A further influence of roughness-induced closure was evident in the work of McCarver and Ritchie (93) who found that the threshold ΔK_0 values for short cracks ($a \sim 20$ to 200 μm) in René 95 nickel-base superalloys were about 60% smaller than for long cracks ($a \sim 25$ mm) at low load ratios, even though no differences were apparent at high load ratios where closure effects were minimal. Tanaka and Nakai (94) measured crack closure loads for small cracks emanating from notches in a low carbon steel and showed that the anomalies between short and long cracks can be eliminated when results are plotted in terms of ΔK_{eff} by factoring out crack closure (Fig. 20). Thus, it may be argued that one of the reasons that short flaws show apparently larger growth rates than long cracks, subjected to the same nominal ΔK level, is because of their limited **wake** and less closure.

Variable-Amplitude Loading

The various crack closure phenomena described in this paper for constant-amplitude fatigue crack growth also have a strong influence on the transient fatigue crack advance following variable amplitude loading in the form of overloads or underloads. It was first suggested by Elber (6) that plasticity-induced closure is the dominant mechanism controlling crack growth retardation due to single or block overloads. Several subsequent studies have attempted to correlate the retarded delay distance with the size of the enlarged plastic zone generated by overloads (e.g., 95-97). Recent work by Suresh (98-100) has shown that in addition to plasticity-induced closure, other forms of closure such as those due to fracture surface oxides or micro-roughness also play an important role in controlling retardation because post-overload crack propagation is effectively governed

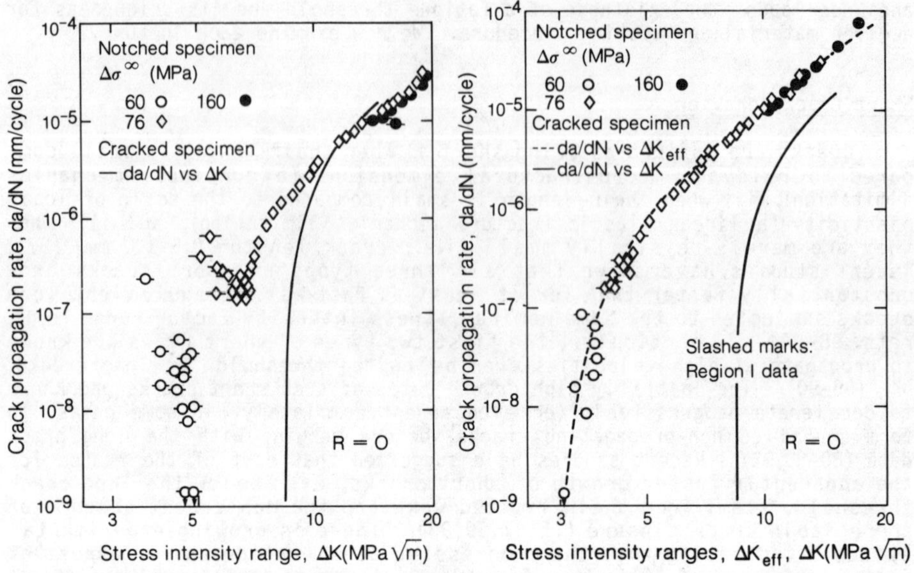

Figure 20 - Variation in fatigue crack propagation rate with a) nominal and b) effective stress intensity range values, ΔK and ΔK_{eff}, respectively, in a 0.17%C steel (σ_y = 194 MPa). Data points represent behavior of short cracks emanating from notches compared with solid line representing conventional long crack da/dN data. After ref. 94.

by the micromechanisms of near-threshold fatigue. Figure 21, for example, shows a highly serrated crystallographic crack profile following an 80% overload in 7075 aluminum alloy at a baseline $\Delta K = 7.7$ MPa$\sqrt{m}$, where a planar fracture surface will be observed in the absence of an overload (100).

Oxide-induced crack closure has also been suggested as a possible explanation for the transient growth characteristics arising from fatigue underloads (101). Suresh and Ritchie (101) demonstrated that even though cyclic underloads with ΔK values smaller than the threshold ΔK_o do not contribute to crack advance or transient growth due to changes in crack tip plasticity, they can result in enhanced oxidation on fracture surfaces. Results in a 2 1/4Cr-1Mo steel indicated (101) that whenever the cyclic CTOD corresponding to the fatigue underload was greater than the maximum thickness of the excess oxide formed prior to the application of the underload, a further enhancement in oxide thickness occurs during underload cycling (such as the case when a fatigue test is "parked" overnight at ΔK values below the threshold ΔK_o).

Crack Growth and Threshold Measurement

The dominant role of crack closure in influencing propagation rates also implies that the conditions in the **wake** of the crack and **prior** loading history can have a bearing on the **current** crack propagation rates. This poses a serious question on the very uniqueness of the crack growth rate at low ΔK levels (where closure effects are important) and on the value of ΔK_o for a given material, environment and testing conditions. Clearly too rapid

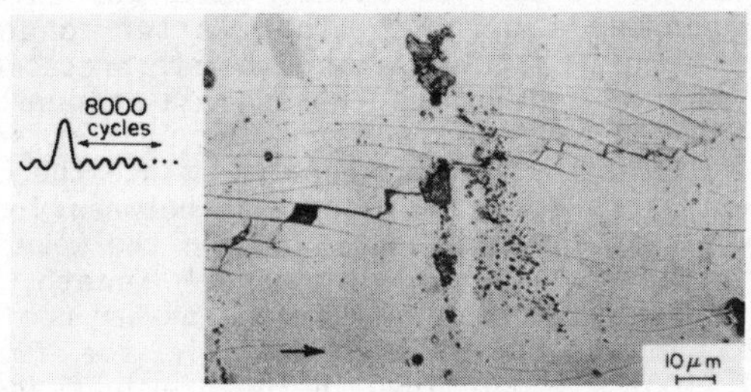

Figure 21 - Crystallographic crack advance following an 80% overload at a baseline ΔK = 5 MPa$\sqrt{m}$ in an underaged 7075 aluminum alloy leading to a highly serrated crack profile and roughness-induced closure; this implies that near-threshold mechanisms can play a significant role in influencing post-overload growth. After ref. 100.

a rate of load shedding will result in premature arrest and an apparently higher threshold ΔK_o value (e.g., 102), presumably from enhanced plasticity-induced closure (akin to retardation following a high-low block loading sequence). Similarly if the rate of load shedding is too slow, it is conceivable that enlarged oxide deposits can form, again leading to an apparently higher ΔK_o from enhanced oxide-induced closure. Evidence for such effects have been reported both for aluminum alloys (102) and steels (103). However, the degree of non-uniqueness for most normal load shedding rates (i.e., < 10% reduction in ΔK for each load step over increments of 1 mm of crack advance) is not as large as might be expected. This follows because the closure mechanisms which dominate near-threshold crack growth behavior are developed in the very near tip region, a short distance behind the crack tip, and thus will only affect subsequent crack extension over a similar distance ahead of the tip. Estimates based on oxide debris measurements suggest that the crack surface contact due to oxide-induced closure to be within roughly 5 μm from the crack tip (18,26,33). More macroscopic experiments where material was removed behind the crack tip similarly indicated that closure within 1 mm of the tip is the most significant in influencing near-threshold behavior (104). However, it is clear that, despite the uncertainty in the location of crack surface contact, the development of a threshold for no fatigue crack growth is intimately linked to the extent of crack closure (25-27), and since the magnitude of such closure will be dependent upon variables such as crack size, load history and so forth, some degree of variability in threshold measurements must be expected.

Concluding Remarks

In this review we have critically examined the many sources of fatigue crack closure and their influence on crack growth behavior, especially in the near-threshold region, under various mechanical, microstructural and environmental conditions. An understanding of the role of such closure processes is also found essential to such phenomena as the behavior of short cracks and the transient crack growth behavior during variable amplitude loading, both for single overloads and block loading, above and below the threshold. In all these instances, closure provides a mechanism whereby the local driving force for crack advance, i.e., the near tip ΔK_{eff} value, differs from the nominally applied values, i.e., ΔK, computed globally in terms of geometry and applied loads. This latter concept clearly is of great importance to the fracture mechanics interpretation of fatigue data since it implies that one can no longer assume a unique dependence of the growth rate, for a given material/environment system, solely in terms of particular values of ΔK and K_{max}. Furthermore, since the closure mechanisms which primarily contribute to this discrepancy between local (near-tip) and global (far field) stress intensities all act in the wake of the crack tip, their effect is critically sensitive to crack length, implying a crack length dependence on the local driving force even when nominal ΔK levels are identical. This constitutes a breakdown in the fracture mechanics similitude concept (106). As outlined in the previous section, it is felt that this latter fact is one of the major causes of the short crack problem where short flaws are observed to propagate below ΔK_o at rates often far in excess of long cracks at the same nominal ΔK levels. However, such "anomalous" behavior of short flaws is only one example, albeit a major one, of problems of uniqueness arising from the current practice of characterizing crack extension in terms of nominal driving forces, e.g., ΔK, ΔJ, etc., when closure mechanisms are clearly present. For example, the varying dependence of ΔK_o values with microstructure and environment at high and low load ratios, the lack of uniqueness in ΔK_o measurements when rates of load shedding are radically changed, the varying response of long crack da/dN tests with classical stress-strain/life behavior, and so forth all follow from the presence of crack closure mechanisms.

In view of this situation, one should conclude that the characterization of fatigue crack propagation data in terms of a nominal characterizing parameter, such as ΔK, should be discouraged since in the presence of closure it does not represent the actual crack driving force. The problem with this approach, however, is what to substitute in its place, since no fundamental continuum mechanics characterizations exist to date for fatigue cracks, which include a consideration of cyclic plasticity, an analysis of the non-stationary crack tip fields, and a modification of these fields due to closure behavior in the wake. Until such analyses are available, the use of ΔK_{eff}, representing closure-adjusted ΔK values, provides probably the most fundamental approach, at least for academic assessment of fatigue behavior. However, in view of the many difficulties in obtaining reproducible K_{cl} data, the use of ΔK_{eff} will clearly present numerous practical difficulties for the standardization of experimentally-determined fatigue crack propagation data for engineering purposes.

Acknowledgements

This work was supported partly by the Director, Office of Energy Research, Office of Basic Energy Sciences, Materials Science Division of the U.S. Department of Energy, under Contract No. DE-AC03-76SF00098 and partly by the Materials Research Laboratory under NSF Grant No. DMR-8216726. Thanks are due to Dr. S. M. Wulf for his continued support, both at the University of California in Berkeley and formerly at M.I.T., to J. L. Tzou and V. B. Dutta for providing us access to their as yet unpublished work, and to Madeleine M. Penton for her help in preparing this manuscript.

References

1. F. A. McClintock, p. 65 in Fracture of Solids, Interscience, New York, 1963.

2. N. E. Frost, p. 1433 in Proc. First Int. Conf. on Fract., T. Yokobori, ed.; Sendia, Japan, 1966.

3. P. C. Paris, Closed Loop Magazine, 2 (1970) p. 11.

4. P. C. Paris, M. Gomez, and W. E. Anderson, Trend in Engineering, 13 (1961) p. 9, University of Washington, Seattle.

5. K. Endo, K. Komai, and Y. Matsuda, Memo. Fac. Eng., Kyoto Univ., 31 (1969) p. 25.

6. W. Elber, ASTM STP 486 (1971) p. 280.

7. T. C. Lindley and C. E. Richards, Mater. Sci. Eng., 14 (1974) p. 281.

8. A. J. McEvily, Met. Sci., 11 (1977) p. 274.

9. R. A. Schmidt and P. C. Paris, ASTM STP 536 (1973) p. 79.

10. R. O. Ritchie, Int. Met. Rev., 20 (1979) p. 205.

11. J. C. Newman, Jr., p. 6-1 in Behavior of Short Cracks in Airframe Components, AGARD Conf. Proceedings No. 328, Advisory Group for Aeronautical Research and Development, France (1983).

12. J. Schijve and W. J. Arkema, Dept. of Aerospace Eng. Report VTH-217, Delft University, The Netherlands (1976).

13. B. Budiansky and J. W. Hutchinson, J. App. Mech., Trans. ASME, Series E, 45 (1978) p. 267.

14. M. Kanninen and J. Atkinson, Int. J. Fract., 16 (1980) p. 53.

15. J. C. Newman, Jr., Ph.D. Thesis, Virginia Polytechnic Inst. and State Univ., 1974.

16. J. C. Newman, Jr., ASTM STP 748 (1981) p. 53.

17. R. O. Ritchie, S. Suresh, and C. M. Moss, J. Eng. Mater. Technol., Trans. ASME, Series H, 102 (1980) p. 293.

18. S. Suresh and R. O. Ritchie, Eng. Fract. Mech., 18 (1983) p. 785.

19. P. C. Paris, R. J. Bucci, E. T. Wessel, W. G. Clark and T. R. Mager, ASTM STP 513 (1972) p. 141.

20. T. C. Lindley, Central Electricity Research Laboratories Report, 1978, Leatherhead, Surrey, U.K.

21. L. K. L. Tu and B. B. Seth, J. Test. Eval., 6 (1978) p. 66.

22. R. P. Skelton and J. R. Haigh, Mater. Sci. Eng., 36 (1978) p. 17.

23. H. Kitagawa, S. Toyohira, and K. Ikeda, in Fracture Mechanics in Engineering Applications, G. C. Sih and S. R. Valluri, eds.; Sijthoff and Noordhof, The Netherlands, 1981.

24. A. T. Stewart, Eng. Fract. Mech., 13 (1980) p. 463.

25. S. Suresh, G. F. Zamiski, and R. O. Ritchie, Met. Trans. A, 12A (1981) p. 1435.

26. S. Suresh, D. M. Parks, and R. O. Ritchie, p. 391 in Fatigue Thresholds, J. Bäcklund, A. F. Blom, and C. J. Beevers, eds.; EMAS Ltd., Warley, U.K., 1982.

27. S. Suresh, J. Toplosky, and R. O. Ritchie, p. I329 in Fracture Mechanics, 14th Symp.; Vol. 1, Theory and Analysis, ASTM STP 791, 1983.

28. J. E. King, Fat. Eng. Mat. Struct., 5 (1982) p. 177.

29. J. L. Yuen and W. Nix, Met. Trans. A, 14A (1983) in press.

30. P. K. Liaw, T. R. Leax, R. S. Williams, and M. G. Peck, Met. Trans. A, 13A (1982) p. 1607.

31. A. K. Vasudévan and S. Suresh, Met. Trans. A, 13A (1982) p. 2271.

32. R. O. Ritchie, S. Suresh, and P. K. Liaw, p. 443 in Ultrasonic Fatigue, J. M. Wells, O. Buck, L. D. Roth, and J. K. Tien, eds.; TMS-AIME, Warrendale, PA, 1982.

33. S. Suresh and R. O. Ritchie, Scripta Met., 17 (1983) p. 575.

34. P. J. E. Forsyth, p. 76 in Crack Propagation, Proc. Symp., Cranfield College of Aeronautics, Cranfield Press, U.K., 1962.

35. K. Minakawa and A. J. McEvily, Scripta Met., 15 (1981) p. 937.

36. E. P. Louwaard, Delft Univ. of Tech., Dept. of Aero Eng. Report LR-243, Delft, The Netherlands (1977).

37. R. O. Ritchie and S. Suresh, Met. Trans. A, 13A (1982) p. 937.

38. S. Suresh, Met. Trans. A, 14A (1983) p. 2375.

39. D. L. Davidson, Fat. Eng. Mat. Struct., 3 (1981) p. 229.

40. S. Purushothaman and J. K. Tien, p. 1267 in Strength of Metals and Alloys, ICSMA5 Conf. Proc., P. Haasen et al., eds.; Pergamon Press, New York, vol. 2, 1979.

41. N. Walker and C. J. Beevers, Fat. Eng. Mat. Struct., 1 (1979) p. 135.

42. R. J. Asaro, L. Hermann, and J. M. Baik, Met. Trans. A, 12A (1981) p. 1133.

43. G. T. Gray, III, A. W. Thompson, and J. C. Williams, Met. Trans. A, 14A (1983) p. 421.

44. F.-S. Lin and E. A. Starke, Jr., Mater. Sci. Eng., 45 (1980) p. 153.

45. S. Suresh, A. K. Vasudēvan, and P. E. Bretz, Met. Trans. A, 15A (1984) p. 369.

46. V. B. Dutta, S. Suresh, and R. O. Ritchie, Met. Trans. A, 15A (1984) in review.

47. S. Suresh and R. O. Ritchie, Met. Trans. A, 13A (1982) p. 1627.

48. K. Endo, T. Okada, and T. Hariya, Bull. JSME, 15 (1972) p. 439.

49. K. Endo, T. Okada, K. Komai, and M. Kiyota, Bull. JSME, 15 (1972) p. 1316.

50. J. L. Tzou, S. Suresh, and R. O. Ritchie, p. 711 in Mechanical Behavior of Materials IV, Proc. 14th Int. Conf. (ICM-4), J. Carlsson and N. G. Olhsen, eds.; Pergamon Press, Oxford, vol. 2, 1983.

51. J. L. Tzou, S. Suresh, and R. O. Ritchie, Lawrence Berkeley Laboratory Report No. LBL-16028, Sept. 1983, University of California, submitted to Acta Met.

52. V. F. Zackay, E. R. Parker, D. Fahr, and R. Busch, Trans. ASM, 60 (1967) p. 252.

53. A. G. Evans, A. H. Heuer, and D. L. Porter, p. 529 in Fracture 1977, Waterloo, Canada (ICF-4), D. R. M. Taplin, ed.; Pergamon Press, Oxford, vol. 1, 1977.

54. E. Hornbogen, Acta Met., 26 (1978) p. 147.

55. R. M. McMeeking and A. G. Evans, J. Amer. Cer. Soc., 65 (1982) p. 242.

56. B. Budiansky, J. W. Hutchinson, and J. C. Lambropoulos, Int. J. Solids Struct., 19 (1983) p. 337.

57. R. O. Ritchie and J. F. Knott, Acta Met., 21 (1973) p. 639.

58. R. J. Cooke, P. E. Irving, G. S. Booth, and C. J. Beevers, Eng. Fract. Mech., 7 (1975) p. 69.

59. R. O. Ritchie, J. Eng. Mater. Technol., Trans. ASME, Series H, 99 (1977) p. 195.

60. E. K. Priddle, Scripta Met., 12 (1978) p. 49.

61. P. Au, T. H. Topper, and M. L. El Haddad, p. 11-1 in Behavior of Short Cracks in Airframe Components, AGARD Conf. Proceedings No. 328; Advisory Group for Aeronautical Research and Development, France (1983).

62. A. Blom, A. Hadrboletz, and B. Weiss, p. 755 in Mechanical Behavior of Materials IV, Proc. 4th Int. Conf. (ICM-4), J. Carlsson and N. G. Olhsen, eds.; Pergamon Press, Oxford, vol. 2, 1983.

63. A. Bignonnet, R. Namdar-Irani, and M. Truchon, Scripta Met., 16 (1982) p. 795.

64. A. K. Vasudēvan and S. Suresh, unpublished research, ALCOA, 1983.

65. V. B. Dutta and R. O. Ritchie, unpublished research, University of California, Berkeley, 1983.

66. I. C. Mayes and T. J. Baker, Fat. Eng. Mat. Struct., 4 (1981) p. 79.

67. H. Kitagawa, H. Nishitani, and J. Matsumoto, in Proc. 3rd Intl. Cong. on Fracture (ICF-3); Dusseldorf, Verein Deutsher Eisenhuttenleute, vol. 5, 1973, paper V-444/A.

68. J. Masounave and J. P. Baïlon, Scripta Met., 10 (1975) p. 165.

69. H. Kitagawa and S. Takahashi, p. 627 in Proc. 2nd Intl. Conf. on Mech. Beh. of Materials, Boston, MA, 1976.

70. G. G. Garrett and J. F. Knott, Acta Met., 23 (1975) p. 841.

71. J. L. Robinson and C. J. Beevers, Met. Sci., 7 (1973) p. 153.

72. G. R. Yoder, L. A. Cooley, and T. W. Crooker, p. 1865 in Titanium 80, H. Kimura and O. Izumi, eds.; AIME, Warrendale, PA, vol. 3, 1980.

73. N. R. Moody and W. W. Gerberich, p. 292 in Fatigue Mechanisms, J. T. Fong, ed.; ASTM STP 675, Amer. Soc. Test. Matls., Philadelphia, PA, 1979.

74. J. E. Allison, Ph.D. Thesis, Carnegie-Mellon University, Pittsburgh, 1982.

75. C. J. Beevers, p. 257 in Fatigue Thresholds, J. Bäcklund, A. F. Blom, and C. J. Beevers, eds.; EMAS Ltd., Warley, U.K., vol. 1, 1982.

76. R. D. Carter, E. W. Lee, E. A. Starke, and C. J. Beevers, Met. Trans. A, 14A (1983) in press.

77. A. K. Vasudévan, P. E. Bretz, A. C. Miller, and S. Suresh, Mater. Sci. Eng., 1984, in press.

78. G. R. Yoder, L. A. Cooley, and T. W. Crooker, Eng. Fract. Mech., 11 (1979) p. 86.

79. H. Suzuki and A. J. McEvily, Met. Trans. A, 10A (1970) p. 475.

80. K. Minakawa, Y. Matsuo, and A. J. McEvily, Met. Trans. A, 13A (1982) p. 439.

81. J. A. Wasynczuk, R. O. Ritchie, and G. Thomas, Mater. Sci. Eng., 62 (1983) in press.

82. T. Kunio, M. Shimizu, K. Yamada, and H. Nakabayashi, p. 409 in Fatigue Thresholds, J. Bäcklund, A. F. Blom, and C. J. Beevers, eds.; EMAS Ltd., Warley, U.K., vol. 1, 1982.

83. K. A. Esaklul, A. G. Wright, and W. W. Gerberich, Scripta Met., 17 (1983)

84. P. K. Liaw, S. J. Hudak, and J. K. Donald, p. II in Fracture Mechanics, 14th Symp., Vol. 2, Testing and Applications, ASTM STP 791, 1983.

85. J. Petit, this volume.

86. B. S. Greenwell and R. N. Parkins, Fat. Eng. Mat. Struct., 5 (1982) p. 115.

87. S. Suresh and R. O. Ritchie, Int. Met. Rev., 25 (1984) in press.

88. S. Pearson, Eng. Fract. Mech., 7 (1975) p. 235.

89. W. L. Morris, M. R. James, and O. Buck, Met. Trans. A, 12A (1981) p. 57.

90. J. Lankford, Fat. Eng. Mat. Struct., 5 (1982) p. 233.

91. R. P. Gangloff, Res. Mechanica Letters, 1 (1981) p. 299.

92. M. M. Hammouda and K. J. Miller, ASTM STP 668 (1979) p. 703.

93. J. F. McCarver and R. O. Ritchie, Mater. Sci. Eng., 55 (1982) p. 63.

94. K. Tanaka and Y. Nakai, Fat. Eng. Mat. Struct., 6 (1983) in press.

95. O. E. Wheeler, J. Basic Eng., Trans. ASME, Series D, 94 (1972) p. 181.

96. J. Willenborg, R. M. Engle, and H. Wood, Technical Report TFR 71-701, North American Rockwell, Los Angeles (1971).

97. E. F. J. von Euw, R. W. Hertzberg, and R. Roberts, ASTM STP 513 (1972) p. 230.

98. S. Suresh, Scripta Met., 16 (1982) p. 995.

99. S. Suresh, Eng. Fract. Mech. 18 (1983) p. 577.

100. S. Suresh and A. K. Vasudĕvan, this volume.

101. S. Suresh and R. O. Ritchie, Mater. Sci. Eng., 51 (1981) p. 61.

102. R. J. Bucci, "Development of a Proposed Standard Practice for Near-Threshold Fatigue Crack Growth Rate Measurement," ALCOA Tech. Rep. No. 57-79-14, Dec. 1979, ALCOA, PA.

103. A. J. Cadman, C. E. Nicholson, and R. Brook, Scripta Met., 17 (1983) in press.

104. K. Minakawa, J. C. Newman, Jr., and A. J. McEvily, Fat. Eng. Mat. Struct., 6 (1983) in press.

105. K. Schulte, H. Nowack, and K. H. Trautmann, Z. Werkstofftech, 11 (1980) p. 287.

106. R. O. Ritchie and S. Suresh, Mater. Sci. Eng., 57 (1983) p. 427.

NEAR-THRESHOLD FATIGUE CRACK GROWTH AND CRACK CLOSURE IN 17-4 PH STEEL AND 2024-T3 ALUMINIUM ALLOY

A.F. Blom

Structures Department
The Aeronautical Research
Institute of Sweden
P.O. Box 11021
S-161 11 Bromma
SWEDEN

and

Department of Aeronautical
Structures and Materials
The Royal Institute of
Technology
S-100 44 Stockholm

An experimental and analytical programme has been carried out to study fatigue crack growth and crack closure at low stress intensity ranges. Experiments were performed on 17-4 PH steel and on aluminium alloy 2024-T3, both materials being used in the aircraft industry, for stress ratios R ranging from - 1 to 0.75. Crack closure was monitored by means of compliance measurements during the tests and it was found that closure loads increased significantly when threshold conditions were approached. These high closure loads are explained in terms of a shear-mode type of crack growth and the formation of oxide debris at the crack tip. The crack growth is transgranular for Al 2024-T3 whereas for 17-4 PH steel crack growth is intergranular at low stress intensites but transgranular at higher ΔK. When crack closure measurements are performed at higher ΔK-values they correlate well with numerically computed values of the Elber type of plasticity-induced crack closure. At R = -1 lower thresholds are found than at small positive R-values. This is explained by smoother crack surfaces due to the compressive loads and consequently less roughness-induced crack closure.

Introduction

During the past decade certification and maintanence of both fighter and civil aircrafts have come to rely on damage tolerance analysis (1). Defects are assumed to exist at highly stressed areas, stress intensity factors as function of crack length are computed by means of linear elastic fracture mechanics and crack length versus number of load cycles or number of flights are calculated by numerical cycle-by-cycle integration of experimentally determined fatigue crack growth rates from constant amplitude tests. Crack growth retardation and/or acceleration due to irregular loading is normally taken into account by means of some semi-empirical method like the Wheeler-model. Once the crack length versus number of load cycles is determined suitable inspection intervals can be defined. A requirement for realistic damage tolerance analyses is a knowledge of the fatigue crack growth properties of the actual material. Recently it has been observed that the linear log da/dN versus log ΔK relationship in the Paris regime (2) ceases to exist when the crack growth rate is decreased below approximately 10^{-6} mm in the so-called near-threshold regime. Much effort is currently spent in investigating the fatigue behaviour in this regime.

This paper presents experimental results from measurements on constant amplitude fatigue crack growth rates and threshold stress intensity levels for two high strength materials currently being used in aeronautical applications, aluminium alloy 2024-T3 and 17-4 PH steel. Emphasis is put on near-threshold behaviour where it is found that the crack growth results, and likely the very existence of a fatigue threshold, may be rationalized in terms of different crack closure mechanisms. Microstructural features at threshold and effects of stress ratio, thickness and specimen geometry are discussed.

Materials

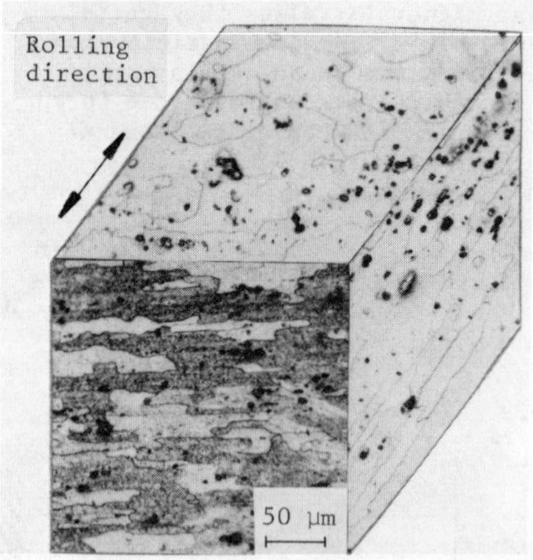

2024-T3 Aluminium Alloy

The aluminium alloy discussed in the present paper is wrought 2024 aluminium in T3 temper, commonly used both in fighter and civil aircrafts in parts subjected to predominantly tensile loads e.g. lower wing skin and fuselage. The specimen material was supplied as a plate of 2 mm and 6 mm thickness with a chemical composition as given in Table Ia. This plate had been given a thermomechanical pretreatment consisting of solution heat treatment, cold working and natural aging to a substantially stable condition resulting in room-temperature mechanical properties as given in Table Ib. The microstructure of the wrought and heat treated Al 2024-T3 consists of a typical pancake-like grain structure, Figure 1, with a grain size of roughly 120×70×20 µm.

Figure 1 - Microstructure of Aluminium Alloy 2024-T3. A typical pancake-like grain structure is seen.

Table I. Properties of Aluminium Alloy 2024-T3

(a) Chemical Composition (Weight per cent)

Cu	Mg	Mn	Fe	Si	Ti	Zn	Cr
4.5	1.5	0.6	0.2	0.1	<0.05	<0.05	<0.05

(b) Room Temperature Mechanical Properties

Yield Strength (MPa)	UTS (MPa)	Elongation (%)	K_{IC} (MPa$\sqrt{m}$)
345	450	17.5	35

17-4 PH Steel

High-strength precipitation hardening (PH) stainless steels are used in aerospace applications as well as in naval structures. The 17-4 family of PH stainless steels is a higly complex alloy with mechanical properties varying widely, depending upon processing and heat treatment (3). Very little information is available on fatigue crack growth behaviour in this material. The author is only aware of the work by Crooker et al (4) which, however, is limited to crack growth rates above 10^{-5} mm/cycle.

Test specimens were machined from a bar, with chemical composition as given in Table IIa, and then heat treated to achieve room-temperature mechanical properties as given in Table IIb. The resulting microstructure is shown in Figure 2. Some elongated non-metallic inclusions parallel to the rolling direction of the bar may be seen in Figure 2 together with traces of prior austenite grains. There is also some moderate banding of the microstructure suggesting slight segregation effects. The general microstructure consists of tempered martensite.

Table II. Properties of Steel 17-4 PH

(a) Chemical Composition (Weight per cent)

C	Si	Mn	P	S	Cr	Ni	Nb	Cu
0.05	1.0	1.0	0.025	0.025	16.5	4.0	0.3	4.0

(b) Room Temperature Mechanical Properties

Yield Strength (MPa)	UTS (MPa)	Elongation (%)	K_Q (MPa$\sqrt{m}$)
1105	1135	14.5	120

Experimental Procedures

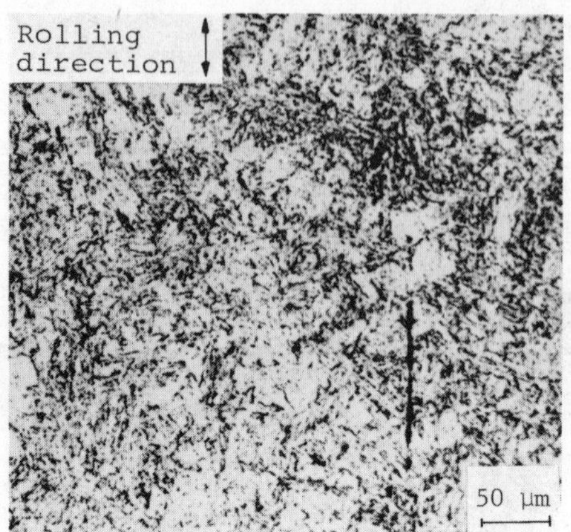

Figure 2 - Microstructure of Steel 17-4 PH. A tempered martensite structure is evident with some small elongated inclusions.

Fatigue crack growth tests in steel were performed on compact tension (CT) specimens, W = 60 mm, 5 and 10 mm in thickness, machined in the C-L orientation. Tests in aluminium were performed both on CT-specimens, W = 60 mm, and on center-cracked-tension (CCT) specimens, W = = 50 mm, 2 and 6 mm in thickness, machined in the L-T orientation, Neither specimen geometry nor thickness affected the rates of near threshold crack propagation. Because of this results will be presented without stating which test combination was used in each case. All experiments were performed with servohydraulic testing machines (MTS and Schenk) at test frequencies between 40 and 80 Hz in laboratory air at ambient temperature. Experiments were performed with stress ratios varying from -1 to +0.75. Several experiments were performed up to five times to check for repeatability and the average values are reported in this paper. Crack length was determined using elastic compliance and visual observation by means of a low-magnification travelling light microscope with a resolution better than 50 µm. For each test specimen the experimental procedure consisted of precracking, manual load shedding down to a threshold ΔK_{th} value defined at a crack propagation rate not ex ceeding 10^{-7} mm/cycle and then increasing the load range and performing fatigue crack propagation measurements at intermediate and high growth rates until fracture occured. The load shedding was done in accordance to the proposed ASTM standard technique (5). Determination of crack closure loads, here defined as the load P_{op} at which the crack opens completely, was per formed with compliance techniques involving either back-face strain (BFS) or crack opening displacement (COD) measurements. Both techniques yielded the same P_{op} values but BFS-measurements were found to be more sensitive to changes in compliance. The relative error in crack closure measurements is estimated to be less than ± 10%. These measurements were performed statically by interrupting the fatigue crack growth experiments.

Numerical Determination of Crack Closure

Although plasticity-induced crack closure has been studied both analytically (6,7) and numerically (8,9) this has sofar only been done for plane stress whereas at near-threshold conditions a state of plane strain prevails. Therefore an elastic-plastic finite element analysis of a growing crack was made, incorporating the residual plastic deformations left in the wake of the crack. Details of the numerical procedure is presented elsewhere (10). Calculations of crack closure loads were done for Al 2024-T3 both at plane strain and plane stress. The results are shown as normalized crack closure loads P_{op}/P_{max} versus stress ratio R in Figure 3. For plane stress an excellent agreement is found with Newman's numerical results (8) and the original experimental results by Elber (11, 12). For plane strain the

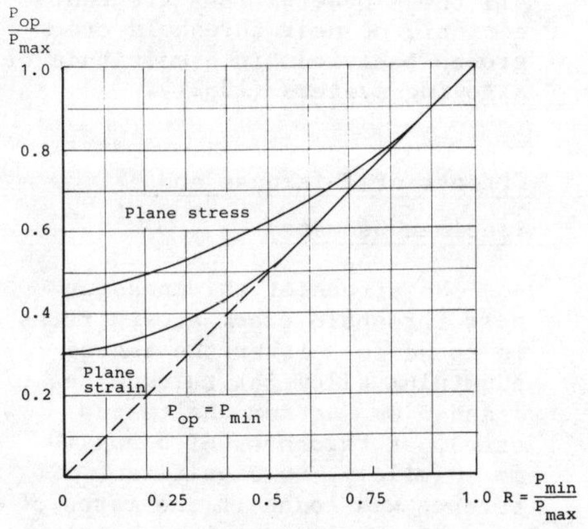

Figure 3 - Normalized plasticity- induced crack closure loads P_{op}/P_{max} as function of load ratio R under plane stress and plane strain. Obtained by elastic-plastic FEM-calculations.

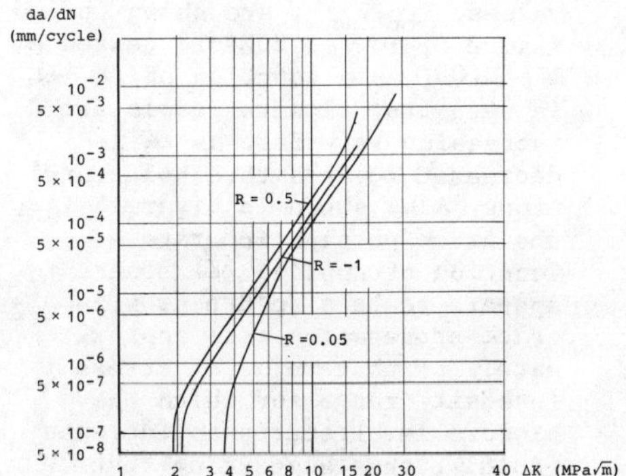

Figure 4 - Effect of R on fcg-rates in Al 2024-T3 at room temperature and laboratory air.

general trend of the results are similar to those for plane stress only with the difference that less crack closure develops, as anticipated, due to more plastic constraint. Thus, the very high crack closure loads found at near threshold crack growth rates, as reported in the pertinent literature and also found in the present investigation, cannot be explained solely in terms of plasticity-induced crack closure but other mechanisms must be considered.

Results

Effect of Load Ratio

The effect of R values on fatigue crack propagation rates in Al 2024-T3 and steel 17-4 PH is shown in Figures 4-5. For both materials an increased load ratio increases the fatigue crack growth rate da/dn and decreases the threshold stress intensity ΔK_{th}. However, this is only true at small positive R values. As seen in Figure 4, for aluminium, a negative load ratio R = -1 produces fatigue crack growth rates which are in between those tested at R equal to 0.05 and 0.5. Also for high load ratios the influence of R is negligible as evidenced in Figure 5 for steel. For those R values where an effect is obvious it is found that this effect increases as the crack propagation rate decreases.

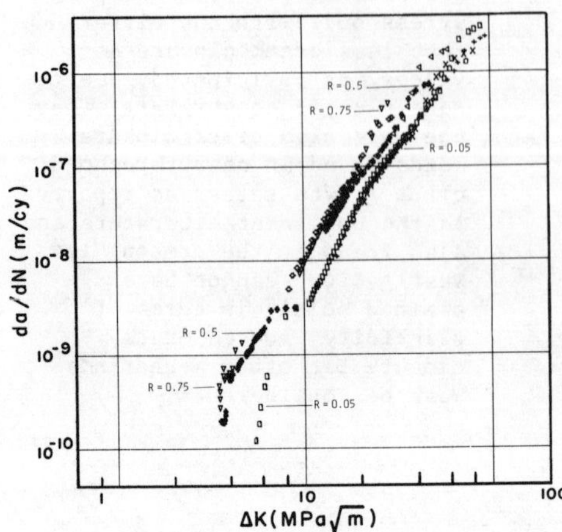

Figure 5 - Fcg-rates in Steel 17-4 PH at room temperature and laboratory environment. Significantly lower near-threshold rates at R = 0.05 than at R = 0.5 and R = 0.75 due to crack closure effects.

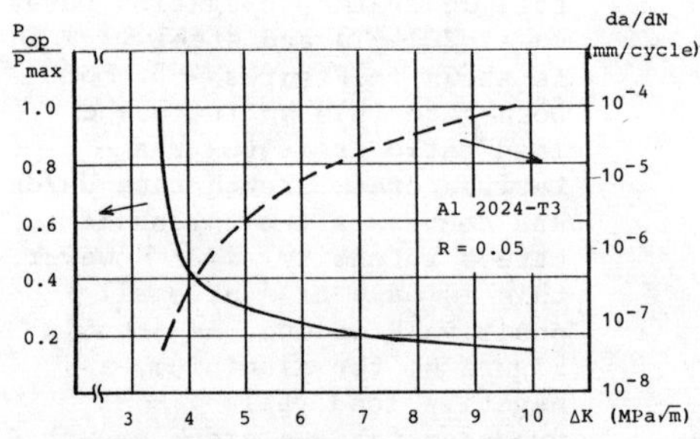

Figure 6 - Variation in crack closure level and crack growth rate versus ΔK for Al 2024-T3 tested at R = 0.05.

All these observations are characteristic of near threshold crack growth behaviour in a multitude of alloying systems (13-21).

Effects of Thickness and Specimen Geometry

No effect of thickness on near threshold crack growth rates is found for either the tested aluminium alloy, at thicknesses of 2 and 6 mm, or for the tested steel, at thickness of 5 and 10 mm. Similary, no significant difference was found in the rates of near threshold crack propagation, for tests on Al 2024-T3, due to different specimen geometry, CT and CCT. Thus, based on concepts of linear elastic fracture mechanics, near threshold fatigue crack growth data obtained in the laboratory may be used in engineering design irrespective of the particular geometry.

Crack Closure

Normalized crack closure values, P_{op}/P_{max}, are shown in Figure 6, for Al 2024-T3 tested at R = 0.05, as a function of ΔK. It is seen that closure levels are increasing very fast as ΔK is decreased towards threshold conditions. Also shown in Figure 6 is the crack propagation rate as function of applied ΔK. There appears to be a sudden drop in crack propagation rate approximately at that value of stress intensity range for which the closure level starts to increase rapidly, thus suggesting that near-threshold crack growth behaviour is essentially due to crack closure. As closure is found to be dependent on ΔK at near threshold growth rates there must be some other mechanism involved than Elber's plasticity-induced crack closure which is, for a given alloy, only assumed to be dependent on stress ratio R (11, 12). Mechanisms that have recently been used to explain the high closure levels at threshold conditions include oxide-induced crack closure (22, 23) and roughness-induced crack closure (24, 25). As will be discussed in some more detail in the subsequent section, both these mechanisms are applicable in the present study both for aluminium and steel. In Figure 6 it may be observed that as ΔK is increased the ratio of P_{op}/P_{max} goes towards a constant value which is less than predicted from continuum considerations based on plasticity-induced crack closure as shown in Figure 3. Similar low closure values at high ΔK levels were also observed by Minakawa and McEvily (26). This is believed to be due to residual plastic deformations building up at the strain gauge, mounted on the back face of the specimen, used for the measurements and thus affecting the measurement results. It seems likely that plasticity-induced crack closure exists as long as continuum considerations are sufficient and that other closure mechanisms become important when the crack growth becomes governed by microstructural features. Assuming that a continuum approach is valid once the plastic zone size is larger than the grain size in the crack growth direction, here approximately 70 μm for Al 2024-T3, it is possible to predict the stress intensity range ΔK at which the ratio P_{op}/P_{max} becomes a constant. Equalizing the monotonic plane strain plastic zone size (27) to the grain size yields a value of $K_{max} = 12.5$ MPa$\sqrt{m}$. A similar calculation for the reversed plastic zone size yields $\Delta K = 25$ MPa$\sqrt{m}$. These values seem to correlate fairly well with the observed trend shown in Figure 6.

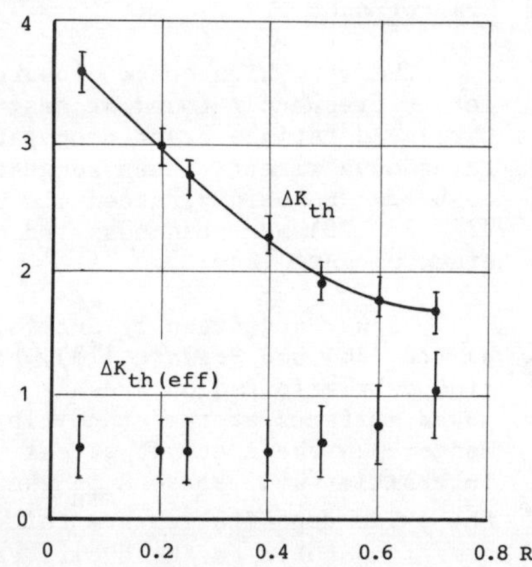

Figure 7 - Comparison of ΔK_{th} and $\Delta K_{th(eff)}$ versus stress ratio for Al 2024-T3 at room temperature and laboratory air.

Although at threshold conditions several closure mechanisms act simultaneously it has been found that the effect of R on ΔK_{th} and on near threshold crack propagation rates may be rationalized in terms of crack closure (22-26), as is done at higher crack growth rates by means of plasticity-induced closure alone (11, 12). In Figure 7 values of ΔK_{th}, for Al 2024-T3, are plotted as function of R. Also shown in Figure 7 are effective threshold stress intensities $\Delta K_{th(eff)}$, determined as $\Delta K_{eff} = = K_{max} - K_{op}$ at threshold conditions, as function of stress ratio. It is deduced from Figure 7 that $\Delta K_{th(eff)}$ is rather independent of stress ratio, although some influence may be seen at high positive R values and at R = -1, and thus it seems that the R-dependence on ΔK_{th} may be explained essentially as a crack closure effect.

The discussion above, based on results for Al 2024-T3, seems equally applicable to the near threshold behaviour of steel 17-4 PH as evidenced from Figures 8-9. In Figure 8 the variation of crack closure levels and fatigue crack growth rates for two different R values, R = 0.05 and 0.75, are plotted versus stress intensity range. It is seen that for R = 0.75, the crack closure level is essentially the same irrespective of the ΔK-value.

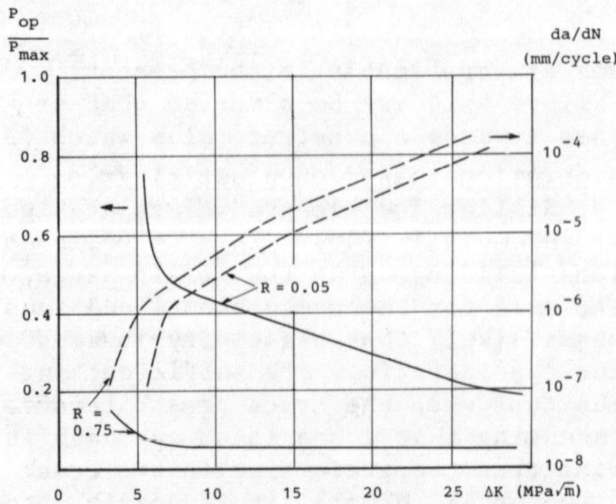

Figure 8 - Variation in crack closure levels and crack growth rates versus ΔK for Steel 17-4 PH tested at R = 0.05 and 0.75.

This means that effects of closure mechanisms, other than plasticity-induced crack closure are minimized at high R-values. Threshold values ΔK_{th} and effective threshold stress intensities $\Delta K_{th(eff)}$ as function of R are shown in Figure 9. The influence of R on ΔK_{th} may partly be explained in terms of crack closure. Another effect of R could be associated with K_{max} rather than ΔK. It is observed that the incidence of intergranular facets increases with R and thus K_{max}, possibly associated with an environmental contribution to crack growth (embrittled grain boundaries). The dependence of $\Delta K_{th(eff)}$ on load ratio in Figure 9 may also partly be due to the uncertainty in crack closure measurements.

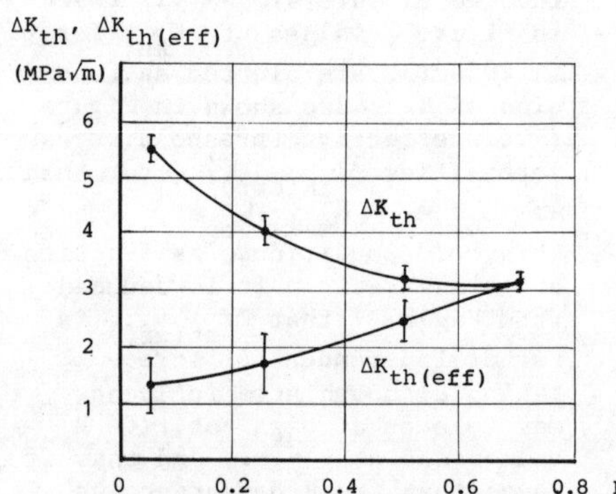

Figure 9 - Comparison of ΔK_{th} and $\Delta K_{th(eff)}$ versus stress ratio for Steel 17-4 PH at room temperature and laboratory air.

Fractography

The very high crack closure levels frequently found at near threshold fatigue crack propagation rates have recently been suggested to be due to oxide-induced closure (22, 23, 28) and roughness-induced crack closure (24-26).

As was suggested by Suresh et al (22, 28) amd Stewart (23), corrosion debris formed on freshly exposed surfaces at the crack tip may wedge-open the crack at stress intensities well above K_{min} when the oxide deposits reach a thickness comparable to the cyclic crack tip opening displacement. This effect is strongest at low stress intensities where fretting oxidation is enhanced by local Mode II displacements (25, 29). Also the effect should be strongest at small positive R-values where the effect of plasticity-induced crack closure is largest as seen in Figure 3. At higher R values crack closure becomes of less importance and fretting oxidation will be supressed. In Figures 10 and 11 fracture surfaces are shown of aluminium and steel specimens, respectively, tested at different R-values. As seen in Figures 10-11 most debris is observed at the lowest stress ratio whereas at high R-values no significant amount of debris can be seen.

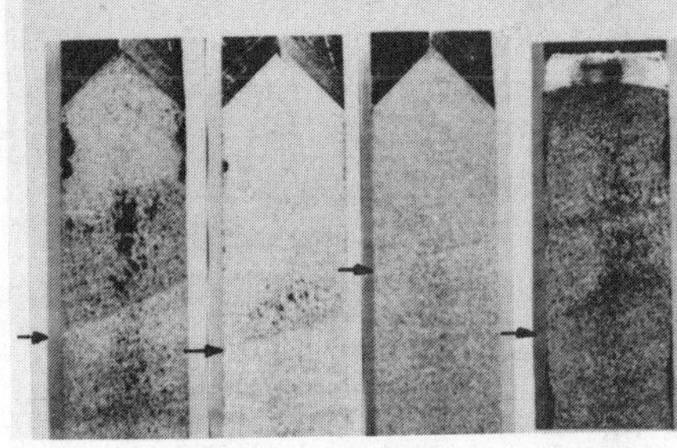

R: 0.05 0.25 0.5 -1
Figure 10 - Fracture surfaces of Al 2024-T3 showing oxide debris at threshold. Threshold indicated by arrows.

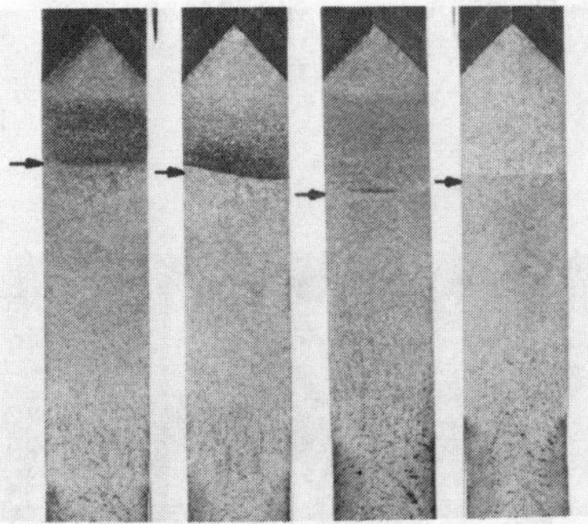

R: 0.05 0.25 0.5 0.75
Figure 11 - Fracture surfaces of Steel 17-4 PH. Note lack of oxide debris for high R-values. Threshold indicated by arrows.

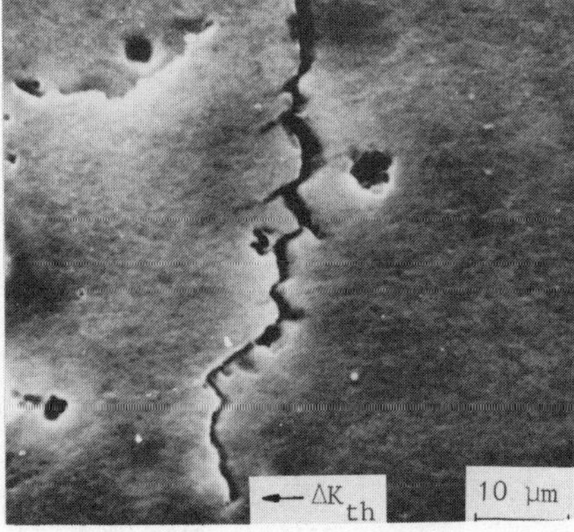

Figure 12 - SEM micrograph showing a serrated crack path in Al 2024-T3 at threshold conditions. R = 0.05.

Besides oxide-induced crack closure, which is believed to play a relatively minor role in the present investigation, high closure levels in the near-threshold regime may be due to roughness-induced closure. Such crack closure is due to a serrated crack path near the threshold ΔK_{th}, Figure 12, in conjunction with local shear displacements which may be deduced from the shear facets shown in Figure 13. The fracture surface topography, for Al 2024-T3, is predominantly transcrystalline as can be seen from the SEM-fractographs shown in Figure 13. Shear facets are seen, in Figure 13, at R = = 0.25 and R = -1, wheras for R = 0.5 the effect of Mode II displacements seems negligible. Similarly for other specimens tested at different R-values an increase in shear facets is found to correlate with a decreased stress ratio. Thus, the effect of roughness-induced crack closure also tends to vanish at high R-values where the effect of plastic closure is negligible, Figure 3. The results discussed above are in agreement with results for other materials presented by Beevers (24) and Minakawa and McEvily (25, 26). As seen in Figure 12 no indications of large amounts of plasticity, such as persistent slip bands, can be seen in Al 2024-T3. This fact was used in a recent paper (20) to explain the lack of any frequency effects, at frequencies between 40 Hz and 20 kHz, on near-threshold crack growth behaviour in this particular aluminium alloy.

For steel 17-4 PH, SEM micrographs of the fracture morphology at threshold are shown in Figure 14. Both for R = 0.05 and R = 0.5 the threshold crack front is clearly visible, Figure 14 (a,b). The fracture surface topography at threshold is a mixture of faceted and fibrous fracture. The faceted mode of fracture appears to be intergranular and forms for R = 0.5 more than 50% of the fracture surface at threshold, Figure 14(b). The highly irregular crack path due to intergranular fracture might give rise to high crack closure load

(a)

levels due to surface asperities blocking each other in conjunction with the observed Mode II displacements at threshold for low R-values, Figure 14(c). Intergranular facets are present both before and after the threshold crack length, together with some secondary cracking, Figure 14(d). Examination of the region immediately after the threshold crack length, i.e. where the load range had been increased, show a few intergranular facets and a rapid return to the fibrous fatigue crack growth mode, Figure 14(e). Well beyond the threshold, at high ΔK, the fracture appearance is fibrous with some secondary branch cracks present, Figure 14(f).

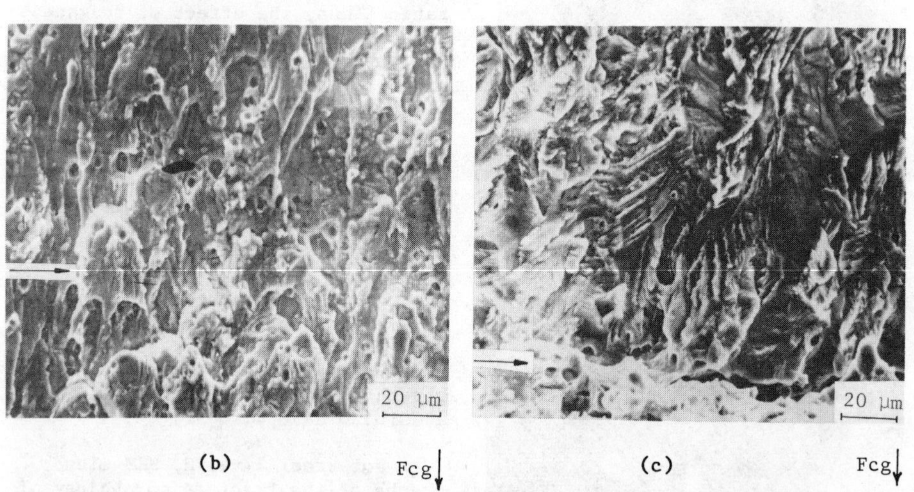

(b) (c)

Figure 13 - SEM micrographs of fracture surfaces of Al 2024-T3 at threshold conditions. Threshold is indicated by arrows. (a) R = 0.25 (b) R = 0.5 (c) R = -1.

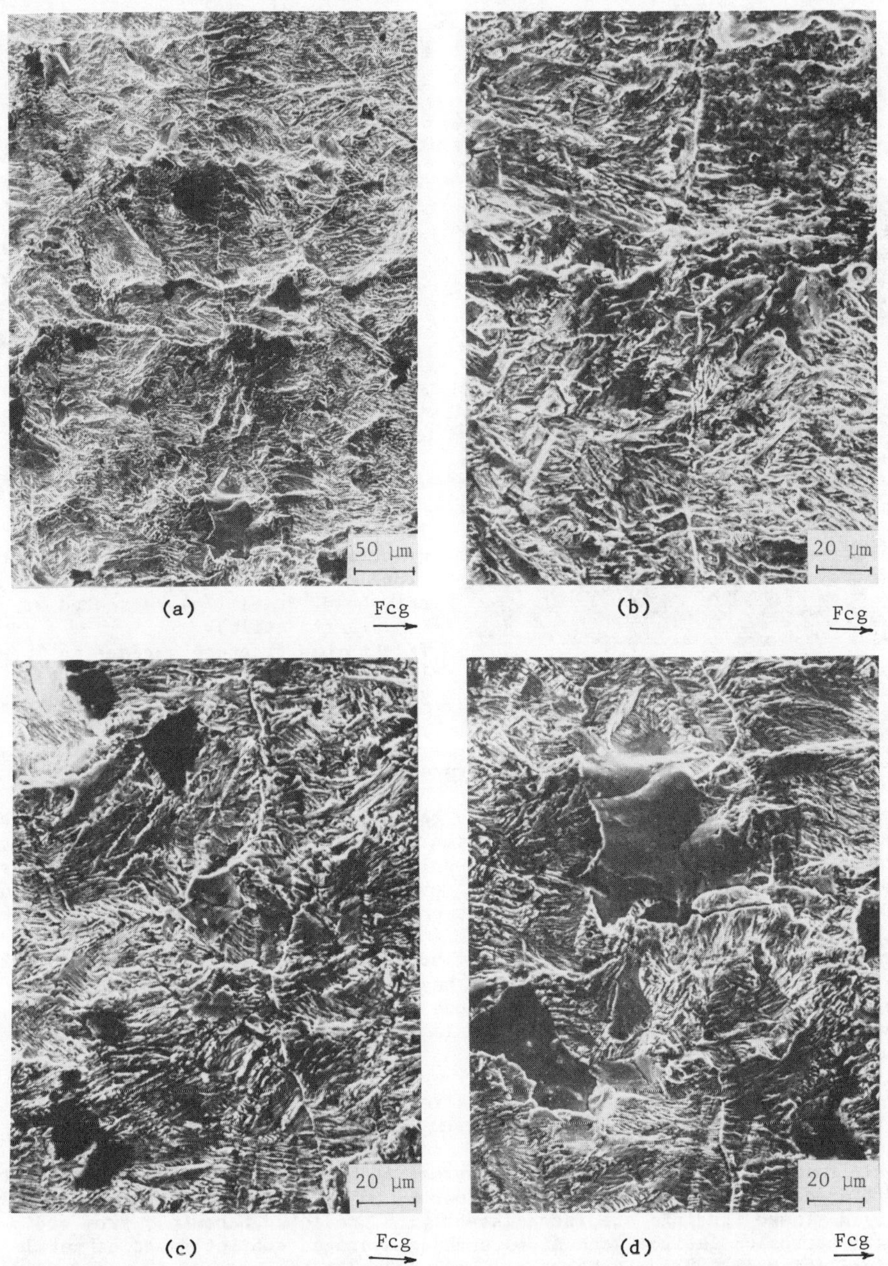

Figur 14 - Showing the threshold crack front in Steel 17-4 PH. Fracture surface topography is a mixture of faceted and fibrous fracture (a) R = 0.05 (b) R = 0.5 (c) Threshold crack front for Steel 17-4 PH tested at R = 0.05. Mode II displacements are observed. (d) SEM micrograph showing intergranular facets before and after the threshold, together with secondary cracking. Steel 17-4 PH tested at R = 0.05

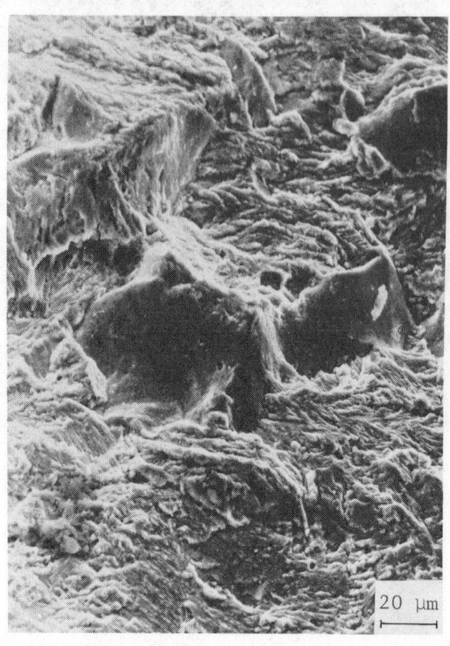

(e)

Fcg↓

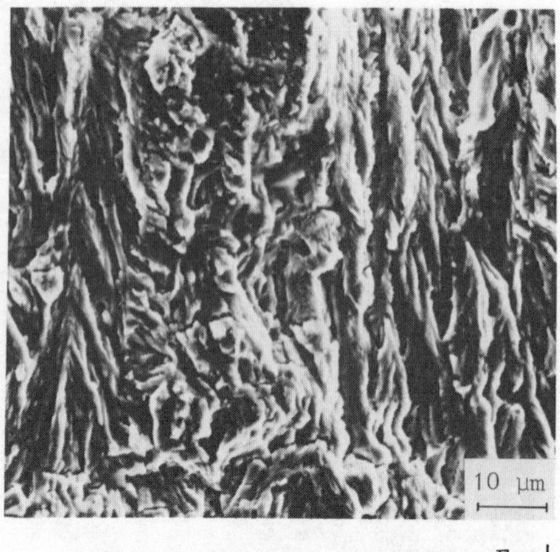

(f)

Fcg↓

Figure 14 - (e) Showing a few intergranular facets and a rapid return to fibrous fracture immediately after threshold. Steel 17-4 PH tested at R = 0.5 (45° tilt).
(f) Fibrous fracture surface in Steel 17-4 PH at high ΔK with R = 0.05. Secondary branch cracks are evident.

Discussion

Near threshold crack propagation behaviour in many different materials have been studied over the past few years. To explain the influence of load ratio R and environment on near threshold behaviour, different crack closure mechanisms have been used by several researchers (22-26). It has been shown that at near-threshold conditions not only plasticity-induced crack closure (11, 12) exists, but also other closure mechanisms such as oxide-induced crack closure (22, 23) and roughness-induced crack closure (24, 26). Crack closure at near threshold conditions may also be due to such phenomena as hydrodynamic wedging effects of viscous fluids within the crack (30) and martensite transformations in steels (31), these mechanisms will, however, not be dealt with here.

Studies in a wide range of low strength steels (22, 23, 28) have clearly shown that near-threshold crack growth rates may decrease in wet environments compared to results for inert environments. This behaviour is only found at low stress ratios R whereas at higher R-values no significant difference, in near threshold behaviour in moist or dry environments, is found. These findings are inconsistent with predicted behaviour from ordinary corrosion fatigue mechanisms such as hydrogen embrittlement or metal dissolution (22, 23). It has been shown (22, 28) by means of scanning Auger spectroscopy measurements that, for lower strength steels, oxide films with a thickness comparable to the crack tip opening displacement may form in the near threshold regime. Such corrosion debris is enhanced at threshold due to fretting between the fracture surfaces arising from plasticity-induced crack closure (11, 12), which is of most importance at low R-values, and Mode II shear displacements (29). Oxide-induced crack closure effects as discussed above seem, however, to be of less importance in high strength steels (32), possibly due to limited plasticity induced crack closure, less fretting or smoother fracture surfaces. This agrees with the results for steel 17-4 PH in the present investigation where the effect of oxide-induced closure is believed to be negligible. For the studied aluminium alloy, Al 2024-T3, it is obvious from Figure 10 that the above discussion on oxide formation applies, much more deposits are seen for low R-values than for high stress ratios. As no attempt was made to measure the thickness of these deposits, their possible contribution to crack closure effects is unclear. In a recent paper by Vasudévan and Suresh (33) oxide thickness measurements on several aluminium alloys, tested at threshold conditions, revealed that oxide-induced crack closure does not play a major role in the near threshold behaviour in Al 2024-T3. Thus, the high closure loads observed at threshold conditions for the two studied materials are supposed to be essentially due to roughness-induced crack closure. Such a closure mechanism may result from surface roughness and an irregular fracture topography (24, 26) in conjunction with Mode II shear displacements at near-threshold crack growth rates (24-26, 29). In the present investigation Mode II displacements are observed for both materials tested, with such shear displacements being more pronounced in the aluminium alloy, at threshold ΔK values when tested at low R-values. In aluminium a serrated crack path was found whereas in steel extensive intergranular fracture causes an irregular crack path.

For aluminium 2024-T3 tests were also performed with a negative stress ratio $R = -1$. At these tests a substantially higher near threshold crack growth rate and much less crack closure were found compared to tests carried out at $R = 0.05$. It is concluded that the compressive part of the load cycles tend to smooth the fracture surface, as evidenced in Figure 13(c), thus reducing the effect of roughness-induced crack closure and consequently giving rise to higher near-threshold crack growth rates. This effect could be of importance to structures subjected to a load spectrum going into compression, e.g. the lower wing skin of an aircraft when landing, where beneficial effects of roughness-induced and also oxide-induced closure may suddenly disappear.

Yet another phenomenon that should be taken into account when interpreting near-threshold fatigue crack growth data is crack deflection, i.e. although the loading may be nominally Mode I the crack may grow in a combination of Modes I and II as evidenced in Figure 12. In a recent paper by Suresh (34) this phenomenon is studied in some detail and it is shown that a crack deflection of 45°, which is the case in Figure 15, leads to a 20 per cent reduction in the Mode I stress intensity factor K_I and a 30 per cent underestimate of the actual growth rate. The serrated crack path shown in Figure 12 was obtained for Al 2024-T3 tested at $R = 0.05$. At higher R-values crack deflection becomes of less importance. This phenomenon may be yet another mechanism to explain the influence of stress ratio on near-threshold crack growth behaviour.

Conclusions

In aluminium alloy 2024-T3 and steel 17-4 PH, an increased stress ratio R increases the near threshold fatigue crack growth rate and decrease the threshold stress intensity ΔK_{th} except for R = -1, at which slightly higher growth rates than at R = 0.05 were observed. The influence of R on near threshold behaviour is explained in terms of crack closure. It is shown that for plane strain, which is prevalent at threshold conditions, plasticity-induced crack closure is not sufficient to explain the high closure loads observed. Instead concepts of oxide-induced crack closure and roughness-induced crack closure are used. Oxide-induced crack closure is, however, believed to play only a minor role in influencing near threshold crack growth behaviour in the two studied materials, particularly the steel, compared to results presented elsewhere for lower strength steels. Roughness-induced crack closure manifests itself by Mode II shear deformations together with a serrated crack path for Al 2024-T3 and an intergranular fracture morphology for steel 17-4 PH. Crack closure effects are most pronounced at small positive R-values where the effect of plasticity-induced crack closure is maximised. At R = -1 the compressive part of the load cycle smoothes the fracture surface, thus reducing the effect of roughness-induced crack closure and resulting in higher near threshold crack growth rates for R = -1 than for R = 0.05. No effect of specimen thickness or specimen geometry could be found. The value of P_{op}/P_{max} decreases when ΔK is increased. This suggests that the very existence of a fatigue threshold largely may be due to crack closure effects. When ΔK is high enough for the plastic zone size to become larger than the grain size, fatigue crack growth is insensitive to microstructure and plastitity-induced crack closure is found to prevail alone.

Acknowledgements

The author is very grateful to J. Byrne, D. Holm and B. Weiss for their help and valuable discussions during the course of this work.

References

(1) MIL-A-83444 (USAF), "Airplane Damage Tolerance Requirements", July 1974.

(2) P.C. Paris, M. Gomez and W.E. Anderson, "A Rational Analytic Theory of Fatigue", Trend in Engineering, 13 (1) (1961), pp. 9-14.

(3) H.J. Rack and D. Kalish, "The Strength, Fracture Toughness and Low Cycle Fatigue Behavior of 17-4 PH Stainless Steel", Metallurgical Transactions, 5 (7) (1974), pp. 1595-1605.

(4) T.W. Crooker, D.F. Hasson and G.R. Yoder, "Micro-Mechanistic Interpretation of Cyclic Crack Growth Behaviour in 17-4 PH Stainless Steels", Fractography-Microscopic Cracking Processes. ASTM STP 600, pp. 205-219, 1976.

(5) Proposed ASTM Test Method For Measurement of Fatigue Crack Growth Rates, Appendix II in ASTM STP 738, pp. 340-356, 1981.

(6) B. Budiansky and J.W. Hutchinson, "Analysis of Closure in Fatigue Crack Growth", ASME Journal of Applied Mechanics, 45, (1978), pp. 267-276.

(7) K.K. Lo, "Fatigue Crack Closure Following a Step-Increase Load", ASME Journal of Applied Mechanics, 47, (1980), pp. 811-815.

(8) J.C. Newman, Jr., "A Finite-Element Analysis of Fatigue Crack Closure", Mechanics of Crack Growth, ASTM STP 590, pp. 281-301, 1976.

(9) M. Nakagaki and S.N. Atluri, "Elastic-Plastic Analysis of Fatigue Crack Closure in Modes I and II", AIAA Journal, 18, (1980), pp. 1110-1117.

(10) A.F. Blom and D.K. Holm", An Experimental and Numerical Study of Crack Closure", Submitted to Engineering Fracture Mechanics.

(11) W. Elber, "Fatigue Crack Closure Under Cyclic Tension", Engineering Fracture Mechanics, 2, (1970), pp. 37-45.

(12) W. Elber, "The Significance of Fatigue Crack Closure", Damage Tolerance in Aircraft Structures. ASTM STP 486, pp. 230-242, 1971.

(13) L.P. Pook, "Fatigue Crack Growth Data for Various Materials Deduced from the Fatigue Lives of Precracked Plates", Stress Analysis and Growth of Cracks, ASTM STP 513, pp. 106-124, 1972.

(14) P.C. Paris, R. J. Bucci, E.T. Wessel, W.G. Clark, Jr. and T.R. Mager, "Extensive Study of Low Fatigue Crack Growth Rates in A533 and A508 Steels", Stress Analysis and Growth of Cracks, ASTM STP 513, pp. 141-176, 1972.

(15) R.J. Bucci, W.G. Clark, Jr. and P.C. Paris, "Fatigue Crack Propagation Growth Rates Under a Wide Variation of ΔK for an ASTM-A517 Grade F(T-1) Steel", Stress Analysis and Growth of Cracks, ASTM STP 513, pp. 177-195, 1972.

(16) R.J. Cooke and C.J. Beevers, "The Effect of Load Ratio on the Threshold Stress for Fatigue Crack Growth in Medium Carbon Steels", Engineering Fracture Mechanics, 5, (1973), pp. 1061-1071.

(17) R.A. Schmidt and P.C. Paris, "Threshold for Fatigue Crack Propagation and the Effects of Load Ratio and Frequency", Progress in Flaw Growth and Fracture, ASTM STP 536, pp. 79-94, 1973.

(18) R.O. Ritchie, "Near-Threshold Fatigue-Crack Propagation in Steels", International Metals Reviews, 20 (5, 6) (1979), pp. 205-230.

(19) T.C. Lindley and C.E. Richards, "Near-Threshold Fatigue Crack Growth in Materials used in the Electricity Supply Industry", pp. 1087-1113 in Fatigue Thresholds, J. Bäcklund, A.F. Blom and C.J. Beevers, Eds.; EMAS Ltd., Warley, U.K., 1982.

(20) A.F. Blom, A. Hadrboletz and B. Weiss, "Effect of Crack Closure on Near-Threshold Crack Growth Behaviour in a High Strength Al-Alloy up to Ultrasonic Frequencies", paper presented at the 4th International Conference on Mechanical Behaviour of Materials, Stockholm Sweden, August 1983.

(21) S. Suresh, A.K. Vasudévan and P.E. Bretz, "Mechanisms of Near-Threshold Fatigue Crack Growth in High Strength Aluminum Alloys: Role of Microstructure and Environment", Metallurgical Transactions A, 15A, (1984) in press.

(22) S. Suresh, G.F. Zamiski and R.O. Ritchie, "Oxide-Induced Crack Closure: An Explanation for Near-Threshold Corrosion Fatigue Crack Growth Behaviour", Metallurgical Transactions A, 12A, (1981), pp. 1435-1443.

(23) A.T. Stewart, "The Influence of Environment and Stress Ratio on Fatigue Crack Growth at Near-Threshold Stress Intensities in Low-Alloy Steels", Engineering Fracture Mechanics, 13, (1980), pp. 463-478.

(24) N. Walker and C.J. Beevers, "A Fatigue Crack Closure Mechanism in Titanium", Fatigue of Engineering Materials and Structures, 1, (1979), pp. 135-148.

(25) K. Minakawa and A.J. McEvily, "On Crack Closure in the Near-Threshold Region", Scripta Metallurgica, 15, (1981), pp. 633-636.

(26) K. Minakawa and A.J. McEvily, "On Near-Threshold Fatigue Crack Growth in Steels and Aluminium Alloys", pp. 373-390 in Fatigue Thresholds, J. Bäcklund, A.F. Blom and C.J. Beevers, Eds.; EMAS Ltd, Warley, U.K., 1982.

(27) J.R. Rice, "The Mechanics of Crack Tip Deformation and Extension by Fatigue", Fatigue Crack Propagation, ASTM STP 415, pp. 247-311, 1967.

(28) S. Suresh, D.M. Parks and R.O. Ritchie, "Crack Tip Oxide Formation and its Influence on Fatigue Thresholds", pp. 391-408 in Fatigue Thresholds, J. Bäcklund, A.F. Blom and C.J. Beevers, Eds.; EMAS Ltd., Warley, U.K., 1982.

(29) D.L. Davidson, "Incorporating Threshold and Environmental Effects Into the Damage Accumulation Model for Fatigue Crack Growth", Fatigue of Engineering Materials and Structures, 3, (1981), pp. 229-236.

(30) J.L. Tzou, S. Suresh and R.O. Ritchie, "Fatigue Crack Propagation in Viscous Environments", paper presented at the 4th International Conference on Mechanical Behaviour of Materials, Stockholm, Sweden, August 1983.

(31) S.Suresh and R.O. Ritchie, "Near-Threshold Crack Propagation: A Perspective on the Role of Crack Closure", paper presented at this symposium.

(32) S. Suresh, J. Toplosky and R.O. Ritchie, "A Mechanism for Environmentally Affected Near Threshold Fatigue Crack Growth in Steels", Fracture Mechanics, ASTM STP 791, pp. I-329-347, 1982.

(33) A.K. Vasudévan and S. Suresh, "Influence of Corrosion Deposits on Near-Threshold Fatigue Crack Growth Behaviour in 2XXX and 7XXX Series Aluminum Alloys", Metallurgical Transactions A, 13A, (1982), pp. 2271-2280.

(34) S. Suresh, "Crack Deflection: Implications for the Growth of Long and Short Fatigue Cracks", Metallurgical Transactions A, 14A (1983), pp. 2375-2385.

INFLUENCE OF R RATIO AND ORIENTATION ON THE FATIGUE CRACK THRESHOLD ΔK_{th}, AND SUBSEQUENT CRACK GROWTH OF A LOW-ALLOY STEEL

A J Cadman, C E Nicholson (Health and Safety Executive, Sheffield UK)
and R Brook (University of Sheffield UK)

The work described within this paper is an investigation into the effects of R and crack orientation on the fatigue threshold (ΔK_{th}), near-threshold growth rate, and the resulting fracture surfaces. The material is a low-alloy free-machining steel (0.31 wt% sulphur). The results show that the effects of R are dependent upon the orientation of the crack with respect to the rolling direction. For both the orientations considered, an increase in the R ratio not only decreases the threshold values but also leads to marked changes in the fracture appearance. These changes are described in detail, along with a consideration of the variations that are seen in the near-threshold crack growth rates for the different R ratios and orientations. The effects observed are explained in terms of the distribution and morphology of the non-metallic inclusions.

Introduction

The fatigue crack growth threshold is an important parameter in the determination of the safety of engineering structures. It is therefore necessary to know how variations in load ratio (R = Kmin/Kmax) and crack orientation (with respect to the rolling direction) affect the value of thresholds, growth rates and fracture surface morphologies. Inclusions can have a marked influence on material properties and so a high sulphur free-machining steel was used for this study to facilitate the easier observation of the effects of inclusions with respect to orientation.

Experimental Procedure

The chemical composition of the material is shown in Table I. The structure consisted of mixed ferrite and pearlite (13% pearlite).

Table I Chemical Composition

Element	C	Mn	Si	Ni	Cr	S	P	Cu
Composition	0.12	1.16	0.03	0.05	0.15	0.31	0.039	0.06

Due to the steel being a free-machining grade its structure also contained 0.92% of manganese sulphide particles. The material had a tensile strength of 504 MPa and a yield strength (σ_y) of 465 MPa. The restrictions in the size of the stock material (44 mm diameter bar) necessitated the use of two different specimen types. To obtain cracks growing transversely to the rolling direction single edge notch three-point-bend specimens were used as shown in Figure 1 (a). Cracks parallel (longitudinal) to the rolling direction were grown in compact tension specimens as shown in Figure 1 (b). Both types of specimens were 12.5 mm in thickness. The other dimensions of the specimens were in accordance with the British Standard for fracture toughness test pieces (1). By using standard specimens and their associated compliance functions the authors expect to have eliminated any effects of specimen geometry from their results.

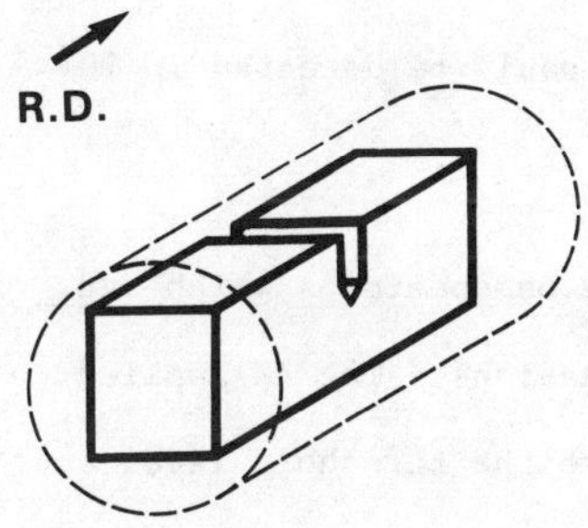

Figure 1 (a) - Orientation of the transverse cracks with respect to the rolling direction

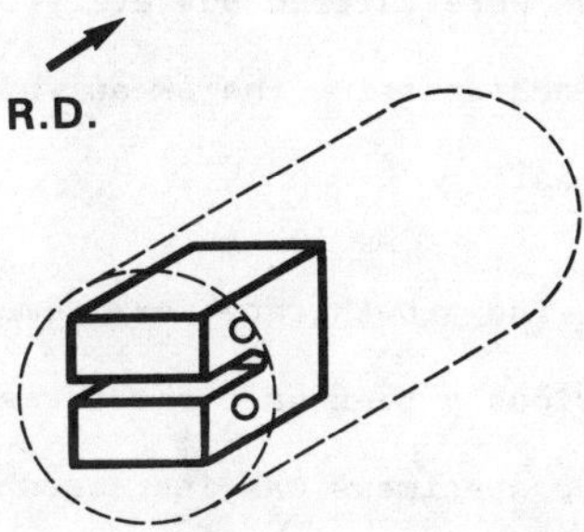

Figure 1 (b) - Orientation of the longitudinal cracks with respect to the rolling direction

Crack length measurement was made using a direct current (30 amp) potential drop technique. The equipment could detect changes in crack length of 0.01 mm and could accurately measure crack lengths to 0.1 mm. A permanent record of the changes in potential difference was made using a chart recorder.

Fatigue testing was carried out on an Amsler Vibrophore at a frequency of 110 ± 10 Hz. No heating of either specimen type was observed at this frequency. The threshold measurements were made using a manual step-down procedure in which the alternating ΔK was reduced by 10% for each increment of crack growth equivalent to the monotonic plastic zone size (r_{pz}) associated with the previous ΔK level.

$$r_{pz} = \frac{1}{6\pi} \frac{Kmax^2}{\sigma y^2}$$

When the plastic zone size became less than 0.2 mm the reductions in ΔK were carried out every 0.2 mm of growth. The threshold was defined as being the ΔK at which no growth could be detected in 10^7 cycles.

The growth rate measurements were made on specimens which had previously been used for threshold determinations. The ΔK applied to these specimens was increased slightly above the threshold level and crack growth was allowed to start. Once the crack had started growing the applied load was kept constant and the crack was grown under increasing ΔK conditions. The crack growth was monitored using the output (V) from the potential difference equipment previously described. The number of cycles (N) at various points throughout the tests were noted. The growth rate curves were calculated from V vs N data using a three point incremental analysis technique (2).

Once the growth rate tests had been completed the specimens were broken open and their fracture surfaces were examined using a scanning electron microscope.

Results

Figure 2 shows the effect of R ratio on the thresholds of the longitudinal and transverse cracks. Both orientations exhibited the

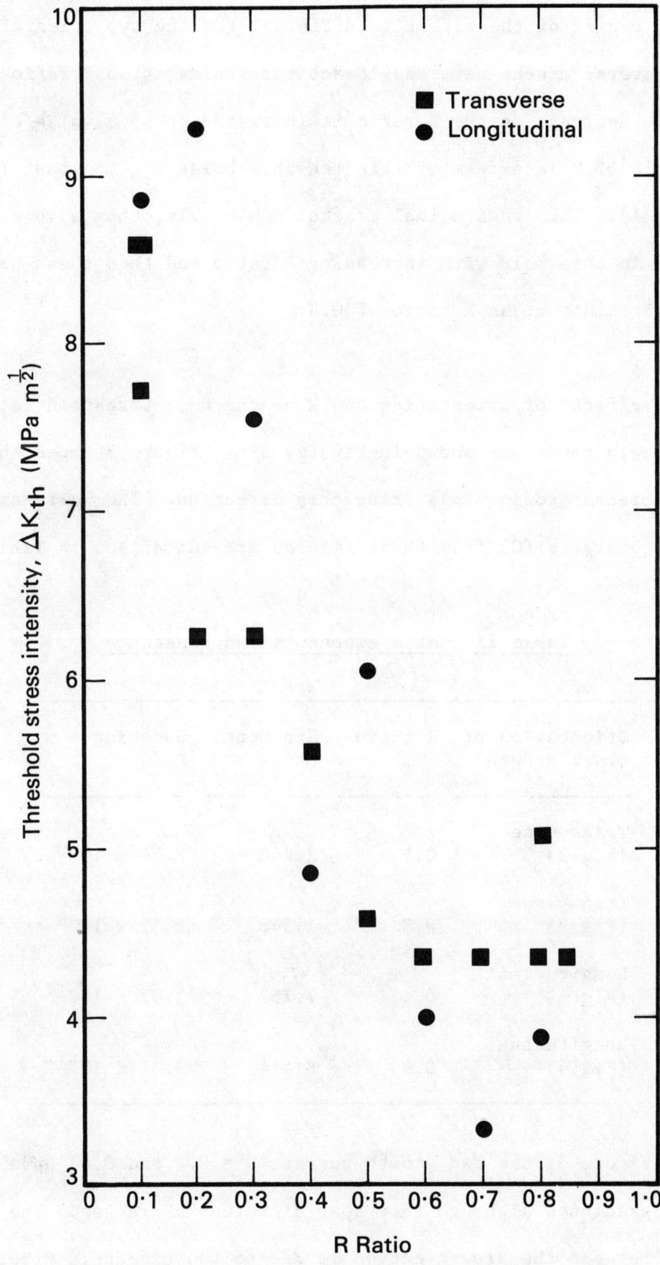

Figure 2 - The variation of threshold with R ratio

well known effect of R on threshold values (3) namely that higher R ratios resulted in lower threshold values. It can be seen, however, that the degree of the effect is different for the two orientations. The transverse cracks have the lowest thresholds at an R ratio of 0.1 and these decrease as the R ratio is increased up to a value of 0.6. For levels of R in excess of this the thresholds are constant (ΔK_{th} = 4.35 $MPam^{\frac{1}{2}}$). The longitudinal cracks, conversely, show a more rapid decrease in threshold with increasing R ratio and thus these have lower threshold values at an R ratio of 0.8.

The effects of orientation and R on the near-threshold fatigue crack growth rates are shown in Figures 3-6. Figure 3 shows the effect of R on cracks growing in a transverse direction. The Paris exponents (n) and constants (C) from these Figures are summarised in Table II.

Table II Paris exponents and constants

Orientation of crack growth	R ratio	Exponent n	Constant C
Transverse (Fig 3)	0.1	2.248	8.54×10^{-9}
Transverse (Fig 3)	0.8	1.548	2.22×10^{-8}
Longitudinal (Fig 4)	0.1	7.752	1.07×10^{-14}
Longitudinal (Fig 4)	0.8	6.414	3.41×10^{-11}

In Figure 3, the two growth curves (R = 0.1 and 0.8) have very similar gradients although they span different ΔK ranges. The lack of overlap between the growth curves is due to two effects. Firstly the

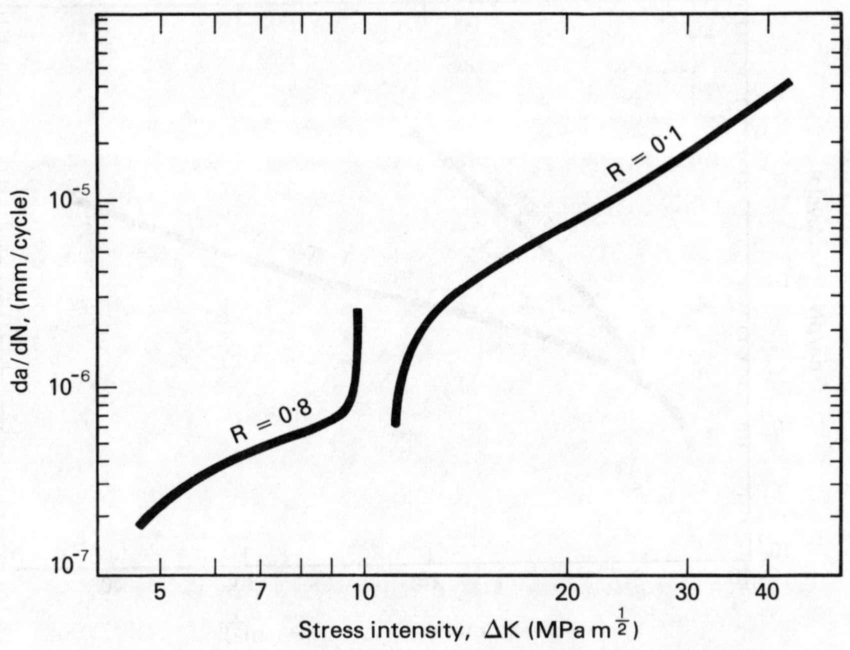

Figure 3 - The effect of R ratio on the transverse cracks

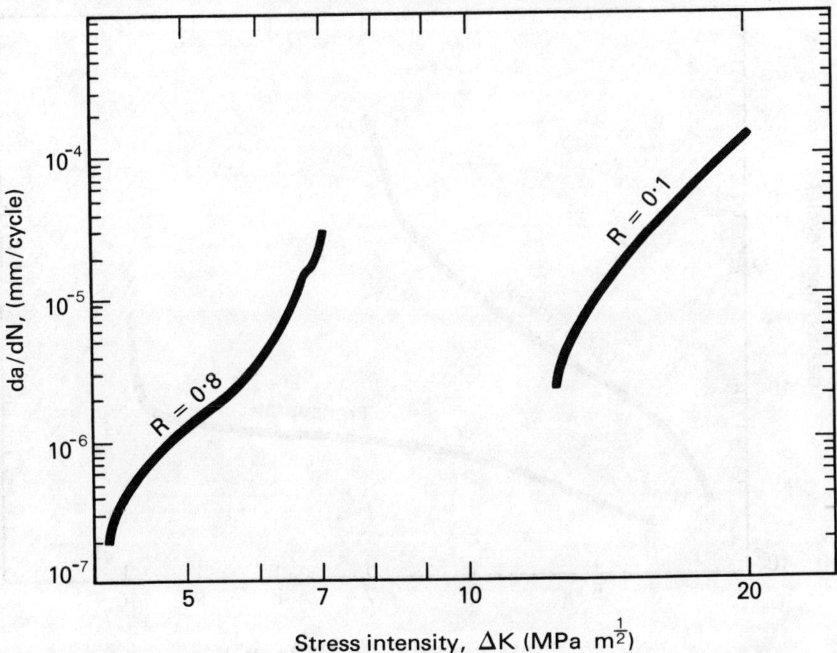

Figure 4 - The effect of R ratio on the longitudinal cracks

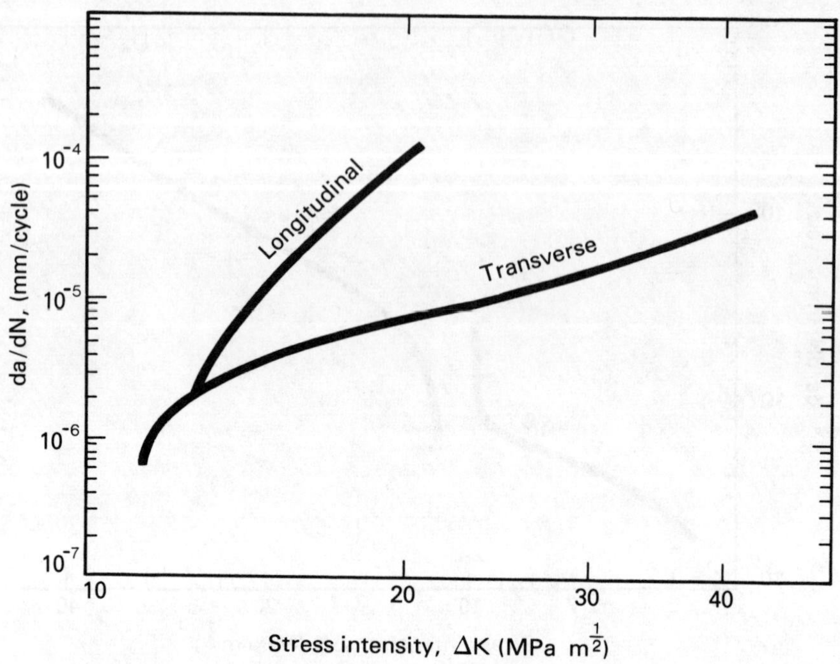

Figure 5 - The effect of crack orientation at an R ratio of 0.1

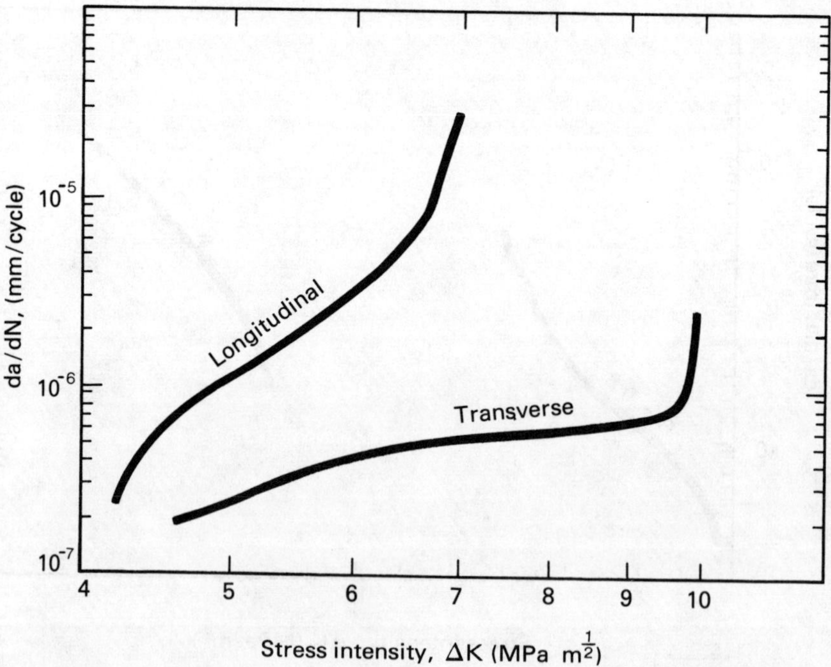

Figure 6 - The effect of crack orientation at an R ratio of 0.8

maximum ΔK to which the high R tests can be carried out is controlled by the ability of the specimen to withstand the high K_{max} levels associated with these tests. The limit of ΔK above which the three-point-bend specimens began to bend was 9 $MPam^{\frac{1}{2}}$, while the compact tension specimens failed by rapid fracture above 7 $MPam^{\frac{1}{2}}$. Secondly the low R tests have relatively high thresholds and so it was not possible to restart crack growth at ΔK's below 10 $MPam^{\frac{1}{2}}$.

The lack of overlap of the ΔK ranges for the longitudinal cracks (Figure 4) can also be explained by the above effects. In this case, however, there is also a large difference in the constants associated with the two growth curves (See Table II). Although the two curves have very similar exponents the tests carried out at an R of 0.8 would have much faster growth rates (in the region of two orders of magnitude), than those at an R of 0.1 if tests could be carried out at the same levels of ΔK.

Figure 5 shows the effect of orientation on growth rates at an R ratio of 0.1. The longitudinal cracks are seen to have much steeper growth curves than the transverse cracks. Thus although their threshold values are similar, the growth rates of cracks at the two orientations are very different, particularly as the ΔK level increases.

At an R ratio of 0.8 (Fig 6) almost identical effects are seen with a difference in growth rate of up to ten times between the linear parts of the growth curves.

Figures (7-12) show the change in fracture morphology for the various combinations of R ratio and specimen orientation. It should be noted that although the fracture surfaces produced by the higher ΔK values are not described here, the effect on the surface morphology of increasing ΔK is only slight when compared with the effects of R ratio and specimen orientation. Figure 7 shows a fracture surface produced by near threshold crack growth of a transverse crack at an R ratio of 0.1.

The fracture surface is very rough and a number of facets can be seen. Some manganese sulphide particles are also visible but these do not seem to have influenced the path of the crack growth.

Figure 8 shows a surface produced at an R ratio of 0.5 by a transverse crack. This fractograph is partially blurred due to the presence of oxide debris which defocusses the electron beam. The surface is flatter than that produced at the lower R ratio. Some facets can still be seen but there are far fewer than were visible at an R ratio of 0.1. Manganese sulphide particles still appear to have no effect upon the path of the crack at this R ratio.

The last micrograph of a transverse crack (Figure 9) shows a surface formed by near-threshold growth at an R ratio of 0.85. This surface is the smoothest of the three examined. No facets were observed but the presence of manganese sulphide particles was more apparent. Voids appear to have been formed around the particles during crack growth, most probably due to the higher mean stress imposed at this high R ratio. The rest of the fracture surface is relatively flat with a number of small furrow-like features which point in the general direction of the macroscopic growth. These features are very similar to those observed in previous work (4) at high R ratios.

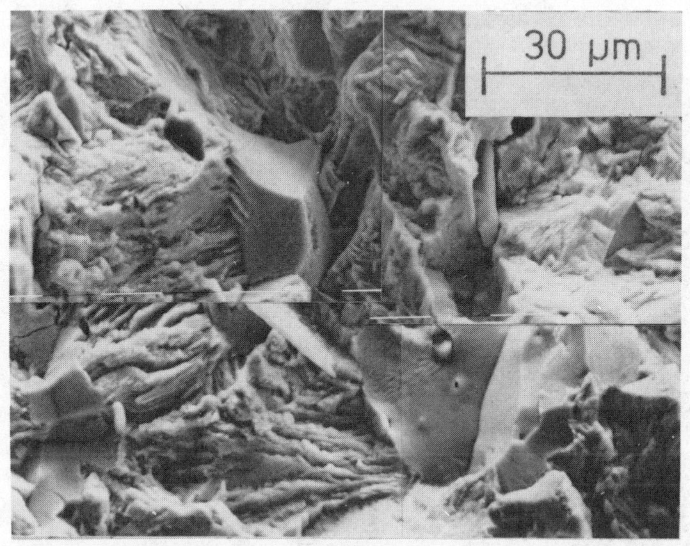

Figure 7 - Fractograph of a surface produced by a transverse crack. R = 0.1 (Direction of crack growth is from right to left)

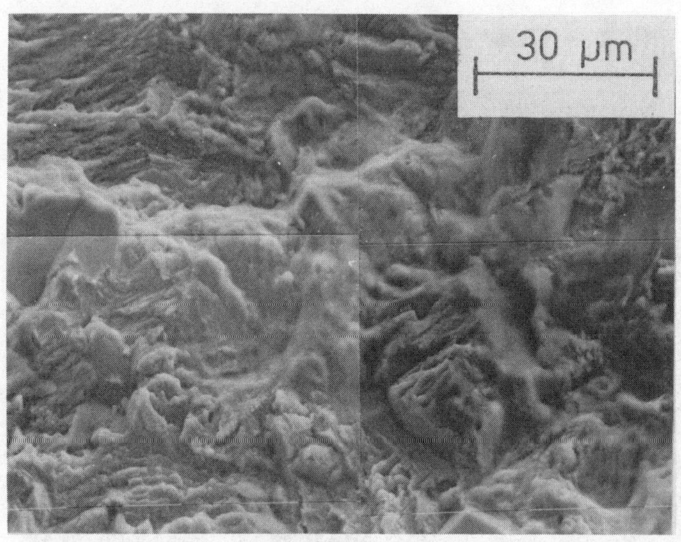

Figure 8 - Fractograph of a surface produced by a transverse crack. R = 0.5 (Direction of crack growth is from right to left)

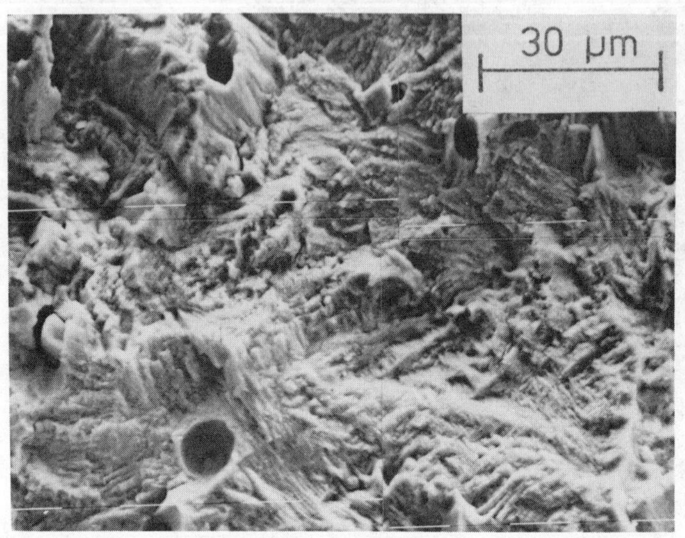

Figure 9 - Fractograph of a surface produced by a transverse crack R = 0.85 (Direction of crack growth is from right to left)

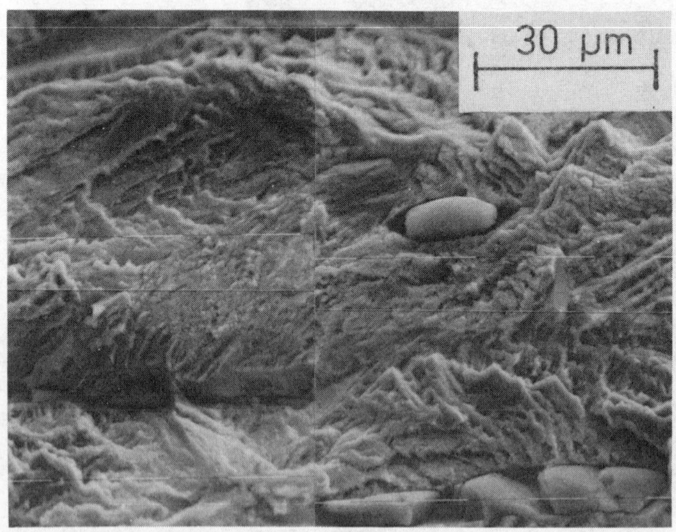

Figure 10 - Fractograph of a surface produced by a longitudinal crack R = 0.1 (Direction of crack growth from right to left)

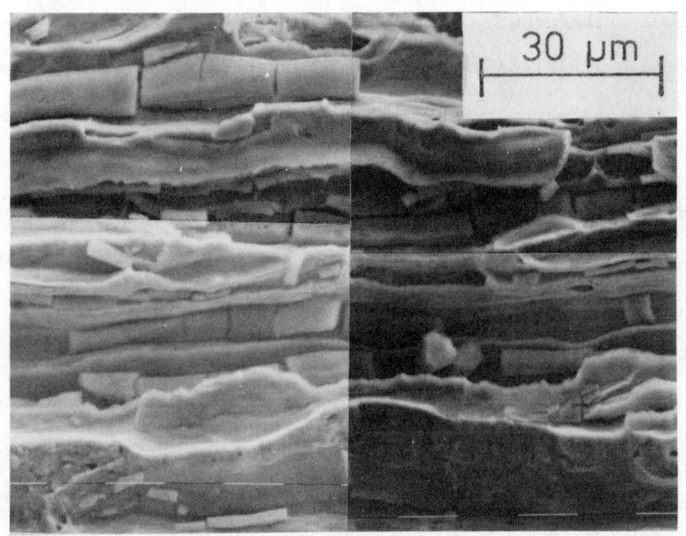

Figure 11 - Fractograph of a surface produced by a longitudinal crack R = 0.5 (Direction of crack growth is from right to left)

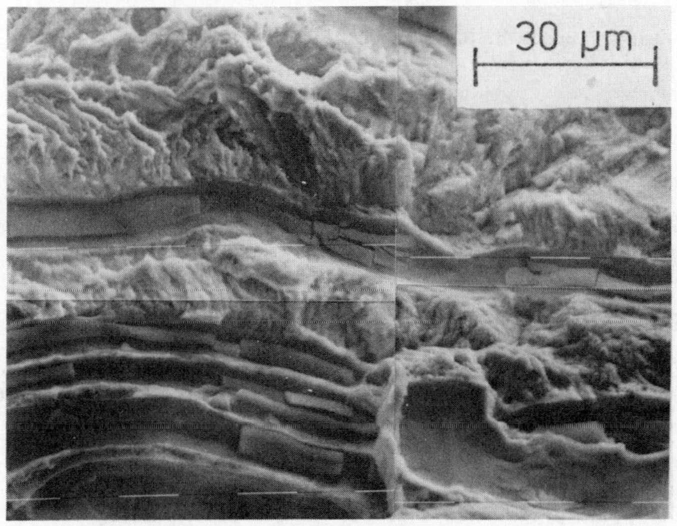

Figure 12 - Fractograph of a surface produced by a longitudinal crack R = 0.8 (Direction of crack growth is from right to left)

The fracture surface produced by longitudinal cracks are shown in Figures 10-12. The surface formed at an R ratio of 0.1 (Figure 10) is highly directional in nature with large ridges parallel to the direction of crack growth. The ridges are fairly smooth features upon which furrowed areas and an occasional manganese sulphide particle can be seen. The valleys between the ridges contain fractured manganese sulphide stringers. The crack grows preferentially along these stringers and the valleys are formed when there is a slight difference in orientation between the stringers and the crack growth direction.

At the higher R ratios, (Figures 11 (R = 0.5) and Figure 12 (R = 0.8)) the manganese sulphides become more prominent. The fracture surfaces appear as a mass of long sulphides with only an occasional ridge that is free of inclusions (Figure 12). At both these R ratios the crack has grown almost exclusively along the manganese sulphide stringers. The effect of the manganese sulphides is so great that the macroscopic appearance of the fracture surfaces (even for the R ratio of 0.1) is very fibrous in nature.

Discussion

Two important factors have emerged from these tests. Firstly cracks grown in the longitudinal direction produce both macroscopically and microscopically rougher fracture surfaces than cracks grown in the transverse direction. The increase in surface roughness would be expected to increase any crack closure (5) that may occur at near-threshold growth rates (4).

Secondly, the preferential growth of the cracks along the manganese sulphide stringers, particularly at higher R ratios, would suggest that these are easier crack growth paths than the metal matrix. Thus the crack growth of the longitudinally oriented cracks would be expected to be faster than that of transverse cracks which must travel through the matrix and cut perpendicularly through the sulphides. The experimental results will now be discussed with respect to these two observations.

The threshold values of the longitudinal cracks are higher than those of the transverse cracks, at low (<0.5) R ratios. The reverse is true at the higher R ratios. One factor influencing the higher thresholds at low R ratios (for both orientations) is mode II crack growth. Because of the small plastic zone size ahead of the crack tip, slip is limited to one slip system (6) and so a rough zig-zag form of crack growth occurs with local mode II displacements. This form of crack growth is responsible for the faceted surface seen in Figure 7. As has already been noted increased roughness causes more crack closure. This has the effect of increasing the threshold at lower R ratios (7). This is particularly true for the longitudinal cracks which produce inherently rougher surfaces. At higher R ratios closure no longer occurs and the resistance of the material to fatigue becomes the dominant factor. Thus the longitudinal cracks which grow along the manganese sulphide stringers are seen to have the lower thresholds. This is because the stringers represent easy paths for crack growth compared with the metal matrix.

Manganese sulphide particles can also be seen to influence the growth rate curves. The growth curves for the longitudinal cracks at different R ratios (Figure 4) have already shown that at the higher R ratio crack growth would be significantly faster (x 100) if the tests had been carried out over the same ΔK range. This difference in crack growth rate is due to the increased effect of the manganese sulphide stringers at the higher R ratio. The high mean stress causes ductile yielding around the stringers such that the crack then follows the resulting voids thus exposing the manganese sulphides on the fracture surface.

The growth rate curves for transverse cracks, Fig 3, show that there is no enhancement of the crack growth rate for an increase in R ratio. The linear parts of both curves are parallel and have similar constants. This effect may have been predicted from the fractographs, since only slight changes in crack growth morphology were observed as the R ratio was increased (Figs 7-9).

When the growth rate curves of the differently oriented cracks at constant R ratios are compared (Figs 5 & 6) the difference between the behaviour is clearly seen. For both R ratios the longitudinal cracks exhibit faster crack growth rates and much higher Paris exponents (Table II). This increase can once again be explained by the easier crack path along the manganese sulphide stringers. It should also be noted that surface roughness does not play a significant role at higher ΔK levels since the crack tip openings and the plastic zone sizes become much larger. Thus growth occurs by conventional mode I mechanisms and crack closure is no longer important.

From these results it can be seen that the factors which influence fatigue threshold do not influence fatigue crack growth in the same manner.

The optimum conditions for a high fatigue threshold can be achieved in a longitudinal crack at a low R ratio, however these conditions do not favour a slow fatigue crack growth rate which is better achieved by a transverse crack whatever the R value.

Conclusions

1 Increasing R ratio decreases the fatigue threshold for both longitudinal and transverse crack orientations.

2 Transversely growing cracks have lower thresholds but also lower growth rates than longitudinal cracks at low R ratios (<0.6).

3 At R ratios above 0.6 transverse cracks have higher thresholds and also lower growth rates when compared with longitudinal cracks.

4 Manganese sulphide stringers have a large influence on the path of a fatigue crack which is growing parallel to them.

5 Conditions which favour a high fatigue threshold do not always favour slow fatigue crack growth rates.

Acknowledgements

The authors are grateful to Professor G J Davies for providing facilities for this research in the Department of Metallurgy at the University of Sheffield. Andrew J Cadman, who was a research student at Sheffield University when this work was carried out, acknowledges financial support from the Science Research Council and the Health and Safety Executive in the form of a CASE Research Studentship. The authors would also like to thank the Research Director of the Health and Safety Executive for permission to publish this paper.

References

1 British Standards Institution "Methods of test for Plane strain fracture toughness (K_{IC}) of metallic materials" BS 5447:1977

2 I M Austen "An assessment of the ability of various methods of fatigue crack growth rate determination" British Steel Corporation Report No SM/PT/7422/8/79/B

3 O Vosikovsky, "The effect of stress ratio on fatigue crack growth rates in steels" Engineering Fracture Mechanics, 11 (1979) pp595-602.

4 A J Cadman, C E Nicholson and R Brook, "Microstructural effects at and near the Fatigue Threshold" in Fatigue Thresholds - Proceedings of an International Conference in Stockholm, June 1981 EMAS, UK, 1982.

5 W Elber, "Fatigue crack closure under cyclic tension", Engineering Fracture Mechanics 2 (1970) pp37-45.

6 S Suresh and R O Ritchie, "A geometric model for fatigue crack closure induced by fracture surface roughness" Metallurgical Transactions 13A (1982) pp1627-1631.

7 R A Schmidt and P C Paris, "Threshold for fatigue crack propagation and the effects of load ratio and frequency" ASTM STP 536 (1973) pp79-94.

AN ASSESSMENT OF INTERNAL HYDROGEN VERSUS CLOSURE EFFECTS ON NEAR-THRESHOLD FATIGUE CRACK PROPAGATION

Khlefa A. Esaklul, Andrew G. Wright and William W. Gerberich

Department of Chemical Engineering and Materials Science
University of Minnesota
Minneapolis, Minnesota 55455

The effect of internal hydrogen on near-threshold fatigue behavior of low strength HSLA steel, ultrahigh strength AISI 4340 steel, iron and three iron-silicon alloys was studied. The objectives were to determine the extent of hydrogen effects on near-threshold crack propagation, threshold stress intensities and crack closure. The results indicate that there is a substantial hydrogen effect on crack growth rates as well as threshold stress intensities. Crack closure plays a major role in determining the extent of hydrogen effects and in several instances it suppressed these effects. However, when closure was subtracted by evaluating the data in terms of effective stress intensity, ΔK_{eff}, hydrogen effects were evident in all cases. The effect of internal hydrogen depended on initial hydrogen concentration and mean stress, i.e., load ratio. Surface roughness induced closure was a major closure mechanism in the presence of internal hydrogen in most of the alloys tested because of hydrogen promoting and increasing intergranular fracture. Results are discussed in relationship to current views on environmental interactions near threshold.

INTRODUCTION

In how many ways does environment affect near-threshold fatigue crack propagation? The true environmental effects, the mechanisms involved and how they can be translated to usable engineering data are still poorly understood. One primary reason is the considerable degree of confusion in the literature as to the hydrogen effect. The points of confusion are a result of inadequate attention to crack-tip closure and/or overlooking the possible mean stress effects when testing at high load ratios to avoid closure. Ignoring closure can lead to premature conclusions. As recently pointed out by Ruppen and Salzbrenner (1) one could be off by an order of magnitude in assessing the true driving force for crack extension and conclude that the material had better resistance in H_2S-brine than in air. Considering the aggressiveness of the H_2S-brine environment, although such data could be used for long cracks near $R \approx 0$, such applications could be disastrous for larger mean stresses. Avoiding closure by testing at high load ratio can also have its setbacks as a result of mean stress effects (2-4). Furthermore, choosing to test at intermediate load ratios leads to the question of at what level of R closure is eliminated. Choosing an arbitrary value might not be the answer particularly when variation in microstructural variables or fracture processes are considered, and could as well lead to premature conclusions. Because of such considerations, understanding of environmental interactions in fatigue-crack growth near threshold has been limited until recently. In fact, almost all data collected prior to about 1977 has little bearing on understanding of environmental effects on near-threshold behavior.

Theoretical modeling has been limited partly because of the complexity of the problem and partly because of the absence of meaningful data. Therefore, any successful models for threshold would necessarily not agree with experimental findings unless the data were taken at very large R-values where closure effects are minimum and possible mean stress affects are considered, or unless closure effects were subtracted out. The later may be accomplished in terms of effective stress intensity, ΔK_{eff}, being defined as K_{max}-K_{op} where K_{op} is the stress intensity value where the crack becomes fully open. This would require the use of suitable crack monitoring and unified guidelines for measuring closure stress intensity.

Most of the work on the effects of environment on near-threshold fatigue behavior have indicated that these effects are vastly dependent on a materials strength (5-19). However, this dependence appears to contradict the general trend of strength on stage II and static crack growth. For example, it is well established that as a result of hydrogen embrittlement gaseous hydrogen environment enhances stage II fatigue crack growth in both low and high strength steels (20-23) even at low pressure (less than 1 atmosphere). On the other hand, near threshold it appears that gaseous hydrogen enhances fatigue crack growth in low strength steels (6,7,9,12-19) but slows it in high strength steels (6,8,13-15) relative to laboratory air. The acceleration in growth in low strength steel was a result of reduced crack closure since similar effects were observed in inert dry environments and there was a total absence of any effects at high load ratios (6,7,9,12-19). Accordingly, low strength steels are relatively immune to hydrogen embrittlement and the effects of hydrogen environment are an artifact due to closure. A very different observation was obtained in high strength steels where gaseous hydrogen had the same deceleration effect at both low and high load ratios. This indicated no influence of closure and an improvement in near-threshold fatigue properites in gaseous hydrogen (6,8,13-15). The later is a result of moist environment (laboratory air) being more aggressive than hydrogen gas (24-25). Most recently, results of Esaklul et al. (3,4,26-28) and Sastry et al. (29) on the influence of internal hydrogen on near-threshold fatigue behavior of low and high strength steels and iron binary alloys indicate that intrinsic hydrogen effects do exist. That is, the presence of internal hydrogen enhances fatigue crack growth rates and lowers threshold stress intensities. Although the effects on threshold were suppressed by closure in the case of high strength steel and iron alloys (4,28,29), the detrimental influence of hydrogen was evident once the data were evaluated in terms of ΔK_{eff}. The variation in closure stress intensities in the materials studied by Esaklul et al. (3,4,27,28) showed that oxide debris was present but that their contribution to closure was not the major effect.

Thus, the influence of environment on near-threshold fatigue crack propagation is controlled mainly by the availability of hydrogen ahead of the crack tip and the rate processes controlling the hydrogen evolution and their influence on closure. This in general can explain the discrepancy in the data in the literature with regard to the various effects of environment. For example, the absence of gaseous hydrogen effects near threshold in low strength steels is a result of low driving force for hydrogen to the crack tip. As ΔK increases, the triaxial state of stress at the crack tip increases enhancing the driving force for hydrogen. Once it exceeds a certain limit sufficient to produce the critical hydrogen concentration for embrittlement a hydrogen effect appears, such as the abrupt acceleration seen by Ritchie and co-workers above $da/dN = 10^{-8}$m/cycle (6,13,14,16). This is also supported by internal hydrogen effects on near-threshold crack growth rates where the effects were enhanced by higher mean stress and higher initial hydrogen concentration (3,28). Differences in fatigue crack growth rates of high strength steels in gaseous hydrogen relative to moist air environments are also controlled by the availability of hydrogen and how hydrogen adsorption appears more effecient in a moist environment than in gaseous hydrogen. Once hydrogen was introduced internally by means of cathodic charging, it was shown that hydrogen has more influence than a moist air environment (4,27-29). Furthermore, the dependence of the degree of intergranular fracture on ΔK seen in hydrogen and moist air (7,15,17,30), is also controlled by the availability of hydrogen. This behavior which appears to be much more pronounced at low R ratios can be rationalized as follows: i) at high values of ΔK, K_{min} is sufficiently high resulting in less fretting action and less hydrogen produced, or the crack growth rate is high enough not to allow sufficient time and hence hydrogen concentration at the crack tip to cause embrittlement; ii) at intermediate ΔK's, high triaxiality, fretting action and lower crack growth rates can operate simultaneously resulting in maximum embrittlement and thus maximum intergranular fracture; iii) as ΔK approaches ΔK_{th}, the driving force for hydrogen to the crack tip being controlled by the triaxial state of stress is reduced and the continuous buildup of corrosion debris on the fracture surfaces might introduce hydrogen into the metal as in moist air or conversely could prevent hydrogen adsorption in a gaseous hydrogen atmosphere. Generally because of the lower stresses, more ductile and transgranular cracking will be found at threshold. At high load ratios, the crack remains open throughout the fatigue cycle preventing the rubbing action between fracture surfaces; hence, less hydrogen is produced in moist environments and less embrittlement results. Also, crack growth rates at high load ratio are higher which may not allow sufficient hydrogen buildup at the crack tip in a given situation. Much of this is still conjecture and needs additional organized studies to document such complex phenomena.

In an attempt to resolve some of the controversial points regarding near-threshold environmental interactions a comprehensive study of the effect of internal hydrogen on three groups of materials with a range of strength making up eight materials conditions has been carried out. The objectives were to determine the intrinsic hydrogen effects, if any, on near-threshold fatigue crack growth, the controlling mechanisms and to characterize the closure processes involved. Testing at the same environment, laboratory air, and at various R values to maintain similar testing conditions and closure contributions was chosen for this study.

Although one might argue that the study of internal hydrogen present in high concentration levels are mainly of academic interest, such contributions are useful in the understanding of environmental effects in general and assist in modeling and prediction of similar effects in other systems. For example, it has been recognized that cathodically over-protected components are suceptible to hydrogen embrittlement due to the ingress of hydrogen evolved (31,32) in both low and high strength steels. This results in degradation of fatigue life (32,33) and it is plausible that large effects on threshold fatigue behavior would be observed although data are lacking in this area.

MATERIALS AND EXPERIMENTAL PROCEDURE

The materials employed in this investigation included a fine grain hot-rolled high-strength-low-alloy (HSLA) steel, an AISI 4340 steel and a sequence of Fe and Fe-Si alloys with different silicon content. The chemical composition and grain sizes are listed in Table I. The HSLA steel, which was available in plate thickness of only 7.5mm was used in both the as-received condition (10 μm) and vacuum heat-treated conditions, the later to obtain larger grain sizes of 60 and 120 μms. The AISI 4340 steel was a vacuum melted steel, austenitized at 1173K for one hour in an inert atmosphere, oil quenched and tempered at 573K for three hours. The Fe and Fe-Si alloys were vacuum induction melted and hot rolled into plates of 13mm thickness. The mechanical properties of these alloys are listed in Table II.

Table II. Mechanical Properties

Alloy	Grain Size (μm)	Yield Strength (MPa)	Tensile Strength (MPa)	Reduction in Area (%)
HSLA Steel	10	365	455	70.3
	60	230	320	74.7
	120	248	327	76.5
	20	1450	1800	-
	103	122	-	-
	73	158	-	-
	63	195	-	-
	93	248	-	-

TABLE I. Chemical Composition (wt.%) and Grain Size (d)

Alloy	C	Mn	Si	P	S	Ti	Al	Nb	Ni	Cr	Mo	Cu	Fe	d(μm)
HSLA Steel	0.070	0.51	0.03	0.01	0.01	<.005	0.10	0.014	0.01	0.01	<.01	<.01	Bal	10
4340 Steel	0.40	0.78	0.27	0.008	<.005	-	-	-	1.80	.077	0.20	0.13	Bal	20
Fe	0.008	-	-	-	-	-	0.017	-	-	-	-	-	Bal	103
Fe 1 at % Si	0.011	0.01	0.45	0.005	0.005	0.09	0.021	-	-	-	-	-	Bal	73
Fe 2.5 at% Si	0.C09	0.01	1.20	0.005	0.005	0.10	0.017	-	-	-	-	-	Bal	63
Fe 4 at % Si	0.020	0.01	1.90	0.005	0.005	0.12	0.018	-	-	-	-	-	Bal	93

All fatigue crack propagation experiments were performed using standard compact tension specimens (CTS) with thicknesses of 6.5mm for the HSLA steel, 12.5mm for Fe and Fe-Si alloys and 19.3mm for the 4340 steel. Hydrogen was introduced by means of cathodic charging prior to testing in a 5% by volume sulfuric acid solution plus As_2O_3 as a recombination poison using two parallel platinum anodes at room temperature. The charging current density and time varied with material and specimen thickness and ranged from $100A/m^2$ and 3 hours for HSLA steel, $100A/m^2$ and 12 hours for Fe and Fe-Si alloys and $20A/m^2$ and 27 hours for 4340 steel respectively. Prior to charging, the specimens were chemically polished and the charging solution was preelectrolized for 24 hours using a dummy specimen. The solution was also deaerated by continuous bubbling of nitrogen gas during preelectrolizing and charging. Immediately after charging, specimens were repolished chemically, electroplated with cadmium in a cyanide bath for 20 minutes and then baked for 30 minutes at 423K to insure uniform hydrogen concentration.

Fatigue crack propagation tests were accomplished by an MTS electro-servo-hydraulic machine under load control in tension-tension cycling at a frequency of 30 Hz. The load ratio, R, being the ratio of the minimum stress intensity, K_{min}, to the maximum stress intensity, K_{max} of the sinusoidal wave form was controlled at 0.1, 0.35 or 0.7. The crack length was monitored by the compliance method with a crack opening displacement gage clipped on the outer surface of the specimen. The load sequence followed a "load shedding" scheme with the cyclic stress intensity range, ΔK, being started at a higher value and progressively decreased until threshold stress intensity, ΔK_{th}, was established. The fatigue threshold was taken as the cyclic stress intensity range at a fatigue crack propagation rate, da/dN, equal to or less than 5×10^{-11}m/cycle. Intermittently, the test was interrupted for crack growth measurements by plotting compliance curves on an x-y recorder with the frequency at 0.1Hz. All tests were run continuously with no interruption except for the crack growth measurements. The closure loads, or more precisely, the opening loads which are defined as the point on the unloading portion of the compliance curve where the curvature of the line started to go negative (34) were measured continuously to determine the effective stress intensity range, ΔK_{eff}, at all values of ΔK. The following terminology was employed

$$\Delta K = \Delta K_{eff} = K_{max} - K_{min} \tag{1}$$

when $K_{min} \geq K_{op}$ where K_{op} is the opening stress intensity measured from the opening load in the compliance curve, or

$$\Delta K_{eff} = K_{max} - K_{op} \tag{2}$$

when $K_{min} < K_{op}$. After the threshold was established, the experiment was terminated on some specimens while others were pulled open for fractographic analysis. Crack surface oxide deposits in the AISI 4340 steel were characterized using Auger spectroscopy and sputter depth profiling. Details of the experimental procedure are described elsewhere (4,28). The testing environment was laboratory air with relative humidity of about 50 percent. Fatigue crack propagation tests in the large grain size HSLA steel were limited to the near-threshold region to avoid excessive plasticity at the crack tip.

RESULTS

HSLA Steel

The presence of internal hydrogen enhanced fatigue crack propagation rates and lowered threshold stress intensities for the as-received (10 μm) and 60 and 120 μm grain size materials. The effect is a true intrinsic hydrogen effect as it is demonstrated in terms of both apparent, ΔK, and ΔK_{eff}, stress intensities shown in Figure 1 for the 10 μm material. The effect was dependent on hydrogen concentration. Increasing the charging current density to 500 A/m^2 resulted in higher crack growth rates and lower threshold stress intensities. Although the effect was apparent in the case of the fine grain size, it was only after subtracting closure that the effects were evident in the other two cases. The degree of hydrogen effects on threshold is shown in Table III. Note that the effects were not due to charging damage and were reversible when hydrogen was removed by baking out the samples (26). The magnitude of the hydrogen effect ranged from a 0.5-1.0 MPa-$m^{1/2}$ decrease in ΔK_{th} to an increase in crack propagation rates of 4-10 times the propagation rates in the absence of internal hydrogen for the 10 μm material. Similarly, the 60 μm and 120 μm materials showed an increase in crack growth rates and a decrease in ΔK_{th}. The effects on crack growth appeared less severe in the 60 μm case relative to the 10 μm and 120 μm cases. Furthermore, the trend in ΔK_{th} in the charged and non-charged cases appeared to follow the general trend of an increase in ΔK_{th} with the increase of grain size in both ΔK_{th} and $\Delta K_{th,eff}$ (5,35-38). In terms of effective threshold stress intensities, it is evident that internal hydrogen has considerable effects which depend on material's strength and hydrogen concentration. For example, $\Delta K_{th,eff}$ dropped by 0.3, 1.4 and 1.8 MPa-$m^{1/2}$ at the lower current density and by 2.0, 1.9 and 2.7 MPa- $m^{1/2}$ at the high current density for the 10, 60 and 120 μm materials, respectively.

Table III. Threshold Stress Intensities for HSLA Steel

Grain Size (μm)	ΔK_{th} MPa-$m^{1/2}$			$\Delta K_{th,eff}$ MPa-$m^{1/2}$		
	no H_2	100 A/m^2	500 A/m^2	no H_2	100 A/m^2	500 A/m^2
10	5.5	5.0	3.5	4.1	3.8	2.1
60	6.5	5.5	5.0	5.0	3.6	3.1
120	8.75	5.5	6.0	6.4	4.6	3.7

Fractographic analysis revealed small differences between the fracture morphology in the presence and in the absence of internal hydrogen. The fracture process was purely transgranular ductile in the non-charged case typical of low strength steels (2,5,16,36) and predominately transgranular with few intergranular facets in the charged case as shown in Figure 2. At ΔK_{th}, the percentage of intergranular fracture ranged from 10 percent in the 10 μm material to 20-30 percent in the 60 and 120 μm materials.

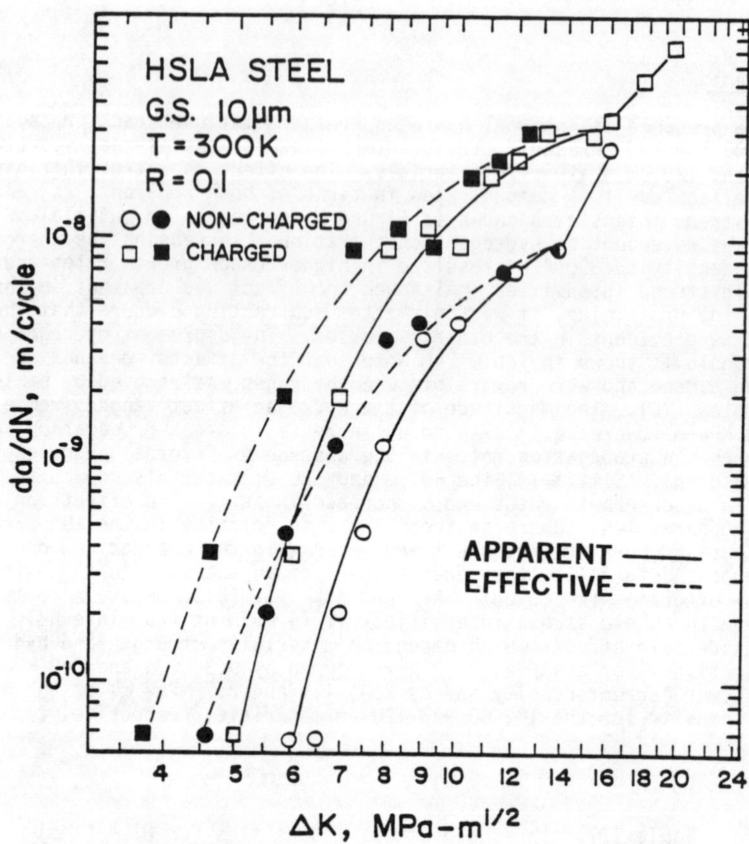

Figure 1 - The effect of internal hydrogen on near-threshold fatigue behavior of HSLA steel in terms of both apparent and effective stress intensities.

Mean stress or load ratio effects were investigated in the presence and absence of internal hydrogen for the lower current density in the 10 μm and 60 μm materials at the three load ratios of 0.1, 0.35 and 0.7. The effect of mean stress is depicted in Figure 3 for the non-charged and charged (HC≈3.5 ppm) 10 μm material. In the absence of internal hydrogen the effect of mean stress is mainly due to closure as shown in Figure 3a. Crack growth rates vs. ΔK of R=0.35 and 0.7 coincided and $\Delta K_{th,eff}$ values were equal for the three load ratios as seen by others (18,35). In the presence of internal hydrogen, crack growth rates were higher at all values of R and, ΔK_{th} and $\Delta K_{th,eff}$ were lower. The later is shown in Figure 4a in terms of $\Delta K_{th,eff}$ vs. R. The 60 μm material behaved similarly to the 10 μm material except for higher closure stress intensities. The relative changes in $\Delta K_{th,eff}$ due to hydrogen are shown in Figure 4b. Nevertheless, no apparent change in the fracture morphology was observed as the mean stress increased in the presence or in the absence of internaly hydrogen. The absence of a measurable mean stress effect in the presence of internal hydrogen for the 60 μm case reflects the presence of sufficient hydrogen concentration ahead of the crack tip at R=0.1. Thus a further increase in the hydrogen concentration by the higher triaxiality at the crack-tip had minor effects. This agrees with the results of concentration dependence where the 60 μm showed the weakest dependence.

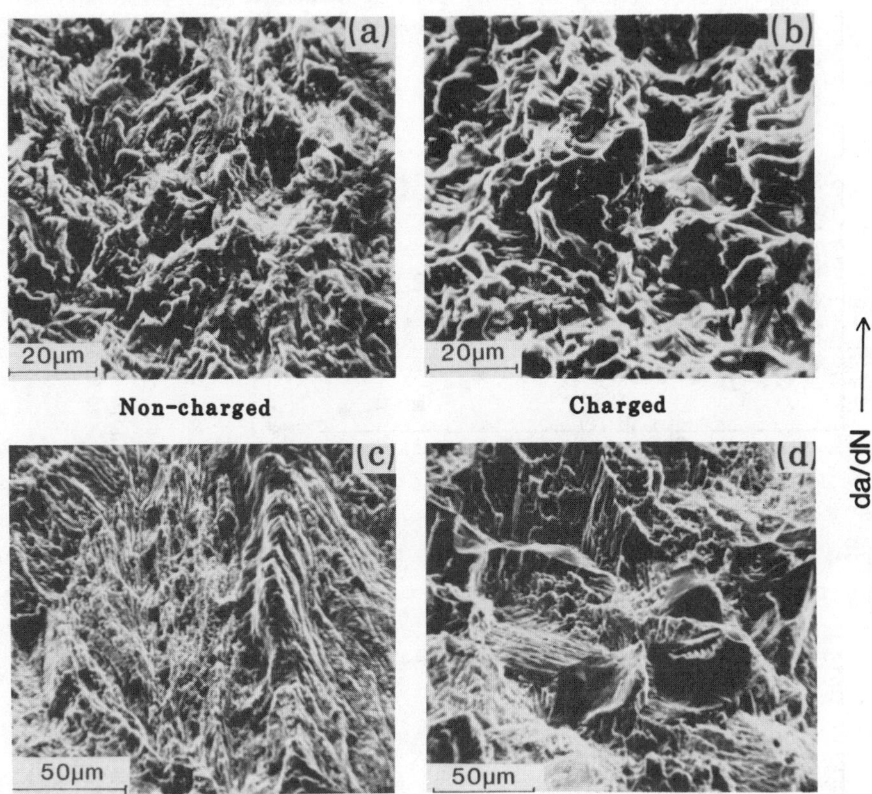

Figure 2 - Fractography of HSLA steel at threshold in the pressence and absence of internal hydrogen. (a & b) 10 μm and (c & d) 60 μm material.

Fe and Fe-Si Alloys

The effect of internal hydrogen on fatigue crack propagation and theshold stress intensities for Fe and Fe-Si alloys was studied only at R=0.1. The behavior of these alloys in the presence and the absence of internal hydrogen is given in Figure 5 in terms of ΔK and ΔK_{eff} for Fe and Fe-2.5% Si. It is evident that internal hydrogen enhances fatigue crack propagation rates and lowers threshold stress intensities in these alloys. The effect appears to be mild near threshold. However, strong hydrogen effects were seen when the data was evaluated in terms of ΔK_{eff} indicating strong closure contributions which were quite evident in the compliance measurements. The effect of hydrogen decreased with the increase in silicon content being the

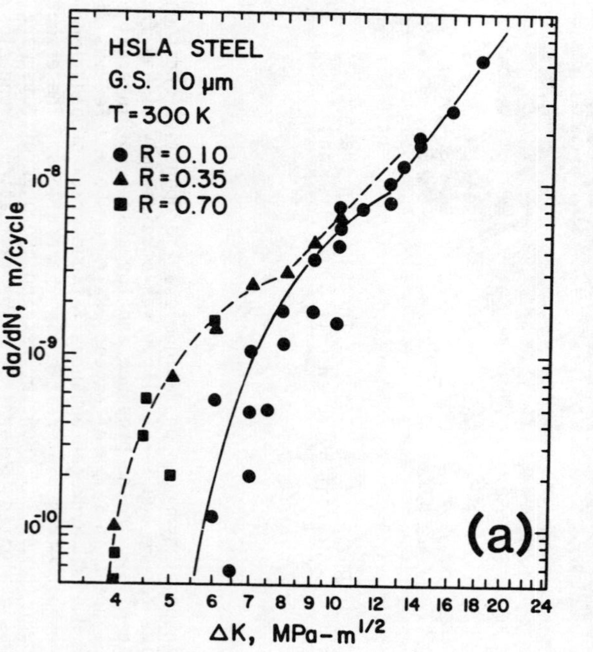

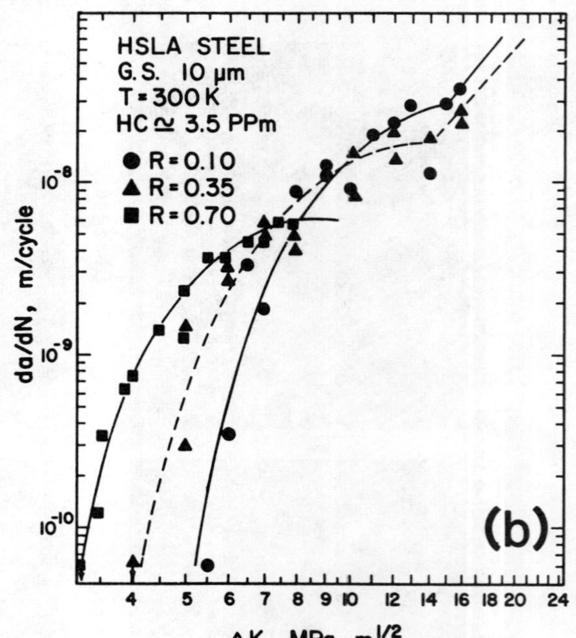

Figure 3 - The effect of load ratio on near-threshold fatigue crack propagation in 10 μm HSLA steel in the non-charged (a) and charged (b) conditions.

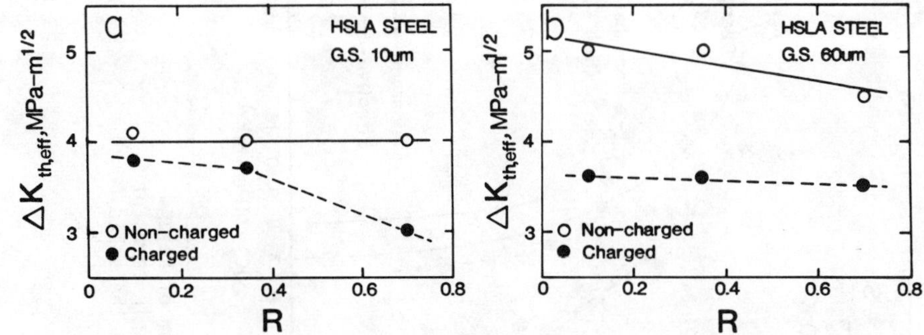

Figure 4 - Effective threshold stress intensity dependence on load ratio, R, for 10 and 60 μm HSLA steel in the presence and absence of internal hydrogen.

strongest in pure Fe and was completely absent in the Fe-4% Si. The Fe-4% Si data showed a large degree of scatter for which the general trend showed no hydrogen effects. The magnitude of internal hydrogen effects ranged from a drop in ΔK_{th} and $\Delta K_{th,eff}$ of 0.5 and 2.0 for Fe, 0 and 1.7 for Fe-1% Si, 0 and 1.6 for Fe-2.5% Si in MPa-m$^{1/2}$ units, respectively, to no effect in Fe-4% Si. Since hydrogen effects increased with a decrease in strength, this behavior tends to disagree with the normally accepted effects of strength. However, it has been shown that hydrogen effects decrease with an increase in silicon content in high strength steels (39). Also, it has been recently shown that silicon contents of this magnitude drastically reduces hydrogen permeation, solubility and trapped diffusivity in iron.*

The fracture process in these alloys varied with silicon content but was primarily of a ductile nature (40). The presence of internal hydrogen appeared to promote intergranular fracture in Fe and Fe-1% Si and increases the degree of intergranular fracture in Fe-2.5% Si down to ΔK_{th}. Fracture morphology of Fe-4% Si did not change with the presence of internal hydrogen in agreement with crack growth data indicating no hydrogen effects in this alloy. The fractography at ΔK_{th} for two of these alloys is shown in Figure 6. Two distinct features were noticed in the presence of internal hydrogen. The first is that intergranular fracture extended to ΔK_{th} and appeared to increase or remain nearly constant as ΔK approached ΔK_{th}. The second is the appearance of a transition region at the early stages of the test where the fracture morphology changed from almost pure transgranular fracture to predominantly intergranular. For example, this sudden change occurred in Fe-2.5% Si at $\Delta K \approx 14$ MPa-m$^{1/2}$ where the initial evidence of crack closure was just observed in the COD measurement and after about 200,000 cycles from the blunt notch root radius. An example of this transition is shown in Figure 7. Comparison of crack growth rates at and before this

*Personal communication from E. Riecke, F. Schambil and K. Bohwenkamp, Max Planck Institut für Eisenforsching, Düsseldorf (1983).

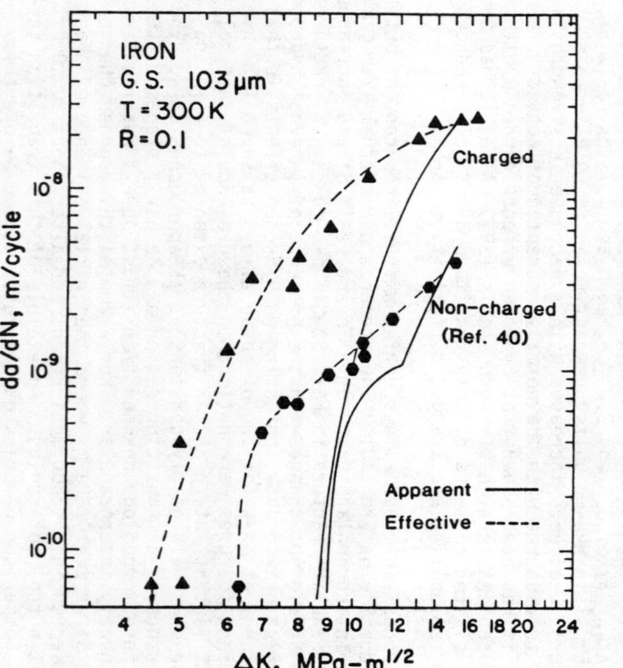

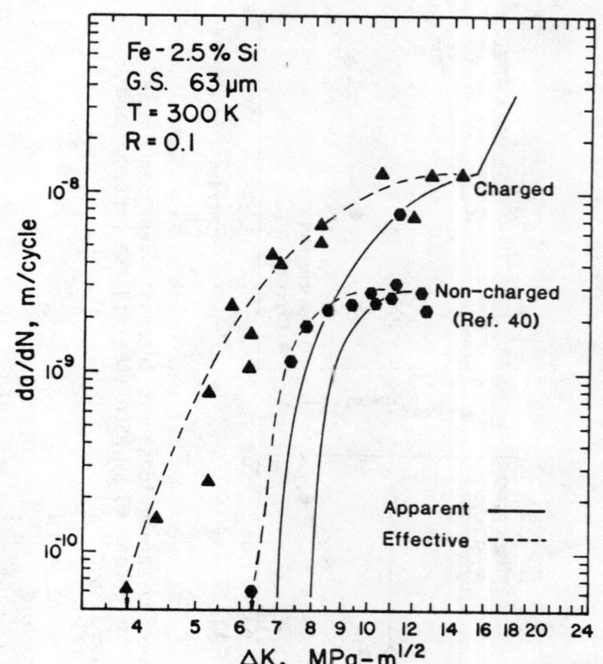

Figure 5 - Near-threshold fatigue crack propagation behavior of Fe and Fe-2.5% Si in the presence and absence of internal hydrogen in terms of ΔK and ΔK_{eff}. For ΔK only the trend of the data is shown by the solid lines for clarity.

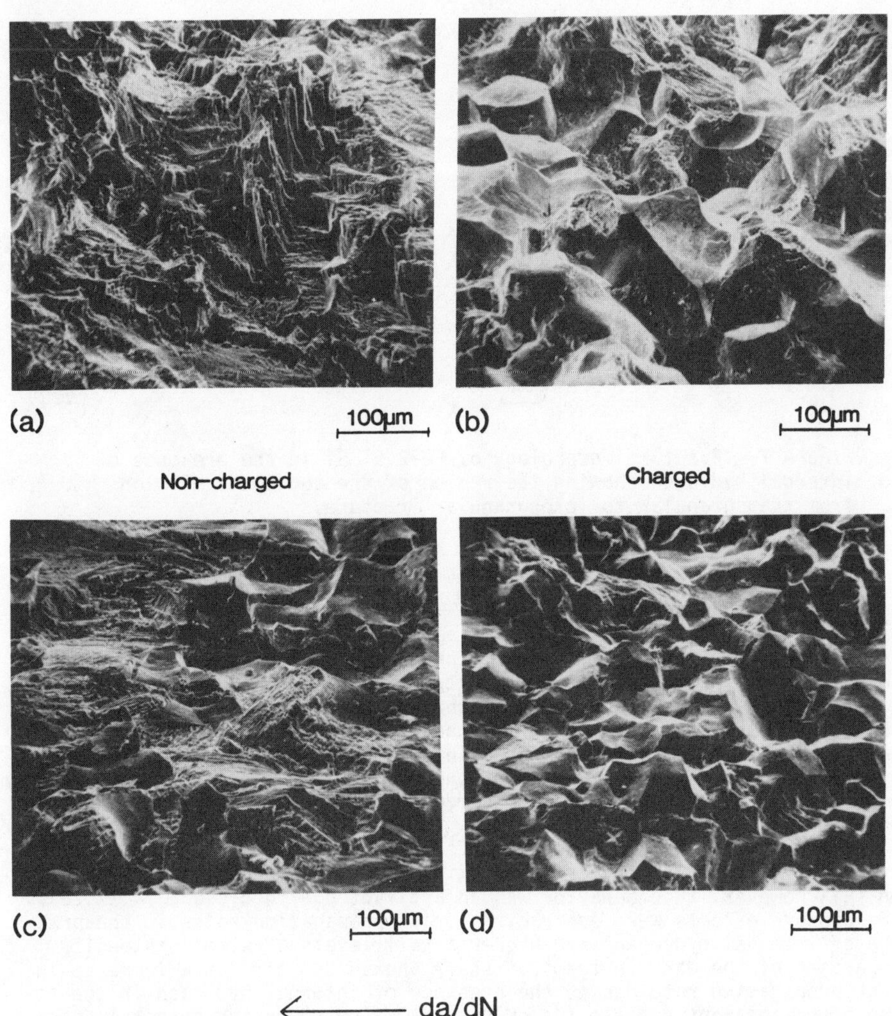

Figure 6 - Fractography of Fe (a & b) and Fe-2.5% Si (c & d) at threshold in the presence and absence of internal hydrogen.

transition relative to growth rates in the absence of internal hydrogen showed a hydrogen enhanced crack growth. This indicates that hydrogen enhanced both modes of fracture and suggests a possible incubation time needed for hydrogen concentration to reach the critical concentration for grain boundary embrittlement. The total change to intergranular fracture rules out any additional environmental effects. First, one might expect any external environmental effect to gradually increase the percent of intergranular fracture. Secondly, at the level of applied ΔK, closure was only 7 percent of the loading cycle, such that the environmental contribution due to fretting-corrosion of the fracture surfaces is minimal. Thus, it appears that the crack propagation process is again controlled by the availability of hydrogen ahead of the crack tip.

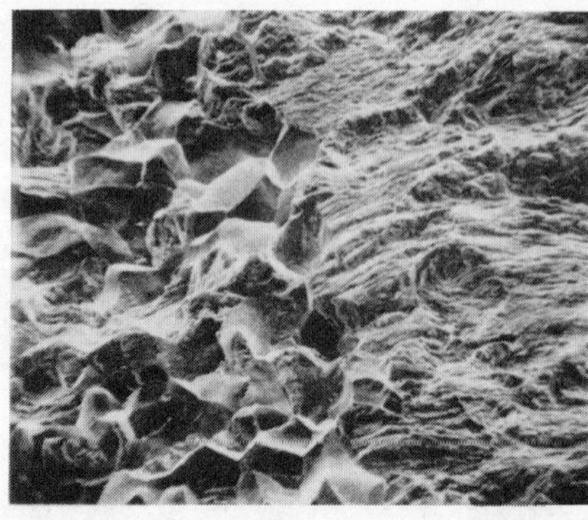

Figure 7 - Fracture morphology of Fe-2.5% Si in the presence of internal hydrogen showing the region of the sudden transition from transgranular to intergranular fracture.

High Strength 4340 Steel

The near-threshold fatigue crack propagation behavior in the presence and absence of internal hydrogen of ultra high strength 4340 steel is shown in Figure 8. It is evident that internal hydrogen enhances crack propagation rates above $\Delta K = 6.0$ MPa-m$^{\frac{1}{2}}$ and that this enhancement depends on initial hydrogen concentration. Decreasing the charging time from 27 to 4.5 hours reduced the crack growth rate considerably but remained higher than in the absence of hydrogen. The enhancement in growth rate increases sharply above $\Delta K = 7.0$ MPa-m$^{\frac{1}{2}}$ to $\Delta K = 8.0$ MPa-m$^{\frac{1}{2}}$. Below $\Delta K = 6.0$ MPa-m$^{\frac{1}{2}}$ it appeared that internal hydrogen has the same effect as gaseous hydrogen where similar or slightly higher resistance to cracking was observed relative to no hydrogen (8); however, this behavior was an artifact due to closure (4,27,28). Once closure effects were subtracted, crack propagation rates in the presence of internal hydrogen were higher at all levels of stress intensity. Comparison of the data in terms of ΔK_{eff} showed 3-4 times the increase in crack propagation rate due to the presence of internal hydrogen in the region of no apparent effects (da/dN vs. ΔK). The effective threshold stress intensity decreased by up to 40 percent from 3.3 MPa-m$^{\frac{1}{2}}$ in the non-charged material to 2.8 and 2.0 MPa-m$^{\frac{1}{2}}$ for the 4.5 and 27 hours charging time, respectively.

The fracture morphology appeared to depend on the level of hydrogen concentration in the specimens. As the samples were broken open a heavily oxidized area appeared near threshold region on all specimens tested at R=0.1. The oxidized area extended the most behind the crack tip in the fully charged specimens, 27 hours, and the least in the non-charged specimens. The oxidized area in the sample which was tested for possible internal damage by charging and baking out to remove hydrogen was similar to the non-charged specimens. Furthermore, a band of dark oxide appeared nearest to threshold which was much wider in the charged specimens and remained slightly behind the crack tip. Variation in the size of the oxidized area in the charged versus the non-charged specimens initially suggested that the identical behavior of charged and non-charged near threshold is a result of

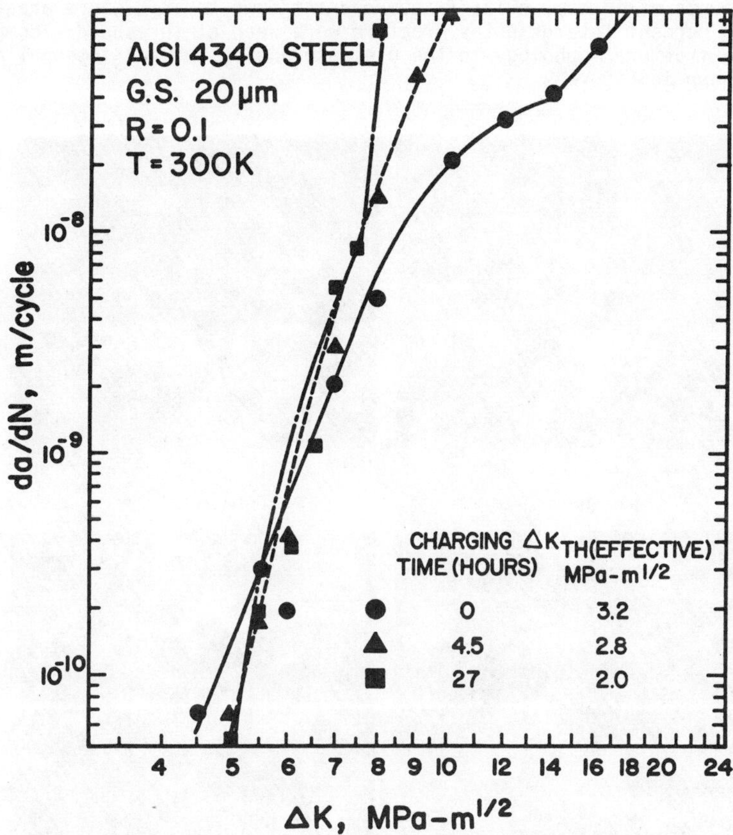

Figure 8 - Near-threshold fatigue crack propagation behavior of AISI 4340 steel in the absence and presence of internal hydrogen at two levels of initial hydrogen concentration represented by the charging time.

the differences in the degree of oxide-induced closure. However, it has been shown that the oxide thickness generated due to the rubbing of fracture surfaces in such high strength steels is not sufficient to produce the closure values observed (4,8,13-15,27).

The fracture process in the hydrogen free case was mixed intergranular and transgranular fracture with the percentage of intergranular fracture varying with the level of ΔK. The highest amount of intergranular fracture $\approx$40-50 percent occurred at $\Delta K \approx$ 10-12 MPa-m$^{\frac{1}{2}}$ and decreased as ΔK increased or decreased from this level. At threshold the fracture morphology ranged from 80-90 percent transgranular with an oxide overlayer. The presence of internal hydrogen increased the degree of intergranular fracture for all values of ΔK and promoted crack branching which was not apparent in the non-charged case. Above ΔK = 7.0 MPa-m$^{\frac{1}{2}}$, the fracture process was predominatly intergranular, 80-90 percent. Below ΔK = 7.0 MPa-m$^{\frac{1}{2}}$ even though the fracture appearance is complicated by deformed asperities, it appears that higher amounts of intergranular fracture exist down to ΔK_{th} where areas as high as 60 percent intergranular fracture were seen at threshold. Comparison of the fracture mophology in the presence and absence of internal hydrogen are shown in Figure 9.

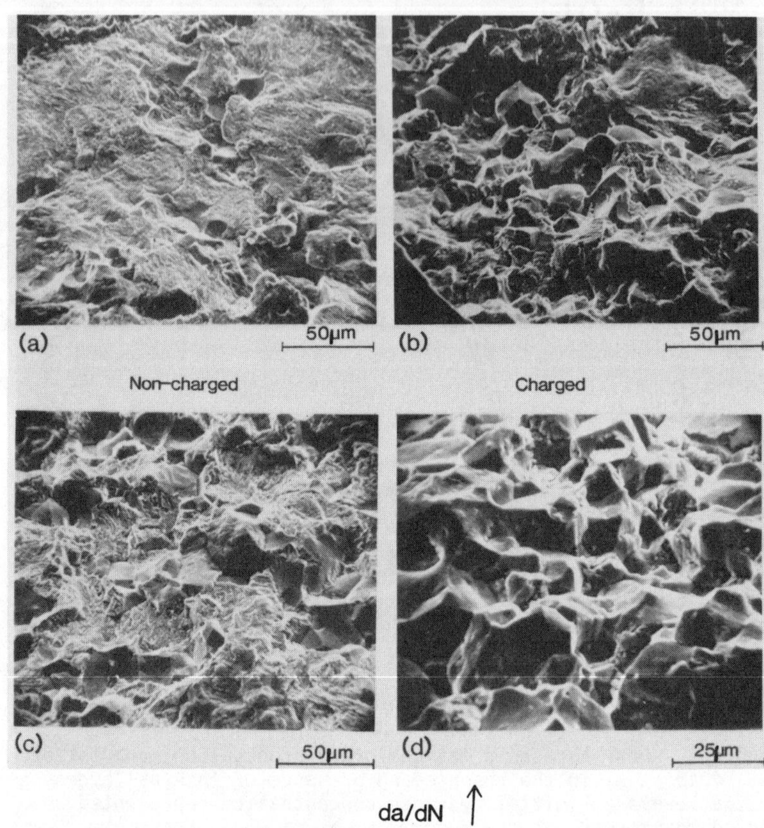

Figure 9 - Fractography of AISI 4340 steel at ΔK_{th} (a & b) and ΔK = 7.0 MPa-$m^{\frac{1}{2}}$ (c & d) in the absence and presence of internal hydrogen for R=0.1.

Increasing the mean stress resulted in higher crack propagation rates and lower threshold stress intensities in the presence and absence of internal hydrogen as shown in Figure 10. In the non-charged condition the load ratio effects appear to be primarily due to closure since crack growth rates were about equal at a ΔK of 8-10 MPa-$m^{\frac{1}{2}}$ and ΔK_{th} at R=0.7 is 10 percent less than $\Delta K_{th,eff}$ at R=0.1. Comparison of da/dN vs. ΔK at R=0.7 and da/dN vs. ΔK_{eff} at R=0.1 showed higher crack growth rates at R=0.7 for ΔK's of 3-8 $MPm^{\frac{1}{2}}$. This suggests that there is some mean stress effect at R=0.7 or the closure contribution at R=0.1 is underestimated. In the presence of internal hydrogen considerable differences between crack growth rates at R=0.1 and 0.7 were observed. For example, at R=0.7 da/dN at ΔK = 2.5 MPa-$m^{\frac{1}{2}}$ was in the order of $1x10^{-8}$ m/cycle. The threshold stress intensity ranged from

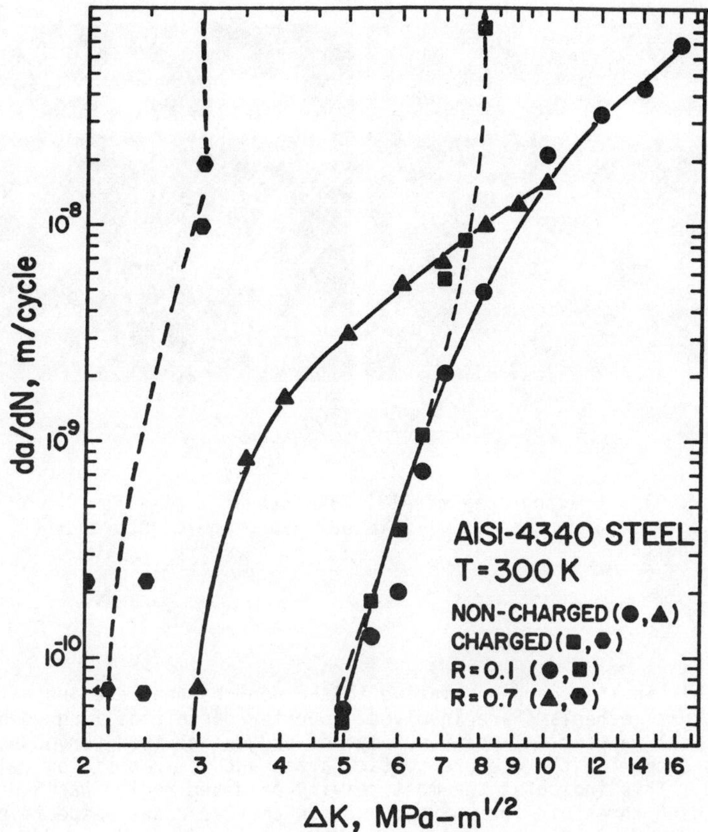

Figure 10 - Load ratio effect on near-threshold crack propagation behavior of AISI 4340 steel in the absence and presence of internal hydrogen.

1.5-2.0 MPa-m$^{1/2}$ which represents a K_{max} of 5-6.6 MPa-m$^{1/2}$. This value is at the same level of K_{max} for ΔK_{th} at R=0.1. However, the fracture morphology at R=0.7 was in complete contrast to that of R=0.1 as shown in Figure 11. The fracture process changed from pure transgranular fracture in the non-charged case to about 90 percent intergranular fracture in the charged case at all levels of ΔK including ΔK_{th}. It is evident from the results above that increasing the mean stress i.e. load ratio in the presence of internal hydrogen accelerates the crack growth process. This increase in crack propagation rate is a result of an increase in K_{max} and the driving force for hydrogen due to higher triaxiality at the crack tip which increases the hydrogen concentration ahead of the crack tip. Thus, intergranular decohesion is easier because of the combination of higher hydrogen concentration and higher σ_{max} which results in a higher degree of intergranular fracture as R increases.

To characterize the extent of the oxide contribution to the closure process, an Auger spectroscopy and sputtering study of the fracture surfaces was performed. The results indicated that the oxide thickness did not vary

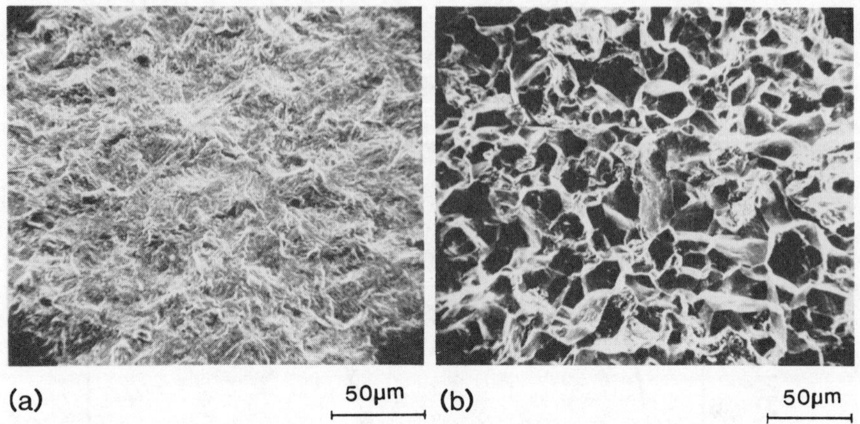

Figure 11 - Fractography of AISI 4340 steel at ΔK_{th} for R=0.7 in the absence and presence of internal hydrogen (a) non-charged, (b) charged.

considerably in the charged relative to the non-charged case indicating that other closure mechanisms are involved. Sputter depth profiling combined with chemical mapping and point surveys of oxygen and iron served as a method to determine the closure contact areas and their position relative to threshold. This indicated the most heavily oxidized region was 300 and 120 μm behind threshold for charged and non-charged cases respectively, and a strip of asperities approximately 100-200 μm wide in the charged case and 75 - 125 μm wide in the non-charged case were overlayed with oxide.

Using a similar approach to the one proposed by Suresh and Ritchie (41) for oxide-induced wedging at a given displacement, K_{op} for geometry-induced wedging at a given load can be calculated using Hartranft and Sih's (42) solution to an edge crack in a semi-infinite plate subjected to concentrated forces. Taking the heavily oxidized area as closure points a distance (a-b) behind the tip of a crack of length a, the closure stress intensity (K_{op}) can be calculated as

$$K_{op} = \frac{2P\sqrt{a}}{\pi B}\left[\frac{1 + f(b/a)}{\sqrt{a^2 - b^2}}\right]$$

where P is the point load wedging the crack faces open, B is the specimen thickness and f(b/a) is a geometric factor that approaches zero for b/a close to unity which is the case here. Assuming that contact area, A_c in a given fatigue cycle involved 50% of the width of the completely overlayed strip then P can be estimated from

$$P = \sigma A_c \tag{2}$$

Since the asperities appeared to have undergone permanent plastic deformation during the repeated closure contact as the crack extended beyond the asperities, then σ can be approximated by the average flow stress, $\bar{\sigma}_{flow}$, of these asperities which could be given by

$$\bar{\sigma}_{flow} = \frac{\sigma_{ys} + \sigma_{uts}}{2} \tag{3}$$

where σ_{ys} and σ_{uts} are yield and ultimate strength. The K_{op} calculated using this approach agree with K_{op} values measured from the compliance measurements as illustrated in Table IV.

Thus, it is evident that the geometric-induced closure is the dominant closure mechanism relative to the oxide-induced closure. The oxide overlay of the geometric asperities which is generated during the fretting fatigue process gives a good estimate of the area of contact and closure position relative to threshold, even though we do not actually know which oxide was formed at which time after crack growth. It is our supposition that the largest band of oxide behind the crack tip is supporting the load at threshold.

DISCUSSION

Separation of closure from the inherent behavior of the materials tested in this investigation in the presence and the absence of internal hydrogen clearly demonstrates the role of crack closure in determining near-threshold crack growth behavior. Testing in the presence and absence of internal hydrogen under similar conditions of external environment (laboratory air) and load ratio in addition to accounting for closure showed the true intrinsic hydrogen effects. These results are in contrast to results in the literature where gaseous hydrogen appeared to have no effect or improved near-threshold fatigue properties relative to laboratory air if the effects of closure were considered (6-18). It is evident at this point that environmental effects on near-threshold fatigue behavior can vary considerably from one material to another and are dependent on several variables. Among these variables are material strength and structure, fracture process, load ratio, frequency, temperature and testing environment. The effect of some of these variables, form the results of this investigation, are summarized in Table V. The data are presented in terms of both ΔK_{th} and $\Delta K_{th,eff}$ to show both the inherent material behavior and the closure contribution to the near-threshold cracking process. The general trend presented in terms of ΔK_{th} at R=0.1 did not show any measurable effects of hydrogen except in the case of the 60 and 120 μm HSLA steel. Nevertheless, comparison of $\Delta K_{th,eff}$ data reveals the degree of hydrogen effects and how closure can contribute to the underestimation of environmental effects (1). Furthermore, results at high load ratios indicate that the closure contribution can be eliminated, but then in the presence of hydrogen, mean stress effects become involved in the fracture process. The closure contributions have varied from one material to another and depended on the fracture morphology and how it was affected by environmental conditions. The least amount of closure

TABLE IV. Calculated Load Opening Stress Intensities for the AISI 4340 Steel

Condition	a	b	B	A_c	P	K_{op}(cal)	K_{op}(obs)
	m	m	m	m^2	Newtons	MPa-$m^{1/2}$	MPa-$m^{1/2}$
Non-charged	.02595	.02576	.01925	$.722 \times 10^{-6}$	1155	1.96*	2.27
	.02595	.02574	.01925	1.203×10^{-6}	1925	3.11+	
Charged	.02640	.02605	.01925	$.9625 \times 10^{-6}$	1540	1.93*	3.42
	.02640	.02600	.01925	1.925×10^{-6}	3080	3.62+	

* based on lower bound of contact band width

+ based on upper bound of contact band width

$\overline{\sigma}_{flow}$ = 1600 MPa

Table V. Threshold Stress Intensities for All Alloys Tested

Alloy	Grain Size	R	No Hydrogen		Hydrogen Charged		ϕ
			ΔK_{th}	$\Delta K_{th,eff}$	ΔK_{th}	$\Delta K_{th,eff}$	
	μm		MPa-$m^{1/2}$	MPa-$m^{1/2}$	MPa-$m^{1/2}$	MPa-$m^{1/2}$	%
HSLA	10	0.1	5.5	4.1	5.0	3.8	7.3
		0.35	4.0	4.0	4.0	3.7	7.5
		0.7	4.0	4.0	3.0	3.0	25
	60	0.1	6.5	5.0	5.5	3.6	28
		0.35	5.0	5.0	5.0	3.6	28
		0.7	4.5	4.5	3.5	3.5	22
	120	0.1	8.8	6.4	5.5	4.6	28
Fe	103	0.1	9.0	6.5	8.5	4.5	31
Fe-1% Si	73	0.1	8.0	5.8	8.0	4.1	29
Fe-2.5% Si	63	0.1	7.0	5.4	7.0	3.8	30
Fe-4% Si	93	0.1	7.0	4.0	7.0	4.2	--
4340	20	0.1	4.75	3.3	5.0	2.0	39
		0.7	3.0	3.0	1.5-2.0	1.5-2.0	33-50

ϕ = Percent change in $\Delta K_{th,eff}$ due to the presence of internal hydrogen.

appeared in the 10 μm HSLA steel due to the fine microstructure and the ductile transgranular mode of fracture. The magnitude of closure increased relatively with the increase in grain size in the same steel particularly in the presence of internaly hydrogen as a result of enhanced plasticity and intergranular fracture. Significant increases in closure appeared when a change in the fracture process occurred both in the presence and absence of internal hydrogen as in the case of Fe, Fe-1% Si, Fe-2.5% Si alloys and 4340 steel. The presence of intergranular fracture enhanced the degree of closure. The closure was further enhanced by the increase in the amount of intergranular fracture in the presence of internal hydrogen. Evidence of the effect of the change in the fracture morphology on closure can be seen when comparing the 60 μm HSLA to the Fe-2.5% Si, two materials of similar strength and grain size. For example, K_{op} ranged from 2.2 and 2.4 MPa-$m^{1/2}$ in the non-charged case to 2.5 and 4.0 in the charged case for HSLA steel and Fe-2.5% Si alloy, respectively. The fracture appearance for the materials in the non-charged condition were predominantly transgranular and the K_{op} values were similar. However, in the charged condition, the nearly 100 percent intergranular fracture in Fe-2.5% Si gave a much larger K_{op} value than the 60 μm HSLA steel which exhibited about 30 percent intergranular fracture. (See Table V.) Other evidence can be seen in the cases of charged versus non-charged 4340 steel, Fe and Fe-Si alloys.

If one compares the effect of yield strength on $\Delta K_{th,eff}$, one finds that in both hydrogen-charged and non-charged conditions, that there is a tendency for threshold to decrease with increasing yield strength. This is shown in Figure 12 where all data except one fall in a relatively narrow scatter band with the non-charged samples tending toward the top and the charged samples toward the bottom of the band. Since in all cases some intergranular fracture was found for the charged specimens at R=0.1, this is additional strong evidence for an embrittlement-enhanced threshold process. The fact that such a process might also manifest itself in the humid air environment simply implies this is a weaker environment with threshold, on the average, being 32% greater than their hydrogen charged counterparts.

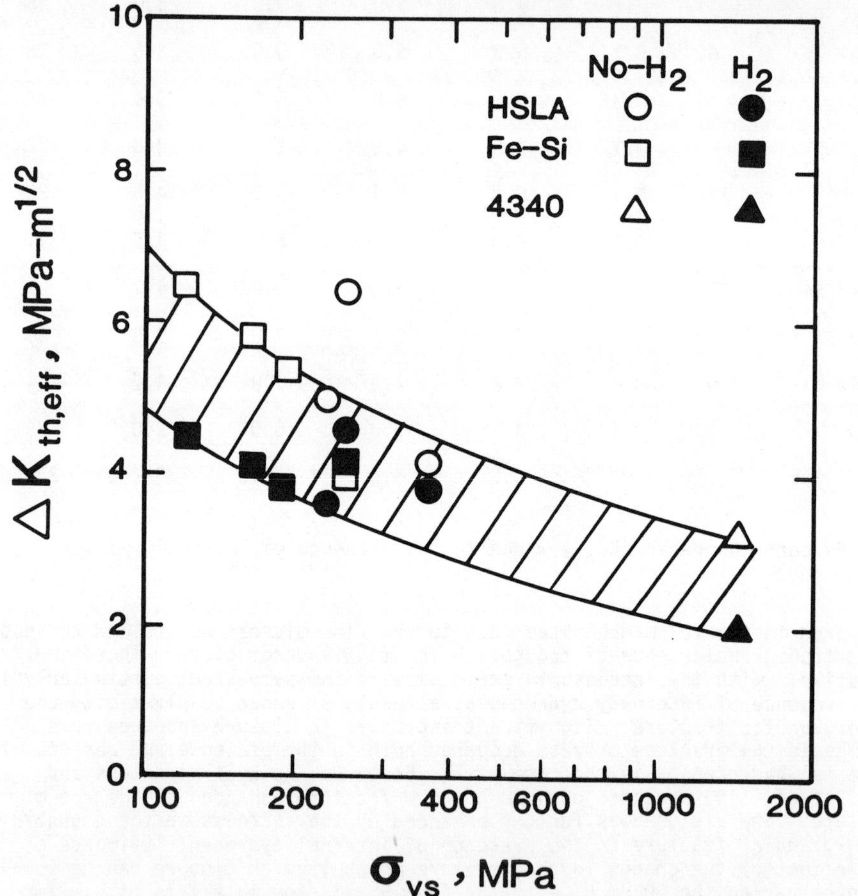

Figure 12 - Dependence of effective threshold stress intensity on yield strength in the absence and presence of internal hydrogen.

Crack closure can be a result of one or more of the following processes: oxide-induced (6,7,10-19,41), geometry or surface roughness induced

(10,30,43-48), or residual stress, reverse slip or plasticity induced (34, 43,49,50). To quantify the contribution of each of these mechanisms to any crack growth process is not easily done but elimination of some or testing under similar conditions to insure similar contributions of others are possible. In this investigation, by testing in laboratory air, the oxide contribution would be similar in the charged and non-charged cases providing the fracture surface morphology is the same. The variations in closure contribution in the charged versus the non-charged cases and from one material to another are mostly a result of the contribution of reverse slip or plasticity and/or surface roughness combined with ModeI-ModeII displacement (51, 52). Since hydrogen did not have any measurable effects on monotonic flow properties, the contribution of plasticity induced closure is expected to remain the same in the presence and absence of internal hydrogen. Thus the main hydrogen effect on closure would tend to be limited to enhanced surface roughness.

As is seen in Figure 13, there is reasonable correlation between K_{op} and ΔK_{th} for both charged and non-charged materials. The least square lines with correlation coefficient, $r \approx 0.8$, demonstrate that for the thresholds of interest, $\Delta K_{th} > 4$ MPa-$m^{\frac{1}{2}}$, the charged materials provided greater opening stress intensities than the non-charged materials. Furthermore, there is a divergence in the curves indicating a more severe effect of hydrogen on closure at higher ΔK_{th} levels. All of these observations are consistent with hydrogen producing a greater geometric closure contribution. First, hydrogen does not appreciably affect the flow properties at room temperature and all the oxide contributions should be similar for a given ΔK-σ_{ys} combination. Thus, at a given ΔK, we propose that the increased K_{op} is due to hydrogen increasing the size and number of geometrical asperities. This is consistent with the ΔK magnitude effect since at higher K_{max} values, more hydrogen-induced intergranular fracture should be expected. A single representation seemed to serve all materials with a yield strength ranging from 122 to 365 MPa. The deviation is with the 4340 steel. This ultra-high strength steel had a more severe effect of hydrogen on the magnitude of K_{op} and may be attributed to its ultra-high strength level of 1450 MPa which makes it more susceptible to intergranular cracking. Thus, it is evident that in the presence of internal hydrogen geometric induced closure is a significant mechanism and the variations in the degree of closure in the charged case relative to the non-charged case are primarily due to more geometric asperities promoted by the change in the fracture morphology.

The mean stress dependence seen in the 10 μm HSLA and 4340 steels in the presence of internal hydrogen reflects the increase in the hydrogen concentration ahead of the crack tip. The increase in concentration is due to the increase in the hydrogen driving force promoted by the higher triaxial state of stress maintained at the crack tip through most of the fatigue cycle at high load ratio. The dependence of near-threshold fatigue behavior on initial hydrogen concentration in the two steels and the total change to intergranular fracture at R=0.7 in the charged 4340 steel are further evidence of the effect of mean stress and the importance of the triaxiality as driving force for cracking under environmental conditions. The domination of intergranular fracture in Fe, Fe-1% Si, Fe-2.5% Si and 4340 steel in the presence of hydrogen at all levels of ΔK including ΔK_{th} strongly suggests that variation in the percent of intergranular fracture in air and gaseous hydrogen (7,15,17,30) are also controlled by the availability of hydrogen and the level of concentration ahead of the crack tip.

In summary, the results of this study clearly indicate that internal hydrogen can have detrimental effects on near-threshold fatigue behavior on low and high strength steels and Fe and Fe-Si alloys. These effects are

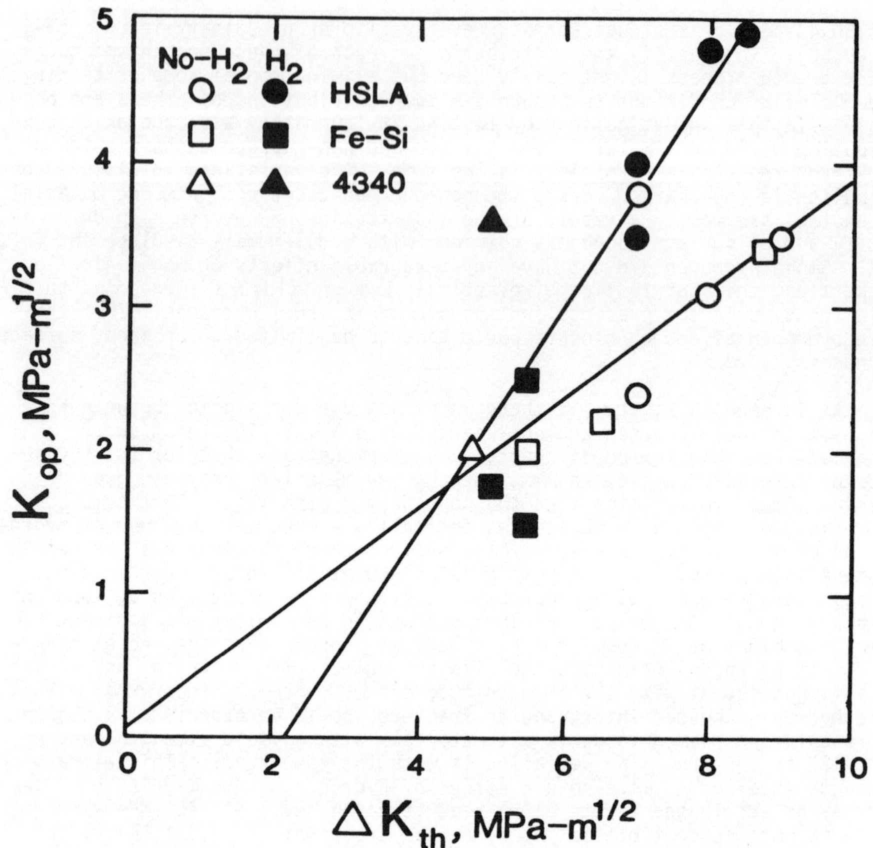

Figure 13 - Variation in K_{op} relative to ΔK_{th} in the presence and absence of internal hydrogen.

strongly controlled by the crack closure processes involved in each material and may be totally suppressed as a result of closure in many instances. Testing at high load ratio to avoid closure can lead to mean stress effects which in some cases enhanced the crack growth process. These results in conjunction with others (14,15,19,24,25,29) indicate that the environmental effects on near threshold fatigue crack propagation are controlled by the availability of hydrogen ahead of the crack tip and crack tip closure mechanisms and how they affect the hydrogen evolution on the surfaces of a growing crack. The apparent immunity of low and high strength steels to gaseous hydrogen embrittlement near-threshold is a result of lower hydrogen availability.

CONCLUSIONS

1) The presence of internal hydrogen enhances near-threshold fatigue crack propagation rates and lowers threshold stress intensities of low strength

HSLA steel, Fe, Fe-1% Si, and Fe-2.5% alloys, and high strength 4340 steel.
2) The effects of internal hydrogen were suppressed by crack closure and were evident only after subtraction of closure.
3) The hydrogen effects depend on the initial hydrogen concentration and mean stress. Increasing the load ratio resulted in higher crack growth rates and lower threshold stress intensities due to the higher driving force for hydrogen promoted by higher triaxiality at the crack tip.
4) Geometry-induced crack closure is an important closure mechanism in Fe and Fe-Si alloys and 4340 steel.
5) Internal hydrogen increased the propensity for intergranular fracture which promoted further crack closure.
6) The influence of environment on near-threshold fatigue behavior is controlled mainly by the availability of hydrogen ahead of the crack tip.

ACKNOWLEDGEMENTS

The authors would like to thank Dr. W. Yu for the use of his iron and iron-silicon data. This work was supported by the Department of Energy, Basic Research Division under contract DOE/DE AC02-79ER10450-A004.

REFERENCES

1. J. A. Ruppen and R. Salzbrenner: "The effect of environment on crack closure and fatigue threshold," accepted for publication in Fatigue of Engineering Materials and Structures, 1983, Vol. 6, pp. 307-314.

2. K. A. Esaklul, W. Yu and W. W. Gerberich: "The effect of geometric closure on threshold stress intensities at low test temperatures," ASTM Symposium on Fatigue at Low Temperatures, 1983, in press.

3. K. A. Esaklul and W. W. Gerberich: "Internal hydrogen degradation of fatigue thresholds in HSLA steel" in Sixteenth National Symposium on Fracture Mechanics, ASTM, 1983.

4. K. A. Esaklul, W. W. Gerberich and A. G. Wright: "Hydrogen effects on roughness-induced closure and near-threshold fatigue cracking in 4340 steel" submitted to Metallurgical Transactions A, 1983.

5. R. O. Ritchie: Intern. Metals Reviews, 1979, Vol. 20, pp.205-230.

6. R. O. Ritchie: in Analytical and Experimental Fracture Mechanics, G. C. Sih and M. Mirabile, eds., pp. 81-108, Sijthoff & Noordhoff, The Netherlands, 1981.

7. A. T. Stewart: Eng. Fract. Mech., 1980, Vol. 13, pp. 463-478.

8. J. Toplosky and R. O. Ritchie: Scripta Met., 1981, Vol. 15, pp. 905-908.

9. C. J. Beevers: in Fatigue Thresholds, J. Backlund et al., eds., pp. 257-275, Engineering Materials Advisory Services LTD., Warley, England, 1982.

10. B. L. Freeman, P. Smith and A. T. Stewart: ibid, pp. 547-561.

11. E. K. Priddle: ibid, pp. 581-600.

12. T. C. Lindley and C. E. Richards: ibid, pp. 1087-1114.

13. R. O. Ritchie, S. Suresh and P. K. Liaw: in Ultrasonic Fatigue, J. M. Wells et al., ed., pp. 443-460, The Metallurgical Society of AIME, Warrendale, PA 1982.

14. S. Suresh, J. Toplosky and R. O. Ritchie: in Fracture Mechanics, ASTM STP 791, 1982, pp. I-329 - I-349.

15. P. K. Liaw, S. J. Hudak, Jr. and J. K. Donald, ibid, pp. II-370-II-388.

16. S. Suresh, C. M. Moss and R. O. Ritchie: Trans. Japan Inst. Metals, 1980, Vo. 21, pp. 481-484.

17. P. K. Liaw, S. J. Hudak, Jr. and J. K. Donald: Metall. Trans. A, 1982, Vol. 13A, pp. 1633-1645.

18. P. K. Liaw, T. R. Leax, R. S. Williams and M. G. Peck: Metall. Trans. A, 1982, Vol. 13A, pp. 1607-1618.

19. A. K. Vasudevan and S. Suresh: Metall. Trans. A, 1982, Vol. 13A, pp. 2271-2280.

20. H. G. Nelson: Effect of Hydrogen on Behavior of Materials, A. W. Thompson and I. M. Bernstein, Eds., pp. 602-611, TMS-AIME, New York, N.Y., 1976.

21. H. G. Nelson: Proceedings of 2nd International Conference on Mechanical Behavior of Materials, Boston, Mass., pp. 690-694, ASM Metals Park, OH., 1976.

22. R. J. Walter and W. T. Chandler: Effect of Hydrogen on Behavior of Materials, A. W. Thompson and I. M. Berstein, eds.,273-286, TMS-AIME, New York, N.Y. 1976.

23. H. F. Wachob and H. G. Nelson: Hydrogen in Metals, I. M. Bernstein and A. W. Thompson, eds.,pp. 703-710, TMS-AIME, Warrendale, PA, 1981.

24. G. W. Simmons, P. S. Pao, and R. P. Wei: Metall. Trans. A, 1978, Vol. 9A, pp. 1147-1158.

25. R. P. Wei, P. S. Pao, R. G. Hart, T. W. Weir and G. W. Simmons: Metall. Trans. A, 1980, Vol. 11A, pp. 151-158.

26. K. A. Esaklul and W. W. Gerberich, Scripta Met., 1983, Vol. 17, pp. 1079-1082.

27. K. A. Esaklul, A. G. Wright and W. W. Gerberich, ibid, pp. 1073-1078.

28. K. A. Esaklul, Ph.D. Thesis, University of Minnesota, 1983.

29. C. N. Sastry, W. E. Wood and K. B. Das, Society for the Advancement of Material and Process Engineering Quarterly, 1981, Vol. 12, pp. 27-32.

30. P. K. Liaw, A. Saxena, V. P. Swaminathan and T. T. Shih: Metall. Trans. A., 1983, Vol. 14A, pp. 1631-1640.

31. W. E. White: Fracture Probelms and Solutions in the Energy Industry, L. A. Simpson, ed., pp. 51-61, Pergamon Press, New York, N.Y. 1981.

32. W. H. Hartt: Design of Fatigue and Fracture Resistant Structures, ASTM

STP 761, P. R. Abelkis and C. M. Hudson, eds., pp. 91-109, ASTM, Philadelphia, PA. 1982.

33. W. W. Gerberich: "Hydrogen fatigue interactions" in Hydrogen Phenomena in Iron and Steels, R. O. Oriani, J. P. Hirth and M. Smialowski, eds., 1983, in press.

34. W. Elber: Damage Tolerance in Aircraft Structures, ASTM STP 486, pp. 230-262, ASTM, Philadelphia, PA. 1971.

35. Y. Nakai, K. Tanaka and T. Nakanishi: Eng. Fract. Mech., 1981, Vol. 15, pp. 291-302.

36. J. P. Lucas and W. W. Gerberich: Materials Sci. Eng., 1979, Vol. 41, pp. 271-280.

37. W. W. Gerberich and N. R. Moody: Fatigue Mechanisms, ASTM STP 675, J. T. Fong, ed., pp. 292-341, ASTM Philadelphia, PA. 1979.

38. J. Masounave and J. P. Bailon: Scripta Met., 1967, Vol. 10, pp. 165-170.

39. I. M. Bernstein and A. W. Thompson: Inter. Metals Review, 1967, Vol. 17, pp. 269-287.

40. W. Yu, Ph.D. Thesis, Univeristy of Minnesota, 1983.

41. S. Suresh and R. O. Ritchie: Scripta Met., 1983, Vol. 17, pp. 575-580.

42. R. J. Hartranft and G. C. Sih: in Methods of Analysis and Solutions of Crack Problems, G. D. Sih, ed., pp. 179-238, Noordhoff, Holland, 1973.

43. A. J. McEvily: Met. Sci., 1977, Vol. 11, pp. 274-284.

44. S. Purushothaman and J. K. Tien: Proc. 5th Inter. Strength of Metals and Alloys, Pergamon Press, 1979, Vol. 2, pp. 1267-1271.

45. N. Walker and C. J. Beevers: Fat. Eng. Mats. Struc., 1979, Vol. 1, pp. 135-148.

46. K. Minakawa and A. J. McEvily: Scripta Met., 1981, Vol. 15, pp. 633-636.

47. K. Minakawa and A. J. McEvily Fatigue Thresholds, J. Backlund et al., eds., pp. 373-390, Engineering Materials Advisory Services Ltd., Warley, England, 1981.

48. G. T. Gary, III, J. C. Williams and A. W. Thompson: Metall. Trans. A., 1983, Vol. 14A, pp. 421-433.

49. R. O. Ritchie, S. Suresh and C. M. Moss: J. Eng. Mat. Technol., 1980, Vol. 102, pp. 293-299.

50. W. W. Gerberich, W. Yu and K. Esaklul: "Fatigue threshold studies in Fe, Fe-Si, and HSLA steel: part I. effects of strength and surface asperities on closure". Accepted for publication in Metall.Trans. A., 1983.

51. D. L. Davidson: Fat. Eng. Mats. Struct., 1981, Vol. 3, pp. 229-236.

52. R. O. Ritchie and S. Suresh: Metall. Trans. A., 1982, Vol. 13A, pp. 937-9.

FATIGUE CRACK CLOSURE AND THE FATIGUE THRESHOLD

C.J. Beevers*, K. Bell* and R.L. Carlson**

* Department of Metallurgy and Materials
University of Birmingham
P.O. Box 363
Birmingham. B15 2TT
U.K.

** School of Aerospace Engineering
Georgia Institute of Technology
Atlanta, Georgia 30332
U.S.A.

The magnitude of ΔK_{th} can vary from 1 to 20 MPa$\sqrt{m}$ depending upon the material and test conditions. One of the reasons for this wide spread is the occurrence and extent of fatigue crack closure. The fatigue threshold stress intensity ΔK_{th} can be considered to have two components, ΔK^c_{th} the closure contribution and ΔK^i_{th} representing the intrinsic crack growth resistance of the material. This paper is concerned with an analysis of the role of the closure contribution in thresholds and to improve the basic understanding of the fatigue closure process a single asperity model has been developed and this will be used to explore the influence of microstructure, yield stress and mean stress on ΔK^c_{th}.

Introduction

The stress intensity range of threshold ΔK_{th} can be considered to have two major components - (a) ΔK^{i}_{th} representing the intrinsic resistance to crack extension and encompassing such variables as microstructure, environment and mean stress and (b) ΔK^{c}_{th} the crack closure component which can be a dominant contributor to ΔK_{th} particularly in cases where a low mean stress state pertains ($R < 0.3$). Thus ΔK_{th} can be expressed as

$$\Delta K_{th\,(measured)} = \Delta K^{i}_{th} + \Delta K^{c}_{th} \qquad (1)$$

In a previous paper (1) the parameters influencing ΔK^{i}_{th} were considered in some detail and this paper will examine the closure contribution to ΔK_{th}. The relevance of the closure contribution to ΔK_{th} can be seen from the results in Table I for a low carbon steel (2), an α/β titanium alloy (3) and nickel base superalloys IN901 and APK-1 (11). The variation in ΔK_{th} for a particular material was associated with changes in grain size, the coarser the grain size the larger the value of ΔK_{th}. Examination of Table I illustrates the major role that closure can play in determining the measured threshold value. To investigate the role of closure in more detail a model has been developed to describe the local stress state in the crack tip region (4).

The Model

Under dominantly plane strain conditions at low stress levels the presence of contact points on the fatigue fracture faces can lead to a wedging open of the fatigue crack in the crack tip region. This non-closure of the fatigue crack faces can occur as a result of the development of microstructural asperities (5) or oxide/corrosion debris asperities (6).

Examination of fatigue fracture faces reveals a non-uniform distribution of asperities which form as a consequence of microstructurally controlled fatigue crack growth. However, if the specimen thickness is very much greater than the spacing of the contact points it is possible to envisage a line of asperities behind the crack tip acting as a load bearing surface. In the case of an oxide layer build up this can be considered simply as creating a relatively uniform ridge across the specimen width. There is therefore the possibility of representing the asperities through thickness by an effective pre-compressed spring which makes line contact through the thickness of the specimen. The line contact concept permits the problem to be treated as a two-dimensional one. Thus the model described treats the case for Mode I opening with a single effective asperity behind the crack tip under dominantly plane strain conditions.

The essence of the crack tip situation is illustrated in Figure 1. The upper half of the crack tip region is represented, an asperity of width b is positioned at a distance C from the crack tip, L represents the magnitude of the interference between the faces produced by the presence of the asperity. The local crack face force P results in a compression of the asperity of magnitude e.

From consideration of the global and local forces on the crack faces the following relationships have been developed (4)

Table I. The intrinsic and closure contributions to ΔK_{th} at low mean stresses for a low carbon steel (2), an α/β titanium alloy (3) and nickel alloys (11).

R	ΔK_{th}	ΔK^{i}_{th}	ΔK^{c}_{th}	Grain Size μm.	Ref.
0	5.5	0.9	4.6	7.8	2
0	6.1	1.2	4.9	20.5	2
0	7.3	1.6	5.7	55	2
0.12	8.1	4.7	3.4	600	3
0.11	11.5	5.3	6.2	730	3
0.1	12.1	4.5	7.6	880	3
0.1	7.8	4.4	3.4	6.8	11
0.1	9.3	5.1	4.2	270	11
0.1	7.8	3.5	4.3	61	11
0.17	16.5	9.0	7.5	600	11

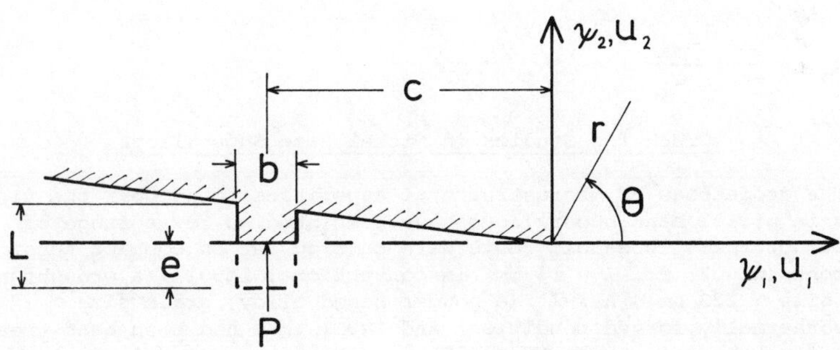

Figure 1 - Upper crack face loading showing terms used in establishing model.

$$K_{CL} = \left(\frac{2}{\pi C}\right)^{\frac{1}{2}} \left[\frac{1}{Eb} + \frac{2(1 - \nu)}{\pi G L}\right]^{-1} \quad (2)$$

$$K_{op} = \frac{LG}{2(1 - \nu)} \left(\frac{2\pi}{C}\right)^{\frac{1}{2}} \quad (3)$$

where K_{op} is the global stress intensity above which there is no contact between the opposing crack faces and K_{CL} is the local stress intensity, which results from the interference with the closing of the faces, produced by the presence of the compressed asperity. Thus K_{CL} is determined by the extent of load transfer at the contact point and as such is sensitive to both the geometry and rigidity of the asperity. In most cases the rigidity of the asperity is close to that of the matrix and hence there are minimal differences in K_{CL} and K_{op} levels. Examination of Equations 2 and 3 reveals that both K_{CL} and K_{op} increase with asperity height and decrease with the distance from the crack tip. With increasing asperity height the local stress intensity will increase and some local yielding of the material at the crack tip could be expected. This can be thought of as a form of relaxation of the local stress intensity and reducing the rate of increase of K_{CL} with asperity height. This feature was incorporated into the model by assuming that local crack tip plasticity would increase the effective value of C. The modified relationships for K_{CL} and K_{op} are presented in Equation 4 and 5.

$$K_{CL} = \left(\frac{2}{\pi (C + Rp)}\right)^{\frac{1}{2}} \left[\frac{1}{Eb} + \frac{2(1 - \nu)}{\pi G L}\right]^{-1} \quad (4)$$

$$K_{op} = \frac{LG}{2(1 - \nu)} \left(\frac{2\pi}{(C + Rp)}\right)^{\frac{1}{2}} \quad (5)$$

where

$$R_p = \frac{1}{6\pi} \left(\frac{K_{max}}{\sigma_y}\right)^2 \quad (6)$$

Crack Tip Studies in Nickel Base Superalloys

The geometries of microstructural asperities found near the tips of cracks in nickel base superalloys have been studied for a range of microstructures. Threshold tests were carried out on compact tension specimens (W = 26 mm, B = 13 mm) in conventional IN901 (a wrought alloy, grain size ∿ 220 μm), APK-1 (a powder based alloy, grain size ∿ 12 μm, in the isothermally forged condition) and IN901 that had been heat treated to increase the grain size to ∿ 1375 μm. Two stage plastic replicas were taken as described elsewhere (4) both when stepping down into threshold and during the grow out in conjunction with back face strain and crack mouth opening displacement measurements of closure.

Crack profiles have been studied on replicas taken at zero load, at K_{min} and at K_{max}. Measurements of the size and position of the contact point nearest to the crack tip (the one considered to affect local crack tip loading) were made from scanning electron microscope examination of the zero load series of replicas. A typical contact point in conventional IN901 is shown in Figure 2.

The principle of using the dimensions of a single surface asperity to describe a through thickness contact line is statistically difficult to justify considering both the types observed in this investigation and the range that must exist through the specimen and yet the inter-relationships of the three parameters defining the geometry of the asperity seem to compensate for this sampling problem. The shapes of the observed asperities did not conform to the idealised case shown in Figure 1 being generally inclined slopes with much longer contact lengths than had been expected - (see Figure 2) -, in both conventional IN901 and APK-1 the average value of b was ∿ 8 μm. The actual contact lengths were influenced by their angle to the crack plane, the extent of Mode II displacement and the distance back from the crack tip.

There was good agreement between measurements of closure using back face strain and crack mouth opening techniques and the calculations of contact point effects using their details in the model. In APK-1, the material with the smallest grain size (an important parameter in determining the range of possible asperity sizes), the mean difference between the measured and the calculated K_{CL} values was 1 MPa$\sqrt{m}$ - the closure stress intensities being of the order of 5.1 (± 0.5) MPa$\sqrt{m}$. In the conventional IN901 where there were less data points K_{CL} was 7.9 (± 0.1) MPa$\sqrt{m}$ with a mean difference of 1.9 MPa$\sqrt{m}$.

Applications of the Model

The model can be used to investigate the role of the following variables on the magnitude of the effect of load transfer at the contact point (K_{CL}): asperity height, distance of the asperity from the crack tip, asperity width, rigidity of the asperity, yield strength of the matrix material and mean stress.

Asperity Height and Position

For an intermediate strength material with microstructural asperities the influence of asperity height L for a range of C values is presented in Figure 3. The magnitude of K_{CL} increases for a given asperity height as the asperity is positioned closer to the crack tip. The plots also illustrate the tendency for K_{CL} to approach a limiting value in the range 6 to 10 MPa$\sqrt{m}$ depending on the value of C. The range of K_{CL}'s predicted are of a similar magnitude to those observed for an α/β titanium alloy with a yield strength of 840 MPa (Table I).

Asperity Width and Position

The influence of asperity width b on K_{CL} as a function of C for a given asperity height of 0.25 μm is illustrated in Figure 4. This represents the behaviour that could be expected from an intermediate strength material (σ_Y = 750 MPa) with microstructural asperities. The magnitude of K_{CL} increases with asperity width tending to a maximum value of 10 MPa$\sqrt{m}$ for C = 10 μm. The variation in b can be envisaged as resulting from changes in the crack path trajectory and the extent of the Mode II component. Increases in both the length of individual crack path elements and the Mode II displacements could be expected to increase the potential value of b. At the same time an increase in C could also be anticipated and hence the disproportionate values of K_{CL} indicated by Figure 4 would not be anticipated, i.e. large Mode II shifts and large

Figure 2 - Typical contact point in conventional IN901.

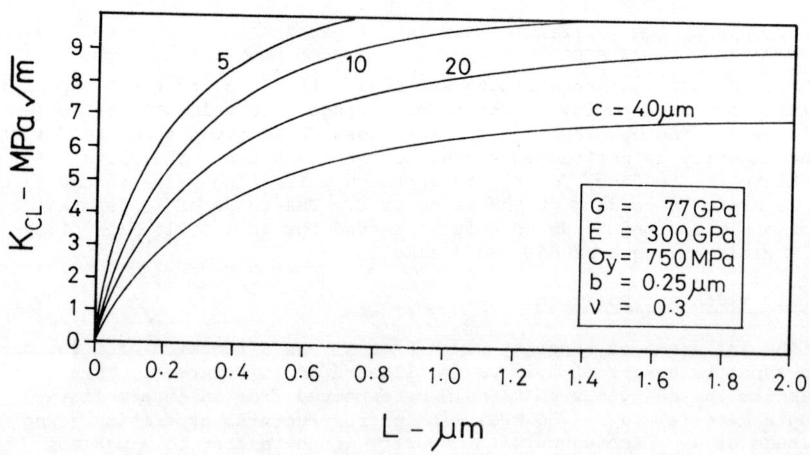

Figure 3 - The influence of asperity height on K_{CL} for a range of distances back from the crack tip.

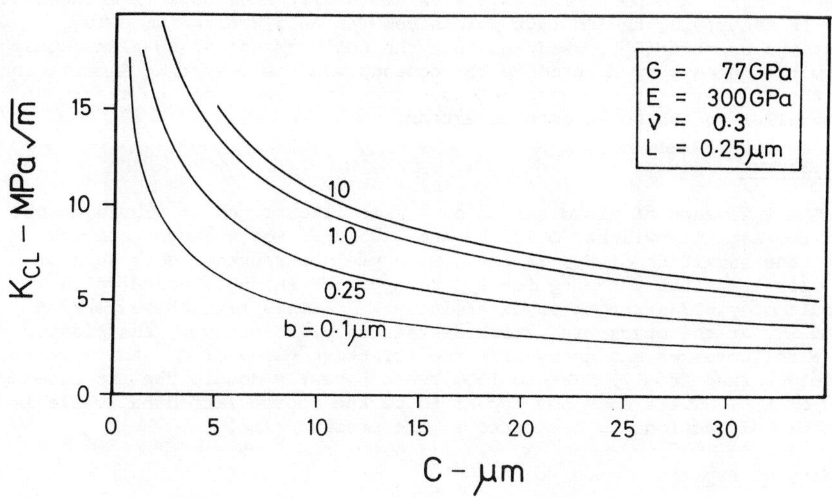

Figure 4 - The influence of the distance back from the crack tip on K_{CL} for a range of contact lengths.

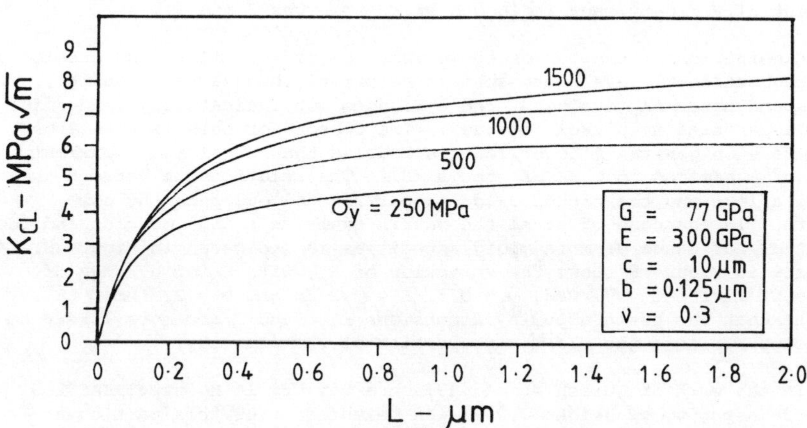

Figure 5 - Variation in K_{CL} with asperity height for a range of yield stresses (b = 0.125 μm).

contact lengths are not compatible with small distances back from the crack tip. In describing the contact points seen in an α/β titanium alloy Walker and Beevers (5) considered that the small amount of in-plane shear deformation necessary to produce the contact was the result of a deviation from the mid plane of the specimen such that an element of the crack was not parallel to the net growth direction.

Yield Stress

The influence of yield stress on K_{CL} is illustrated in Figure 5 and 6 for two asperity widths, 0.125 μm and 0.25 μm. For a given asperity height the magnitude of K_{CL} increase with yield strength. Both sets of data illustrate the tendency for K_{CL} to approach an upper bound value. The role of yield strength in determining K_{CL} arises from the extent of plasticity at the crack tip. With decreasing yield strength the plastic zone size increases and hence also the effective value of C. An increase in yield stress from 250 MPa to 1500 MPa can nearly double the projected value of K_{CL}. There does not appear to be the appropriate data available for these projections to be tested at the present time.

Rigidity of Asperity

The rigidity of the asperity can have a significant influence on the magnitude of K_{CL}. The particular significance of asperity rigidity is included in Figure 7 which projects the behaviour of a range of engineering alloys, details of the parameters used to compute these curves are included in Table II. For a high strength aluminium alloy (curves 1 and 2) the presence of an alumina asperity raises the potential upper bound K_{CL} from ∿ 2.5 $MPa\sqrt{m}$ to 7.5 $MPa\sqrt{m}$. In some high strength aluminium alloys (7) ΔK^{i}_{th} is ∿ 1 $MPa\sqrt{m}$ and ΔK^{c}_{th} is ∿ 2 to 3 $MPa\sqrt{m}$, clearly in this instance the closure was more likely to be due to microstructural rather than oxide asperities. For titanium alloys the presence of TiO_2 asperities also has a potentially significant influence in K_{CL} (curves 3 and 4).

Examination of the curves shows that the nickel alloys and steels with high strength levels can achieve relatively high upper bound K_{CL} levels and hence ΔK_{th} values. Table II does not indicate any values for oxide asperities in nickel or steel. The reason for this is that iron oxide have in general a lower Young's Modulus than steel and a maximum value of E similar to that of steels (8). The relationship between the nickel alloys and the nickel oxide properties follows much the same pattern. In the case of steel the matrix exhibits a high modulus and the potential influence of more rigid asperities is indicated in Figure 8. The data in Figure 8 shows the variation of K_{CL} with C for a range of values of E with G = 77 GPa, ν = 0.3, L = 0.2 μm and b = 0.5 μm. It is evident that the basic asperity dimensions L, b and C are more likely to influence K_{CL} than the elastic properties of the asperity.

In the work of Suresh et al. (9) on a 2¼% Cr, 1% Mo steel SA542-3 an oxide asperity of height 0.2 μm was measured. For this particular steel K_{CL} can be estimated from the thresholds at R = 0.05 and R = 0.7 to be ∿ 4.5 $MPa\sqrt{m}$. The lower curve in Figure 8 may be considered to approximate to the case of a steel with oxide asperities. An asperity of height 0.2 μm would result in a K_{CL} of 4.5 $MPa\sqrt{m}$ if situated ∿ 7 μm from the crack tip. There are no direct measurements of C available but the dimensions involved appear eminently reasonable.

The data available for nickel oxide indicates a maximum value of 120 GPa for Young's Modulus (10). Thus in nickel alloys the closure

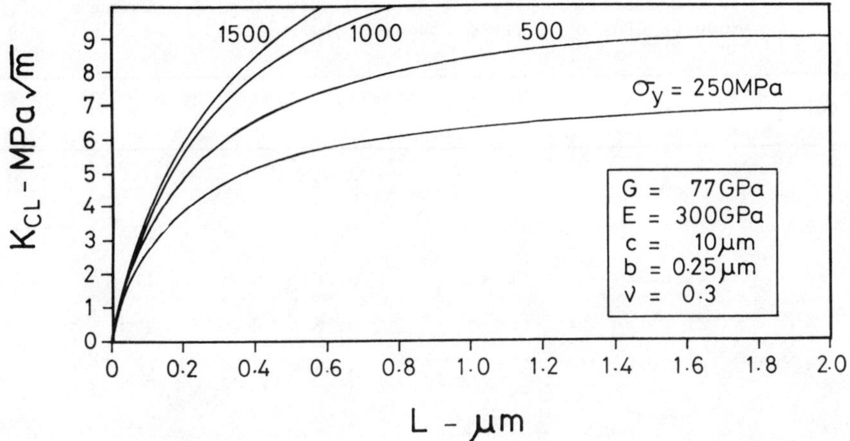

Figure 6 - Variation in K_{CL} with asperity height for a range of yield stresses (b = 0.25 μm).

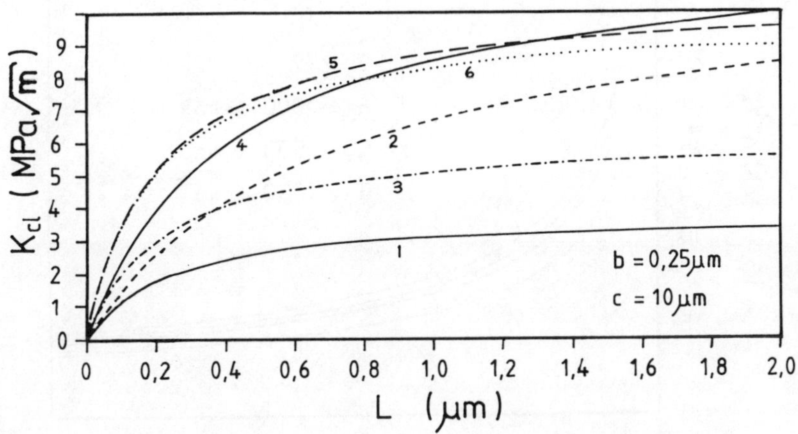

Figure 7 - Variation in K_{CL} with asperity height for a range of engineering alloys (see Table II for material properties).

Table II. The values used in the computation of the curves in Figure 7. (ν - Poisson's Ratio, G - Shear Modulus GPa, E - Young's Modulus GPa, σ_y - Yield Stress MPa).

Curve	Matrix	Asperity	ν	G	E	σ_y
1	Al	Al	0.34	26	70	500
2	Al	Al_2O_3	0.34	26	370†	500
3	Ti alloy	Ti alloy	0.36	45	110	850
4	Ti alloy	TiO_2	0.36	45	270†	850
5	Ni alloy	Ni alloy	0.375	76	200	900
6	Steel	Steel	0.29	82	210	1000

† Data from 'The Oxide Handbook', 2nd Edn. G.V. Samsonov ed., IFI/Plenum, New York (1981).

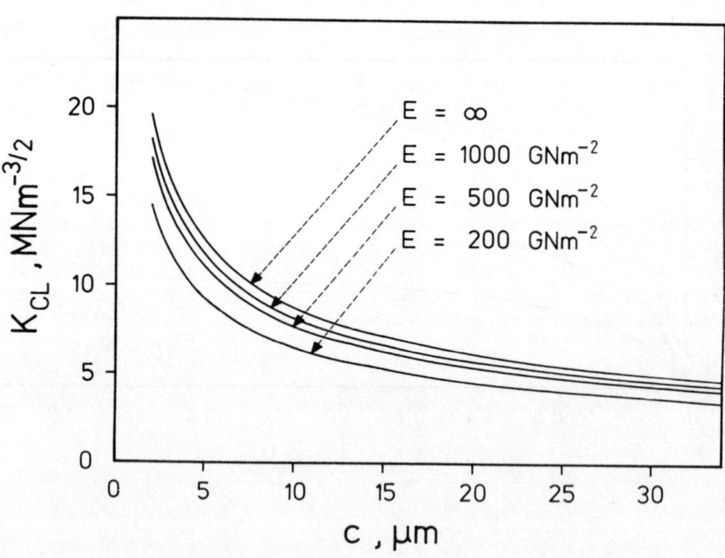

Figure 8 - The influence of the distance back from the crack tip and the rigidity of the asperity on K_{CL} (L = 0.2 μm, b = 0.5 μm, G = 77 GPa, ν = 0.3).

contribution both at room temperature and elevated temperature can be expected to derive from the formation of microstructural asperities. The upper bound value of 8 to 9 MPa$\sqrt{m}$ for K_{CL} is in reasonable agreement with that observed for coarse grained IN901 (I1) (Table I).

Grain Size

The influence of grain size d on ΔK_{th} has been expressed in the form

$$\Delta K_{th} = \Delta K_{th(o)} + kd^{\frac{1}{2}} \tag{7}$$

that is the threshold stress intensity range increases as $d^{\frac{1}{2}}$. In Table I it is evident that the value of ΔK^{c}_{th} exhibits a stronger grain size dependence than ΔK^{i}_{th}. It is reasonable to assume that the asperity height and the distance from the crack tip increase in proportion to the grain size for a dominantly transgranular failure mode i.e. $L = k_1 d$ and $c = k_2 d$. Thus, with $K_{op} \propto \frac{L}{c^{\frac{1}{2}}}$, we obtain $K_{op} \propto d^{\frac{1}{2}}$. Whilst this approach is somewhat simplistic it does emphasize the role that closure has to play in explaining the grain size dependence of ΔK_{th}.

Mean Stress

The role of mean stress in threshold determination has been extensively studied and is well documented (e.g. 1,3,5,7,11). The reason for this mean stress dependence is not completely understood, however at least two major contributions have been identified, namely, crack closure and the influence of K_{max}. The results in Figure 9 show that in conventional IN901, for the geometries found, closure in both the elastic and elastic/plastic cases ceases to have an influence for R > ∿ 0.5. Computations with the ranges of yield stress, contact length and distances from the crack tip observed show the closure contribution falling to zero at R values of 0.3 - 0.5 for the plasticity corrected condition.

Clearly this decrease in K_{CL} can explain the decrease in ΔK_{th} at low R ratios however the obvious and rather dramatic changes in ΔK_{th} that take place at high R ratios must be attributed to some other feature. The combination of high mean stress and environmental factors is a contribution in this regime of marked ΔK_{th} variation at high R's.

By use of the definition of ΔK an expression for ΔK_{CL} may be developed That is

$$\Delta K_{CL} = K_{CL} - K_{min} = K_{CL} - \frac{\Delta K}{(1/R - 1)} \tag{8}$$

For given loading and geometry parameters ΔK is given and K_{CL} may be computed from Eq. 4. A plot of ΔK_{CL} versus R is presented in Fig. 9 for the given parameter values.

It should be noticed that whilst Figure 9 illustrates the case for a constant ΔK and increasing R ratio at a fixed crack length the same pattern of behaviour is observed during the grow out from threshold at a constant load and R ratio with increasing crack length - that is that ΔK_{CL} progressively decreases.

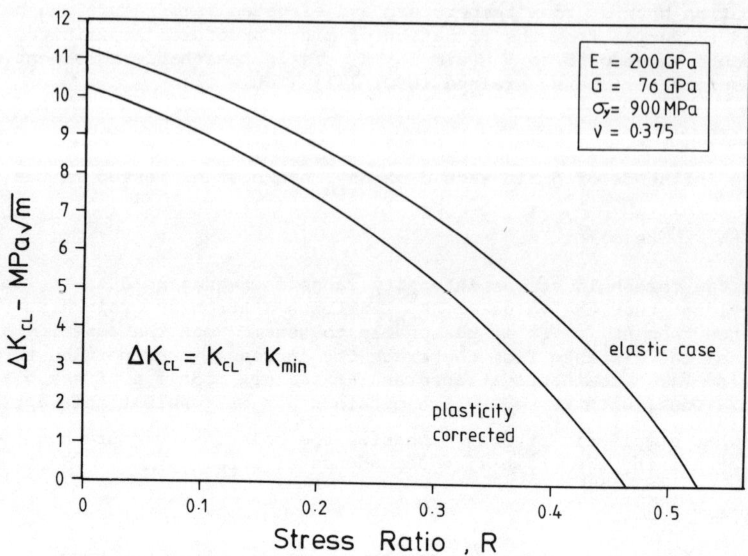

Figure 9 - Variation in closure contribution with stress ratio for both purely elastic and elastic + plastic cases. The data used are taken from the study of conventional IN901 asperities (ΔK = 10 MPa$\sqrt{m}$, C = 34.5 μm, b = 8.8 μm, L = 0.25 μm).

Conclusions

(1) At low stress ratios there is a major contribution to ΔK_{th} from ΔK^c_{th}, the closure component.

(2) Asperities - resulting from non-planar crack growth or enhanced oxide formation - can be seen near to the crack tip.

(3) The presence of these asperities with load transfer at the contact points leads to a modification of the stress state at the crack tip.

(4) A two dimensional model has been developed which treats the case of a single asperity in the crack tip region; based on the geometry of the asperity calculations of K_{CL} and K_{op} can be made.

(5) For a range of materials e.g. mild steel, titanium and nickel alloys crack closure can make significant contributions to the level of the fatigue threshold at low mean stress levels, the extent of this contribution has been shown to increase with increasing grain size. Results from the nickel alloy study have shown that the model is relatively consistent and realistic in its estimates of K_{CL}.

(6) The model has been used to demonstrate the influence of mean stress and the rigidity of the asperity and the role of microstructure in determining the closure contribution for a range of engineering materials.

References

(1) C.J. Beevers, "Some Aspects of the Influence of Microstructure and the Environment on ΔK Threshold", pp. 257-275 in Fatigue Thresholds (Proc. Int. Conf.), J. Backlund, A. Blom and C.J. Beevers, eds.; Stockholm, Sweden, 1982.

(2) S. Taira, K. Tanaka and M. Hoshina, "Grain Size Effect on Crack Nucleation and Growth in Long-Life Fatigue of Low-Carbon Steel", pp. 135-173 in Fatigue Mechanisms ASTM STP 675, J.T. Fong, ed.; 1979.

(3) M. Hicks, R.H. Jeal and C.J. Beevers, "Slow Fatigue Crack Growth and Threshold Behaviour in IMI 685", Fat. Engng. Mat. Struct., 6, (1983), pp. 51-65.

(4) C.J. Beevers, K. Bell, R.L. Carlson and E.A. Starke, "A Model for Fatigue Crack Closure", Engng. Fract. Mech., 19, (1984), pp 93-100.

(5) N. Walker and C.J. Beevers, "A Fatigue Crack Closure Mechanism in Titanium", Fat. Engng. Mat. Struct., 1, (1979), pp. 135-148.

(6) R.O. Ritchie, S. Suresh and C.M. Moss, "Near-Threshold Fatigue Crack Growth in 2¼Cr - 1 Mo Pressure Vessel Steel in Air and Hydrogen", J. Eng. Mat. Tech. Trans. ASME 'H', 102, (1980), pp. 293-299.

(7) B.R. Kirby and C.J. Beevers, "Slow Fatigue Crack Growth and Threshold Behaviour in Air and Vacuum for Commerical Aluminium Alloys", Fat. Engng. Mat. Struct., 1, (1979), pp. 203-215.

(8) "The Swelling of Steam Grown Oxide from Superheat and Reheat Tube Steels" EPRI FP686 (1978).

(9) S. Suresh, G.F. Zamiski and R.O. Ritchie, "Oxide Induced Crack Closure: An Explanation for Near Threshold Corrosion Fatigue Crack Growth Behaviour", Met. Trans., 12A, (1981), pp. 1435-1443.

(10) N.H. Van Vllack, "Nickel Oxide", The International Nickel Co. Inc.

(11) K. Bell, "Fatigue Crack Growth and Arrest in Nickel Base Superalloys", Ph.D. Thesis, University of Birmingham.

Appendix

The governing equations for the single asperity model described previously (4) are obtained from considerations of vertical displacements and asperity deformation.

The vertical displacement at the asperity is given by

$$U_2(C,\pi) = \frac{2(1-\nu)}{G}\left(\frac{C}{2\pi}\right)^{\frac{1}{2}} K_{I(local)} + \frac{2(1-\nu)}{G}\left(\frac{C}{2\pi}\right)^{\frac{1}{2}} K_{I(global)} \qquad \text{(A1)}$$

$K_{I(global}$ depends on the external loading

$$K_{I(local)} = \left(\frac{2}{\pi C}\right)^{\frac{1}{2}} \frac{P}{B} \qquad \text{(A2)}$$

The deformation of the asperity by the crack face forces

$$e = \frac{PL}{EBb} \qquad \text{(A3)}$$

The crack face displacement in terms of the deformation of the asperity is

$$U_2\ (C,\pi) = L - e \qquad \text{(A4)}$$

Through algebraic manipulations these equations can be used to describe the variation of the total stress intensity factor with external loading and by considering the special cases of zero external loading and zero asperity pressure equations 2 and 3 are developed.

INFLUENCE OF STRESS RATIO ON FATIGUE THRESHOLDS

AND STRUCTURE SENSITIVE CRACK GROWTH IN Ni-BASE SUPERALLOYS

R. A. Venables*, M. A. Hicks+ and J. E. King*

*Department of Metallurgy and Materials Science
University of Nottingham, University Park
Nottingham NG7 2RD, UK

+Rolls-Royce Ltd., PO Box 31, Derby DE2 8BJ, UK

The load or stress-ratio (R-ratio) dependence of threshold stress intensity values and structure-sensitive (near-threshold) fatigue crack propagation rates has been investigated in two nickel-base superalloys relevant to turbine disc applications. Both fine and coarse grained microstructures have been examined at room temperature, where near-threshold crack growth occurs along slip bands on {111} planes, producing highly faceted fracture surfaces.

In the coarse grained microstructures thresholds fall markedly with increasing R-ratio, from values of 8 to 12 $MNm^{-3/2}$ at R = 0.1 to between 5 and 6 $MNm^{-3/2}$ at R = 0.8, whereas in the fine grained material, where the fracture surfaces are relatively flat, there is little change in threshold with R. Measurements of crack mouth opening during a single fatigue cycle, made using a clip gauge, indicate that crack closure plays an important part in determining structure-sensitive crack propagation rates. The results are interpreted in terms of surface roughness induced closure.

Introduction

Over the last few years a considerable body of evidence has been building up which suggests that crack closure processes have a very significant effect on threshold stress intensity values (ΔK_{th}) for fatigue crack propagation, and on structure-sensitive, near-threshold crack growth rates. Indeed, it now seems possible to explain many of the effects of microstructure, stress ratio (R-ratio) and environment on threshold behaviour in terms of closure arguments (1-14).

Closure refers to the closing of the tip of a crack before the minimum load in a fatigue cycle has been reached, with the effect that the crack tip actually experiences a stress intensity range which is smaller than that which is being applied, i.e.:

$$\Delta K_{applied} = K_{max} - K_{min},$$

but $\Delta K_{effective} = K_{max} - K_{cl},$

where $\Delta K_{effective}$ is the range of stress intensity seen by material at the crack tip and K_{cl} is the stress intensity at which closure occurs. Where this effect is due to plasticity in the wake of the crack it is generally termed "plasticity induced closure" (15,17).

The same effect of reducing the stress intensity range at the crack tip can also be achieved in a rather different way, by propping or wedging the crack open before K_{min} is reached. This can occur when a fretting oxidation process produces oxide debris at the crack tip (9-11) or because of the imperfect mating of rough fracture surfaces (1-8,12-14), and these effects are referred to as "oxide induced closure" and "surface roughness induced closure" (15). A summary of these effects has been given by Suresh and Ritchie (15).

Plasticity induced closure has been shown to be important in plane stress where there is a large plastic zone around the crack tip, but not in plane strain (18), and it is unlikely, therefore, to play a significant role in near-threshold crack growth in high strength alloys.

In steels (8-10,12,14) and aluminium alloys (6-8,11) both oxide and surface roughness induced closure have been used successfully to interpret the effects of environment, microstructure and R-ratio on fatigue behaviour, but the results of work on titanium (1-3) and nickel-base (13) alloys, in which highly faceted fracture surfaces can be produced during structure-sensitive crack growth, are of most relevance to this work.

In many Ni (13,19-22) and Ti (1-3) base alloys, near-threshold crack growth proceeds by propagation along slip bands on active slip planes. The sharp changes in crack path which this necessitates on reaching grain boundaries produce rough fracture surfaces, the roughness of which is strongly dependent on the grain size. The mode of growth involves both tensile opening (Mode I) and shear displacement (Mode II) components in the slip band in which the crack is propagating. A small amount of irreversibility of plastic deformation at the crack tip, brought about perhaps by cross slip of dislocations in the slip band to accommodate the tensile opening, leads to wedging open of the crack as the load is reduced, so that the crack tip is screened from the effect of the full range of applied load. In three dimensions the process will be considerably more complex than that depicted in figure 1, giving rise to a greater incidence of this type of closure.

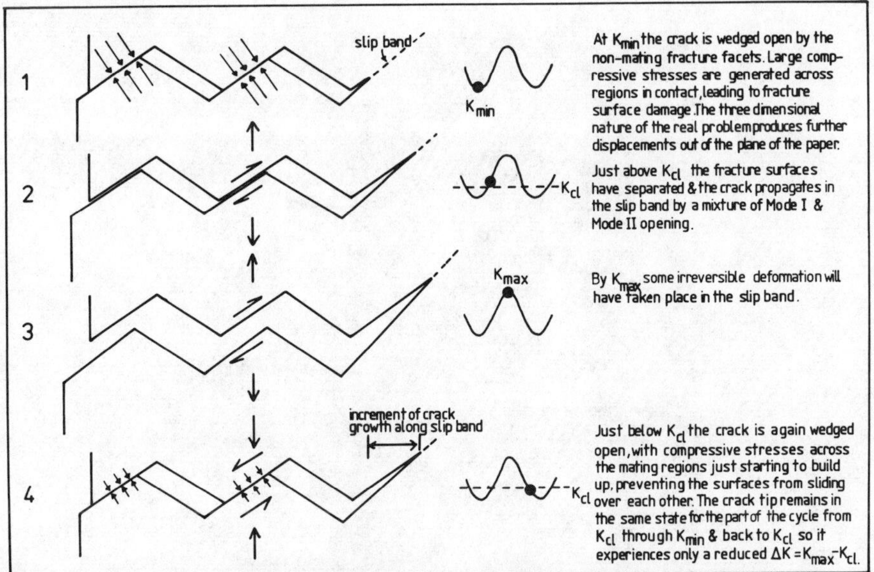

Figure 1. Generation of closure during faceted crack growth.

Surface roughness induced closure has been observed in titanium and titanium alloys both by metallographic techniques, which show non-mating fracture facets and fracture surface damage, and by compliance measurements (1,3). These measurements have been used to explain many of the effects of R-ratio and microstructural features, which control the crack path, on near-threshold behaviour. A compliance technique was also found to be the most satisfactory method for detecting closure in the nickel-base alloy, Nimonic 105 (13). The influence of crack geometry on near-threshold crack growth rates in this alloy has been interpreted in terms of changes in closure behaviour (13).

In this paper, closure measurements are used to interpret the R-ratio and grain size dependence of threshold in two nickel-base alloys, Nimonic 901 (N901) and Astroloy.

Materials

The nominal compositions of the two alloys used, N901 and Astroloy, are given in Table I. The N901 was conventionally cast and wrought and the Astroloy was produced by a powder route.

Table I. Chemical compositions (wt.%)

	Cr	Mo	Ti	Al	B	Zr	C	Co Ni	Fe
N901	12	6	3	0.3	0.1	-	0.04	40-45	balance
Astroloy	15	5	3.5	4	0.025	0.03	0.03	17 bal.	-

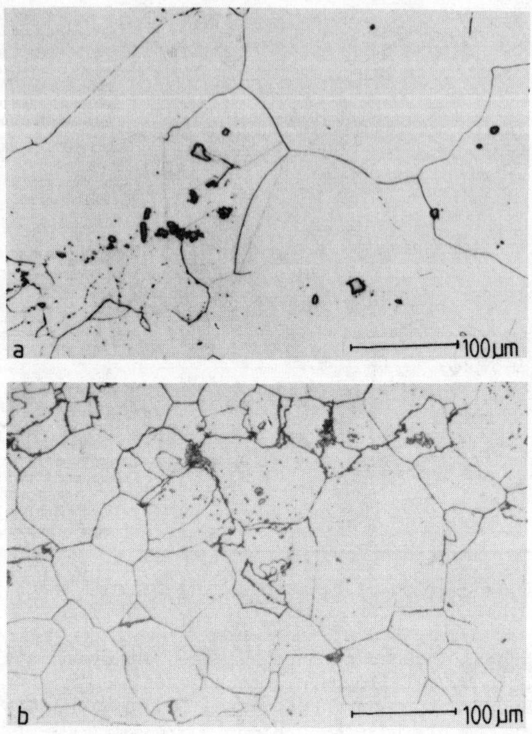

Figure 2. Microstructures (a) C901 (b) M901

Two forms of N901 were examined, a very coarse grained structure (C901) with a grain size varying between 200 and 400 µm and a finer grained form (M901) with a grain size of about 50 µm. The alloy is solution treated and aged to produce a strengthening precipitate of γ' ($Ni_3(Ti,Al)$) in the form of fine spheres. Optical micrographs of C901 and M901 are shown in figure 2.

In Astroloy two grain sizes were also investigated, fine: FG, 11-13 µm, and coarse: CG, ∿50 µm. The heat-treatment again involved solution treatment and ageing. The FG material was given a partial solution treatment below the γ' solvus, resulting in a bimodal γ' distribution in the final microstructure, with coarse γ' particles pinning the γ grain boundaries. Solution treatment above the γ' solvus for the CG material produced a microstructure containing only fine γ' particles. Figure 3 shows optical micrographs of Astroloy in the FG and CG conditions. The greater Ti + Al content in Astroloy compared with N901 results in a much higher volume fraction of γ' in the microstructure, typically 60-70% γ' in Astroloy (23) and only 15-25% in N901 (24).

Room temperature yield stress values for both materials are given in Table II.

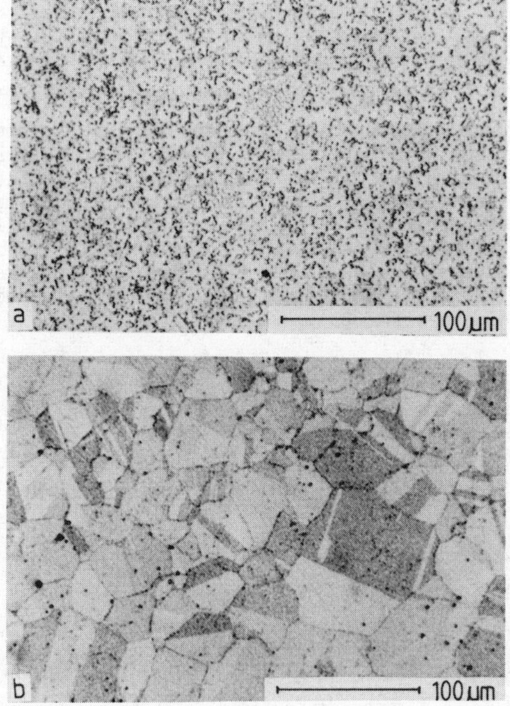

Figure 3. Microstructures (a) FG Astroloy
(b) CG Astroloy

Table II. Room temperature yield stress values (MNm^{-2})

N901	M901	900
	C901	810
Astroloy	FG	1021
	CG	954

Experimental Techniques

Threshold testing was carried out in air, at room temperature and R-ratios of 0.1 and 0.8, on a Mayes servo-controlled, electro-hydraulic testing machine, operating at 50 Hz. Crack length was monitored continuously using a DC potential drop technique. The test-pieces were of the single edge notch type, with dimensions 72.5 x 12.5 x 12.5 mm, loaded in three point bending (19).

A load shedding procedure was applied in order to obtain the threshold stress intensity value, ΔK_{th}, taken as that corresponding to a crack growth rate of approximately 10^{-11} m/cycle. The magnitude of the load reductions was continually decreased as the growth rate become slower, until the final reductions were less than 5% of ΔK. After reaching threshold the loads were increased and the crack allowed to grow to an $^a/W$ of ~ 0.6. Crack growth rate data were obtained from both the decreasing and increasing load stages of the test.

Load versus crack mouth opening displacement traces were plotted through a fatigue cycle for the C901 material at $\Delta K \simeq 12.5$ $MNm^{-3/2}$ for R = 0.1 and R = 0.8 and $\Delta K \simeq 24$ and 35 $MNm^{-3/2}$ for R = 0.1 only. The traces were obtained by mounting a clip gauge on two knife edges either side of the notch, and at least six traces were recorded for each set of conditions. This was done for both the conventional bend specimens and a version with 1.5 mm deep side grooves on either side of the crack, to eliminate any effects associated with the plane stress regions of the test-piece. For the side grooved specimens K values were calculated from (25):

$$K = K_{nom} \left(\frac{B}{B_N} \right)^{\frac{1}{2}},$$

where K_{nom} is the stress intensity factor for a specimen of uniform thickness, B, and B_N is the reduced thickness of the side grooved specimen between the grooves, i.e.:

$$B_N = B - 2 \times \text{side groove depth.}$$

Etching of metallographic specimens was carried out in Kalling's reagent or 2% Bromine in Ethanol. Fracture surfaces from the threshold tests were examined using a JEOL 35C scanning electron microscope.

Results

1. Nimonic 901

Figure 4(a) is a plot of da/dN versus ΔK for C901 at R = 0.1 and R = 0.8. The data show a strong R-ratio dependence of threshold, with ΔK_{th} falling from 10 $MNm^{-3/2}$ at R = 0.1 to about 6 $MNm^{-3/2}$ at R = 0.8. In the finer grained microstructure, M901 (fig. 4(b)) there is a slightly reduced dependence, ΔK_{th} is 7.8 $MNm^{-3/2}$ at R = 0.1, falling to 5.1 $MNm^{-3/2}$ at R = 0.8. At R = 0.1 and R = 0.8 threshold values in C901 are higher than in M901, with the greatest difference at the lower R-ratio. For both conditions the da/dN vs ΔK curves start to come together above threshold ($\simeq 15$ $MNm^{-3/2}$), but the high K_{max} values for the R = 0.8 tests, and consequent early acceleration towards final fracture, mean that this trend is not shown as markedly as it would have been at intermediate R values.

In the near-threshold regime C901 and M901 show faceted, structure-sensitive crack growth, as can be seen in figure 5. A very similar fracture morphology is produced at both R = 0.1 and R = 0.8, crack growth occurs along {111} slip bands, but with frequent direction changes between intersecting slip bands, such that instead of producing planar, grain sized facets, much smaller facets result. These appear slightly coarser in C901 than in M901 (fig. 5(a) to (c)). At higher ΔK levels much flatter fracture surfaces are produced, for example figure 5(d).

Figures 6, 7 and 8 show experimental traces of crack opening vs. load, obtained in C901 using conventional and side-grooved bend specimens. In all cases the same behaviour was observed in both the loading and unloading halves

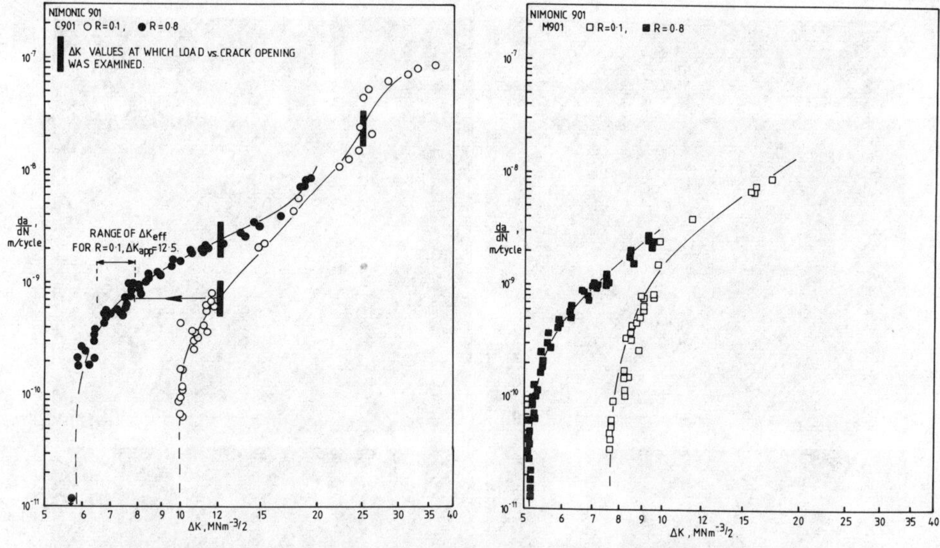

Figure 4 Crack growth rate versus stress intensity range
(a) C901 (b) M901

of the fatigue cycle. At R = 0.1 and ΔK of about 12.5 $MNm^{-3/2}$ (12.2 $MNm^{-3/2}$ in the conventional specimen and 12.9 $MNm^{-3/2}$ in the side-grooved one) the load vs. crack opening trace had a steep slope near minimum load and a shallower slope at high loads. Figure 6 shows that the change in slope occurs over a range of K from 6.7 to 7.9 $MNm^{-3/2}$. A very similar range of values was measured for all twelve traces from conventional and side-grooved specimens.

Figure 7 shows the traces of load against clip gauge displacement at ΔK = 24 $MNm^{-3/2}$ and R = 0.1 for both side-grooved and conventional test-pieces. The trace for the conventional test-piece appears to show a slightly steeper slope close to the minimum load, although the side-grooved trace does not.

A straight trace, typical of those obtained from the measurements made at R = 0.8 is shown in figure 8.

2. Astroloy

The da/dN vs. ΔK curves for FG and CG Astroloy at R = 0.1 and R = 0.8 are shown in figure 9. The coarse grained material (CG) shows a very marked R-ratio dependence of threshold and near-threshold crack growth rates, with ΔK_{th} dropping from 12.0 $MNm^{-3/2}$ at R = 0.1 to 6.7 $MNm^{-3/2}$ at R = 0.8. By contrast ΔK_{th} in the fine grained material (FG) is about 5.8 $MNm^{-3/2}$ at both R values and the crack growth rates are also very similar.

In the CG condition near-threshold crack growth occurs in a faceted manner along {111} planes (21), producing a fracture surface consisting of

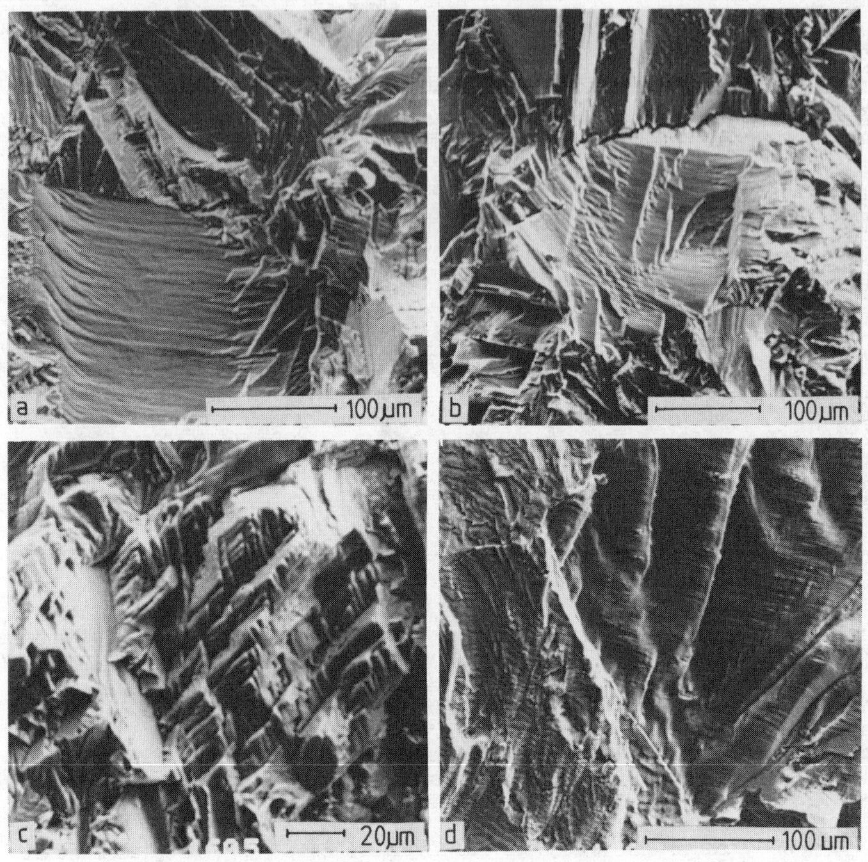

Figure 5 N901 fracture surfaces

(a) C901, R = 0.1. Near threshold.

(b) C901, R = 0.8. Near threshold.

(c) M901, R = 0.1. Near threshold.

(d) C901, R = 0.1. $\Delta K = 35\ MNm^{-3/2}$.

flat, angular facets. However, they differ from those observed in N901 in that their size is of the order of the grain size (figure 10(a)). In the R = 0.1 test the faceted growth persisted up to a ΔK of about 30 $MNm^{-3/2}$. At higher ΔK values a flatter, striated mode of growth is dominant, as shown in figure 10(b).

In both CG and FG material the near-threshold growth mode was independent of R in the range investigated. Figure 10(c) shows the fracture surface produced during near-threshold crack growth in FG Astroloy. In the fine grained material only occasional small facets are produced.

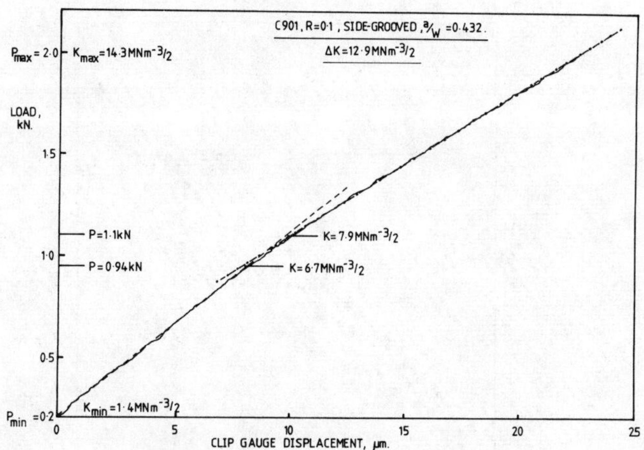

Figure 6 Load versus clip gauge displacement.
C901, side-grooved, R = 0.1, ΔK = 12.9 $MNm^{-3/2}$

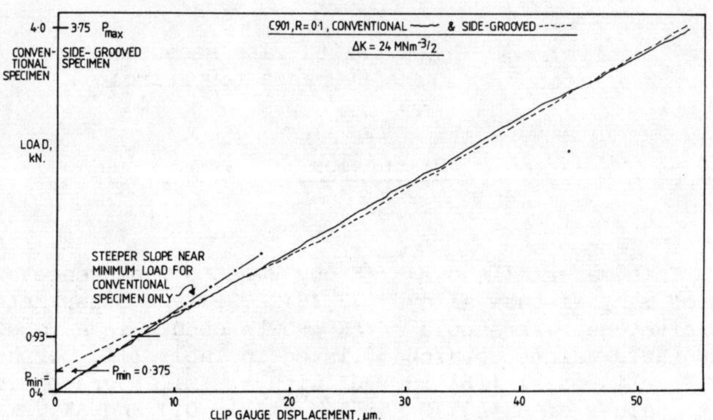

Figure 7 Load versus clip gauge displacement.
C901, conventional and side-grooved, R = 0.1, ΔK = 24 $MNm^{-3/2}$

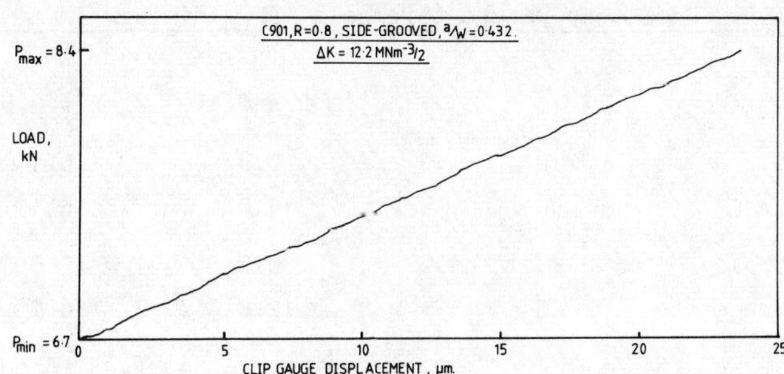

Figure 8 Load versus clip gauge displacement.
C901, side-grooved, R = 0.8, ΔK = 12.2 $MNm^{-3/2}$

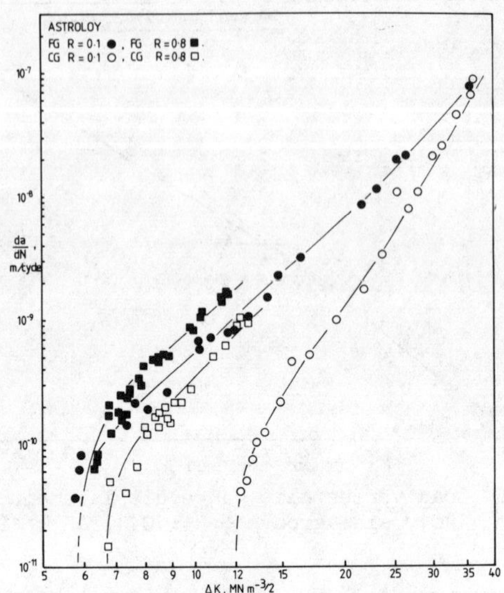

Figure 9 Crack growth rate versus stress intensity range for Astroloy.

Discussion

Nimonic 901

(a) C901. This material shows a strong R-ratio dependence of threshold characteristic of many Ni-base alloys (13,19-22,26) and other materials where structure-sensitive, near-threshold crack growth occurs in a faceted manner (1-3). The threshold values obtained (listed in Table III) compare well with those shown by Scarlin et al (26) in N901 with a similar grain size. Their results indicate values of ΔK_{th} ∿10 $MNm^{-3/2}$ at R = 0.1 and ΔK_{th} ∿7.2 $MNm^{-3/2}$ at R = 0.5, compared with ΔK_{th} = 10.0 $MNm^{-3/2}$ at R = 0.1 and ΔK_{th} ∿6 $MNm^{-3/2}$ at R = 0.8 in the present work.

Table III. Room temperature threshold (ΔK_{th}) values ($MNm^{-3/2}$)

		R = 0.1	R = 0.8
N901	M901	7.8	5.1
	C901	10.0	∿6
Astroloy	FG	5.8	5.8
	CG	12.0	6.7

A satisfactory explanation of these results can be given in terms of surface roughness induced closure. Figure 11 shows traces of near-threshold fracture surfaces, taken at the mid-section of the test-piece, illustrating

Figure 10 Astroloy fracture surfaces.

(a) CG. Near threshold.

(b) CG. $\Delta K = 40\ MNm^{-3/2}$.

(c) FG. Near threshold.

the degree of fracture surface roughness involved in this type of faceted crack growth. The trace of crack opening versus load, shown in figure 6 for a side-grooved specimen at R = 0.1 and $\Delta K = 12.9\ MNm^{-3/2}$, shows that at low loads a steep trace characteristic of a short crack is obtained but as the load increases the gradient becomes shallower, i.e. the crack behaves as if it were significantly longer. The change from the steeper to the shallower gradient is interpreted as the range of load over which the crack tip itself opens and starts to experience the full range of applied ΔK. This gives a value of K_{cl} somewhere between 6.7 and 7.9 $MNm^{-3/2}$, so the alternating stress intensity actually experienced by the crack tip ($\Delta K_{eff} = K_{max} - K_{cl}$) becomes $\Delta K_{eff} = 6.4$ to 7.6 $MNm^{-3/2}$. When this is plotted on the graph of da/dN vs. ΔK (figure 4(a)) it shifts the R = 0.1 data over the data obtained at R = 0.8, in a range which extends beyond it, to a slightly lower ΔK. This

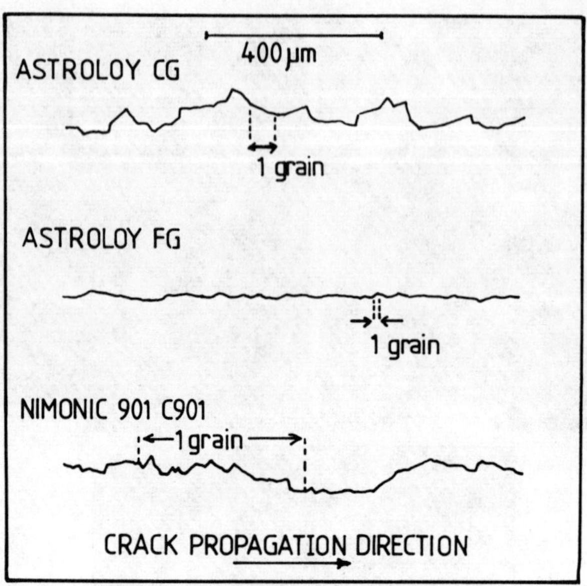

Figure 11 Near-threshold fracture surface profiles. (From centre section of specimen).

suggests that there may be a small closure contribution operative, even at R = 0.8, at this ΔK value (∿7.5 $MNm^{-3/2}$). By ΔK = 12.2 $MNm^{-3/2}$ in the R = 0.8 test (and at higher ΔK values) no closure effect was measurable, as can be seen from the straight plot of load against clip gauge displacement shown in figure 8.

At ΔK ∿12.5 $MNm^{-3/2}$ there was no difference in the compliance behaviour of the side-grooved and conventional specimens in terms of their load vs. crack opening behaviour. In contrast, at K = 24 $MNm^{-3/2}$, the conventional specimen showed a steeper slope near minimum load, from which a K_{cl} value of ∿6.2 $MNm^{-3/2}$ can be calculated, whereas the side-grooved specimen did not (figure 7). At ΔK = 24 $MNm^{-3/2}$ the fracture surface still appears similar to the near-threshold regime, so this K_{cl} value of 6.2 $MNm^{-3/2}$ (close to the range of K_{cl} = 6.7 to 7.6 $MNm^{-3/2}$ obtained at ΔK ≃12.5 $MNm^{-3/2}$) seems a reasonable value. The behaviour of the side-grooved specimen, however, suggests that the measured closure load could be due to plane stress closure effects, although it seems unlikely that these would be so strong when the maximum plane stress plastic zone radius, calculated from (27):

$$r_y = \frac{1}{2\pi}\left(\frac{K_{max}}{\sigma_y}\right)^2,$$

is only 173 μm. This is an area requiring further investigation.

(b) M901. The finer grained form of N901, M901 produced smaller threshold values at both R-ratios than the coarse grained C901 (Table III). This is the commonly observed effect of grain size on threshold: (3,19,28,29) increasing the grain size gives higher threshold values, and is in keeping with surface roughness induced closure arguments, in that a smaller grain size will give rise to smaller deviations of the crack path and therefore less closure. The R-ratio dependence of threshold is also reduced when compared with C901, consistent with the reduced closure contribution. However, the grain size effect observed here is not very strong, a six-fold reduction in grain size (from 300 to 50 μm) bringing about only a 20% reduction in threshold at R = 0.1. By way of comparison, in Astroloy, reducing the grain size from 50 to ∿12 μm produces a 50% reduction in threshold.

An explanation for the small effect of grain size on threshold in N901 can be obtained from observation of the near-threshold fracture path. Figure 5 shows the morphology of the fracture surfaces produced during near-threshold crack growth in N901. This material does not show the planar, grain-sized facets seen in some other Ni-base alloys (19,20,22), but shows facets which are generally heavily stepped and, within a single grain crack, growth often occurs alternately along two sets of intersecting slip bands. This has the effect of reducing the scale of the fracture surface roughness from that which would be expected if grain-sized facets were produced, and so reduces closure effects. The scale of roughness in M901 appears only slightly less than in C901, despite the large difference in grain size, thus explaining the relatively small reduction in threshold value.

Astroloy

In Astroloy the CG material shows a strong R-ratio dependence on threshold, whereas the FG material shows only a weak one, and there is a large grain size dependence of threshold at R = 0.1, although only a much reduced one at R = 0.8, as can be seen from the threshold values in Table III.

In the CG material grain-sized planar facets are produced during near-threshold crack growth (figure 10(a)), giving rise to a fracture profile which shows a similar scale of surface roughness to the C901, despite the large difference in grain size. This can be seen in the fracture profiles in figure 11. Indeed the threshold behaviour in C901 and CG Astroloy is very similar, with CG Astroloy showing values of 12 $MNm^{-3/2}$ at R = 0.1 and 6.7 $MNm^{-3/2}$ at R = 0.8, whilst C901 gives 10 $MNm^{-3/2}$ at R = 0.1 and 6 $MNm^{-3/2}$ at R = 0.8. If it is assumed, by analogy with the behaviour of C901, that the removal of the closure element will bring the CG Astroloy R = 0.1 crack growth data over the R = 0.8 curve, then at R = 0.1 and ΔK = 12.9 $MNm^{-3/2}$ an approximate value of K_{cl} = 7 $MNm^{-3/2}$ can be obtained. This is in the range K_{cl} = 6.7 to 7.9 $MNm^{-3/2}$, measured in C901, suggesting that similar surface roughness levels give rise to similar closure contributions in these two materials.

The difference in the near-threshold faceted growth in N901 and CG Astroloy may be associated with two factors. The high retained work levels in N901 (30,31) mean that the crack may be propagating into material which already contains slip bands on a number of different {111} planes, introduced during the forging process. Thus, even at low ΔK levels it may be relatively easy to activate slip on intersecting slip bands, where they already exist ahead of the crack, and so the crack can propagate along a less tortuous path by moving from slip band to slip band within a single grain.

Astroloy has higher yield stress levels and a much larger γ' volume fraction than N901 and this may also contribute to the increased size and planarity of the fracture facets, with the greater volume of shearable precipitates producing more effective hindrance of the activation of secondary slip (32).

Comparison of the near-threshold fracture surfaces of FG and CG Astroloy (micrographs in figure 10 and profiles in figure 11) shows that a much flatter, less faceted fracture surface is produced in the fine grained material. This is associated with the increased difficulty of forming well defined slip bands (19) within small grains, because of the accommodation effects required where slip bands meet grain boundaries. The "geometrically necessary" dislocations which must be introduced around the boundaries (33) have the effect of dispersing the slip and therefore making it more homogeneous, an effect which can become dominant when the grain size is small. Thus shear (Mode II) displacements at the crack tip are restricted and a flatter fracture surface is produced, more characteristic of Mode I crack propagation. Closure effects are reduced and only a small R-ratio dependence of threshold values and near-threshold crack growth rates results.

At R = 0.8 there is little difference in crack growth rates and threshold values between CG and FG Astroloy, indicating that there is only a small effect of grain size on intrinsic threshold. Indeed the R = 0.8 threshold values for both Astroloy and N901 fall into a narrow range between 5.1 and 6.7 $MNm^{-3/2}$ despite considerable microstructural variations and up to 20% differences in yield stress.

These results suggest that a number of earlier interpretations of microstructural effects on crack propagation behaviour, for example differences between under and overaged precipitates (19,34-37), which attributed improvements to increased reversibility of slip at the crack tip, in microstructures deforming inhomogeneously, should now be looked at again in terms of the effect of the nature of the slip on the fracture surface roughness.

One feature of roughness induced closure frequently observed in titanium alloys (1,2), but which is not clearly demonstrated in nickel alloys fatigued in air, is fracture surface damage at closure points. Despite the large compressive stresses which must be generated at contact points when the crack is wedged open, obvious fracture surface damage of the type seen in titanium alloys has not been reported in Ni-base alloys and was not observed in this work. However, in vacuum, evidence of re-welding at contact points has been demonstrated, in a Ni-base superalloy (20).

Comparison with model of Suresh and Ritchie (Ref. 15)

Suresh and Ritchie (15) have developed a model for roughness induced closure which predicts:

$$\frac{K_{cl}}{K_{max}} = \sqrt{\frac{2\gamma x}{1+2\gamma x}}$$

where γ is a non-dimensional fracture surface roughness factor, equal to the fracture surface asperity height divided by the length of the base of the asperity, and x is the ratio, U_{II}/U_{I}, of Mode II to Mode I displacement, at the crack tip. The predictions of the model are shown in figure 12. Points for Astroloy and N901, determined from the present work, have been added to the figure, although there is some difficulty in determining γ for a real fracture surface. For the coarse grained structures they indicate a considerable Mode II component of crack tip deformation, with a much smaller one in the FG Astroloy, in keeping with the ease of accommodation of large shear displacements in slip bands in fine and coarse grained structures.

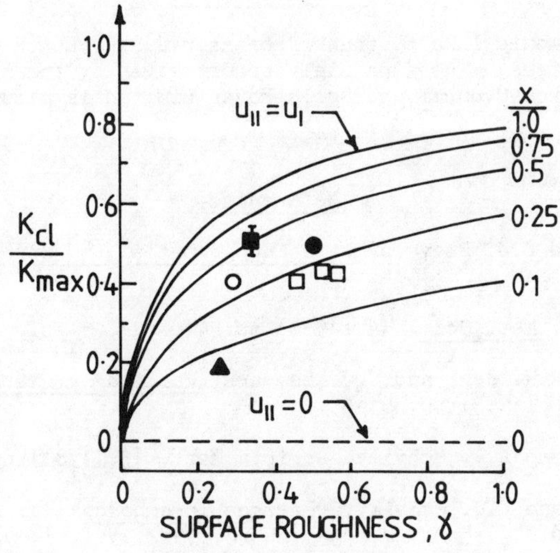

Figure 12 Predictions of model of Suresh and Ritchie (15) for roughness induced closure.

Conclusions

1. The room temperature R-ratio and grain size dependence of threshold in two Ni-base alloys, Nimonic 901 and Astroloy, can be satisfactorily explained in terms of surface roughness induced closure.

2. In these alloys threshold behaviour and closure stress intensities correlate closely with surface roughness.

3. Surface roughness is related to grain size, but the relationship is not a simple one, and it varies from material to material even in alloys deforming by similar mechanisms. The exact nature of the crack tip deformation behaviour can affect whether or not facets are of the order of, or significantly smaller than, the grain size.

4. In high strength alloys, near-threshold closure measurements can be made satisfactorily in conventional fatigue specimens, using a clip gauge mounted across the crack mouth, as any plane stress contributions are negligible in this regime.

Acknowledgements

The authors would like to thank Professor J.S.L. Leach for provision of laboratory facilities. The financial support given by the Science and Engineering Research Council and Rolls-Royce Limited is also gratefully acknowledged.

References

1. N. Walker and C.J. Beevers, Fat. Engng. Matls. and Structures, 1 (1979) pp. 135-148.

2. C.J. Beevers, Met. Sci., 14 (1980) pp. 418-423.

3. M.A. Hicks, R.H. Jeal and C.J. Beevers, Fat. Engng. Matls. and Structures, 6 (1983) pp. 51-65.

4. K. Minakawa and A.J. McEvily, Scripta Met., 15 (1981) pp. 633-636.

5. K. Minakawa and A.J. McEvily, "Fatigue Thresholds", J. Backlund, A.F. Blom and C.J. Beevers, eds.; EMAS, 1982, Vol. 1, pp. 373-390.

6. A.J. McEvily and K. Minakawa, "Strength of Metals and Alloys", (ICSMA6) R.C. Gifkins, ed.; Pergamon Press, 1982, pp. 845-850.

7. K. Minakawa, J.C. Newman Jr. and A.J. McEvily, Fat. Engng. Matls and Structures. To be published.

8. S. Purushothaman and J.K. Tien, Proc. 5th International Conference on Strength of Metals and Alloys (ICSMA5), Aachen, August 1979. Vol. 3, pp. 1267-1271.

9. R.O. Ritchie, S. Suresh and C.M. Moss, J. Engng. Matls. and Technology, 102 (1980) pp. 293-299.

10. S. Suresh, G.F. Zamiski and R.O. Ritchie, Met. Trans., 12A (1981) pp. 1435-1443.

11. A.K. Vasudevan and S. Suresh, Met. Trans., 13A (1982) pp. 2271-2280.

12. I.C. Mayes and T.J. Baker, Fat. Engng. Matls. and Structures, 4 (1981) pp. 79-96.

13. J. Byrne and T.V. Duggan, "Fatigue Thresholds", J. Backlund, A.F. Blom and C.J. Beevers, eds.; EMAS, 1982, Vol. 2, pp. 759-776.

14. G.T. Gray III, J.C. Williams and A.W. Thompson, Met. Trans., 14A (1983) pp. 421-433.

15. S. Suresh and R.O. Ritchie, Met. Trans., 13A (1982) pp. 1627-1631.

16. W. Elber, Engng. Fract. Mech., 2 (1970) pp. 37-45.

17. W. Elber, ASTM STP 486, 1971, pp. 230-242.

18. T.C. Lindley and C.E. Richards, Mat. Sci. and Engng., 14 (1974) pp. 281-293.

19. J.E. King, Met. Sci., 16 (1982) pp. 345-355.

20. J.E. King, Fat. Engng. Matls. and Structures, 5 (1982) pp. 177-188.

21. J.E. King, Fat. Engng. Matls. and Structures, 4 (1981) pp. 311-320.

22. M.A. Hicks and J.E. King, Int. J. Fatigue, 5 (1983) pp. 67-74.

23. A. Porter and B. Ralph, J. Mat. Sci., 16 (1981) pp. 707-713.

24. Rolls-Royce, Derby. Private communication.

25. C.N. Freed and J.M. Krafft, J. Matls., 1 (1966) pp. 770-790.

26. R.B. Scarlin, K.N. Melton, W. Hoffelner, "Fracture and the Role of Microstructure", F.E. Matzer and K.L. Maurer, eds.; EMAS, 1982, Vol. II, pp. 689-700.

27. J.F. Knott, "Fundamentals of Fracture Mechanics", Butterworths, 1973.

28. R.O. Ritchie, Int. Met. Revs., 5 and 6 (1979) pp. 205-230.

29. C.J. Beevers, Met. Sci., 11 (1977) pp. 362-367.

30. F. Turner, "The development of Gas Turbine Materials", Applied Science Publishers, 1981, pp. 177-205.

31. G.W. Meetham, The Metallurgist and Matls. Technologist, (Sept. 1982) pp. 387-391.

32. D.A. Koss and K.S. Chan, Acta Met., 28 (1980) pp. 1245-1252.

33. M.F. Ashby, Philos. Mag., 21 (1970) pp. 399-424.

34. E. Hornbogen and K-H. ZumGahr, Acta Met., 24 (1976) pp. 581-592.

35. E.A. Starke, Jnr. and G. Lütjering, "Fatigue and Microstructure", ASM. 1970, pp. 205-243.

36. J. Lindigkeit, A. Gysler and G. Lütjering, Met. Trans., 12A (1981) pp. 1613-1619.

37. A. Gysler, J. Lindigkeit and G. Lütjering, Proc. ICSMA5, Aachen, 1979, Vol. 2, pp. 1113-1118.

Applications and Special Techniques

APPLICATION OF FATIGUE THRESHOLD CONCEPTS TO VARIABLE AMPLITUDE CRACK PROPAGATION

S. SURESH

Division of Engineering, Brown University
Providence, Rhode Island 02912, U.S.A.

and

A. K. VASUDEVAN

Aluminum Company of America
Alcoa Technical Center
Alcoa Center, Pennsylvania 15069, U.S.A.

The application of mechanisms associated with fatigue thresholds to variable amplitude crack propagation is examined in detail. It is shown for some high strength aluminum alloys that post-overload retarded growth is effectively governed by the mechanisms of near-threshold crack propagation. Such concepts are used to explain many apparent anamolies in microstructurally influenced variable amplitude fatigue crack propagation in 7075 and 2020 aluminum alloys. Experimental results and possible mechanistic reasons are discussed to rationalize the observation that heat treatments, which promote resistance to constant amplitude fatigue crack propagation, lead to detrimental fatigue response under the more realistic conditions of variable amplitude cycling.

Introduction

The mechanisms underlying crack growth attenuation or arrest during variable amplitude fatigue have long been a topic of considerable academic interest as well as engineering importance. The classical models, proposed to rationalize transient fatigue behavior, can be nebulously categorized into concepts based on changes in crack-tip yielding (1-3), blunting (4) or strain hardening (5,6) induced by load interactions. Although a precise understanding of such mechanisms is still lacking, it recently has become evident (7,8) that the near-threshold fatigue characteristics of a material observed under constant amplitude loading conditions can have a strong influence on its response to variable amplitude fatigue at nominally higher stress intensity range values. This is in view of the observations that retarded crack growth rates following overloads correspond to the near-threshold regime ($da/dN \lesssim 10^{-6}$ mm/cycle), even though the nominal post-overload driving force may be large enough to induce crack growth corresponding to the intermediate regime ($10^{-6} \lesssim da/dN \lesssim 10^{-3}$ mm/cycle) (7-10). Extensive correlations between the micro-mechanisms of near-threshold and post-overload crack propagation behavior have seldom been attempted (7-10) because of the fact that several aspects of fatigue behavior in the near-threshold regime have long remained unclear even for constant amplitude fatigue cycling. In the past few years, however, many novel mechanistic interpretations of near-threshold crack advance, particularly with respect to the role of environmental, microstructural and closure processes, have begun to emerge for a wide range of alloy systems (for review, see ref. 11). It is, therefore, appropriate at this time to examine the retardation behavior of fatigue cracks subjected to peak tensile overloads in light of such recent developments in the conceptual understanding of near-threshold fatigue crack propagation.

Review of Prior Work

Figure 1 shows the variable amplitude fatigue crack propagation behavior of the peak-aged (standard -T651 temper) and over-aged (standard -T7351 temper) conditions of 7075 aluminum alloy, from the work of Bucci et al. (12). The application of a single overload after every 4000 constant amplitude cycles results in a more pronounced retardation for the peak-aged condition than for the over-aged alloy, with increasing values of overload ratio (12). Results derived solely from constant amplitude fatigue tests at the same nominal (post-overload) ΔK level, however, show exactly the opposite trend, with the over-aged alloy have a more favorable microstructure for fatigue crack growth resistance (as evidenced from the value of da/dN for the two conditions at OLR = 1.0, in Fig. 1). The work of Knott and Pickard (6) also suggested that the under-aged microstructure of Al-4.5Zn-2.5Mg high strength alloy leads to greater retardation and delay period than the over-aged condition of comparable strength level, although it gives rise to faster constant amplitude crack propagation than the over-aged temper at ΔK values nominally identical to those in the retarded growth regime (Fig. 2).

Previous explanations of the above effects of aging treatment on post-overload retardation in high strength aluminum alloys focused on crack-tip plasticity and cyclic work hardening characteristics (6,12-14). It was postulated (6), for example, that the higher flow stress at the crack tip arising from the larger cyclic strain hardening exponent in the under-aged alloy (as compared to the over-aged heat treatment) leads to a smaller subsequent crack tip plastic strain and a more pronounced retardation. It is known, however, that cyclic strain hardening can occur due to local plasticity even for constant amplitude crack growth where such interpretations would lead to erroneous predictions of the effects of aging treatment

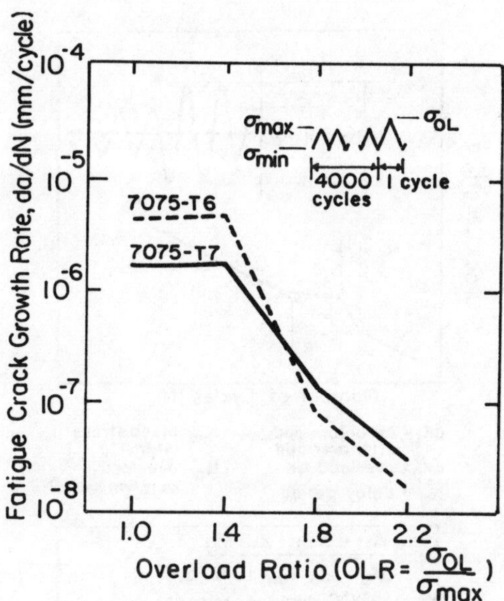

Fig. 1 Variation of fatigue crack growth rates with overload ratio in the peak-aged (-T651) and over-aged (-T7351) conditions of 7075 aluminum alloy tested in the intermediate regime of crack growth; tensile overloads applied after every 4000 constant amplitude cycles in an environment of room temperature humid air (> 90% RH) (after ref. 12).

on crack growth rates. Furthermore, recent studies (7,8) have pointed out that although such changes in crack tip plasticity, induced by the overloads, play a role in influencing retardation, they cannot account for the influences of various mechanical, microstructural and environmental factors observed in the post-overload regime. For example, the earlier interpretations based on changes in crack tip plastic zone or cyclic work hardening do not convincingly rationalize such typical retardation phenomena as striationless crack advance, fracture surface abrasion, retardation well beyond the overload elastic-plastic boundary, plane strain growth attenuation, increase in microstructure-sensitivity of fatigue crack growth and environmental interactions (7,8). It, thus, becomes necessary to explore additional processes related to microstructural effects on variable amplitude fatigue crack growth. In the present paper, the sequential changes in crack profile and in the fracture surface features are examined as a function of post-overload cycles in two high strength aluminum alloys: 7075 and 2020, and the implications of the experimental results are discussed in terms of the micro-mechanisms of near-threshold fatigue crack propagation.

Materials and Experimental Procedures

The 2020 and 7075 aluminum alloys investigated in this study were both obtained in the form of plate material, the nominal compositions of which are given in Table I. The 2020 alloy was fatigue-tested in the standard peak-aged (-T651) temper, whereas 7075 aluminum was investigated

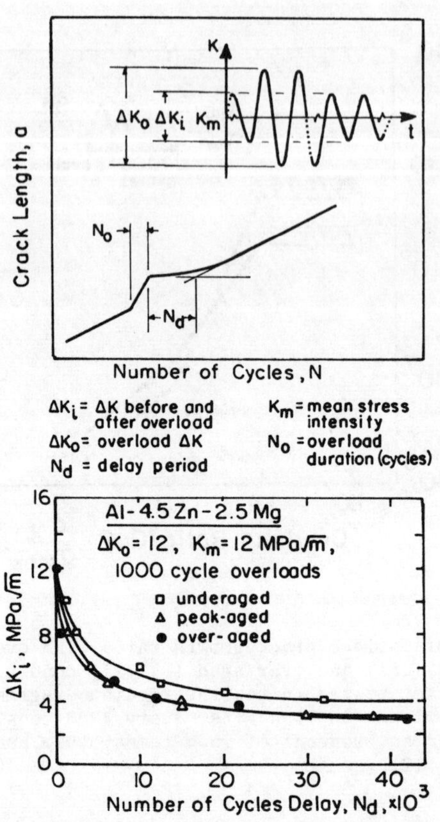

Fig. 2 Influence of overloads on delay period for various baseline ΔK_i values in the under-aged, peak-aged and over-aged conditions of Al-4.5Zn-2.5Mg alloy tested in laboratory air (after ref. 6).

Table I. Chemical Composition (in wt. Percent) of Alloys Investigated

Material	Cu	Li	Mn	Si	Fe	Cd	Ti	Al
2020	4.46	1.08	0.52	0.09	0.20	0.20	0.02	Balance

Material	Zn	Mg	Cu	Cr	Fe	Si	Al
7075	6.25	2.26	1.51	0.23	0.28	0.23	Balance

in the under-aged (heat treated at 120°C for 90 mins following solution treatment and a 2% stretch), peak-aged (-T651 temper) and over-aged (-T7351 temper) conditions. The under-aged and over-aged conditions were designed to produce similar yield strength values. The room temperature mechanical properties of all the alloys are provided in Table II.

Table II. Room Temperature Mechanical Properties of Alloys Investigated

Alloy	Yield Strength (MPa)	U.T.S. (MPa)	Elongation (%)	K_{Ic} (MPa$\sqrt{m}$)
2020-T651	517	547	5.3	23.6
7075-UA	450	570	17.2	--
7075-T651	530	589	11.5	28.7
7075-T7351	454	505	13.0	32.0

Constant amplitude fatigue crack growth studies were carried out at a frequency of 25 Hz and load ratio of 0.33 in an environment of room temperature humid air (∿ 95% RH) on 6.25 mm thick compact specimens and slow bend Charpy specimens machined in the L-T orientation. Both specimen geometries resulted in identical crack propagation rates. The crack growth retardation behavior following single peak tensile overloads was monitored using the bend specimens. This was in view of the fact that the geometry of the bend specimen facilitated sequential observations of crack profiles prior to and following the application of the overload. An overload ratio (defined as the ratio of the maximum stress intensity factor during the overload to that during the constant amplitude cycle) of 1.8 was used for all the overload tests. The extent of fracture surface oxidation was characterized in a secondary ion mass spectrometer using procedures described in detail elsewhere (15).

Results and Discussion

Effects of Microstructure and Crack Morphology

7075 Aluminum Alloy. Figure 3 shows the influence of microstructure on the constant amplitude fatigue crack growth rates for the 7075 aluminum alloy tested in a humid air environment at a frequency of 25 Hz and load ratio of 0.33. In the intermediate range of constant amplitude fatigue crack propagation (above 10^{-6} mm/cycle), the 7075 alloy shows a greater resistance to cyclic crack growth in the over-aged condition than in the under-aged or peak-aged tempers. In the near-threshold region of constant amplitude fatigue crack growth, however, exactly the opposite trend is observed. Here, increased aging causes a substantial increase in near-threshold growth rates and a decrease in threshold ΔK_o values (Fig. 3). These changes in microstructural influences on crack growth with decreasing ΔK are analogous to the variations in microstructural effects previously observed between constant amplitude and post-overload crack growth in the intermediate regime (Figs. 1 and 2). Such typical microstructural influences on constant and variable amplitude fatigue crack growth behavior of high strength aluminum alloys are summarized in Table III.

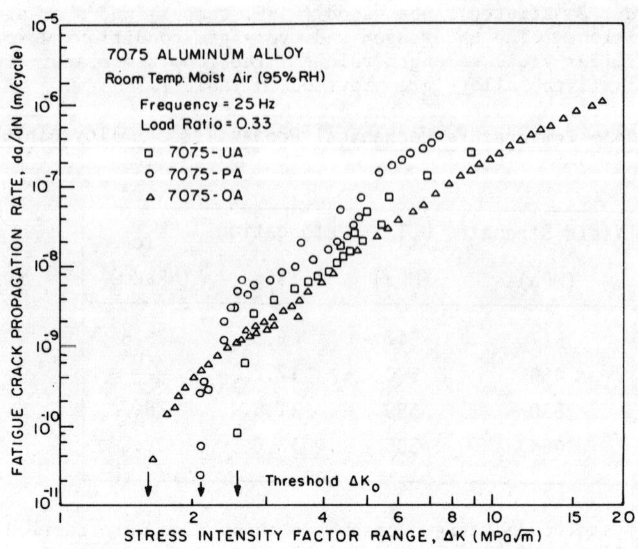

Fig. 3 Constant amplitude fatigue crack propagation data for the 7075-UA-PA and -OA tempers tested in ∿ 95% RH room temperature air at a load ratio of 0.33.

Table III. VARIATIONS IN RELATIVE FATIGUE CRACK GROWTH RATES WITH HEAT TREATMENT TYPICALLY OBSERVED IN HIGH STRENGTH ALUMINUM ALLOYS

HEAT TREATMENT	CONSTANT AMPLITUDE GROWTH RATES		VARIABLE AMPLITUDE* GROWTH RATES
	intermediate range	near-threshold	
underaged	high	low	low
peak-aged	high	medium	low-medium
overaged	low	high	high

*for an overload ratio typically greater than 1.5; complete crack arrest can occur at low baseline growth rates and large overload ratios.

Figure 4 shows the mode of crack advance and the resulting fracture profile immediately following and 8000 constant amplitude cycles after the

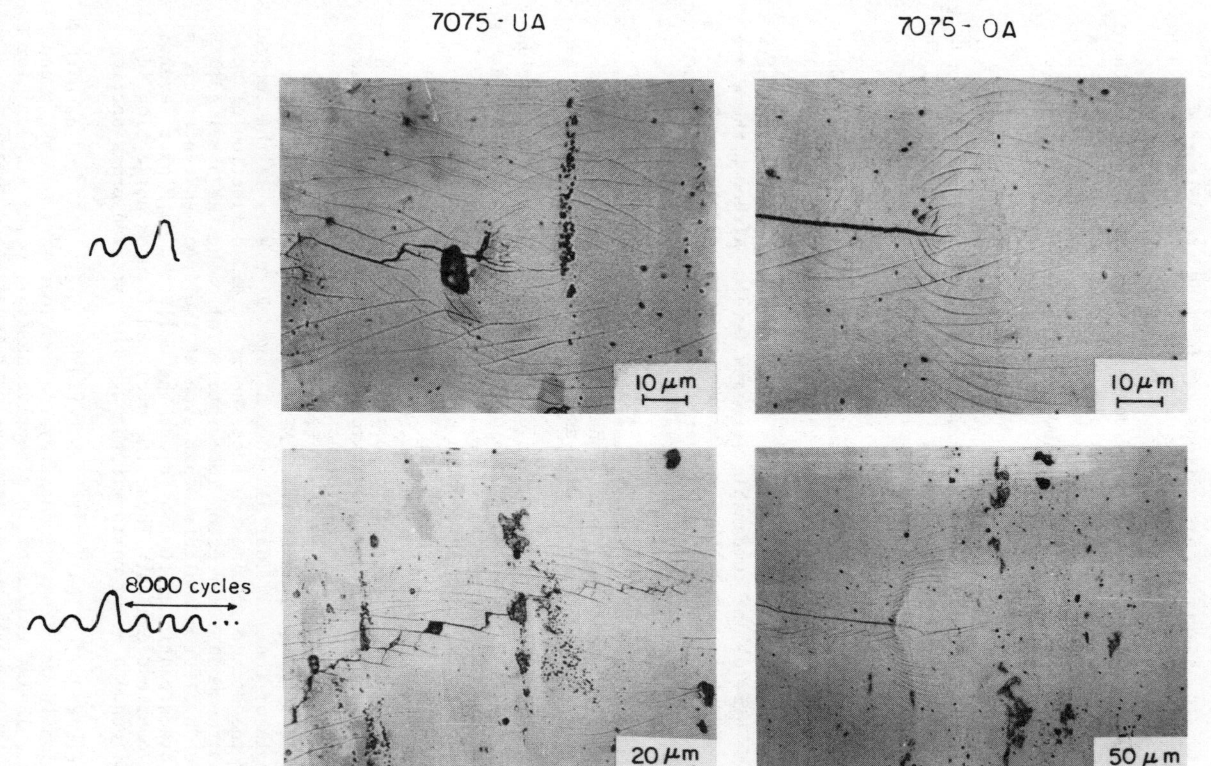

Fig. 4 Changes in fatigue crack profile immediately following and 8000 constant amplitude cycles after the application of an 80% single overload at baseline $\Delta K = 5\ MPa\sqrt{m}$ ($R = 0.33$) in the under-aged and over-aged tempers of the 7075 aluminum alloy.

application of an 80% overload at a baseline ΔK = MPa$\sqrt{m}$ (R = 0.33) in the under-aged and over-aged conditions of 7075 aluminum alloy. Although conventional models for retardation behavior, such as Elber's plasticity-induced crack closure model (1) or the Wheeler (2) and Willenborg (3) models, would predict identical transient behavior for these two microstructures because of their comparable strength (and hence similar calculated plastic zone size values), an examination of Figure 4 clearly reveals completely different retardation characteristics for the UA and OA materials. The following specific observations can be made from Figure 4:

i) In the under-aged microstructure, the mode of crack advance is highly crystallographic as evidenced by the non-linear growth along intense shear bands visible on etched sample surfaces;

ii) In the over-aged condition, however, the crack path is highly linear both prior to and after the overload;

iii) Significant crack deflection ($\sim$ 45°), away from the nominal Mode I growth direction, occurs on application of the overload in the under-aged temper. In the over-aged temper, however, the crack continues to propagate along the pre-overload (Mode I growth plane).

Figure 5 shows the constant amplitude fatigue crack propagation data for the 2020-T651 aluminum alloy tested in room temperature 95% RH air at a frequency of 25 Hz and load ratio of 0.33. The overall fatigue crack growth behavior of the 2020-T651 alloy is found to be substantially superior to that of 7075-T651 of comparable strength level. Although there are inherent differences in composition and microstructure between the two alloys, recent work (16,17) has shown that a partial reason for the beneficial fatigue response of the 2020-T651 alloy stems from its highly crystallographic and tortuous crack path (as compared to 7075-T651), attributed primarily to the shearability of its θ' (Al_2Cu) precipitates. It is also found (17) that the serrated nature of crack profile in this alloy is promoted as the K values decrease, approaching the threshold ΔK_o.

Figure 6 shows the sequential changes in crack profile as a function of the number of cycles following an 80% overload in 2020-T651 aluminum alloy at a baseline ΔK = 4.5 MPa$\sqrt{m}$. The application of the overload (corresponding to the crack location denoted by the smaller arrow) leads to the deflection of the crack by about 45° from the nominal Mode I growth direction (larger arrow), in addition to the formation of a secondary crack. Figure 6 shows that no crack growth occurs up to 1000 cycles after which the retarded post-overload crack propagates with a non-linear and zig-zag profile (Fig. 6d), which is typical of near-threshold crack advance, in this material. The deflection of the fatigue crack, induced by an 80% overload applied at a baseline ΔK = 7.7 MPa$\sqrt{m}$, is shown in Figure 7 for a duplicate sample of the 2020-T651 aluminum alloy. Figure 7a is the optical micrograph of the etched specimen surface showing a 45° deflection of the crack tip from the Mode I direction immediately following the overload, whereas Figure 7b clearly indicates that even at the specimen center cross section, the crack continues to grow along the deflected path over several thousand post-overload cycles. Thus, the overload-induced deflection in crack path is found to occur uniformly through the specimen cross-section.

Recent crack deflection models presented by Suresh (18) indicate that a 45° deflection in crack path induced by the overload (e.g., Figs. 6 and 7) gives rise to a 15-19% reduction in effective ΔK ($K_{eff} \approx \{k_1^2 + k_2^2\}^{1/2}$ for co-planar growth, where k_1 and k_2 are the local tensile and shear stress

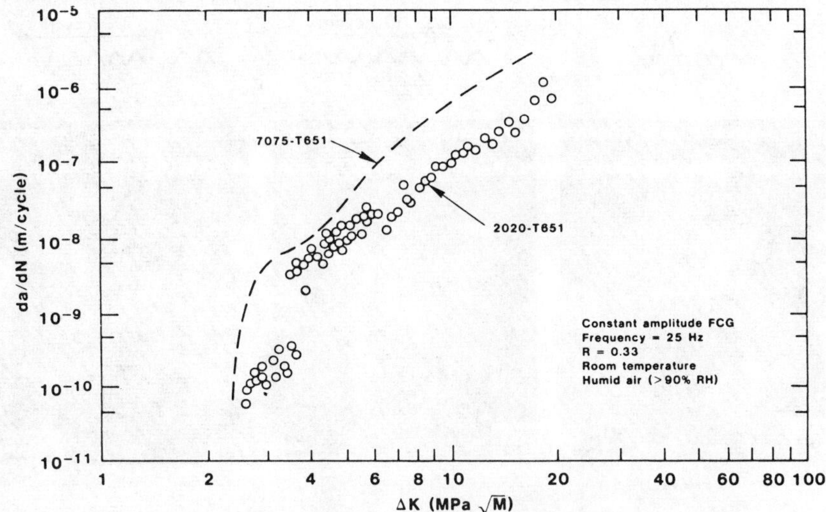

Fig. 5 Constant amplitude fatigue crack propagation data for 2020-T651 and 7075-T651 aluminum alloys tested in room temperature moist air (> 90% RH) at a load ratio of 0.33.

intensities for a deflected crack). Additionally, crack deflection can lead to a 30% underestimate in the retarded crack growth rate, if the changes in crack profile are not considered in the fatigue data analyses. Furthermore, it has also become evident that mechanistic processes typical of near-threshold crack growth such as serrated and mixed mode crack advance as well as various sources of crack closure are activated in the post-overload regime since crack deflection and changes in crack tip plasticity due to the overload reduce the effective ΔK to the near-threshold regime although the nominal post-overload ΔK may correspond to the intermediate growth regime (7,8). We now examine the relevance of near-threshold mechanisms to post-overload crack growth.

Effects of Plasticity-Induced Crack Closure

Conventional models for fatigue crack growth retardation have centered around changes in crack-tip plasticity arising from the overload. It was first suggested by Elber (1) that the enhanced residual plastic deformation due to the overload cycle results in crack face contact in the wake of the advancing crack tip which accounts for not only the growth attenuation, but also delayed retardation. The argument that such plasticity-induced closure (1,19) or residual compressive stress models (2,3) alone can rationalize retardation has recently been questioned (7,8) for the following reasons: i) Plasticity-induced closure is considered to play a dominant role only under plane stress loading conditions, whereas considerable crack growth retardation occurs even in plane strain; ii) While residual compressive stresses are normally considered active over a distance comparable to the size of the cyclic plastic zone generated by the overload (20), marked retardation is known to occur over distances well beyond the greatest prior elastic-plastic boundary generated by the overload; iii) Models based on changes in crack-tip plasticity do not fully rationalize (even when cyclic strength properties are considered) such post-overload characteristics as absence of striations, abraded crack growth, as well as increase in microstructure and environment-sensitivity. Thus, plasticity-induced crack

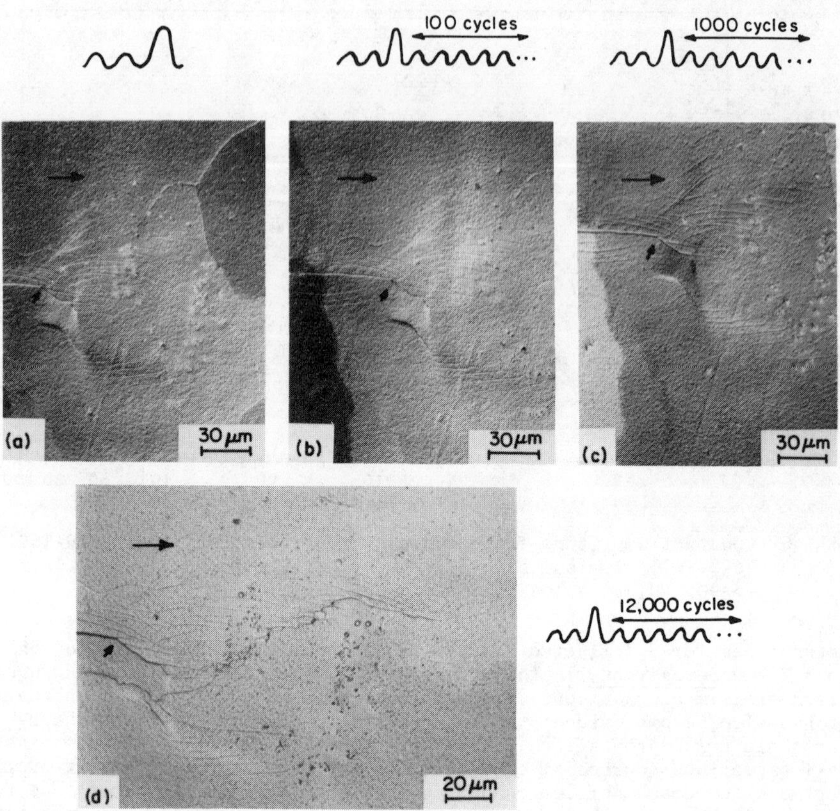

Fig. 6 Sequential changes in fatigue crack profile following (a) an 80% single overload and (b) 100 cycles, (c) 1000 cycles and (d) 12000 cycles after the overload in 2020-T651 aluminum alloy at baseline $\Delta K = 4.5\ MPa\sqrt{m}$; room temperature moist air (~ 95% RH).

closure, which is considered to play a minimal role in influencing near-threshold crack growth, does not also account for a number of the transient effects arising from overloads. It therefore becomes necessary to examine other sources of crack closure which affect crack growth in plane strain.

Effects of Oxide-Induced Crack Closure

The concept of oxide-induced crack closure (e.g., ref. 21) has recently been suggested as a possible explanation for the role of environment in influencing near-threshold fatigue crack growth behavior. Such closure arises as a result of enlarged oxide formation on the fracture surfaces due to repeated contact (fretting) at low ΔK levels and leads to premature contact between the crack faces thereby reducing effective ΔK values (21). Studies of post-overload retardation (7,8) first revealed that oxide-induced closure is a viable mechanism for transient crack growth following overloads in certain microstructures capable of forming copious oxide layers in moist environments.

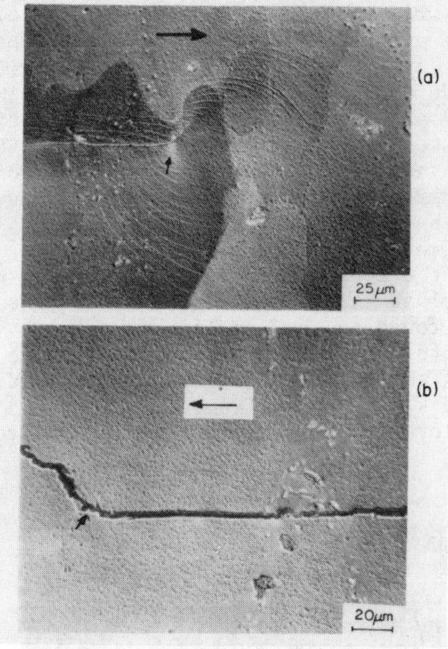

Fig. 7 a) Surface profile of deflection of a Mode I fatigue crack immediately after the application of an 80% overload in 2020-T651 aluminum alloy at baseline $\Delta K \approx 7.7$ MPa$\sqrt{m}$. b) Crack profile at specimen center thickness at 9000 cycles of constant amplitude loading following the overload in 2020-T651 alloy; smaller arrows indicate crack tip locations at which overload was applied. Larger arrows indicate nominal Mode I growth direction.

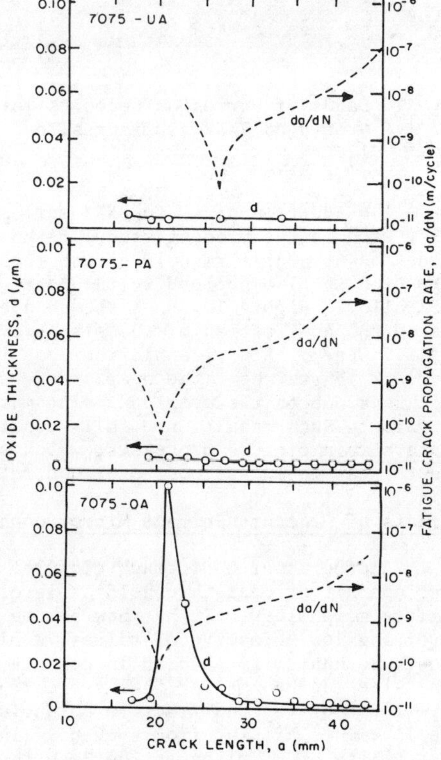

Fig. 8 Variation of approximate excess oxide thickness values with crack length and constant amplitude growth rates for 7075-UA, -PA and -OA alloys tested in 95% humid air at R = 0.33.

The importance of oxide-induced closure processes to the near-threshold fatigue behavior of high strength aluminum alloys is now known to be a strong function of alloy composition and aging treatment. Figure 8 shows the variation of estimates of fracture surface oxide deposits of the UA, PA and OA microstructures of the 7075 aluminum alloy as a function of crack length (as well as constant amplitude crack growth rates) for tests performed in 95% RH air at a load ratio of 0.33. At near-threshold crack growth rates, the over-aged 7075 alloy forms up to 0.1 μm thick oxide layers. The fracture surface excess oxide thickness for both the under-aged and peak-aged tempers, however, is only about 0.005-0.01 μm at all growth rates. Such effects of microstructure on crack face oxidation in the near-threshold regime are also similar to those observed in the post-overload regime. The present study as well as previous Alcoa Research (12) have revealed that negligible oxide formation occurs in the 7075-UA microstructure following an overload, while macroscopically visible, thick oxide bands can be observed in the over-aged structure. Figure 9 shows that marked crack face oxidation occurs in the over-aged condition of a 7XXX aluminum alloy whenever an overload is applied (after ref. 12).

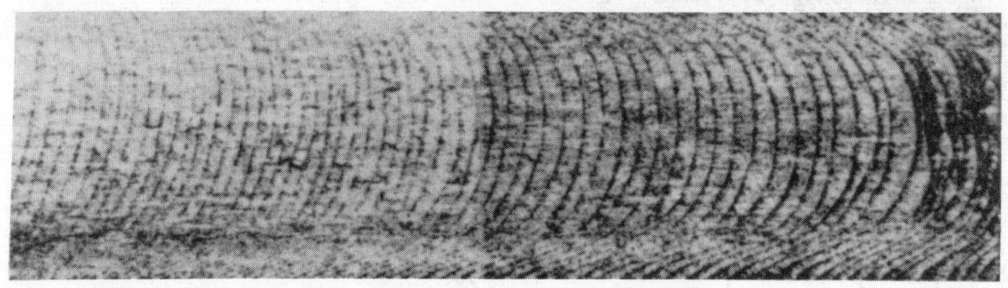

Fig. 9 Bands of corrosion deposits formed on the fracture surface of an over-aged 7XXX aluminum alloy after each overload cycle (after ref. 12).

The addition of Li to 2XXX series aluminum alloys (such as the 2020 alloy used in the present study) is known to increase the fracture surface oxidation at near-threshold growth rates (17). Such influence of oxide formation is also noticed in the post-overload growth region of the 2020-T651 alloy. Figure 10 shows the changes in fracture surface oxide thickness due to the application of 80% single overloads at baseline ΔK values of 4.5 and 8.7 $MPa\sqrt{m}$. A noticeable increase in oxidation is found at the location of crack length where the overload is applied, with the increase in oxide thickness due to the overload being more pronounced at the lower baseline ΔK value. Such results are fully consistent with the hypothesis (7,8) that plane strain closure processes, which are typical of near-threshold crack growth, can influence post-overload retardation.

Effects of Fracture Surface Micro-Roughness

Another source of crack closure, which leads to an increase in crack face contact at near-threshold ΔK levels, is that due to rough fracture surface morphology. It has now become clear that in fatigue situations involving low effective ΔK values, a highly crystallographic (Stage I) growth mechanism is favored in certain microstructures because the plastic zone size can be smaller than typical grain size (22,23). This results in a serrated or faceted fracture morphology. Given the presence of Mode II displacements arising from such non-linear crack advance (21) and the possibility of misalignment between the fracture surfaces during the

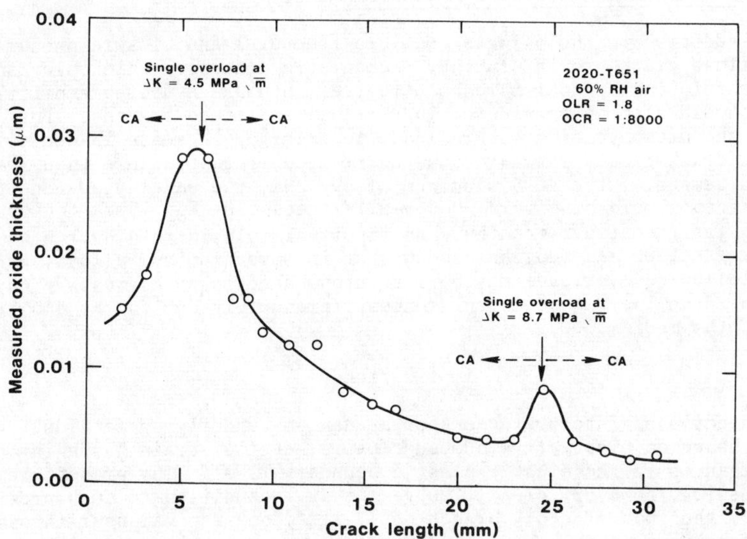

Fig. 10 Estimates of fracture surface oxidation from secondary ion mass spectroscopy analyses showing an increase in oxide thickness due to 80% overloads at crack length locations corresponding to baseline ΔK values of 4.5 $MPa\sqrt{m}$ and 8.7 $MPa\sqrt{m}$.

unloading portion of the fatigue cycle due to oxidation of slip steps and irreversible crack tip deformation, the rough fracture morphology readily provides a mechanism for additional closure through the so-called roughness-induced closure process (21-23). Such closure is of greater significance at lower <u>effective</u> ΔK levels where the scale of crack tip opening displacements is comparable to the size of fracture surface asperities (21-23).

Closure arising from fracture face micro-roughness has also been suggested as a possible mechanism for <u>enhanced</u> retardation following overloads (7,13,14) and simple quantitative micro-roughness models (8) are available for rough estimates of such closure. Post-overload closure due to micro-roughness is dominant in microstructures containing highly shearable/coherent secondary particles, such as the under-aged conditions of high strength aluminum alloys (e.g., Figs. 4a and 4c) and certain aluminum alloys containing lithium such as 2020-T651 (Fig. 6d). The latter figure, for example, shows the change in fracture mode (to a crystallographic growth mechanism) induced by the overload, as evident from the advance of the crack along the intense slip lines visible on the specimen surface, and the possibility of enhanced closure due to micro-roughness. Arguments based on the micro-roughness model (7,8,13) also rationalize such transient features as fracture surface abrasion and absence of striations. Fractography and closure results obtained by numerous independent studies indicate that crack face asperity contact due to micro-roughness is an important mechanism for overload effects (12-14,19) although in most cases the underlying mechanism was (erroneously) attributed to plasticity-induced closure.

Effects of Other Near-Threshold Mechanisms

In addition to the effects of microstructural and closure phenomena on post-overload crack growth, further mechanistic processes which play an important role in the near-threshold regime (such as "hydrogen embrittlement"), should also be considered in variable amplitude loading. For example, the presence of moisture leads to a larger increase in near-threshold crack growth rates for the under-aged microstructure than for the over-aged temper of the 7075 aluminum alloy (25), the under-aged condition gives rise to a more beneficial "over-all" threshold ΔK_0. Such conventional corrosion fatigue effects can have an important role in retardation behavior. The work of Wei et al. (26) has shown that in some aluminum alloys, delay effects following overloads may be less pronounced in an aggressive environment than in an inert gaseous environment, presumably due to the embrittling effect of the medium.

Implications

Conventional interpretations of mechanisms underlying retardation have all been based on plasticity-induced closure, cyclic strain hardening, blunting and changes in the elastic-plastic boundary (1-6). The present results and discussion, however, clearly suggest that <u>in addition</u> to such previous arguments, the role of crack branching (7,8,27) and typical near-threshold growth mechanisms (such as oxide-induced and roughness-induced closure (21-23) should be considered in order to rationalize fully transient retardation phenomena. A careful examination of numerous previous papers on this topic (6,9,10,12-14,19) reveals that although there is ample experimental evidence supporting the occurrence of such additional mechanisms following overloads, the significant majority of earlier studies have erroneously ascribed <u>all</u> the retardation characteristics only to conventional mechanisms. Furthermore, studies (e.g., 2,3,19) correlating <u>nominal</u> crack growth distance during retardation to the overload plastic zone must be regarded as questionable since, as demonstrated here and in recent studies (7,8,27), considerable differences exist between the <u>nominal</u> and <u>actual</u> post-overload crack growth rates (and delay distance) due to the occurrence of crack branching in many aluminum alloys.

The implications for post-overload growth of various <u>additional</u> mechanistic processes detailed in this work can be summarized as follows. Pronounced crack branching and highly serrated retarded crack growth occur after an overload in aluminum alloy microstructures, such as 7075-UA or 2020-T651 alloys, where crystallographic crack advance is favorable. (A review of overload-induced crack branching of aluminum alloys indicates that up to 35% decrease in effective ΔK can be caused <u>solely</u> by crack deflection and that such reductions are comparable to those induced by other conventional mechanisms (8). Furthermore, a neglect of changes in crack path produced by the overload can result in up to a 30-40% <u>underestimate</u> of post-overload growth rate due to errors in crack length measurement (8,18).) The microstructure dependence of fracture surface oxidation and highly faceted crack growth in the post-overload region of 7075 and 2020 alloys appear similar to those observed in the near-threshold regime. Such typical microstructural effects on near-threshold (15,25) and post-overload crack propagation for high strength aluminum alloys such as 7075 are given in Table IV.

The present argument that near-threshold mechanisms effectively govern retarded crack advance following overloads provides an explanation for such typical post-overload characteristics as striationless crack growth and enhanced microstructural effects in aluminum alloys. The observations of oxide bands and abraded crack faces during overload tests can also be interpreted in terms of such threshold phenomena as oxide-induced and roughness

Table IV. Typical Microstructural Effects on Near-Threshold and Post-Overload Fatigue Crack Growth in High Strength Aluminum Alloys

Microstructure	Fatigue Behavior
Under-Aged	Reduction in effective ΔK due to crack deflection and roughness-induced crack closure, embrittlement due to moisture.
Over-Aged	Embrittlement due to moisture, substantial fracture surface oxidation and oxide-induced crack closure; possible enhancement in embrittlement due to hydrogen produced concomitantly with fracture surface oxidation.

Fatigue behavior in the peak-aged condition is similar to that in the under-aged temper even though the degree of influence of these mechanisms is significantly less than in the under-aged temper.

induced crack closure, respectively. Furthermore, such arguments are consistent with the occurrence of delayed retardation since considerable oxide build-up or faceted growth and the resulting additional closure processes are activated only after the effective ΔK in the post-overload region is reduced to near-threshold levels (7,8). This also implies that if overload-induced changes in crack-tip plasticity and crack branching reduce the effective ΔK values below the threshold ΔK_o, complete crack arrest would result, an inference fully supported by experimental observations (8).

Finally, the application of threshold concepts to variable amplitude loading is found to have certain interesting implications for the design and ranking of aluminum alloys in terms of superior fatigue resistance. Over-aged aluminum microstructures have traditionally been favored for corrosion fatigue applications because of their slower constant amplitude crack growth in the intermediate regime (12) whereas it is shown in this work that the under-aged temper has a far superior resistance to fatigue crack growth in the near-threshold regime (which accounts for a major portion of the crack growth life) and in variable amplitude loading situations (which are typical of in-service conditions). Although it is premature to draw conclusions based on the limited extent of experimental data, the present results for conventional high strength ingot aluminum alloys suggest that designing microstructures to produce superior near-threshold crack growth response may translate into better fatigue propagation characteristics under spectrum loading conditions.

Conclusions

Detailed analyses of the retardation characteristics following single peak overloads in 7075 and 2020 aluminum alloys reveal that in addition to the conventional mechanisms based on changes in crack tip plasticity, microstructural and environmental factors significantly influence post-overload crack growth. Pronounced crack branching occurs in the 7075-UA and 2020-T651 alloys on application of the overload. The present results substantiate previous arguments (7,8) that the post-overload retarded crack advance is strongly influenced by the micromechanisms of near-threshold fatigue. Specifically, typical near-threshold mechanisms of crack closure due to oxide layers and fracture face microroughness are also shown to be of importance in the retarded post-overload region. Such concepts are utilized to rationalize the observation that the under-aged microstructure, which offers less resistance to constant amplitude fatigue crack growth (as compared to the over-aged structure) in the Paris regime can lead to far superior retardation following an overload. The possible applications of the present interpretations are examined for the ranking of aluminum alloys in terms of fatigue resistance.

Acknowledgments

This work was supported partly by Brown University's Materials Research Laboratory, funded by the National Science Foundation, and partly by Alcoa Internal Research funds. Thanks are due to Drs. P. E. Bretz, R. O. Ritchie, R. R. Sawtell and J. T. Staley for their help and support. One of the authors (S.S.) acknowledges the support provided by the University of California, Berkeley, under Grant No. AFOSR-82-0181 for preparing some of the figures and the help of Miss M. M. Penton and Mrs. L. Gray in typing this manuscript.

References

1. W. Elber, "The Significance of Fatigue Crack Closure", pp. 230-242, ASTM STP 486, American Society for Testing and Materials, Philadelphis, PA 1971.

2. O. E. Wheeler, "Spectrum Loading and Crack Growth", Journal of Basic Engineering, Trans. ASME, Series D, 94 (1972) pp. 181-186.

3. J. Willenborg, R. M. Engle and H. Wood, "A Crack Growth Retardation Model Using an Effective Stress Intensity Concept", Technical Report TFR 71-701, North American Rockwell, Los Angeles Division, 1971.

4. R. H. Christensen, Metal Fatigue, McGraw-Hill, Inc., New York, NY, 1976.

5. R. E. Jones, "Fatigue Crack Growth Retardation After Single Cycle Peak Overload in Ti-6 Aℓ-4V Titanium Alloy", Engineering Fracture Mechanics, 5 (1973) pp. 585-604.

6. J. F. Knott and A. C. Pickard, "Effects of Overloads on Fatigue Crack Propagation: Aluminum Alloys", Metal Science, 11 (1977) pp. 399-404.

7. S. Suresh, "Crack Growth Retardation due to Micro-Roughness: A Mechanism for Overload Effects in Fatigue", Scripta Metallurgica, 16 (1982) pp. 995-999.

8. S. Suresh, "Micromechanisms of Fatigue Crack Growth Retardation Following Overloads", Engineering Fracture Mechanics, 18 (1983) pp. 577-593.

9. P. J. Bernard, T. C. Lindley and C. E. Richards, "The Effect of Single Overloads on Fatigue Crack Propagation in Steels", Metal Science, 12 (1977) pp. 390-398.

10. S. W. Hopkins, C. A. Rau, G. R. Leverant and A. Yuen, "Effect of Various Program Overloads on the Threshold for High Frequency Fatigue Crack Growth", pp. 125-141, ASTM STP 595, American Society for Testing and Materials, Philadelphia, 1976.

11. S. Suresh and R. O. Ritchie, "Near-Threshold Fatigue Crack Propagation: A Perspective on the Role of Crack Closure", this volume.

12. R. J. Bucci, A. B. Thakker, T. H. Sanders, R. R. Sawtell and J. T. Staley, "Ranking 7XXX Aluminum Alloy Fatigue Crack Growth Resistance Under Constant Amplitude and Spectrum Loading", pp. 41-78, ASTM STP 714, American Society for Testing and Materials, Philadelphia, 1980.

13. J. M. Baik, L. Hermann and R. J. Asaro, "Fatigue Crack Growth and Overload Retardation in 2048 Aluminum", pp. 33-51, in _Mechanics of Fatigue_, T. Mura, ed., ASME, New York, NY, 1981.

14. H. Nowack, K. H. Trautmann, K. Schulte and G. Lütjering, "Sequence Effects on Fatigue Crack Propagation: Mechanical and Microstructural Contributions", pp.32-43, ASTM STP 677, American Society for Testing and Materials, Philadelphia, 1979.

15. A. K. Vasudevan and S. Suresh, "Influence of Corrosion Deposits on Near-Threshold Fatigue Crack Growth Behavior in 2XXX and 7XXX Series Aluminum Alloys", _Metallurgical Transactions_, 13A (1982) pp. 2271-2280.

16. E. J. Coyne, T. H. Sanders and E. A. Starke, "The Effect of Microstructure and Moisture on the Low Cycle Fatigue and Fatigue Crack Propagation of Two Aℓ-Li-X Alloys", pp. 293-307, in _Aℓ-Li Alloys_, T. H. Sanders and E. A. Starke, eds., The Metallurgical Society of AIME, Warrendale, PA, 1980.

17. A. K. Vasudevan, P. E. Bretz, A. C. Miller and S. Suresh, "Fatigue Crack Growth Behavior of Aluminum Alloy 2020", _Materials Science and Engineering_, (1984) in press.

18. S. Suresh, "Crack Deflection: Implications for the Growth of Long and Short Fatigue Cracks", _Metallurgical Transactions_, 14A (1983) pp. 2375-2385.

19. E. F. J. vonEuw, "Effect of Overload Cycle(s) on Subsequent Fatigue Crack Propagation in 2024-T3 Aluminum Alloy", Ph.D. Thesis, Lehigh University, 1971.

20. J. R. Rice, "Mechanics of Crack Tip Deformation and Extension by Fatigue", pp. 247-311, ASTM STP 415, American Society for Testing and Materials, Philadelphia, 1967.

21. S. Suresh, G. F. Zamiski and R. O. Ritchie, "Oxide-Induced Crack Closure: An Explanation for Near-Threshold Corrosion Fatigue Crack Growth Behavior", _Metallurgical Transactions_, 12A (1981) pp. 1435-1443.

22. K. Minakawa and A. J. McEvily, "On Crack Closure in the Near-Threshold Region", _Scripta Metallurgica_, 15 (1981) pp. 633-636.

23. S. Suresh and R. O. Ritchie, "A Geometric Model for Fatigue Crack Closure Induced by Fracture Surface Morphology", _Metallurgical Transactions_, 13A (1982), pp. 1627-1631.

24. D. L. Davidson, "Incorporating Threshold and Environmental Effects Into the Damage Accumulation Model for Fatigue Crack Growth", Fatigue in Engineering Materials and Structures, 3 (1981) pp. 229-236.

25. S. Suresh, A. K. Vasudevan and P. E. Bretz, " Mechanisms of Slow Fatigue Crack Growth in High Strength Aluminum Alloys: Role of Microstructure and Environment", Metallurgical Transactions, 15A (1984) pp. 369-379.

26. R. P. Wei, N. E. Fenelli, K. D. Unangst and T. T. Shih, "Fatigue Crack Growth Response Following a High-Load Excursion in 2219-T851 Aluminum Alloy", Journal of Engineering Materials and Technology, Trans. ASME, 102 (1980) pp. 280-292.

27. J. Lankford and D. L. Davidson, "The Effect of Overloads Upon Fatigue Crack Tip Opening Displacement and Crack Tip Opening/Closing Loads in Aluminum Alloys", pp. 899-906 in Advances in Fracture Research, D. François, ed., Pergamon Press, Oxford, 1981.

ON THE INTERACTION OF OVERLOAD AND METALLURGICAL VARIABLES ON FATIGUE CRACK GROWTH IN THE THRESHOLD REGIME

C. H. Newton, R. S. Vecchio, R. W. Hertzberg, and R. Jaccard[1]

Metallurgy and Materials Engineering Department
Lehigh University
Bethlehem, PA 18015
USA

Fatigue crack growth characteristics in the near-threshold regime in an extruded aluminum alloy are examined as a function of prior loading history and microstructural variables. Previous studies have shown that the crack growth delay N_d can be divided into three regimes according to the ΔK level prior to and following the overload (ΔK_{base}). Overloads near threshold cause increasing delay with decreasing ΔK_{base} while fatigue crack growth at intermediate ΔK_{base} is attenuated only modestly by overloads. At higher ΔK_{base}, N_d increases sharply, thereby leading to the development of a U-shaped curve of N_d vs. ΔK_{base}. Grain size effects, including the relationship between the plastic zone, grain size, and overload-induced delay, are examined in the threshold regime. The concept of an effective overload ratio to normalize delay effects is explored, along with the effect of metallurgical variables on constant amplitude FCP threshold behavior.

[1]Swiss Aluminium Ltd., Zurich, Switzerland

Introduction

The application of fracture mechanics principles to fatigue permits separate consideration of crack initiation, propagation, and final failure. In many applications, crack-like defects exist when the component is put into service, thereby eliminating the initiation portion of fatigue life. The crack propagation behavior in the low crack growth rate regime then becomes the most important criterion in determining fatigue life. Crack growth rates in this near-threshold regime increase very rapidly as the cyclic stress intensity (ΔK) increases from a threshold value to a value corresponding to the transition to a region described by a power law relationship between crack growth rate and ΔK.

The fatigue threshold is characterized by a cyclic value of the stress intensity factor at which the crack growth rate becomes vanishingly small(ΔK_{th}). The threshold value of ΔK and the crack growth rate in the near-threshold regime are dependent on the microstructure as shown by Waldron _et al_. in experiments on plain carbon steels [1] and by Yoder _et al_. in experiments on titanium alloys [2]. For example, most experiments have shown that as grain size increases, ΔK_{th} increases which results in greater fatigue resistance in the near-threshold regime [3, 4, 5, 6]. Carlson and Ritchie, however, found that ΔK_{th} was inversely proportional to the grain size in a high strength Fe-Cr-C steel [7] though Yoder _et al_. questioned whether the critical microstructural feature for this study should have been the martensite lath spacing rather than the prior austenite grain size [2]. The threshold for copper and Cu-Al alloys was also found to decrease with increasing grain size [8]. Mutoh and Radhakrishnan [9] assumed that the cell spacing was the critical parameter for these copper alloys but no study was attempted to confirm this hypothesis. The influence of yield strength on ΔK_{th} has been examined and the results found to be conflicting. Namely, the threshold value of ΔK was found to increase with decreasing yield strength in steels [5, 9, 10], to increase with the square root of the yield strength [6] in titanium alloys and pearlitic steels, and to have very little dependence on the yield strength in other titanium alloys [11]. Obviously, the effects of grain size and yield strength on threshold behavior vary markedly and are not well understood. Another microstructural feature which influences the near-threshold regime is the spacing and coherency of the precipitates [12]. Cadman _et al_. found that precipitates could accelerate the crack growth rate by starting cracks ahead of the crack front or slow down crack growth by pinning the crack front [13].

Though many fatigue tests are limited to constant load amplitude, fatigue analyses more directly related to actual structures must consider delays and accelerations in crack growth due to varying loads. The resulting changes in growth rate are complex and have been attributed to plasticity-induced closure [14], oxide-induced closure [15], roughness-induced closure [16], crack blunting, and microcracking [16].

Simple experiments which involve load excursions periodically superimposed onto constant amplitude load fluctuations are useful in studying complex load interactions and the associated micromechanisms of failure. Such simple overload interactions have been used in several studies, e.g. [17, 18, 19]. More recently, the trends in overload-induced delay have been observed to follow a U-shaped curve as shown in Figure 1 [20]. The increase in delay at the upper end of the curve correlates with the transition from plane strain to plane stress conditions. It has been hypothesized that the increase in delay at the lower end of the curve is due to the influence of metallurgical variables, such as grain size, on the fatigue fracture process. One of the purposes of this study is to evaluate this hypothesis in an aluminum alloy. In addition, an attempt was made to evaluate the amount of cyclic delay as a function of the effective overload ratio, the latter being dependent on the effective ΔK_{base} level.

Experimental Procedure

Two materials prepared by the Swiss Aluminium Company were used in this investigation. Four different thermomechanical treatments of alloy AC050 were used in an attempt to obtain a wide range of grain sizes . The composition of this alloy and the mechanical properties resulting from each heat treatment are shown in Tables 1 and 2. Overload experiments, based on an effective value of ΔK_{base}, were conducted with Swiss Aluminium Company alloy AC062/61 which was subjected to two heat treatments; the properties and composition of this material are also listed in Tables 1 and 2.

Metallography

Each material was examined metallographically on three orthogonal planes. Large grain sample surfaces were polished and etched using Tucker's reagent. The surfaces of samples with small grains were etched with 0.5% hydrofluoric acid. The microstructures obtained from the strain anneal procedure resulted in grain sizes which were considerably larger than those originally sought (see Table 3). Volumetric average diameters of 0.1 and 18 mm were used in calculations involving the small and large grain materials. Subgrain diameter measurements were also made for AC050-2 using Keller's reagent.

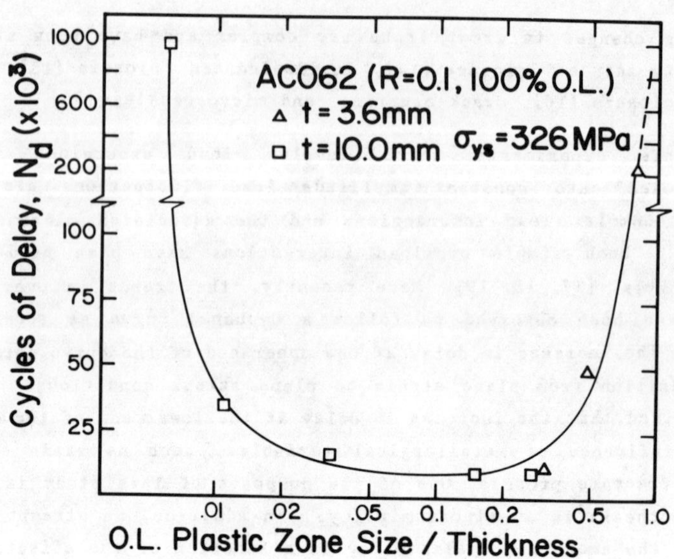

Figure 1: Overload delay normalized with respect to sheet thickness for an aluminum alloy [27].

Table I: Composition

Alloy	Si	Fe	Cu	Mn	Mg	Cr	Zn	Ti
AC050[1]	0.5-0.6	0.1-0.3	0.10	0.1	0.45-0.6	0.05	0.15	0.1
AC062/61	0.63	0.21	0.17	0.03	0.54	0.11	0.08	0.01

[1]Specification

Table II: Mechanical Properties

Alloy	Prestraining	Solution Treatment	Aging	σ_{ys} (MPa)	σ_{uts} (MPa)
AC050-1	10%	290h/530°C		220	238
AC050-2	10%	290h/530°C	1h/510°C	125	170
AC050-12	5%	170h/530°C	10h/165°C	160	175
AC050-13	5%	290h/530°C		170	190
AC062/61-H	–	12h/520-580°C	14h/160°C	326	357
AC062/61-L	5%	170h/530°C	14h/160°C	264	292

Table III: Grain Dimensions

Alloy	Grain Size (mm)		
	L	T	S
AC050-1	0.12	0.12	0.075
AC050-2	7-70	7-70	2-5
AC050-12	7-70	7-70	2-5
AC050-13	7-70	7-70	2-5

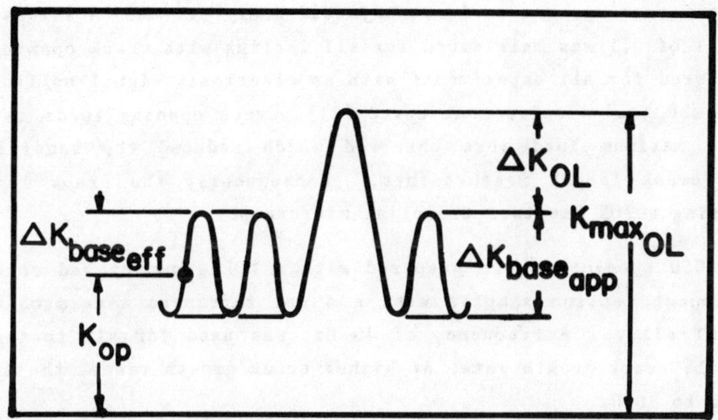

Figure 2: Overload nomenclature.

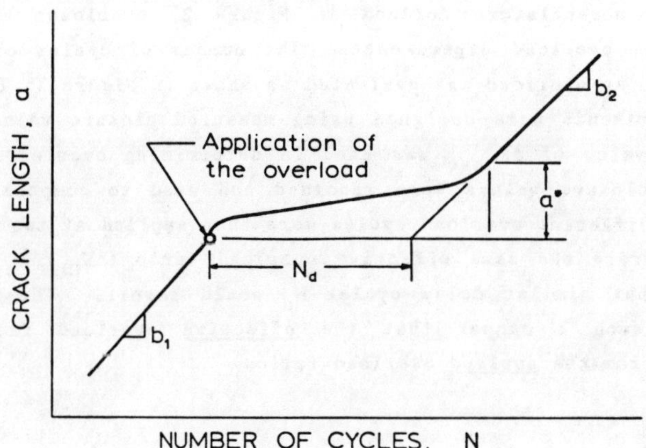

Figure 3: Method of determining N_d.

Fatigue testing

Fatigue tests were conducted on a servohydraulic fatigue testing machine under manual control. Values of ΔK_{th} were determined using K-decreasing test methods based on a constant rate of decrease C = −0.06 /mm where

$$C = \frac{1}{K} \cdot \frac{dK}{da} \tag{1}$$

During manual testing, the load was adjusted every 0.1 mm. A stress ratio($R = K_{min}/K_{max}$) of 0.1 was maintained for all testing with crack opening levels being monitored for all experiments with an electronic signal nulling scheme applied to the load-displacement curve [21]. Crack opening loads as high as 94% of the maximum load were observed which reduced the sensitivity of compliance crack length measurements. Consequently the crack length was measured using a 30X Gaertner traveling microscope.

The AC050 specimens were prepared with a WOL geometry and were 10 mm. thick. Compact tension samples with a 4 mm. thickness were machined from the AC062/61 alloy. A frequency of 40 Hz. was used for all testing below 10^{-5} mm/cycle crack growth rate; at higher crack growth rates, the frequency was lowered to 30 Hz.

The parameter used in designing the overload experiments was the applied overload ratio defined in the following equation.

$$\%OL_{app} = \Delta K_{OL}/\Delta K_{base} \tag{2}$$

Based on the nomenclature defined in Figure 2, overloads of 100% were applied in the overload experiments. The number of cycles of delay (N_d) resulting from an overload was evaluated as shown in Figure 3. The effective overload experiments were designed using measured closure values such that an effective value of ΔK_{base} was used in determining overloads. In these experiments, closure values were recorded and used to compute ΔK_{eff}(i.e., $K_{max} - K_{op}$). Different overload cycles were then applied at two ΔK_{eff} levels so as to generate the same effective overload ratio ($\Delta K_{OL}/\Delta K_{eff}$) with the expectation that similar delay cycles N_d would result. These tests were designed in such a manner that the <u>effective</u> overload ratio differed dramatically from the <u>applied</u> overload ratio.

Results and Discussion

Near-threshold fatigue results

Constant amplitude fatigue crack growth rate data were determined for the small and large grain materials with primary interest in the near-threshold regime. These low stress results (R = 0.1) are shown in Figure 4. Additional data for R = 0.3, 0.5, and 0.8 were also obtained for the small-grain material. Transients, such as the one that is obvious at $\Delta K = 17\ MPa\sqrt{m}$ for AC050-2, were observed for all three large-grain materials. The transients in fatigue crack growth rates probably result from the change in crystallographic direction at the grain boundaries, which produced the facetting. The location of grain boundaries and secondary cracks correlated with the crack length associated with the transients in fatigue crack growth rates. The appearance of the fracture surfaces observed in an ETEC scanning electron microscope showed large-scale facets which could be associated with individual grains. In addition, secondary cracking was observed on the polished face of the large-grain samples during the fatigue experiments as well as on the fracture surfaces as noted with the aid of the scanning electron microscope.

Four threshold experiments on the small-grain material gave threshold values which increased with decreasing stress ratio as expected [22]. Figure 5 shows threshold results from Stofanak _et al_. for cast and extruded aluminum alloys and reveals that ΔK_{th} decreases with increasing stress ratio (R) [23]. The threshold experiments on the three large-grained materials provided threshold values between 4.8 and 5.2 $MPa\sqrt{m}$ whereas a threshold value of 3.1 $MPa\sqrt{m}$ was observed at low stress ratio for the fine-grained material. It can be seen that these data fit the trend lines reported earlier. The threshold values increase with grain size which agrees with the trends noted by Yoder _et al_. [2] and others [3, 24, 25]. However, Higo _et al_. have presented results for copper and copper-aluminum alloys for which ΔK_{th} increases with decreasing grain size. Beevers [6] noted that these copper and Cu-Al alloys have low yield stresses (40-180 MPa) and large grain sizes (20-400 µm). The threshold stress intensity increased with increasing monotonic yield stress (as did the results of this report). Beevers attributed this discrepancy to a reverse plastic zone which was much smaller than the grain size. Similarly, the AC050 large grain materials have a grain size which is much larger than the reverse plastic zone, yet the threshold values increased with increasing grain size. It is suggested that the relevant barrier to fatigue crack growth near threshold for these large-grain materials may well involve a smaller microstructural dimension than the actual grain boundary.

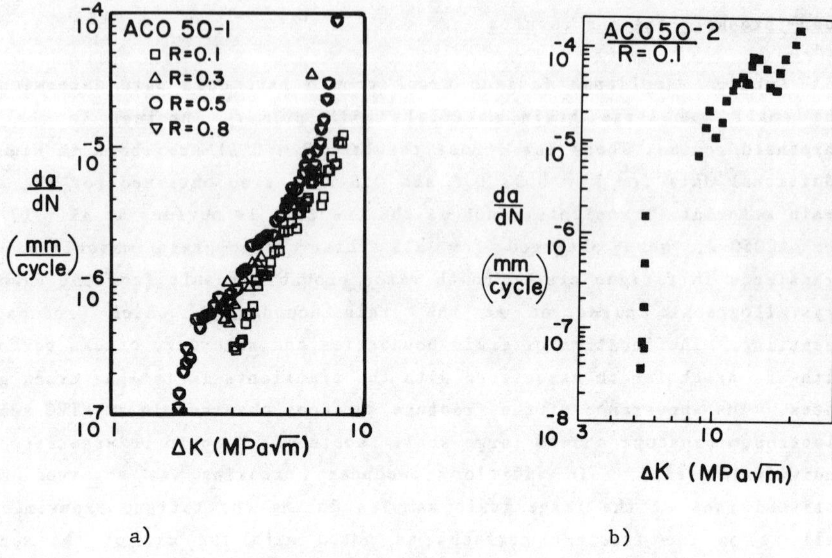

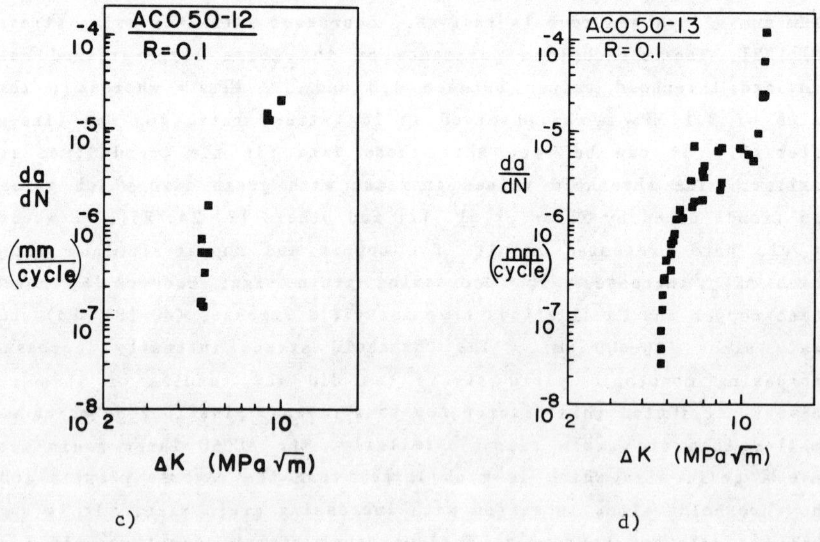

Figure 4: Fatigue crack propagation results: a) AC050-1, b) AC050-2, c) AC050-12, and d) AC050-13.

Yoder et al. developed a correlation for ΔK_{th} as a function of microstructural parameters with the following form:

$$\Delta K_{th} = k \cdot \sigma_{ys}\sqrt{\bar{\ell}} \quad (3)$$

where $\bar{\ell}$ is the relevant microstructural parameter [2]. Yoder noted the importance of choosing the correct microstructural parameter. Using this type of correlation and taking the ratio of σ_{ys} d for the two grain sizes of AC050, a factor of twenty between values of ΔK_{th} is expected. The difference between the observed threshold values for the two grain sizes is actually 1.7. Again a smaller microstructural parameter may be more relevant in comparing test data from the large-grained material with data from the relatively fine-grained samples.

Metallographic examination showed no organized substructure in the fine-grained material while subgrains with a mean diameter (d_s) of 0.4 mm were observed in the AC050-2 coarse-grained material. Foils of slightly-deformed fine-grained material were also examined in a transmission electron microscope and revealed a homogeneous dislocation distribution along with no evidence for a subgrain structure. Since subgrain boundaries may be viewed as locked arrays of dislocations, the subgrain boundary was chosen to represent the effective barrier to dislocation movement in the coarse-grained material. On the other hand, the major grain boundary was believed to represent the effective slip barrier in the fine-grained material since no subgrain boundaries were observed in the latter material. On this basis, a prediction of the ratio of ΔK_{th} for these two materials based on $\sigma_{ys}\sqrt{d}$ and $\sigma_{ys}\sqrt{d_s}$ for the fine and coarse-grained materials, respectively, gave an expected ratio of ΔK_{th} values of 1.4, which is much closer to the observed ratio of 1.7.

In order to explore further the correlation between threshold values and associated yield strengths and grain sizes, data were collected from the literature and compared with the AC050 results. These data for copper, copper-aluminum, steels, titanium, and nickel-base alloys [3, 8, 11, 23, 24, 25, 26] are plotted in Figure 6. Data from the transition (ΔK_T) between the near-threshold and power-law regimes were included as they have been associated with the threshold values. Two branches can clearly be seen in this figure. A closer examination of the data reveals that the upper branch is almost entirely composed of ΔK_T transition values, whereas the lower branch contains only threshold (ΔK_{th}) values. The copper and Cu-Al data of Higo et al. form a short U-shaped region encompassing the lower ends of both branches. The AC050 data which are the circled points, fit in the threshold branch of the curve. The two independent branches of the curve suggest two ideas. First, a correlation

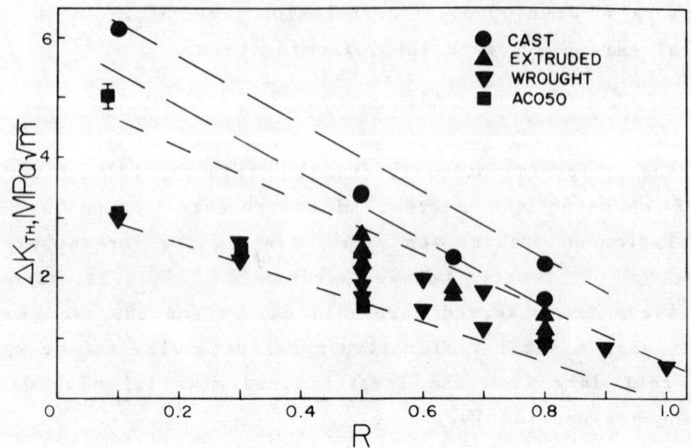

Figure 5: Threshold values as a function of R for cast, wrought and extruded aluminum alloys after Stofanak et al. [23]. Threshold values for AC050 are included.

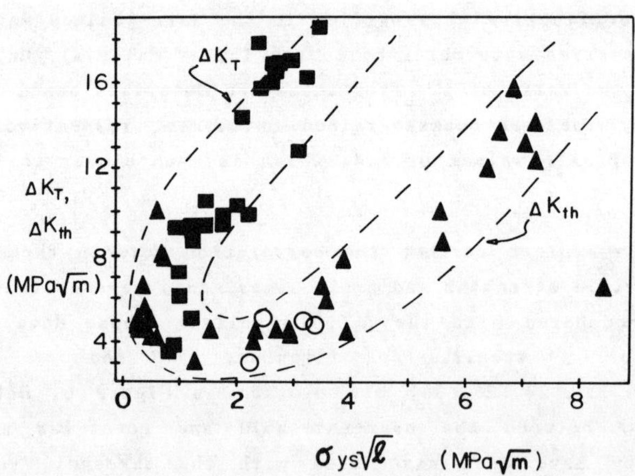

Figure 6: Threshold and transition values as a function of $\sigma_{ys}\sqrt{\ell}$ for steels, copper, copper-aluminum, titanium, and aluminum alloys [3,8,11,23,24,25,26]. Threshold results for AC050 are included (O).

similar to Yoder's for the transition regime may be reasonable also for threshold data. A linear least squares fit to the data gives a correlation of the following form:

$$\Delta K_{th} = 2.4\ \sigma_{ys}\sqrt{\ell} - 3.8 \qquad (4)$$

Secondly, the data shown in Figure 6 suggest that the slopes of da/dN - ΔK plots in the threshold region between ΔK_{th} and ΔK_T are sensitive to mechanical properties and microstructure as described by the parameter $\sigma_{ys}\sqrt{\ell}$.

Overload-induced delay

The results for the overload experiments are presented in Table 4 and reveal the number of cycles of delay (N_d) resulting from applied overloads as a function of ΔK_{base} and grain size. Since the plastic zone size resulting from the maximum stress intensity due to the overload ($r_{y_{OL}}$) is a relevant parameter for evaluating overload-induced delay, the number of cycles of delay is plotted as a function of plastic zone size in Figure 7. For a given plastic zone size, it is seen that longer delays occur in the large-grained samples than in the finer-grained specimens; consequently, a single correlation between $r_{y_{OL}}$ and cyclic delay is inadequate. Vecchio et al. [27] noted that a good correlation of delay cycle data could be obtained when $r_{y_{OL}}$ is normalized with respect to the some relevant microstructural parameter. Accordingly, the delay data obtained in this study were normalized with the relevant microstructural parameter for each material (i.e., subgrain size in the large-grained samples and the grain size in the small-grained samples)and the data shown in Figure 8.

With the exception of the AC050-1 data points on the right-hand side of this plot, the cyclic delay increased with decreasing $r_{y_{OL}}/\ell$ in agreement with the earlier findings for several steel and aluminum alloy reported by this laboratory. The data points on the right-hand side of the figure are influenced by the plane strain/plane stress transition as discussed by Vecchio et al. [20] and Mills and Hertzberg [17].

Based on the recent results of Vecchio et al. [20] and the supportive findings of the present study, attempts were made to analyze overload results reported by others. In so doing, the generality of the influence of grain size and yield strength on delay cycles after an overload appears to have been established. For example, Antolovich and Jayaraman [28] noted that the amount of cyclic delay in Waspaloy following a 50% overload decreased with increasing ΔK level (plane strain controlled behavior) and varied with grain size and γ' particle size as shown in Figure 9. The data from this

Table IV: Overload-Induced Delays

Sample	ΔK_{base} (MPa$\sqrt{m}$)	$\%OL_{app}$	N_d (10^3)	$r_{y_{OL}}$ (mm)	$r_{y_{OL}}/d$	$r_{y_{OL}}/d_s$
1/6	5.0	100	> 300	0.36	1.06	
1/6	7.0	100	65	0.72	3.8	
1/5	10.0	100	10-15	1.5	15.4	
1/5	12.0	100	20	2.1	22.2	
1/6	15.0	100	175-200	3.3	35.0	
2/13	12.5	100	59	2.4	0.15	6.0
2/13	15.0	100	80	3.4	0.21	8.5
12/34	10.0	100	520	0.82	0.051	0.21
12/11	12.5	100	> 294	1.3	0.080	3.3
13/25	6.0	100	> 3 085	0.33	0.020	0.83
13/25	7.0	100	2 900	0.45	0.028	1.1
13/25	8.25	100	360	0.63	0.039	1.6

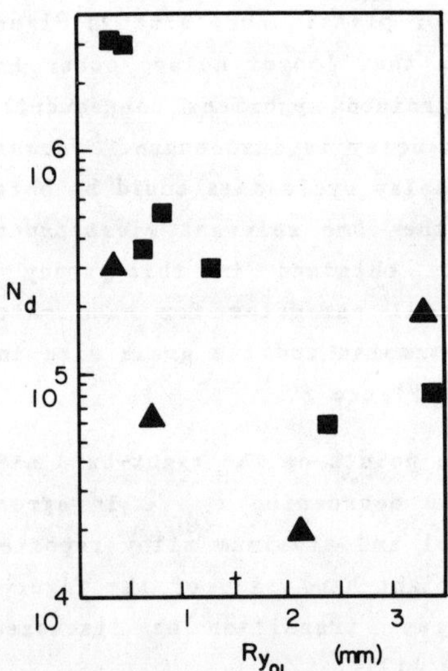

Figure 7: Overload-induced delay as as a function of the overload plastic zone size. Different data points correspond to results with various grain sizes.

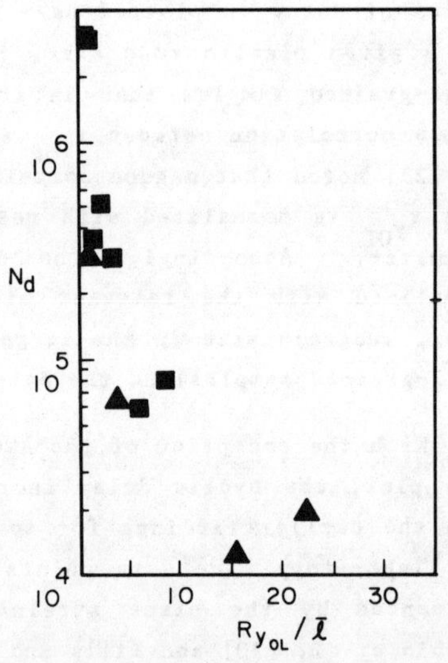

Figure 8: Overload-induced delay as a function of the ratio of plastic zone size to relevant microstructural parameter.

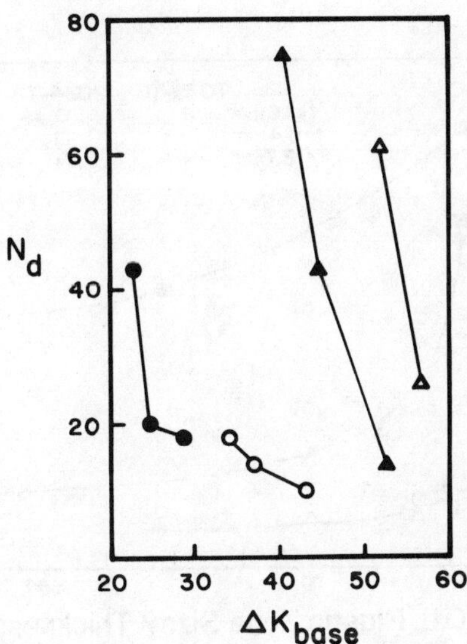

Figure 9: Tensile overload data for Waspaloy after Antolovich and Jayaraman [29]. The overload ratio is 1.5.

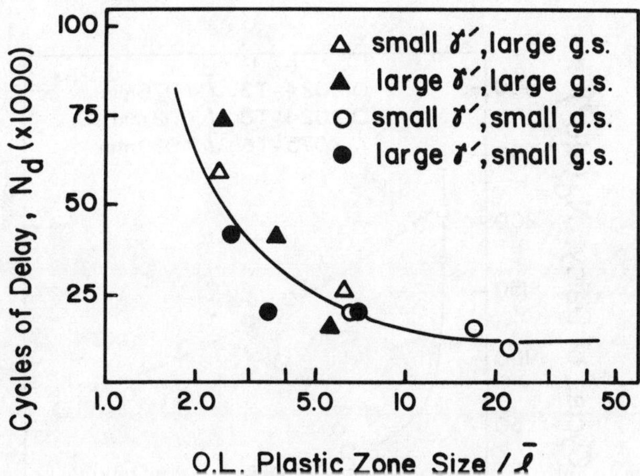

Figure 10. Cyclic delay following 50% overload. Data from Figure 9 normalized with respect to overload plastic zone size/grain size ratio.

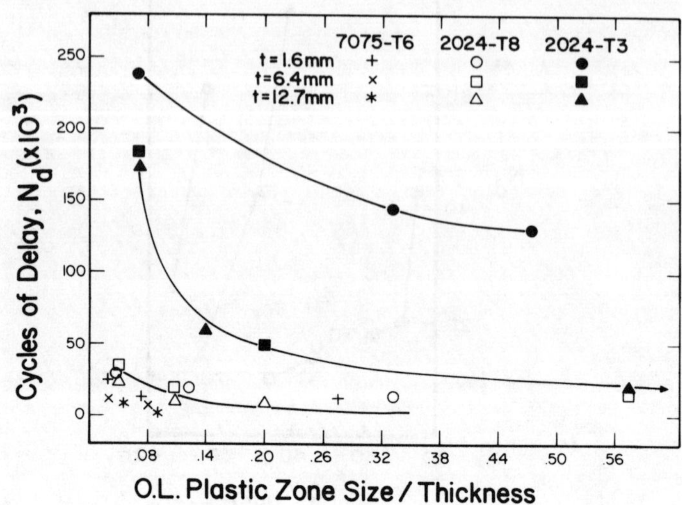

Figure 11: Cyclic delay versus the overload plastic zone size to sheet thickness ratio for several aluminum alloys (low to intermediate ΔK levels). (Data from Chanani [30].)

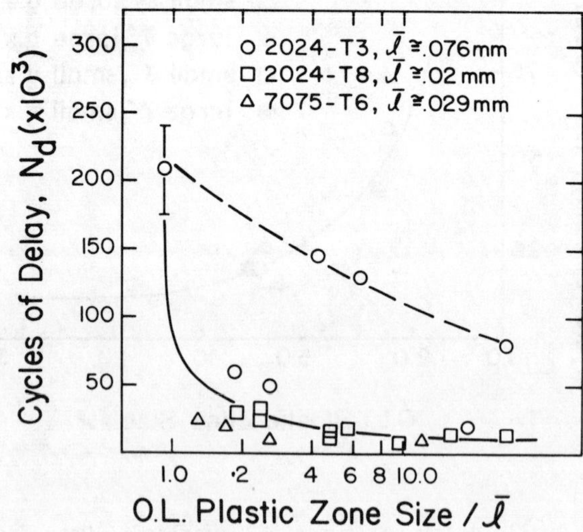

Figure 12: Cyclic delay as a function of the overload plastic zone size normalized with respect to grain size. (Data from Figure 11.)

paper were analyzed with the cyclic yield strength (approximated by the flow stress) the latter measure of yield strength having been chosen in recognition of the significant amount of cyclic strain hardening exhibited by materials of this type. The overload plastic zone was then computed and normalized by the grain size with the results shown in Figure 10. The normalization of these data is most encouraging in that it supports our fatigue model as it relates to the combined influence of microstructure and overload cycles on the fatigue of engineering alloys.

To further illustrate the influence of microstructure on cyclic delay in the low growth rate regime, the data from Chanani [29] were re-examined. Figure 11 shows the cyclic delay behavior for 2024-T3, 2024-T8, and 7075-T6 as a function of $r_{y(OL)}/t$. It is apparent that there is no unique curve that describes all the data. If these data are now plotted as a function of $r_{y(OL)}/\bar{\ell}$ where $\bar{\ell}$ is the mean grain diameter, a shift toward a single curve is observed (Figure 12). Note however, that there is some lack of correlation for the results obtained from specimens with very thin sections (i.e. t = 1.6 mm). It is believed that this lack of correlation is related to the enhanced plasticity effects in the very thin sections (i.e. contributions from plane stress or tensile displacement effects) as discussed elsewhere [20].

Finally, the current overload data as well as the data of Antolovich and Jayaraman, Chanani, and Vecchio et al., are plotted in Figure 13. We conclude that overload-induced delay is normalized by the ratio of overload plastic zone size to the relevant microstructural barrier, the latter often though not always being the grain size.

Effective Overloads

Based on the preceeding results, an attempt was made to rationalize the increasing amounts of delay with decreasing ΔK_{base} as one approaches ΔK_{th}. We noted previously [27] that this trend could be attributable to important changes in the effective overload ratio in the region where crack closure changes significantly. Accordingly, it was speculated that similar amounts of delay would occur in conjunction with different applied overload ratios, so long as the effective overload ratio was the same. As defined by Elber [14], the effective ΔK value that drives a fatigue crack is given by $\Delta K_{eff} = K_{max} - K_{op}$. The effective overload ratio may then be defined as:

$$\%OL_{eff} = \Delta K_{OL}/\Delta K_{base_{eff}} \tag{5}$$

The results of the three effective overload experiments are shown in Table 5. Two of the experiments were designed so that 100% effective overloads

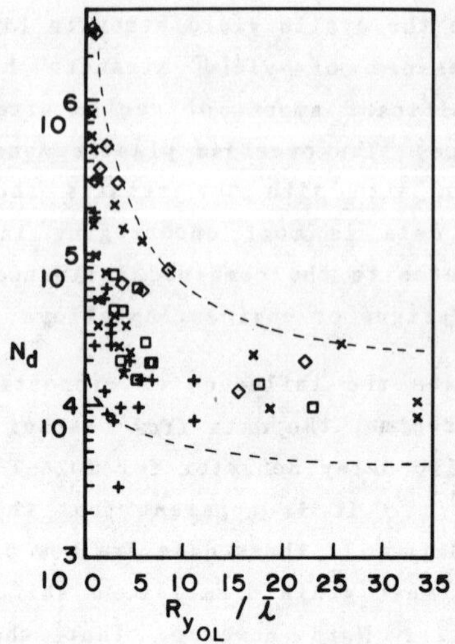

Figure 13: Overload-induced delay normalized with respect to grain size. AC050 data and data from Antolovich and Jayaraman, Chanani, and Vecchio et al. Data points at right are believed to be influenced by the plane strain-plane stress transition effects.

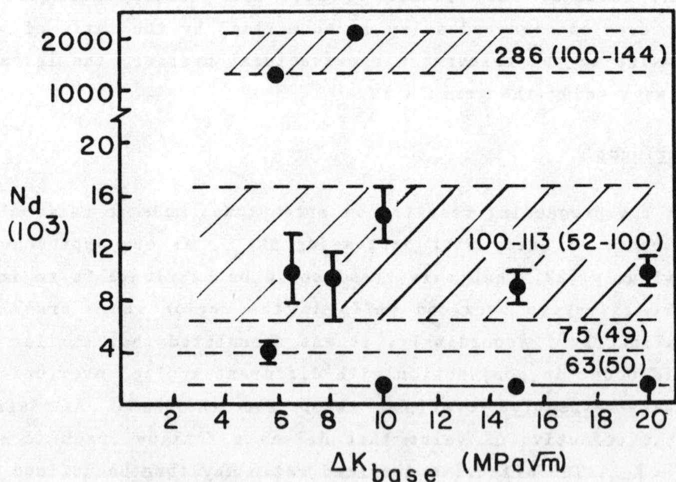

Figure 14: Cyclic delay from effective overload experiments plotted vs. ΔK_{base}. Effective overload ratios are noted with the applied overload ratios indicated in parentheses.

Table V: Effective Overload Results

Alloy	ΔK_{base} (MPa m)	$\%OL_{app}$	$\%OL_{eff}$	N_d (cycles)	$\frac{K_{op}}{K_{max\ base}}$
AC062/61-H	6.3	54	100	7600	50%
	6.3	54	100	12900	50%
	6.3	54	100	10000	50%
	8.1	82	100	11700	26%
	8.1	82	100	10000	26%
	5.6	46	75	3000	45%
	5.6	46	75	4700	45%
	8.1	92	100	6200	17%
	8.1	92	100	10400	17%
AC062/61-L	6.0	100	286	$1.3x10^6$	72%
	9.0	144	286	$2.1x10^6$	50%
Other AC062/61 results after Vecchio *et al*.[20] and Stofanak *et al*.[23]					
AC062/61-H	10	100	113	16500	15-25%
	10	100	113	12200	15-25%
	10	100	113	15300	15-25%
	10	100	113	13700	15-25%
	15	100	113	10100	15-25%
	15	100	113	8400	15-25%
	15	100	113	9600	15-25%
	15	100	113	7600	15-25%
	20	100	113	11100	15-25%
	20	100	113	8800	15-25%
	10	50	63	1350	15-25%
	15	50	63	1700	15-25%
	15	50	63	1040	15-25%
	20	50	63	1600	15-25%
	20	50	63	1400	15-25%
	20	50	63	1200	15-25%

were applied at two different values of ΔK_{base}. The third experiment involved a 100% applied overload at a ΔK_{base} of 6.3 $MPa\sqrt{m}$ and an equivalent effective overload at an applied ΔK_{base} of 9.0 $MPa\sqrt{m}$. Reasonable agreement is found between the overload-induced delay in each experiment.

Additional overload experiments from Vecchio _et al_. are also reported in Table 5. These results were not originally intended to be experiments evaluating effective overload phenomena however, as noted by Stofanak _et al_. K_{op}/K_{max} is approximately the same for ΔK_{base} greater than 8 $MPa\sqrt{m}$. Therefore, the effective and applied overload ratios could be computed from these data. All of these results are plotted in Figure 14 as a function of applied ΔK_{base}. It is seen that changes in delay cycles result from changes in the _effective_ overload ratio. Furthermore, the same amount of delay results from different _applied_ overload ratios so long as the _effective_ overload ratio remains essentially unchanged. The extent of cyclic delay is compared in Figure 15 as a function of both effective and applied overload ratios. Much less scatter in results is associated with the _effective_ overload ratios as compared to the applied overload ratios. Whereas the linear fit between cyclic delay and the effective overload ratio may be fortuitous, this relationship certainly identifies certain factors responsible for overload-induced delay phenomena. The effective overload experiments show that the influence of grain size on overloads below the plane strain-plane stress transition can be explained in terms of closure considerations and the effective overload ratio.

Conclusions

The following conclusions can be drawn from the present investigation:

1. Overload-induced delay is effectively normalized by the ratio of the overload plastic zone size to the relevant microstructural parameter. This parameter is frequently the grain size but for the large grain materials involved in these experiments, the subgrain size appears to be the effective barrier to fatigue crack growth.

2. The increase in delay with decreasing ΔK_{base} values that results from an overload cycle can be rationalized in terms of an increasing _effective_ percentage overload.

3. Constant amplitude threshold results increase with grain size. For the small-grained materials the relevant microstructural parameter was the grain size, while the subgrain size was more relevant for the large-grained samples.

Acknowledgments

Financial support from the Swiss Aluminium Company and from AFOSR Grant No. 83-0029 are gratefully acknowledged. Alusuisse also provided tensile test results. Professional Services Group - Ocean City Research Company provided

the electronic nulling scheme for evaluating closure effects. Thanks are also due to Dr. Jeffrey Crompton for performing the TEM analysis.

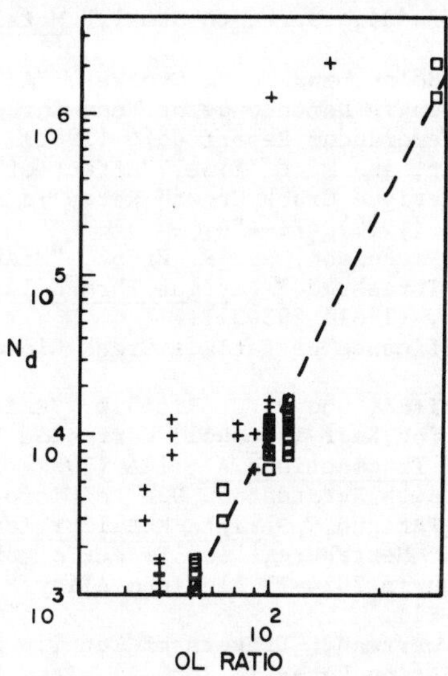

Figure 15: The effect of applied (+) and effective (□) overload ratios on cyclic delay. Note better correlation with effective overload ratio data.

References

1. G. W. J. Waldron, A. E. Inkle, and P. Fox, "Application of SEM to the Study of Surface Topography of Fatigue Fracture," Proceedings of the Third Annual Scanning Electron Microscope Symposium, (1970) 297-312.

2. G. R. Yoder, L. A. Cooley, and T. W. Crooker, "Quantitative Analysis of Microstructural Effects of Fatigue Crack Growth in Widmanstatten Ti-6Al-4 and Ti-8Al-1Mo-1V, Engineering Fracture Mechanics, 11 (1979) 805-816.

3. G. M. Lin and M. E. Fine, "Effect of Grain Size and Cold Work on the Near Threshold Fatigue Crack Propagation Rate and Crack Closure in Iron," Scripta Metallurgica, 16 (1982) 1249-1254.

4. J. Masounave and J.-P. Bailon, "Effect of Grain Size on the Threshold Stress Intensity Factor in the Fatigue of a Ferritic Steel," Scripta Metallurgica, 10 (1976) 165-170.

5. G. T. Gray, III, J. C. Williams, and A. W. Thompson, "Roughness-induced Crack Closure: An Explanation for Microstructurally Sensitive Fatigue Crack Growth," Metallurgical Transactions A, 14A (1983) 421-433.

6. C. J. Beevers, "Some Aspects of the Influence of Microstructure and Environment on ΔK Thresholds," Fatigue Thresholds, Fundamentals and Engineering Applications, (1981) 257-275.

7. M. F. Carlson and R. O. Ritchie, "On the Effect of Prior Austenite Grain Size on Near-Threshold Fatigue Crack Growth," Scripta Metallurgica, 11 (1977) 1113-1118.

8. Y. Higo, A. C. Pickard, and J. F. Knott, "Effects of Grain Size and Stacking Fault Energy on Fatigue Crack Propagation Thresholds in Cu-Al Aluminum Alloys," Metal Science, 15 (1981) 233-239.

9. Y. Mutoh and V. M. Radhakrishnan, "An Analysis of Grain Size and Yield

Stress Effects on Stress at Fatigue Limit and Threshold Stress Intensity Factor," J. Engineering Materials and Technology, 103 (1981) 229-233.

10. R. O. Ritchie, "Influence of Microstructure on Near-Threshold Fatigue-Crack Propagation in Ultra-High Strength Steel," Metal Science, 11 (1977) 368-381.

11. G. R. Yoder, L. A. Cooley, and T. W. Crooker, "A Critical Analysis of Grain Size and Yield Strength Dependence of Near-Threshold Fatigue-Crack Growth in Steels," NRL Memorandum Report 4576 (1981).

12. M. Zedalis, L. Filler, and M. E. Fine, "Effect of Purity and Dispersoid Type on Near Threshold Fatigue Crack Growth Rates in Al-Zn-Mg-Cu Alloys," Scripta Metallurgica, 16 (1982) 471-474.

13. A. J. Cadman, C. E. Nicholson, and R. Brook, "Microstructural Effects at and Near the Fatigue Threshold," Fatigue Thresholds, Fundamentals and Engineering Applications, (1981) 293-311.

14. W. Elber, "The Significance of Fatigue Crack Closure," ASTM STP 486, (1971) 230-246.

15. S. Suresh, G. F. Zamiski, and R. O. Ritchie, "Oxide-Induced Crack Closure: An Explanation for Near-Threshold Corrosion Fatigue Crack Growth Behavior," Metallurgical Transactions A , 12A (1981) 1435-1443.

16. S. Suresh, "Crack Growth Retardation Due to Micro-Roughness: A Mechanism for Overload Effects in Fatigue," Scripta Metallurgica, 16 (1982) 995-999.

17. W. J. Mills and R. W. Hertzberg, "The Effect of Sheet Thickness on Fatigue Crack Retardation in 2024-T3 Aluminum Alloy," Engineering Fracture Mechanics, 7 (1975) 705-711.

18. R. D. Brown and J. Weertman, "Effects of Tensile Overloads on Crack Closure and Crack Propagation Rates in 7050-Aluminum," Engineering Fracture Mechanics, 10 (1978) 867-878.

19. R. P. Wei and T. T. Shih, "Delay in Fatigue Crack Growth," International Journal of Fracture, 10 (1974) 77-85.

20. R. S. Vecchio, R. W. Hertzberg, and R. Jaccard, "Overload Induced Crack Growth Rate Attenuation Behavior in Aluminum Alloys," Scripta Metallurgica, 17 (1983) 343-346.

21. A. Saxena, S. J. Hudak, Jr., J. K. Donald, and D. W. Schmidt, "Computer-Controlled Decreasing Stress Intensity Technique for Low Rate Fatigue Crack Growth Testing," J. Testing and Evaluation, 6 (1978) 167-174.

22. R. W. Hertzberg, Deformation and Fracture Mechanics of Engineering Materials, John Wiley and Sons, (1983), 2nd edition.

23. R. J. Stofanak, R. W. Hertzberg, G. Miller, R. Jaccard, and J. K. Donald, "On the Cyclic Behavior of Cast and Extruded Aluminum Alloys, Part A: Fatigue Crack Propagation," Engineering Fracture Mechanics, 17 (1983) 527-539.

24. J. E. King, "Effects of Grain Size and Microstructure on Threshold Values and Near Threshold Crack Growth in Powder-Formed Ni-base Super-Alloy," Metal Science, 16 (1982) 345-355.

25. Y. Nakai, K. Tanaka, and T. Nakanishi, "The Effects of Stress Ratio and Grain Size on Near-Threshold Fatigue Crack Propagation in Low-Carbon Steels," Engineering Fracture Mechanics, 15 (1981) 291-302.

26. G. R. Yoder, F. H. Froes, and D. Eylon, "Effect of Microstructure, Strength, and Oxygen Content on Fatigue Crack Growth Rate of Ti-4.5Al-5.0 Mo-1.5Cr (Corona 5)," submitted to Metallurgical Transactions A.

27. R. S. Vecchio, R. W. Hertzberg, and R. Jaccard, "On the Overload-Induced Fatigue Crack Propagation Behavior in Aluminum and Steel Alloys," submitted to Fatigue of Engineering Materials and Structure.

28. S. D. Antolovich and N. Jayaraman, "The Effect of Microstructure on the Fatigue Behavior of Ni Base Superalloys," Fatigue, Environment, and Temperature Effects, (1982) 119-143.

29. G. R. Chanani, "Effect of Thickness on Retardation Behavior of 7075 and 2024 Aluminum Alloys," ASTM STP 631 (1977) 365-387.

CONCEPTS OF FATIGUE CRACK GROWTH THRESHOLDS

GAINED BY THE ULTRASOUND METHOD

S. E. Stanzl* and H. M. Ebenberger

University of Vienna, Institute of Solid State Physics, A 1090 Wien, Austria

* Swiss Federal Institute of Technology Zurich, Institute of Metallurgy

CH 8092 Zürich, Switzerland, during 1983/84

The ultrasound resonance method has shown to be very useful for fracture mechanics research and especially advantageous in studying fatigue crack growth in the near-threshold regime, as high numbers of cycles are gained within short testing times. Experimental technique and results are reviewed. In addition new measurements of the environmental influence on fatigue crack propagation at ultrasonic frequency are discussed in the light of fracture-surface roughness-induced and viscous fluid-induced crack closure. Obviously these phenomena are effective besides hydrogen embrittlement, anodic dissolution and pronounced environment-grain boundary interaction.

Introduction

Whether a small defect or an already existing short crack in some structural component will degenerate into a growing crack under fatigue loading is a problem of high practical importance and therefore has been considered extensively for many years. The emphasis has shifted from a defect-free design philosophy to a defect-tolerant one, as fracture mechanical principles became better known. However, the problem of a quantitative characterization of "short cracks" and a correlation of endurance limit and threshold stress intensity is still lacking.

Measurements of the endurance limit and threshold values need extremely long testing times. Therefore raising the testing frequency has been discussed and tried since fatigue tests were performed; the extreme case of using 20 000 Hz ultrasound for measuring S-N curves was first published by Neppirras in 1960 (1). The use uf the ultrasound method for fracture mechanical measurements has not been reported before 1973, when Mitsche et al. (2) gave a detailed description of the ultrasound method for measuring $\Delta a/\Delta N = f(\Delta K)$ curves and presented first crack growth data, with the characteristic deflection of the crack growth curve at low ΔK values pointing to the existence of a threshold value. At the same time, Purushothaman et al. (3) also pointed out the advantages of the ultrasound method for threshold measurements and published crack growth curves in 1978 (4). Since that time several other investigators adopted the method for their own measurements (5,6).

Until now the following information has been gained with the ultrasound method (2): The method is appropriate for rapid measurements of fatigue crack growth rates with time and energy savings. It is especially advantageous for measurements in the threshold regime, as high numbers of cycles are obtained within short testing times and therefore make possible to study such low crack growth rates that cannot be gained in conventional tests.

In addition, it was found that no pronounced frequency effect in crack growth exists (7) which is of high practical importance, as ultrasound tests therefore can complete low frequency tests or may even replace them: the crack growth curves of mild steel and chromium steel are the same for 200 Hz and 20 kHz, if non-corrosive environment is used (7). A similar result was gained by Hoffelner (6) who compared the crack growth rates of some heat-resisting Ni-alloys at 2.3, 30 and 20 000 Hz. On the other hand, Puskar (8) found that the crack growth rates of low-carbon steel were about 10 times smaller for 22 kHz than those for 70 Hz. In studies with various metallic materials (7) a series of further similarities in the behavior of these materials at 200 Hz and 20 kHz loading were found, for example: ductile transcrystalline fracture in non-corrosive environment at room-temperature, ledgelike crack propagation at low stress intensities and the existence of a threshold stress intensity in non-corrosive environment at 293 and 77 K for polycrystalline materials and for copper single-crystals with a defined orientation (7). On the other hand, Purushothaman and Tien (4) observed that no threshold existed in a monocrystalline Ni-alloy.

Experimental Procedure

Details of using the ultrasound method for fracture mechanical measurements have been described in (2) already. Essentially the same testing technique is applied today by other authors also (5,6).

The loading technique is the same as used for smooth specimens by Mason since 1958 (9). Specimens with a length of $\lambda/2$ of the longitudinal 20 kHz wave are vibrating in resonance, thus producing a standing sinusoidal wave with strain and stress maximum in the centre of the specimen, and dispacement maxima at its ends. Loading is of fully reversed push-pull type (R = -1), recently however, efforts have been undertaken to verify stressing with changed load ratios (10-12).

For fracture mechanical studies, flat specimens are single-edge (2,7) or double-edge or centre (5) notched; a new specimen shape is being developed now (13). Crack growth is recorded optically by watching the polished specimen surface with a camera, which is attached to a microscope (2,7). In previous works (e.g. 2), a high-speed camera with frame-speeds up to 6000 frames per second was used. Nowadays video-cameras are applied, which work with exposure times of 1/30 of a second. Observation of the specimen surfaces is possible with magnifications between 50 and 3000. This allows measuring not only crack increments of a few micrometers, but also a study of slip band generation, crack initiation processes and short crack behavior. In addition measurements of displacement amplitude, number of cycles and other parameters are introduced into the video-record synchronously, as no exact correlation of these data with the crack length is otherwise possible at this high loading frequency. Therefore, tests which were performed later (5), using only a travelling microscope and without applying a camera, do not render reliable results.

Improvements of the testing technique have been reported recently (14). With the aid of a computer-control device variable-amplitude tests can be performed. The scheme of this equipment is shown in Fig. 1.

The control unit works with an electrodynamic amplitude gauge consisting of a permanent magnet and a coil, which allows measuring mechanical vibrations without touching the specimen or ultrasound horn. In order to obtain adequate accuracy, this assembly is attached to the objective-lens of a microscope so that the distance between horn and gauge is kept constant (10). The signal of the amplitude gauge is magnified by a factor of 50 up to several Volts and used twice. First it renders the displacement amplitude, which is observed by an oscilloscope and which yields $\Delta\varepsilon$ and $\Delta\sigma$ values after optical calibration. Second, it is used for feedback control: The signal is rectified and serves as control signal for the PID (proportional integral differential) controller. It is digitized with an A/D (analogous-digital) converter and then fed to the computer. The PID controller is designed in such a way, that a compromise between quick control response and small overshoot of the vibration amplitude is attained. The computer sends a specified signal in Volts as preset value. Both values, i.e. actual voltage measured with the magnet-coil gauge and preset value are displayed in digits. Control of the displacement amplitudes however also is possible by using a constant-voltage source and a potentiometer instead of the computer.

In 20 kHz tests one cycle lasts 50 μ sec only, and a pulse of 1000 cycles (which is mostly used in order to avoid overheating of the specimens due to damping) lasts 50 m sec. Therefore a PLL-(phase lock loop)frequency counter is applied (15). The measurements are read by a 7-segments-display and received in binary form by the computer. An assembly of a 4-digits counter, which may be initiated by appropriate software and a 20 kHz oscillator make it possible to start the measuring process after a defined frequency is attained and also to stop the test, when a certain frequency or frequency-change between two or more pulses has been reached. This device allows to define a certain amount of "damage", as the resonance frequency of a specimen changes with increasing damage; similarly it may be used as a crack-detector.

The cycle counter consists of two 4-digit pre-adjustable counters. An 1 : 10 division counter registers numbers of cycles up to 10^9. Another 4-digit pre-counter together with a 20 kHz oscillator, serves as time-set for the pause lengths during the pulses. On the other hand, this oscillator predetermines the moment of initiation of the counter during the transient oscillation at the start of each pulse. Furthermore, this device sends a signal to the computer for an interruption, when a defined number of cycles is attained.

The measurements are recorded with a differential-input consisting of 16 channels and a 12-bit A/D converter, which works at input-voltages of 0 - 10 Volts. By means of a programmable pre-amplifier, a sensitivity of 1 mV is attained, thus allowing the recording of all needed values. Furthermore, these signals can be used for defining an interruption of the tests. The computer is a PDP 11/03 with a double-floppy discette assembly. The results are printed out and can be watched as well on a monitor during the test.

Further important steps in changing the test technique have been recently developed by Tschegg (13) using an electric device for measuring crack growth instead of the optical equipment. With the aid of this it is possible

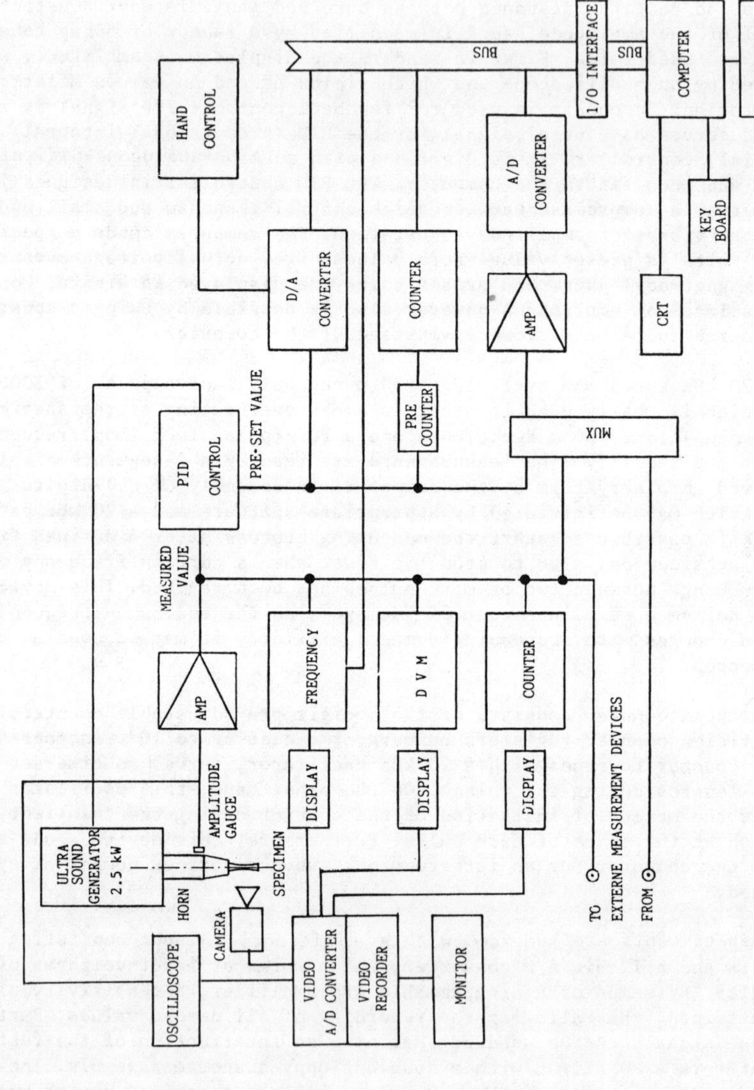

Figure 1 - Ultrasound testing machine for measuring fracture mechanical data with video equipment, computer control and automatic data recording to test not only flat specimens but also specimens of any shapes, and also meet the problems with inhomogenous crack growth. Another variant in performing ultrasonic fracture mechanics tests has been made possible by superimposing static loads (6,11), changing the mean stress (12) or by performing bending-tests (11).

Ultrasound tests are usually displacement controlled and a magnet-coil assembly is used for measuring displacements as mentioned above (10). From the displacements Δu the maximum strains $\Delta\varepsilon = 2\pi\Delta u/\lambda$ in the centre of the specimens are calculated for the unnotched case, and from this the stresses are $\Delta\sigma$, according to Hookes law: $\Delta\sigma = E\cdot\Delta\varepsilon$, assuming that the overall plastic deformation is negligible. For calculation of the cyclic stress intensity ΔK, only the tensile part of $\Delta\sigma$ is introduced, assuming that crack growth cannot take place during the compression period. This means that the effect of not fully closed cracks at zero stress is neglected. Thus $K_{max} = \sigma_{max}\sqrt{a}\cdot Y$ results, with the correction function $Y = f\ (a/W)$ taken from the literature for stress-controlled tests (16). For crack-lengths (a/W) of SEN-specimens up to about 0.4 these Y values have been found to be a useful approximation in the case of the rather long resonance specimens with $(\ell/W) \geqq 12$ (a being the crack length, ℓ the specimen length, and W the specimen width). For longer cracks or broader specimens however, the function Y of the stress-controlled test does not approximate the displacement-controlled case sufficiently and Schoeck (17) has developed correction functions for displacement controlled ultrasound tests.

Further correction factors have been suggested by Purushothaman et al. (4) to account for the resonance in ultrasonic testing and by Hoffelner et al. (18) to account for dynamic effects. A correction factor, suggested by Weiss et al. (19) for ultrasonic tests where the ligament strain is kept constant, is not further considered here in light of the improvements introduced by Schoeck (17).

Corrosion Fatigue at 20 kHz Ultrasound Loading

Fatigue cracking in corrosive environments is of high practical importance, and numerous studies have been performed in order to explain the mechanisms of corrosion fatigue cracking. According to Speidel (20), three groups of theoretical concepts may explain the environmental influences; these mechanisms are film rupture and anodic dissolution, hydrogen embrittlement and the effect of absorption of specific ions or molecules on cracking. Recently, Ritchie and Suresh (21) have discussed the concept of oxide-induced crack closure as being responsible for near-threshold crack growth behavior in air environment. It is thought that the ultrasound method would provide an elegant way for studying this question; as the various mechanisms discussed in the literature display different time-dependencies, it is expected that comparing 20 kHz results with low frequency measurements would provide some new information.

In earlier studies (22) crack propagation was compared in dehumidified silicone iol as inert environment and 3.5 % aqueous NaCl solution. It was found that the crack growth rates differ only at rates below about 10^{-9}m/cycle. As rate-limiting steps, surface diffusion (for liquids) and build-up of a gas-monolayer at the crack tip were considered. Calculations confirmed that at the very high frequency of 20 kHz an influence by the environment cannot become effective at crack growth rates above approximately 10^{-9}m/cycle.

In this work, 60 % Cu - 40 % Zn brass was loaded with the refined ultrasonic technique described above in vacuum of 10^{-3} and 1 Pa, as well as in air, distilled water, 3.5 % NaCl-water solution and silicone oil (23).

The specimen geometry was: length ℓ = 100 mm, width W = 10 mm and thicknesses d were 1.5, 2 and 3 mm; a single edge notch of wedge shape with notch radius 0.5 mm and a depth of 1 mm was applied prior to annealing in the place of maximum strain. Annealing was performed in vacuum at 650°C during one hour plus furnace cooling. The resulting grains were elongated in rolling direction and were about 140 μm long and 35 μm broad. The specimens were polished mechanically in order to produce optimal condition for identification of crack paths.

Ultrasound stressing was performed as described above. The specimens were immersed into the liquid, or put into a vacuum chamber with a roughing pump with or without a diffusion pump; the tests in air were performed with forced air at room temperature. The tests were done without interruptions, since it was desired to study time-dependent effects during the 20 kHz test. However, it is not possible to keep the temperature constant at room-temperature during the tests in vacuum without interruptions. Therefore it was measured with thermocouples in the area of maximum strain of dummy specimens. The results are compared in Fig. 2 for 1.5 and 3 mm stick specimes for ultrasound tests in vacuum of 10^{-3} Pa and air respectively.

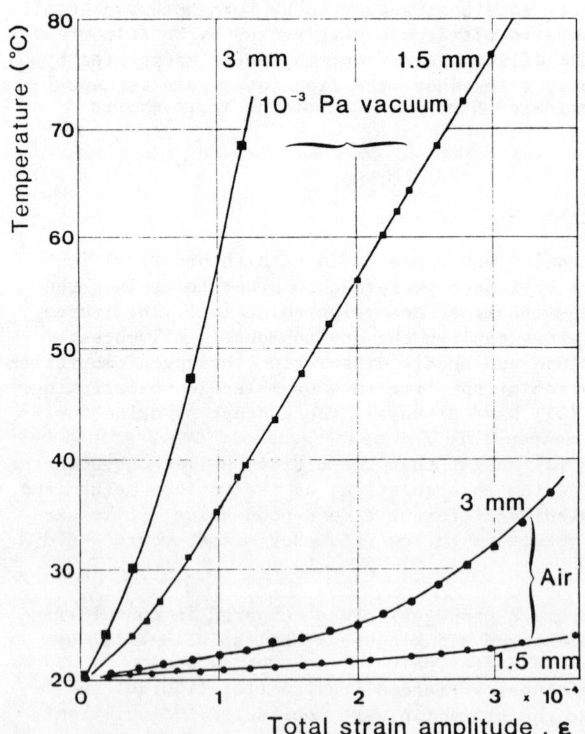

Figure 2 - Temperatures in maximum stressed area in centre of λ/2 specimens of 1.5 and 3 mm thickness during ultrasound loading in air and 10^{-3} Pa vacuum, respectively, vs. total strain amplitudes

It shows that in vacuum 50°C are not exceeded, if 1.5 mm thick specimens are stressed with strain amplitudes up to 1.7×10^{-4}, and for 3 mm thick specimens 0.7×10^{-4} is the limiting strain amplitude. For tests in air much higher amplitudes may be applied without exceeding 50°C, as shown in Fig. 2. With the aid of these curves, the crack growth experiments were performed in such a way, that the temperature was kept below 50°C, neglecting local heating at the crack tip.

In order to obtain $\Delta\sigma$ values, the displacement amplitudes Δu were measured at the specimen end as described above, and from these the $\Delta\varepsilon$ values in the centre of the specimens were calculated. Additionally $\Delta\varepsilon$ was measured directly with micro-straingages. It was found that in liquid environments and to some smaller extent even in air, the measured strains are smaller than the calculated values. Since this effect could not be observed in vacuum, it is attributed to damping of the specimens at their free end by the hardly compressible liquids. The following results consider this effect.

Crack growth rates were obtained from crack lengths between 0.2 a/W and 0.6 a/W. K_{max} was calculated according to

$$K_{max} = \sigma_{max} \cdot Y_s \cdot \sqrt{a} \cdot Y$$

where Y = f (a/W) is the correction function for finite sample width, reported in the literature for stress controlled testing (16), Y_s being Schoeck's correction function (17), which accounts for the displacement controlled test and an effectively strained sample length in the ultrasound resonance test. The Young's modulus was experimentally found by the Förster resonance technique; its magnitude is 98.1 GPa. Crack growth rates $\Delta a/\Delta N$ were determined from the monitor, which displayed the specimen surfaces, 140 times magnified. The resolution limit is 10 µm in this case. Crack increments Δa of 500 µm were evaluated. The data points do not originate from a smoothed Δa vs. ΔN curve but from individual measurements. Thus scattering is more extensive. The resulting fracture surfaces were oriented normal to the stress axis indicating plane strain conditions. Measurements of the plastic zone sizes with the recrystallization technique confirmed this assumption.

Fig. 3 shows the curves of crack propagation in three liquid environments, i.e. silicone oil, distilled water and 3.5 % NaCl water solution. They reflect a pronounced influence of the environment. The curve of crack growth in distilled water is shifted to lower K_{max} values with respect to oil at "high" K_{max} values. The curve of crack growth in NaCl is shifted to still smaller values. The threshold stress intensity is 2.7 MN $m^{-3/2}$ for both oil and distilled water. It is 1.8 MN $m^{-3/2}$ for NaCl-solution. Thresholds were found by lowering the stress intensity stepwise until no crack growth could be observed during 10^8 cycles and by then increases the stress intensity stepwise from a level below the already determined threshold value. The data points with arrows in Figs. 3, 5, 7, 9 and 10 were obtained by assuming a maximum increase of the crack length of 20 µm (which is approximately the scale of optical resolution) and dividing this by the applied number of cycles.

In Fig. 4 three typical SEM pictures of the fracture surfaces are reproduced. Fig. 4a shows the area of crack initiation in oil environment, where K_{max} was 4.2 MN $m^{-3/2}$ and $\Delta a/\Delta N = 1.7 \times 10^{-10}$ m/cycle. In Fig. 4b the environment was distilled water, K_{max} was 3.4 MN $m^{-3/2}$ and $\Delta a/\Delta K = 2.4 \times 10^{-10}$ m/cycle. In Fig. 4c with salt water, K_{max} was 2.5 MN $m^{-3/2}$ and $\Delta a/\Delta N = 3.6 \times 10^{-10}$ m/cycle. The fracture surface of loading in oil is

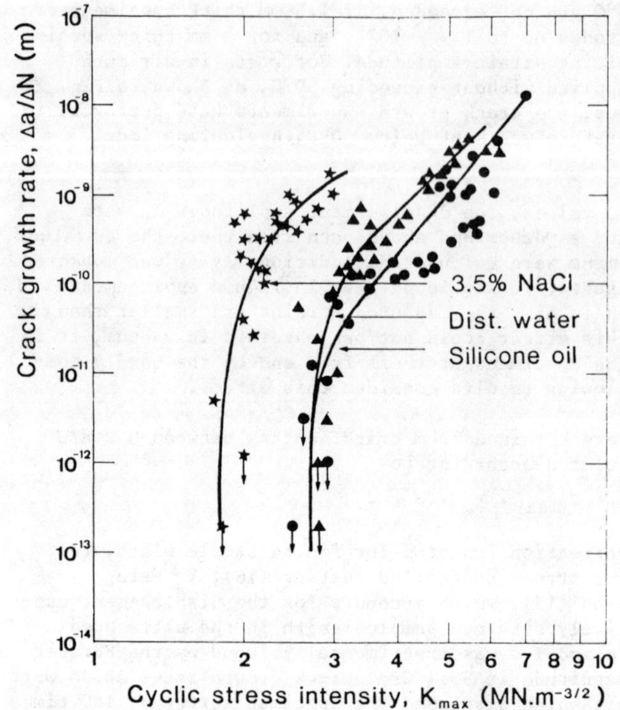

Figure 3 - Crack growth rates and thresholds of brass, which was fatigued with 20 kHz ultrasound in silicone oil, distilled water, and 3.5 % NaCl solution

charaterized by a transcrystalline, smooth appearance, whereas specimens stressed in water, display an embrittled structure and corrosion pits. In specimens stressed in a NaCl solution these features are more pronounced, and in addition intercrystalline cracking is visible. These are fracture surfaces which are characteristic for the threshold regime; at higher cyclic stress intensities however, the fracture appearance is not as different, and fatigue striations are eventually visible.

In Fig. 5 the crack propagation curves for oil and air as environments are compared. Down to stress intensities of about 4 MN $m^{-3/2}$, where the crack growth rates are about 10^{-10} m/cycle, the data points of the two curves are roughly within the same scatterband, whereas below this value the curve for air-environment points to a higher threshold stress intensity. The effect of specimen thickness on crack growth rate in air was also tested. 1.5, 2 and 3 mm thick specimens were tested and no influence was found, which means that the data points in Fig. 5 were obtained with all three specimen thicknesses. A similar result was reported in earlier measurements on mild steel (24) where 0.8 mm to 10 mm thick specimens were tested and no thickness effect was found. This contrasts the opinion of Weiss et al. (19) who dimensioned their specimens according $d \geq 2.5\ (K_{max}/R_p)^2$, which is the ASTM recommendation for static loading.

The fracture surface of a specimen loaded in air is shown with low magnification in Fig. 6a. Crack propagation direction was from bottom to top. At the bottom and in the centre two horizontal bands are clearly visible, showing

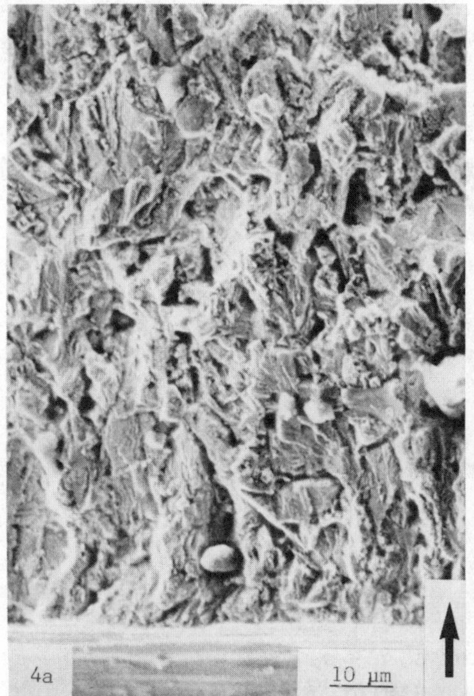

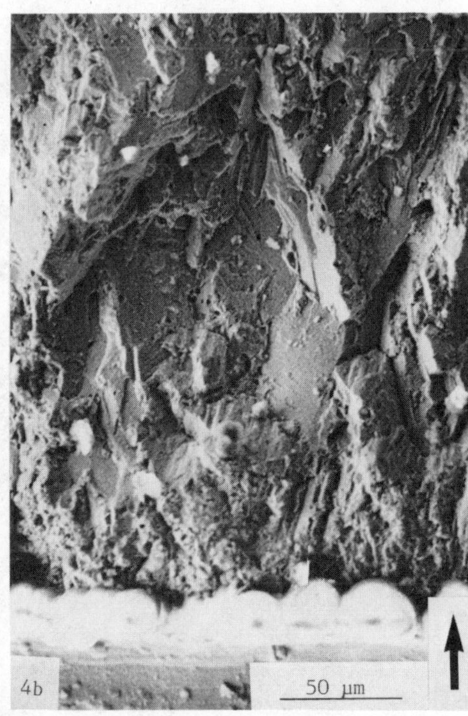

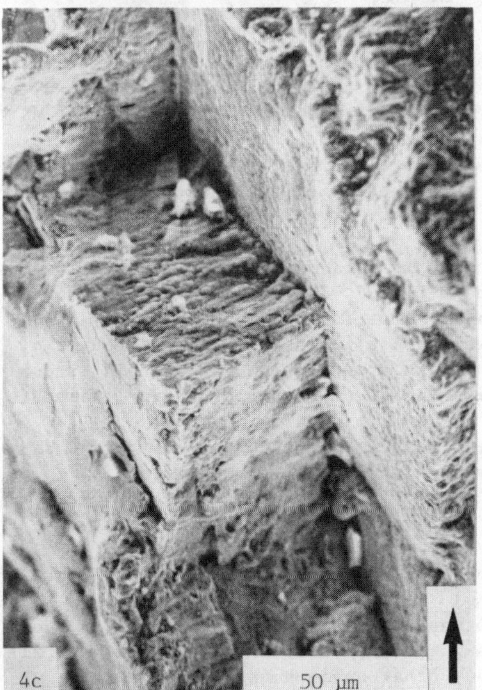

Figure 4 - Fracture surfaces after ultrasonic fatigue. Crack growth direction is indicated by an arrow.

Figure 4a - Area of crack initiation in silicone oil environment; $K_{max} = 4.2$ MN $m^{-3/2}$, $\Delta a/\Delta N = 1.7 \times 10^{-10}$ m/cycle.

Figure 4b - Area of crack initiation in distilled water, $K_{max} = 3.4$ MN $m^{-3/2}$, $\Delta a/\Delta N = 2.4 \times 10^{-10}$ m/cycle.

Figure 4c - Fracture surface after loading in 3.5 % NaCl-water solution; $K_{max} = 2.5$ MN $m^{-3/2}$, $\Delta a/\Delta N = 3.6 \times 10^{-10}$ m/cycle.

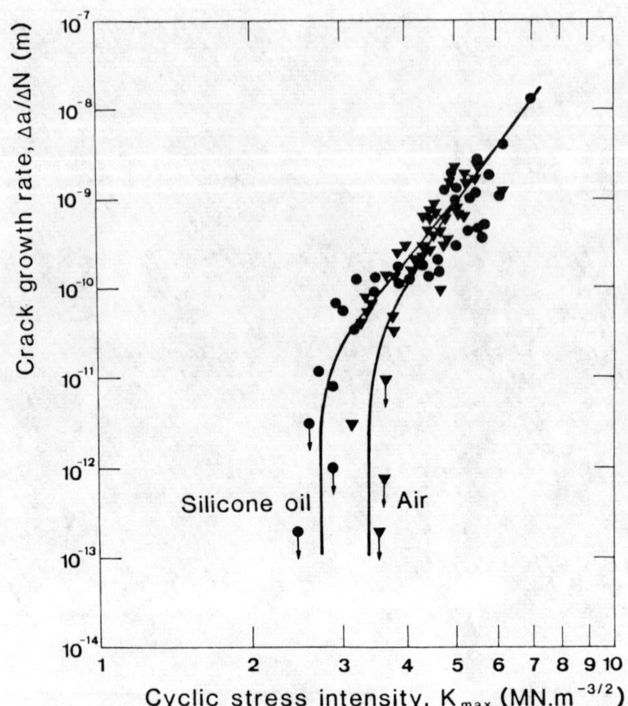

Figure 5 - Crack propagation and thresholds after 20 kHz ultrasonic loading in oil and air

extensive intergranular cracking. These are regions where the cyclic stress intensity was very low, approximately in the threshold regime. In the middle band e.g., it was 3.9 MN$^{-3/2}$; the crack growth rate was 9.1 x 10^{-11}m/cycle. A detail of this area is shown in Fig. 6b, where the intergranular fracture is quite obvious in contrast to Fig. 4b (distilled water environment, where the fracture surface is characterized by an essentially transcrystalline brittle structure).

Additional tests have been performed in a vacuum of 1 Pa and 1 x 10^{-3}Pa. The crack growth curves are reproduced in Fig. 7. The resulting crack growth rates are approximately the same for both vacua. Compared with air, the crack growth curve is shifted to somewhat lower stress intensities in the regime of "higher" cyclic stress intensities and shows a marked lower threshold stress intensity, which is in contrast with most measurements reported in literature.

The fracture surfaces of specimens tested in vacuum of 10^{-3} Pa are characterized by a smooth, transcrystalline structure; for example, in Fig. 8 K_{max} was 3.5 MN m$^{-3/2}$ and $\Delta a/\Delta N$ = 2.2 x 10^{-10} m/cycle. Loading in vacuum of 1 Pa revealed fracture surfaces which are essentially similar to Fig. 8.

Furthermore, it could be observed that in vacuum of 10^{-3} Pa the crack growth rates are an order of magnitude higher at short crack lengths, between 0.1 a/W and 0.3 a/W (notch depth was 0.1 a/W), (Fig. 9). There is even crack growth at such small K_{max} values where longer cracks do not grow at all. This effect could not be observed in any other environments until now.

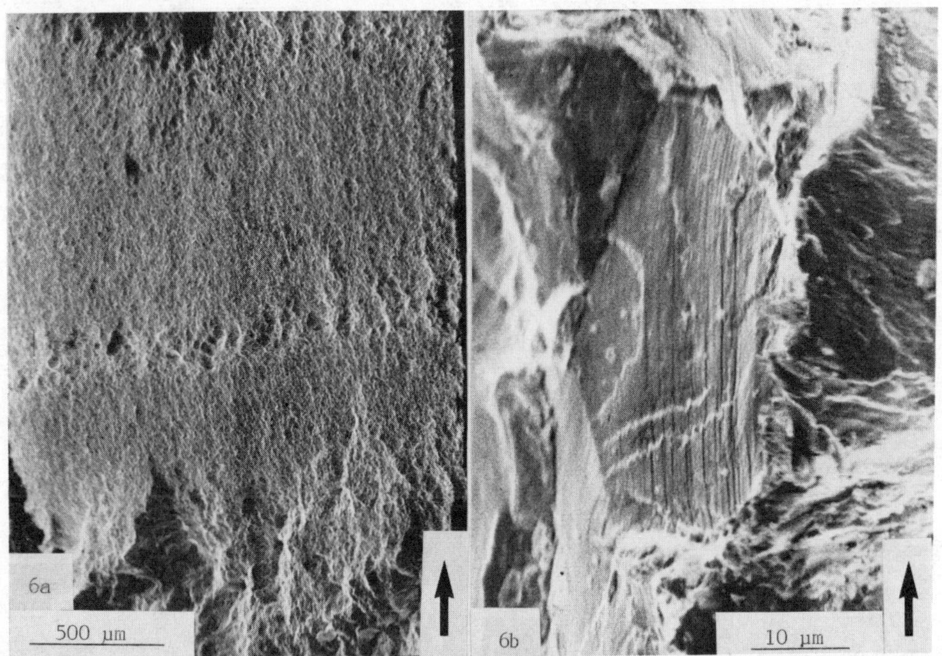

Figure 6a - Fracture surface after ultrasonic loading in air. Crack growth direction: From bottom to top. The two horizontal bands showing extensive intergranular fracture were generated at near-threshold stress intensities. In the middle band K_{max} = 3.9 MN $m^{-3/2}$ and $\Delta a/\Delta N$ = 9.1 x 10^{-11}m/cycle

Figure 6b - Detail of Fig. 6a. Intergranular fracture after loading with a near - threshold K_{max} = 3.9 MN $m^{-3/2}$ in air environment; $\Delta a/\Delta N$ = 9.1 x 10^{-11} m/cycle

In Fig. 10 all curves of the previous figures are summarized. It shows increasing K_{max} values for same crack growth rates in the series NaCl solution,water, oil though with identical threshold stress intensities for oil and distilled water. Secondly a slight increase of the K_{max} values needed to obtain same $\Delta a/\Delta N$ is visible in the series: vacuum, oil and air; and the curves for vacuum and water cross at crack growth rates below about 10^{-11} m/cycle. Thirdly the crack growth curves for silicone oil and vacuum are the same down to approximately 10^{-11} m/cycle; at lower rates however, higher threshold stress intensities were found for the oil environment than for vacuum. In vacuum, a steep transition leading to a threshold cyclic stress intensity occurs at the very low crack growth rate of approximately 10^{-12} m/cycle whereas this transition takes place at about 10^{-11}m/cycle in silicone oil and air and between 10^{-11} and 10^{-10} m/cycle in water and NaCl solution.

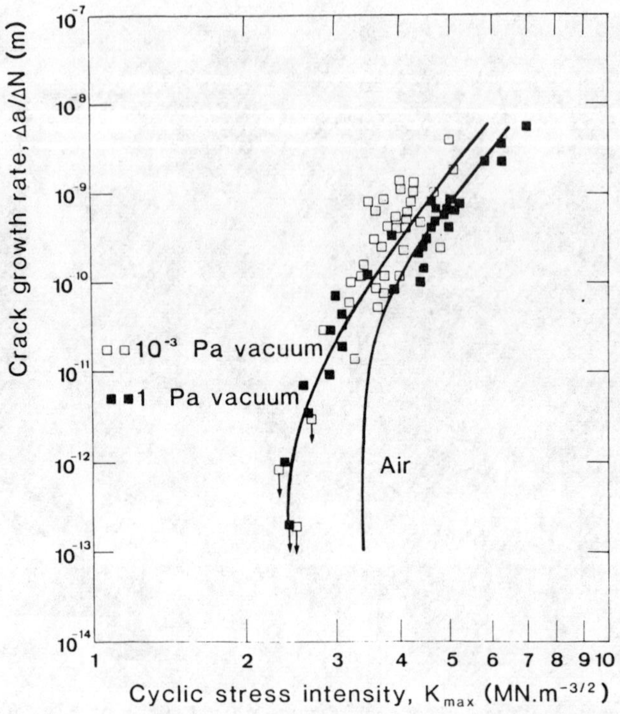

Figure 7 - Crack growth curves for ultrasound stressing in vacuum of 10^{-3} Pa, 1 Pa and air

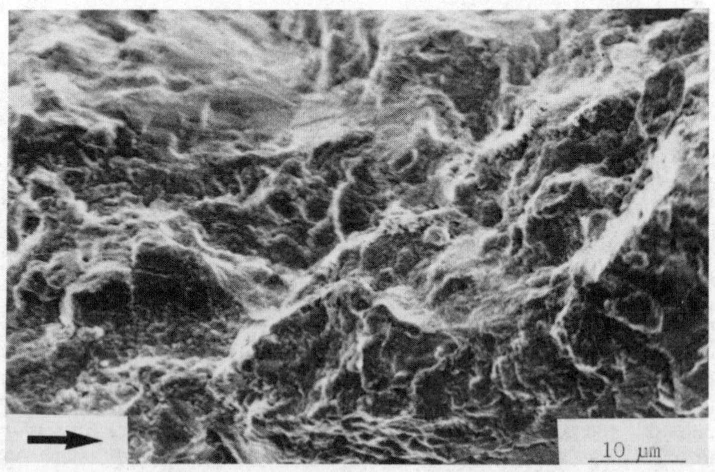

Figure 8 - Fracture surface after ultrasound stressing in vacuum of 10^{-3} Pa with K_{max} = 3.5 MN $m^{-3/2}$, $\Delta a/\Delta N$ = 2.2 x 10^{-10} m/cycle

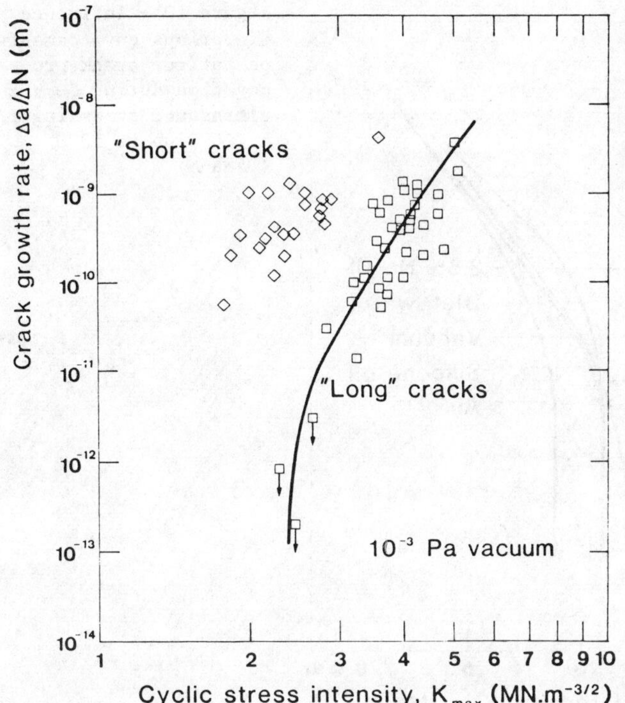

Figure 9 - Crack growth rates of " short" cracks, when specimens are stressed with ultrasound in vacuum of 10^{-3} Pa. Comparison with growth of "long" cracks

Discussion

Higher crack growth rates in water than in oil environment and higher rates in NaCl solution than in water may be explained by anodic dissolution and hydrogen embrittlement respectively. Scanning microscope pictures, displaying faceted and intergranular structures with eventual corrosion pits support this assumption. Furthermore, surface spectroscopy revealed that decomposition of brass into copper and tin had taken place close to the fracture surfaces and close to the specimen surfaces. It is assumed that this process has accelerated crack growth additionally, as differences of the crack growth rates could not be detected in copper, iron and steel at crack growth rates above 10^{-9} m/cycle in earlier works (22). Spectroscopical measurements further showed noticeable (3 %) coverage of specimens with Cl^-ions after stressing in NaCl solution, which points to extensive anodic dissolution. Fracture surfaces of specimens tested in oil were almost completely covered by an oil layer, which could not be removed chemically. Thus it may be concluded that during crack propagation oil serves as protective film against chemical attack by the surrounding environment, which is moist air usually. The striking result of the same threshold stress intensities for distilled water and oil may be explained by the much rougher fracture surfaces of the specimens loaded in water (Fig. 4b). This means, that in the regime of very low crack growth rates in water at least two mechanisms are controlling the magnitude of the threshold: anodic dissolution which lowers the threshold and fracture-surface roughness-induced crack closure, which raises it.

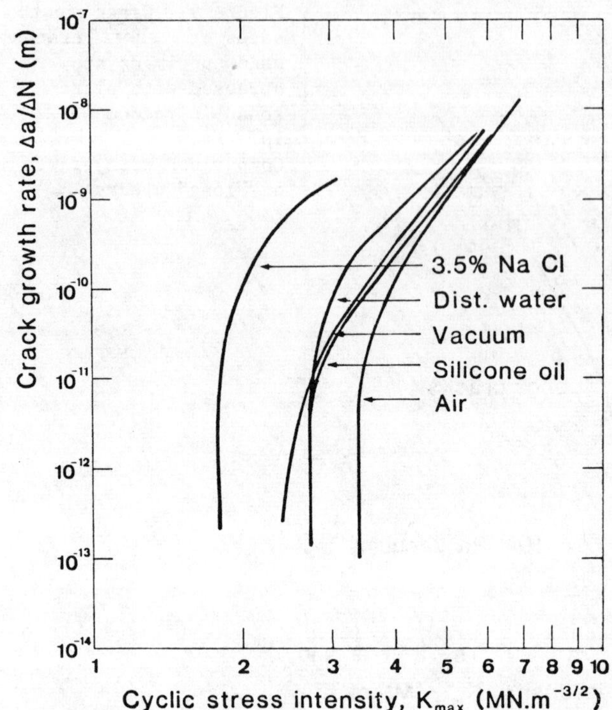

Figure 10 - Influence of various environments on fatigue crack propagation during 20 kHz ultrasound stressing at 20°C up to 50°C (vacuum)

In comparison specimens cracked in oil and air environment display almost identical crack growth curves down to K_{max} values of about 4 MN $m^{-3/2}$ and crack growth rates of about 1 x 10^{-10} m/cycle; below these values, however, the steeper curve for air environment indicates that in the low threshold regime, crack propagation in air obviously needs higher nominal cyclic stress intensities. This result is similar to Tzou's et al. finding on 2¼ Cr-1 Mo steel (25), where the curves even cross each other, which means that the crack growth rates in moist air are higher at high K_{max} compared to silicone oil and drop to crack growth rates below the values in oil at low stress intensities. In Tzou's et al. work, oxide induced crack closure is made responsible for this effect. In a 50 Hz test, 0.2 μm thick oxide layers were found at crack growth rates of about 10^{-11} m/cycle in air, which results in a closure stress intensity of 4.5 MN $m^{-3/2}$ according $K_{c\ell} = dE/4\sqrt{\pi\ell}$ with an assumed distance 2ℓ = 4 μm for the maximum oxide wedge thickness d behind the crack tip; the applied stress intensity is reduced by this amount.

If it is assumed that the build-up time of the oxide layer is the rate-controlling process, 400 times more cycles would be needed at 20 kHz than at 50 Hz to obtain same exposure times. For the ultrasound test, therefore only layer-thicknesses in the range of 10^{-3} μm can be expected at a crack growth rate of about 10^{-11} m/cycle, which would cause a $K_{c\ell}$ of about 10^{-2} MN $m^{-3/2}$ only. This result is in accordance with considerations of the build-up time of gas-monolayers according Langmuir's equation and is additionally confirmed by spectroscopic studies, which did not reveal any oxide-layers in the range of 10^{-1} μm. In addition the change in slope of the curve for air, leading to a higher K_{th}, takes place at a crack growth rate of approximately 10^{-11} m/cycle, which is the same range as reported by Tzou et al. (25). For the much shorter testing times in the ultrasound test, the curves should show this change at lower crack growth rates, if oxide-induced crack closure is assumed. Therefore it is concluded that oxide-induced crack closure does not occur in ultrasound fatigue. On the other hand, the pronounced intercrystalline fracture at low stress intensities in air, as shown in Fig. 6, is strong evidence for the mechanism of crack closure induced by fracture surface roughness as responsible mechanism for increased threshold stress intensity values. In this case the contact points behind the crack tip act as "wedges" (26,27).

The phenomenon of intercrystalline fracture may be explained by corrosive interaction of the environment with grain boundary matter, causing grain boundary embrittlement and possibly also metal dissolution. The resulting fracture surface in this work is very rough due to the rather large grains (the grain diameter of the tested material is about 35 μm in the plane of crack growth and 140 μm normal to it). Studies with different grain sizes and mean stresses should therefore help to further clarify the problem of environmental induced crack closure and threshold behavior.

In this work less intercrystalline cracking has been observed for water environment compared to air, but more embrittled transgranular features (Fig. 4b), which corresponds to lower threshold values for water environment (due to less roughness induced closure) than for air. Irrespectively of this, brass-decomposition as mentioned above, may explain in addition lower threshold stress intensity values in water.

Comparing the curves for water and vacuum, it shows that crack growth at "high" cyclic stress intensities is slower in vacuum, whereas in the threshold regime, the crack growth curve for vacuum points to lower K_{th} values, indicating that crack growth still takes place at such low ΔK values where crack propagation is not longer observed in water (or oil or air). This result indicates that crack closure due to fracture surface roughness may be the reason for raised K_{th} values in water, air and oil compared to vacuum. In addition, wedging due to partial penetration of a liquid into the crack (25) from the crack mouth and the specimen sides seems to be a plausible explanation. In situ high-speed filming of the specimen surfaces during the ultrasound test (28) verified penetration of the used liquids into the crack and extensive pumping and squeezing-out of it. It has been shown by Tzou et al. (25) that the resulting hydrodynamic pressure will lower the cyclic stress intensity.

Dehumidified oil has been used in numerous studies instead of vacuum as inert environment. As the viscosities of oil and vacuum or air are different, it was thought necessary to perform tests in vacuum in order to exclude eventual crack-closure effects due to a wedging action of the liquids within the crack and also possible oil-film induced effects. Indeed, the curves in Fig. 6 and 10 demonstrate that the crack growth curve for oil is shifted to slightly higher K_{max} values in the threshold regime. Wedging due to partial penetration of the oil into the crack seems to be responsible. At all other stress intensitites however, silicone oil simulates the vacuum environment quite well.

Two other interesting results are enhanced crack propagation rates in vacuum and higher stress intensities in the threshold regime compared with air. Speidel (20) discussed crack branching or crack tip blunting by oxidation as possible explanations for higher threshold values in air than in vacuum, when Nickel-base alloys were fatigued at 850°C. However in this work crack branching may be excluded as it could not be observed. Both above mentioned results also might be explained, based on the work of Suresh et al. (21), by oxide-induced crack closure. However, as shown above, oxide layers, which are generated during the ultrasound test, are much too thin to cause crack closure. Therefore it is concluded that the difference of K_{th} values for air and vacuum by a factor of 1.25 originates from the very pronounced fracture-surface roughness produced by stressing in air.

Further influences, like possible capillary or pumping effects of the crack certainly play some additional role, but have not been considered here.

Enhanced crack growth rates of short cracks found in vacuum may be understood again in the light of roughness-induced crack closure (27). Up to crack lengths of one or some few grain diameters the fracture surface obviously is smooth enough (Fig. 8) to prevent any roughness-induced crack closure effect. The remarkable result, that enhanced crack growth of short cracks was observed in vacuum of 10^{-3} Pa only, confirms this assumption. In all other environments either intercrystalline or embrittled transcrystalline fracture surfaces with considerable roughness have been generated and may have (see e.g. Fig. 4, 6) caused crack closure at small crack lengths already. In liquid environments viscous fluid-induced crack closure (25) may in addition be responsible for lack of enhanced growth of short cracks at very low cyclic stress intensities (in oil e.g.).

Conclusions

Ultrasound loading is a very useful method for studying corrosion fatigue cracking, especially at very low stress intensities.

The results of this work show that dominating processes are probably anodic dissolution and hydrogen embrittlement, which may take place preferentially at grain boundaries during crack growth in distilled water or 3.5 % NaCl solution. The mechanisms enhance crack propagation compared to intert silicone oil environment.

Higher thresholds for air and distilled water than for vacuum indicate a reduction of the applied stress intensity by surface roughness-induced crack closure. SEM pictures of the fracture surfaces, which show extensive intergranular and embrittled features of high roughness in the threshold regime are strong evidence for this assumption. The slightly higher threshold for oil compared with vacuum is attributed to a reduction of the applied stress intensity by hydro-dynamic wedging.

Furthermore, enhanced crack growth rates for "short cracks", which were found in 10^{-3} Pa vacuum only, are also attributed to the absence of surface roughness-induced crack closure. In liquids, in addition viscous fluid-induced crack closure may be responsible for lack of enhanced crack growth rates of short cracks.

Acknowledgement

The measurements of this work are part of the Ph.D. thesis of H.M. Ebenberger. Support by the "Fonds zur Förderung der wissenschaftlichen Forschung", Vienna, contract No. 4107 is gratefully acknowledged.

References

1. F.A. Neppirras, "Very High Energy Ultrasonics," British Journal of Applied Physics, 11 (1960) pp. 143 - 150

2. R. Mitsche, S. Stanzl and D.G. Burkert, "Hochfrequenzkinematographie in der Metallforschung," Wissenschaftlicher Film, 14 (1973) pp. 3-10

3. S. Purushothaman, J.P. Wallace and J.K. Tien, "High-Power Ultrasonic Fatigue," pp. 244-249, in Ultrasonics International Conf. Proc. IPC Press, Surrey, England, 1973

4. S. Purushothaman and J.K. Tien, "Slow Crystallographic Fatigue Crack Growth in a Nickel-Base Alloy," Metallurgical Trans., 9A (1978) pp. 351-355

5. B. Weiss, R. Stickler, J.F. Femböck and K. Pfaffinger, "High Cyclic Fatigue and Threshold Behavior of Powder Metallurgical Mo and Mo Alloys," Fatigue of Eng. Mater. and Struct., 2 (1979) pp. 73-84

6. W. Hoffelner, "The Influence of Frequency on Fatigue Crack Propagation of Some Heat-Resisting Alloys using Ultrasonic Fatigue," pp. 461-472 in Ultrasonic Fatigue, J.M. Wells et al., eds., The Metall. Soc. of AIME, Warrendale Pennsylvania, 1982

7. S.E. Stanzl and E.K. Tschegg, "Fatigue Crack Growth and Threshold Behavior at Ultrasonic Frequencies," in Fracture Mechanics: ASTM STP 791, J.D. Lewis and G. Sines, eds. 1983 Vol. 2. pp. II 3 - II 18

8. A. Puskar, "Fatigue Properties of Steels at Low and High Loading Frequences," pp. 223-227 in Ultrasonic Fatigue, J.M. Wells et al., eds., The Metallurgical Soc. of AIME, Warrendale, Pennsylvania 1982

9. W.P. Mason, Piezoelectric Crystals and Their Application in Ultrasonics, Van Nostrand, New York, 1950

10. W. Kromp, K. Kromp, H. Bitt, H. Langer and B. Weiss, "Technique and Equipment for Ultrasonic Fatigue," pp 238-243, Ultrasonics International Conf. Proc., IPC Press, Surrey, England, 1973

11. W. Hoffelner, "Fatigue Crack Growth at 20 kHz - a New Technique," J. Phys. E.: Sci. Instr., 13 (1980) pp. 617-619

12. W. Hollanek, "A Method for Generating Quick Mean Load Changes in Ultrasound Fatigue," to be published in Ultrasonics

13. E.K. Tschegg, "Fully Automatic Measurements of Fatigue Crack Growth during Ultrasound Stressing," to be published in Scientific Instruments

14. S.E. Stanzl, W. Hollanek and E.K. Tschegg, "Fatigue and Fracture under Variable-Amplitude Loading at Ultrasonic Frequency," submitted to ICF 6, India, Dec. 1984

15. A. Lindner, "Frequenzmessungen bei Ultraschallbeanspruchung," Ph.D. thesis at the University of Vienna, 1981

16. W.F. Brown and J.E. Srawley, "Plane Strain Crack Toughness Testing of High Strength Metallic Materials," ASTM STP 410 (1966)

17. G. Schoeck, "Calculation of the Stress Intensity in Ultrasonic Resonance," Zeitschrift für Metallkunde, 73 (9) (1982) pp. 576-578

18. W. Hoffelner and P. Gudmundson, "A Fracture Mechanics Analysis of Ultrasonic Fatigue," Eng. Fracture Mech., 16 (3) (1982) pp. 365-371

19. B. Weiss, H. Bildstein, R. Stickler, K. Pfaffinger and J. Femböck, "Bestimmung des Schwellwertes des Spannungsintensitätsfaktors von Mo und Mo-Legierungen mit einer 20 kHz Methode, " Metall, 34 (1980) pp. 636-641

20. M.O. Speidel, "Influence of Environment on Fracture," pp. 2685-2704 in Advances in Fracture Research, D. Francois, ed., ICF 5, Cannes, France, 1981, Pergamon Press 6, 1982

21. S. Suresh, G.F. Zamiski and R.O. Ritchie, "Oxide-Induced Crack Closure: A Mechanism for Near-Threshold Corrosion Fatigue Crack Growth," Metall. Trans., 12 A (1981) pp. 1435-1443

22. S.E. Stanzl and E.K. Tschegg, "Influence of Environment on Fatigue Crack Grwoth in the Threshold Region," Acta Metallurgica, 29 (1981) pp. 21-32

23. H.M. Ebenberger, "Rissausbreitung bei 20 kHz Ultraschall-Wechselbeanspruchung," Ph. D. thesis at the University of Vienna, 1983

24. S.E. Stanzl, "Ueber den Einfluss der Probendicke auf die Wachstumsgeschwindigkeit von Ermüdungsrissen," Z. Metallkunde, 71 (1980) pp.195-202

25. J.L. Tzou, S. Suresh and R.O. Ritchie, "Fatigue Crack Propagation in Viscous Environment," paper presented at Fourth Int. Conf. on Mechanical Behavior of Materials (ICM 4), Stockholm, Sweden, Aug. 1983

26. N. Walker and C.J. Beevers, "A Fatigue Crack Closure Mechanism in Titanium," Fat. of Eng. Mat. and Structrues, 1 (1979) pp. 135-148

27. S. Suresh and R.O. Ritchie, "A Goemetric Model of Fatigue Crack Closure Induced by Fracture Surface Roughness," Metall. Trans., 13 A (1982) pp. 1627-1631

28. S.E. Stanzl and R. Mitsche, "Ermüdungsrissausbreitung in metallischen Werkstoffen bei hochfrequenter Wechselbeanspruchung," 16 mm movie No. 1533, produced at BHWK 1974 and published 1976 in Vienna, A 1050 Wien, Schönbrunnerstrasse 56

AN ASSESSMENT OF THE FATIGUE THRESHOLD AS A DESIGN PARAMETER

R. Brook
Department of Metallurgy
University of Sheffield
Sheffield
S1 3JD
U.K.

In view of the complex nature of the threshold effect and its sensitivity to numerous external variables the value currently of ΔK_{th} as a design parameter is questionable. This problem is examined from the viewpoint of (a) threshold measurement, (b) the effect of R ratio, (c) material variables and (d) fatigue life prediction.

Introduction

The threshold stress intensity, ΔK_{th}, reflects the conditions under which an existing crack of known size in a prescribed material just remains dormant under the combined influence of applied cyclic loading and environment. As such, the threshold parameter *per se* potentially presents a powerful design parameter to help avoid the premature failure of structures and components exposed to high cycle fatigue loading.

When the Paris relationship (1) is used to estimate fatigue lives on the basis of the defect-tolerant approach it is conservative i.e. the remnant life will be under-estimated because of the many cycles accumulated in the low growth rate, near-threshold region. More accurate estimations should be possible using extensions of the Paris approach which take into account the non-linear dependency of crack growth rate found at each end of the ΔK spectrum (e.g. 2). From this point of view the fact that the growth rate just above ΔK_{th} can be several orders of magnitude lower than that predicted by the Paris expression is particularly attractive. The benefit to be gained in added fatigue life should more than compensate for the extra cost incurred in threshold testing. However, the situation is complicated by the fact that the threshold phenomenon can be attributed to one, or more, of several possible mechanisms; in view of this it is not surprising that the effect of factors influencing threshold may be difficult to predict or rationalise, even in the laboratory. One must seriously question, therefore, the value of much of the data available generally to engineers who seek to add this new dimension into design against fatigue. This paper seeks to highlight features of threshold behaviour about which the designer should be aware before a ΔK_{th} value is transferred from a material data sheet into design calculations to determine loads and lifetimes.

ΔK_{th} Measurement

The foundation for confident use of a property in engineering is that the test procedure and specimen provide the designer with an accurate value which can be used reliably to predict the response of the component to service loadings. First of all, therefore, let us examine the crack growth threshold from the viewpoint of a characterising test parameter as determined in the laboratory.

A well-controlled test can yield threshold values that are reproducible to, say, ±5% and this can mask the inherent sensitivity of ΔK_{th} to many of the detail differences in test procedure and environment which will be discussed later. This sensitivity can be manifest in quite different ways, such as (a) the significant variation that may be found between threshold values from different sources on nominally similar materials (e.g. 3,4) and (b) the inability to provide a definitive expression relating ΔK_{th} and R which is applicable generally (5). The fact that threshold testing is costly in terms of time and facilities does not always encourage or allow a thorough appraisal of all the important variables to be made.

Load-shedding Procedures

Although no standard procedure exists, as yet, it is usual to measure ΔK_{th} by a load-shedding technique. This involves adjusting loads to reduce the applied ΔK and hence the crack growth rate.

Control of the load-shedding procedure has, for the most part, been achieved by a step-wise reduction in load made after a certain amount of crack growth: the crack extension being sufficient to take the tip beyond the material deformed plastically under the influence of the higher load associated with the previous step. A wide range of estimates of the effect of crack-tip plasticity have been used in justifying different load-shedding procedures, many of these are close in value to Irwin's original estimate (6). A recent study (7) by interferometry of deformation at the tips of fatigue cracks in a carbon-manganese steel showed close agreement with Irwin's (plane stress) estimation of plastic zone, r_p,

$$r_p = \frac{1}{2\pi}\left(\frac{K_{max}}{\sigma_y}\right)^2 \tag{1}$$

where σ_y = yield strength

following a step-down sequence at R = 0.1, but at R = 0.6 the measured zone size (Figs. 1 and 2) was some ten times larger than predicted. A first simple consideration of the data suggests it is reasonable to assume that yielding occurs between the first and second (outer) fringes of the pattern; a more detailed analysis is in progress. Nevertheless, the observations provide additional evidence to support earlier criticism (8) of load-shedding based on current estimations of nominal crack-tip plasticity; attention

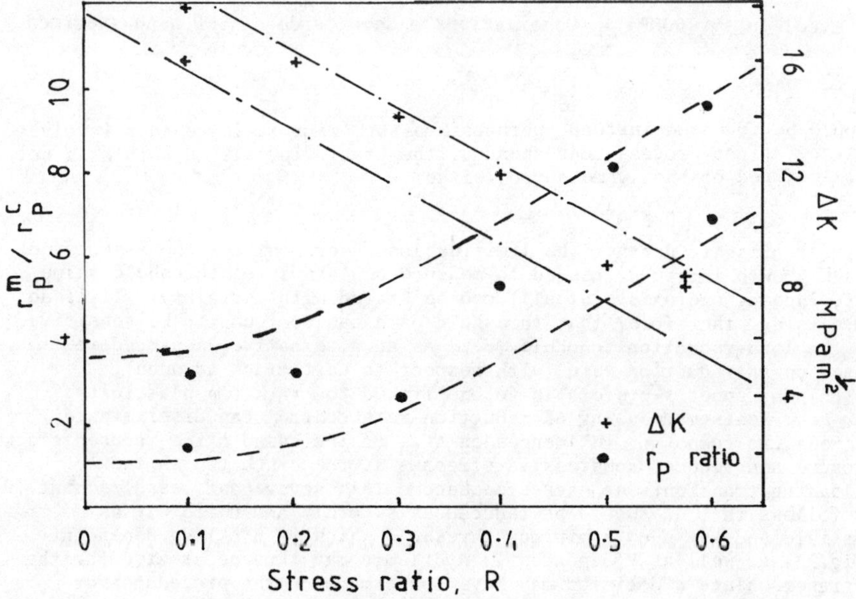

Figure 1 - The dependency of the ratio of measured and calculated (6) plastic zone sizes, r_p^m / r_p^c, and the stress intensity range, ΔK on the stress ratio, R, for a 0.15%C 1.5%Mn steel.

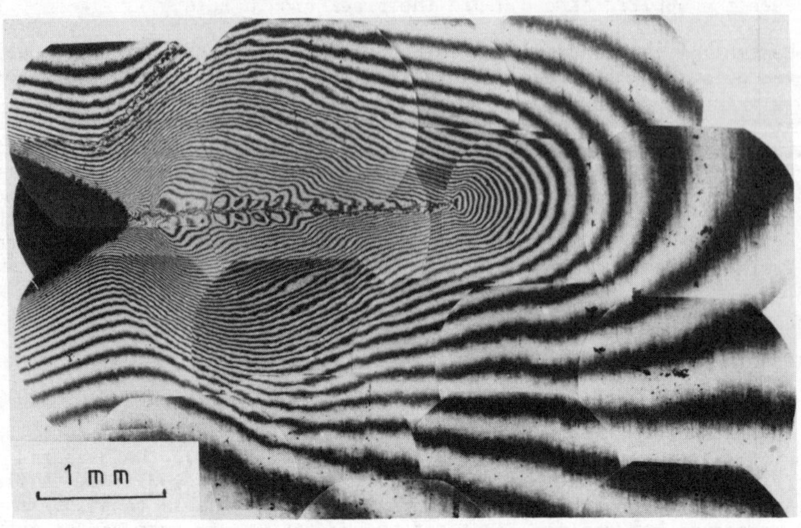

Figure 2 - Fringe pattern developed by crack growing at a ΔK reduction rate of $5MPa.m^{-\frac{1}{2}}mm^{-1}$ between ΔK levels of 20 and 12 $MPa.m^{\frac{1}{2}}$ at R = 0.6 (Height interval between fringes $\sim$ 0.27μm). 0.15%C, 1.5%Mn steel (σ_y = 460MPa). Observations made on side of SEN bend specimen.

should be focussed instead, perhaps, on the 'microscale' events involved in the fatigue-process zone ahead of the crack tip, although this is not well-defined or easily measured, either.

At present, neither the limiting load decrement nor the associated crack growth increment needed to measure an 'absolute' threshold value (if, indeed, one exists at all) can be stated with certainty. It is not surprising, therefore, that threshold data can be found to be sensitive to the load-reduction schedule (8). As an alternative, a procedure based on ΔK-reduction rate (with respect to unit crack advance), d (ΔK)/da, appears preferable to one linked to crack tip plasticity, *per se*. Tests made using ΔK reduction-rate control can discriminate between the competing influences on ΔK_{th} of increased oxide-induced crack closure and reduced compressive stresses at the crack tip as the unloading transients at each step become less severe and less frequent (9). It follows that ΔK should be reduced as smoothly and uniformly as possible and may show a minimum threshold which is d(ΔK)/da dependent (Fig. 3) as well as R dependent. A dilemma may thus be created for the designers since a decision may have to be made on the procedure for threshold measurement, depending on the use to be made of the data generated. If the test procedure is optimised to provide the minimum ΔK_{th} value measurable then this should give the best basis for comparison between different materials or production and fabrication routines. However, such a test and its ΔK_{th} may be quite different to in-service

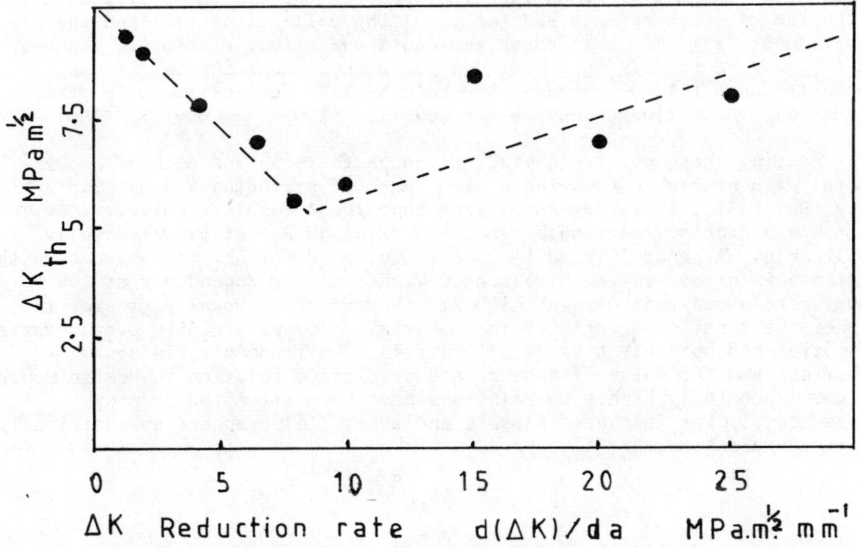

Figure 3 - Effect of ΔK - reduction rate on ΔK_{th} at R = 0.1 (6)
0.15%C, 1.5%Mn steel; Initial ΔK, 20MPam$^{\frac{1}{2}}$; Growth step 0.1mm.

conditions. Thus a choice may have to be made between data ranging from that of a routine, 'standard', test that may be 'ultra-safe' to that from a 'customised' test which reproduces (if practicable) the fluctuating load routines of either normal service or most-severe overload conditions appropriate to the component, so securing more appropriate, higher, thresholds. The influence is such that, when designing against fatigue, based on a threshold criterion, the smoothly changing loads may be more critical than rapid transients, whereas the reverse could be expected for other failure criteria.

R-dependency of Threshold

Threshold tests are generally made under constant R conditions to eliminate the systematic variation of ΔK_{th} with R usually observed, particularly when R is less than 0.5.

The threshold state can also be approached in a number of ways under changing R, by keeping one of these following factors constant, K_{max}, K_{min}, ΔK, σ_{mean}. There appears to have been no comprehensive study of the effect of prior variations in R on the eventual threshold value. However it is well-established that crack closure can account for much of the R-dependency shown by threshold. The stress intensity seen to be applied (ΔK_{th}) being the sum of an effective component (ΔK_o) and that lost as a result of closure mechanisms (ΔK_{cl}).

Closure is known to be influenced by plasticity in the wake of the crack (10), the degree of surface roughness of the fracture (11), the mechanism of crack advance and the resulting deflections (12) and the environment (13). In some cases the cause and effect of closure changes on ΔK_{th} may be readily apparent: an increase in humidity of 35% was recently found (7) to increase threshold by 15%, for instance, in other situations the influence may be more subtle.

Bearing these points in mind, although there is a wealth of experimental data available covering a wide range of strengths and materials (e.g. Ref (14)), it is not surprising that these results, collectively, do not show a precise relationship between ΔK_{th} and R that is universally applicable. A general trend is usually followed for ΔK_{th} to decrease with increase in stress ratio, frequently with a strong dependency at low R tending to a constant ΔK_{th} at high R: the effect is usually greater the higher the tensile strength of the material. To measure ΔK_{th} over a range of ratios and possibly a range of controlled environments, is again laborious and the value of an accurate predictive relation in design would be correspondingly high. Correlations have been attempted by many researchers. For instance, Klesnil and Lukas (15) proposed an empirically-derived expression,

$$\Delta K_{th} = \Delta K_{th}^{R=0} (1 - R)^{\gamma} \quad (2)$$

where γ is a constant $0 < \gamma < 1$. Schmidt and Paris (16) based their model on crack closure concepts,

$$\text{At low R,} \quad \Delta K_{th} = (K_{cl} + \Delta K_{o}) (1 - R) \quad (3)$$

$$\text{At high R,} \quad \Delta K_{th} = \Delta K_{o} \quad (4)$$

K_{cl} and ΔK_{o} being provided by experiment.

Certain relationships predict that $\Delta K_{th} = 0$ as $R \to 1$, whereas experimental data in many instances suggests that a finite value of ΔK_{th}, typically in the range 2-4 $MPam^{\frac{1}{2}}$ is approached. If this is really the case then materials of relatively low monotonic toughness could in certain circumstances fail catastrophically before threshold could be established. For example, an alloy of $K_{Ic} = 40$ $MPam^{\frac{1}{2}}$ and $\Delta K_{th} = 4$ $MPam^{\frac{1}{2}}$ would experience brittle failure before any fatigue crack growth could occur at $R \geqslant 0.9$.

Austen (17) has developed a general relation which enables the different forms of ΔK_{th} / R expressions to be compared easily. Its basis lies in a survey by Garwood (18) on mild steel which suggested a linear relationship,

$$\Delta K_{th} = \Delta K_{th}^{R=0} - \text{constant} \times R \quad (5)$$

Austen modified this equation, making the constant equal $(\Delta K_{th}^{R=0} - \Delta K_{o})$ so that $\Delta K_{th} \to \Delta K_{o}$ as $R \to 1$, hence,

$$\Delta K_{th} = \Delta K_{th}^{R=0} (1 - R) + R\, \Delta K_{o} \quad (6)$$

By substituting the identity

$$\Delta K_{th}^{\ o} = K_{cl} + \Delta K_o \qquad (7)$$

from the Schmidt and Paris relation (equation 3) into equation (6) and exponentiating the (1 - R) term, Austen provided a general relation

$$\Delta K_{th} = K_{cl}\ (1 - R)^{\gamma'} + \Delta K_o \qquad (8)$$

where γ' is determined experimentally.

Figure 4 illustrates the sensitivity of the ΔK_{th} / R relation to the value of γ', using equation (8). In this example the limits are represented by tests in vacuum and in hydrogen since the environment will influence the value taken by γ', as it will K_{cl}, of course.

Figure 4 also shows the replacement of fatigue crack growth by fast fracture as K_{max} becomes equal to K_c at ΔK_{th} when R values approach unity.

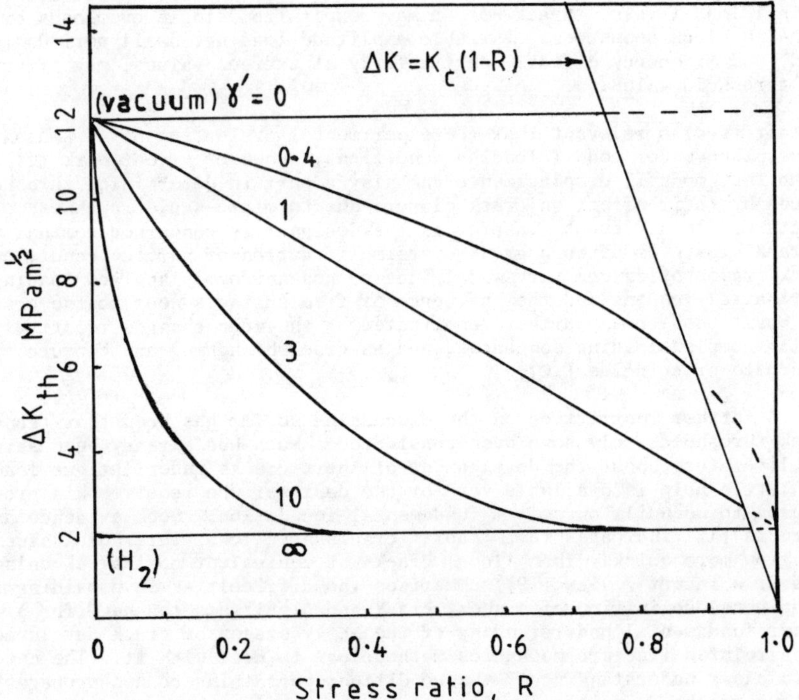

Figure 4 - The influence of γ' on the form of ΔK - R relation based on Eq. (8) and applied to an alloy assumed to have the following properties: $K_{Ic} = 50MPam^{\frac{1}{2}}$, $K_{cl} = 10MPam^{\frac{1}{2}}$.

Austen's proposal in no way disguises the fact that an absolute generalisation to describe the stress ratio dependency of ΔK_{th} seems to be out of the question. The cumulative effects of the variables influencing the experimental measurement of the individual ΔK_{th} values are superimposed and distributed over a range of R values. Thus once more the conclusion must be drawn that only (a) when the service conditions are known to be reproduced precisely by the conditions of threshold measurement or (b) when, for the case in question, the ΔK_{th} is known to be insensitive to other variables likely to be encountered in service, can predictions of R dependency in design be made with confidence.

Other Mechanical Variables

The designer must also bear in mind that there are other variables, not accounted for in the normal laboratory determination of ΔK_{th} that may have a significant effect on the threshold value appropriate to service. Data on these effects tends to be scarce and in some instances contradictory evidence can be found. The ΔK-decreasing routines that have been described so far are rarely found in practice, the usual service conditions being those in which ΔK increases as the crack grows when cycled between nominal load levels. Crack arrest may result from single overloads or high-low block sequences. Variable amplitude loading, dwell periods (off load) and frequency effects (particularly at extreme values) may affect the threshold value.

It is also relevant that the experimental evaluation of ΔK_{th} usually takes place under mode I loading conditions. However recent work (11) has shown that mode II displacements can play a part in determining threshold values by their effect on crack closure due to micro-scale roughness of the fracture. In any event, insofar as the designer is concerned, nominal mode I crack growth is often a gross oversimplification of service conditions. Crack trajectories can vary widely due to non-uniformly applied loadings, multi-axial loading and the influence of free surfaces near to the crack. The position becomes further complicated as the mode changes relatively during complex loading sequences, and as crack branching and closure effects alternate or coincide (12).

A further restriction on the discussion so far has been that 'long' crack thresholds only have been considered. Much has already been said in the literature about the development of short cracks under fatigue loading but little help is available yet for the designer who requires his procedures to be built on a firm fundamental foundation. Such evidence as there is (12) indicates that 'short' cracks have lower threshold values and grow more quickly than 'long' cracks at equivalent nominal ΔK values. However a recent review (19) summarises the difficulties in providing an unequivocal definition of a short crack and highlights the need for a better fundamental understanding of the early stages of crack development and a related fracture mechanics methodology to deal with it. The need for a clear understanding of the conditions pertaining to non-propagating fatigue cracks is painfully apparent, to both researcher and designer.

Material Factors

It has been well established that the ΔK_{th} is sensitive to changes in microstructure (e.g. 20). Although this fact is not generally acknowledged in modelling threshold behaviour, a recent proposal (28) by Taylor is capable of taking the influence of microstructural differences into account, including the unusual effect due to grain size. Taylor uses the balance between energy gained in increasing compliance and that lost in extending crack tip plasticity and creating new surface. The estimates of 'absolute' threshold which this model provides for a variety of metals and alloys lie between 1 and 2 MPa $m^{\frac{1}{2}}$ and agree reasonably well with the lowest values determined experimentally for these materials.

A better understanding of the influence of microstructure on ΔK_{th} is required, however. This could be helpful, both at the material selection stage and when trying to evaluate the fatigue properties of components with heterogeneous microstructures or stress distributions (e.g. textures, residual cold work (21)) due to prior processing. The problem is perhaps highlighted if one considers the extreme difficulty of providing laboratory threshold data that relates accurately to a structural weld and its heat-affected-zone.

Threshold and Life Prediction

A further role of threshold is to be found in fatigue life predictions incorporated within a defect-tolerant approach. It has been indicated already that the threshold value to be used must be selected or measured with care. Nevertheless the threshold regime can form a very significant part of a fatigue life-time and it may be possible to justify the cost of experiments and integrate the growth of 'long cracks' over the whole ΔK spectrum.

Assuming then that a reliable and appropriate estimation of ΔK_{th} is available, how useful would it be in calculating fatigue lives? The wealth of fatigue crack growth data available has resulted in a substantial number of fatigue crack propagation "laws" being established empirically, together with a few based on theoretical models of fatigue crack growth. However, only a few of the relationships proposed (e.g. 2, 15, 22) take into account ΔK_{th} and R.

The simple and obvious approach is to integrate the Paris relationship between ΔK limits imposed by threshold and final fracture, the latter, like ΔK_{th}, being R dependent. Although this method will provide a conservative estimate it is often overly so.

An evaluation of thirty-nine formulae was made by Hoeppner and Krupp (23) and a modified Forman equation

$$\frac{da}{dN} = \frac{C(\Delta K^n - \Delta K_{th}^{\ n})}{(1 - R)K_c - \Delta K} \tag{9}$$

was proposed as the best approximation for engineering application. More recently, Austen and Walker (24) have evaluated some eighteen relationships. Only six of these take into account ΔK_{th} and early stage crack growth. The inability to consider this regime implies fatigue crack propagation at infinitely small ΔK values based on extrapolation of "Paris-type" behaviour. Comparisons were based on experimental data collected for a low alloy steel, σ_y 480 MPa, over a range of R, 0.05 to 0.85. Austen derived a predictive relation (25),

$$\frac{da}{dN} = C\,(\Delta K)^n \left[\frac{(\Delta K - \Delta K_{th})\,\Delta K.K_c}{K_c - \Delta K/(1 - R)}\right]^p \tag{10}$$

where C, n, p are determined experimentally or assigned values of,

$$C = \frac{1}{4\pi\ \sigma_y\ E} \quad ; \quad n = 2; \quad p = 0.25$$

based on theory.

In the form given the same shaping exponent, p, applies to both ends of the ΔK range, but it is possible to assign different values of exponent to each extremity.

Austen found the expression provided a sigmoidal form of the log $\frac{da}{dN}$ versus log ΔK plot that was in reasonable agreement with experimental data and accounted for the influence of R.

Davenport (26) has recently reviewed methods available to estimate the residual fatigue life of materials and components used in the U.K. power-generating industry. His recommendations involve summing upper-bound crack growth data computed for the three regions which generally characterise crack growth and which include a lower-bound estimate of ΔK_{th}. These threshold values are estimated as follows:

For ferritic steels ($\sigma_y \leq 620$ MPa)

$$\Delta K_{th} = 5.0\ (1 - R),\ 0 < R < 0.6 \tag{11}$$

$$\Delta K_{th} = 2.00, \qquad 0.6 < R < 1.0 \tag{12}$$

Steels of higher strength are always assumed to have ΔK_{th} equal to 2.00 MPa$\sqrt{m}$. To ensure a conservative value, the near-threshold crack growth is best estimated by relating ΔK_{th} to K_{min} rather than R. Thus combining equations (11) and (12) and using

$$R = \frac{K_{min}}{\Delta K + K_{max}} \tag{13}$$

gives $$\Delta K_{th} = (5 - K_{min}), \quad 0 \leq K_{min} \leq 3.00 \tag{14}$$

$$\Delta K_{th} = 2.00\ , \qquad K_{min} > 3.00 \tag{15}$$

Equations (14) and (15) can then be incorporated into an equation proposed by Priddle (27) for near-threshold crack growth,

$$\frac{da}{dN} = \lambda \times 10^{-11} (\Delta K - \Delta K_{th})^n \quad (16)$$

giving

$$\frac{da}{dN} = \lambda \times 10^{-11}\left(\frac{\Delta K}{(1 - R)} - 5\right)^n, \; 0 \leqslant K_{min} \leqslant 3.00 \quad (17)$$

and
$$\frac{da}{dN} = \lambda \times 10^{-11} (\Delta K - 2.00)^n, K_{min} > 3.00 \quad (18)$$

Substituting values of $\lambda = 8$ and $n = 3$ ensures pessimistically high values of growth rate. This is then linked with a third power law

$$\frac{da}{dN} = 10^{-11} (\Delta K)^3, \; \Delta K < 0.7 K_c (1 - R) \text{ and } K_{max} \leqslant 0.7 K_c \quad (19)$$

to give an upper bound to the Paris regime and

$$\frac{da}{dN} = \frac{0.604 \times 10^{-11} \Delta K^3}{[1 - (\frac{K_{max}}{K_c})^2]^{3/4}} \quad (20)$$

provides a similar bound to the final stage of crack growth.

There is insufficient evidence available currently to judge the relative merits of the proposals by Austen and by Davenport.

Conclusions

A ΔK_{th} value for a specific material and condition that can be applied generally, and with precision, in component design and service cannot be guaranteed, as yet, from a simple test procedure or a predictive relation.

The options available to the designer (and listed in a suggested order of increasing preference) are therefore to either

(a) ignore threshold, in effect, and assume $\Delta K_{th} \approx 0$

(b) use a ΔK_{th} measured under 'normal' service loading history

(c) seek an 'absolute' (minimum) threshold value and take this to be the 'worst case.'

or (d) use a ΔK_{th} appropriate to the most damaging fatigue loading envisaged.

Although a number of fatigue crack growth laws exist which account for threshold, no one has found general acceptance yet.

Acknowledgements

The provision of unpublished data by Dr. I. M. Austen (British Steel Corporation) and Mr. D. Bristow (University of Sheffield) is gratefully acknowledged.

References

1. P. C. Paris and F. Erdogan, "A Critical Analysis of Crack Propagation Laws," Jnl. Basic Eng. (Trans. ASME, D), 85, (1963) pp. 528-534.

2. C. E. Nicholson, "Influence of Mean Stress and Environment on Crack Growth," pp. 226-243 in Proc. Conf. "Mechanics and Mechanisms of Crack Growth", Cambridge, British Steel Corporation Publication (1973).

3. O. Vosikovsky, "Frequency, Stress Ratio and Potential Effects on Fatigue Crack Growth of HY130 Steel in Salt Water," Jnl. Test Evaluation 6, (3), (1978) pp. 175-182.

4. S. I. Kwun and M. E. Fine, "Fatigue Macro-Crack Growth in Tempered HY80, HY130 and 4140 Steels: Threshold and Mid-K Range," Fat. Eng. Mat. Struct., 3, (1980) pp. 367-382.

5. R. T. Davenport and R. Brook, "The Threshold Stress Intensity Range in Fatigue," Fat. Eng. Mat. Struct. 1, (1979), pp. 151-158.

6. G. R. Irwin, "Structural Aspects of Brittle Fracture," App. Mat. Res. 3, (1964), pp. 65-81.

7. D. Bristow, Private Communication, University of Sheffield, July 1983.

8. A. J. Cadman, R. Brook and C. E. Nicholson, "Effect of Test Techniques on the Fatigue Threshold (ΔK_{th})" pp. 59-75 in Proc. Int. Symp. "Fatigue Thresholds," Vol. 1, Stockholm, Sweden, EMAS Ltd., U.K. 1981.

9. A. J. Cadman, R. Brook and C. E. Nicholson, "Factors Affecting the Meausrement of the Fatigue Threshold (ΔK_{th})", Scripta Met. 17, (1983) pp. 1053-1056.

10. W. Elber "The Significance of Fatigue Crack Closure in Damage Tolerance in Aircraft Structures," pp. 230-243 in STP.486, ASTM, 1971.

11. S. Suresh, "Micro-mechanisms of Fatigue Crack Growth Retardation following Overloads," Eng. Fract.Mech. 18, (1983), pp. 577-594.

12. S. Suresh, "Crack Deflection: Implications for the Growth of Long and Short Cracks," Met. Trans. 14A, (1983) pp. 2375-2385.

13. S. Suresh, G. F. Zamiski and R. O. Ritchie, "Oxidation and Crack Closure. An Explanation for Near-Threshold Corrosion Fatigue Crack Growth Behaviour," Ibid., 12A, (1981), pp. 1435-1443.

14. T. C. Lindley and C. E. Richards, "Near-Threshold Fatigue Crack Growth in Materials used in the Electricity Supply Industry," pp. 1087-1113 in Proc. Int. Symp. "Fatigue Thresholds" vol. 2, Stockholm, Sweden, EMAS Ltd., U.K. (1981).

15. M. Klesnil and P. Lukas, "Effect of Stress Cycle Asymmetry on Fatigue Crack Growth," Mat. Sci. Eng. 9, (1972), pp. 231-240.

16. R. A. Schmidt and P. C. Paris, "Threshold for Fatigue Crack Propagation and the Effects of Load Ratio and Frequency," pp. 79-94 in STP 536, ASTM, (1983).

17. I. M. Austen, "Measurement of Fatigue Crack Growth Threshold Values for the Use in Design," B.S.C. Technical Report SH/EM/9708/2/83/B British Steel Corporation, (1983).

18. S. J. Garwood and C. F. Boulton, Cumulative Damage to Welded Steel Structures," Final Report EUR 7635. Commission of the European Communities, Luxembourg, (1982.)

19. S. Suresh and R. O. Ritchie, "The Propagation of Short Fatigue Cracks," Report No. UCB/RP/83/1014, Department of Materials Science and Minerals Engineering, University of California at Berkeley, (1983)

20. C. J. Beevers, "Fatigue Crack Growth Characteristics at Low Stress Intensities of Metals and Alloys," Metals Science, 11, (1977), pp. 362-367.

21. J. Blacktop, C. E. Nicholson, R. Brook and R. T. Towers, "The Effect of Cold Deformation on the Fatigue Threshold, pp. 629-638 in Proc. Int. Symp. "Fatigue Thresholds" Stockholme, Sweden, EMAS, Ltd., U.K. (1981).

22. A. Hartman and J. Schjive, "The Effects of Environment and Load Frequency on the Crack Propagation Law for Macro Fatigue Crack Growth in Aluminium Alloys," Eng. Fract. Mech., 1, (1970) pp. 615-631.

23. D. W. Hoeppner and W. E. Krupp, "Prediction of Component Life by Application of Fatigue Crack Growth Knowledge," Eng. Fract. Mech, 6, (1974), pp. 47-70.

24. I. M. Austen and E. F. Walker, "An Analysis of the Applicability of Empirical Fatigue Crack Growth Relationships," B.S.C. Report SH/PT/6795/9/79/B. British Steel Corporation, (1979).

25. I. M. Austen, "Representation of Fatigue Crack Growth Information for use in Remnant Life Calculations," B.S.C. Technical Note GSP/EM/6/83/B, British Steel Corporation, (1983).

26. R. T. Davenport, "A Review of Fatigue Crack Growth Laws and Some Recommendations for Calculating Residual Fatigue Life," Report No. NW/SSD/SR/69/81. Central Electricity Generating Board (1982).

27. K. Priddle, "Some Effects of Temperature, Vacuum and CO_2 Environments on Fatigue Crack Growth and Threshold for En3A Mild Steel and Weld Metal," Report RD/4990/N81, Central Electricity Generating Board, (1981).

28. D. Taylor, "A Model for the Estimation of Fatigue Threshold Stress Intensities in Materials with Various Different Microstructures," Fatigue Thresholds, V. I, 455-470, EMAS, Cradley Heath, UK, 1982.

Small Cracks

THE EFFECTS OF TEXTURE AND GRAIN SIZE ON THE SHORT FATIGUE CRACK GROWTH RATES IN Ti-6Al-4V

C. W. BROWN* AND D. TAYLOR**

* MATERIALS ENGINEERING GROUP
ROLLS-ROYCE LTD, DERBY, ENGLAND

**DEPARTMENT OF MECHANICAL ENGINEERING
TRINITY COLLEGE, DUBLIN, IRELAND

The influence of test piece orientation and material microstructure on the propagation rates of short surface thumbnail cracks have been examined in Ti-6Al-4V using cellulose acetate replicas. In addition, attempts were made to determine the threshold values for propagation of these cracks by both an annealing and a normal load shedding technique. The large degree of scatter found in the data restricted the quantitative conclusions that could be reached, however it was evident that variation in short crack growth rates could be achieved through microstructural control. The results are discussed in terms of the well established influence of such parameters on near threshold long crack propagation behaviour.

Introduction

The discrepancy between the propagation rates of "long" and "short" cracks when analysed in terms of conventional linear elastic fracture mechanics (LEFM) is now well established. These differences have been attributed to a number of factors, notably microstructural size (1), and stress level (2), but all of which are linked through the inability of plane strain LEFM to model the actual deformation processes occurring at a crack tip in a real material. It is clear that no fatigue crack propagates as a planar, infinitely sharp defect in an elastic continium and therefore that the acknowledged success of LEFM in quantifying fatigue crack growth has been in spite of this. Indeed it has been suggested (3), this success has only been due to the material under test responding in the same way when a sufficiently large volume is subjected to identical stressing conditions. It is therefore highly likely that the rationalisation of long crack behaviour trends with changes of material, microstructure, environment, loading sequence, etc. cannot necessarily be read across to the short crack case where individual units of the microstructure are controlling.

The effects of grain size (4,5) and test-piece orientation (6,7) have been the topics of many long crack studies. There is a general concensus that increasing grain size reduces crack growth rates, particularly at low stress intensities, and results in a higher threshold for fatigue crack propagation. Thus effect has been rationalised by crack face "non-closure" (8) or "roughness induced" closure (9) which is increased in the coarse grained material due to the structure sensitive nature of the crack extension when the crack tip plastic zone is smaller than the grain size.

There is less agreement, however, with regard to the influence of test-piece orientation on the fatigue behaviour of textured titanium alloys. It has been shown with other materials (10) that many effects can be accounted for by normalising the applied stress intensity with respect to Young's modulus. However, in titanium alloys, changing orientation can also result in a change of fracture mechanism and thereby mask the underlying behaviour. Indeed, Bowen (6) has proposed several such modes of crack advance which correlate well with the fractography of his tests performed in laboratory air at room temperature. However, Wanhill (7) has shown that these effects of test piece orientation can either be exaggerated or even reversed dependant upon environment. It is therefore clear that if any of these long crack results are to be read across to the field of short crack growth then the micromechanisms of crack extension in titanium alloys become of prime significance.

Data on the growth rates of short cracks in titanium alloys is relatively scarce, although that which does exist suggests their propagation rates to be very fast when compared to long crack data in identical material (11). This comparison is critically dependant upon accurate measurement of the long crack behaviour, however, it would appear that microstructure rather than modulus represents the better method of characterising short crack behaviour in the first instance.

Specifically, for aluminium alloys, Morris (12) and Lankford (13) have examined the interaction between the growth rate of a propagating fatigue crack and the proximity of the nearest grain boundry. Both agree that the boundary can cause arrests in the crack extension and thereby increase the overall fatigue life if a finer grained material is used. However, Morris (12) showed that this benefit was lost to some extent by the acceleration which occurred immediately prior to the crack tip reaching the boundary due to a decrease in plasticity induced crack closure.

Few such detailed experiments have been performed on titanium alloys however data for different microstructures of IMI 685 (11) showed that an increase in the proportion of high angle grain boundaries led to a significant reduction in the overall crack propagation rate. Indeed, Morris (14) has also performed work to show that this overall average growth rate can be correlated with long crack growth rate data through the use of closure concepts. The degree and type of closure however depends critically on material grain size, yield strength, R ratio, applied maximum stress, environment and proximity of certain microstructural features to the crack tip and hence analytical treatment of the problem has so far been limited to particular conditions where detailed microscopical measurements have been made (15).

Bearing in mind the critical application for these materials, improvements in short crack growth resistance are obviously desirable and represent the only true way of increasing the defect tolerance of highly stressed components, where the critical crack size at fracture is very small.

The paper describes an investigation into two possible microstructural variables, grain size and texture, in a widely used titanium alloy to compare and contrast the apparent benefits separately manifest in long and short crack growth resistance.

Materials and Experimental Techniques

Material, in the form of 100mm square billet, was supplied by IMI with a standard mill annealed microstructure (Figure 1). A coarsened grain size but still consisting of equiaxed α and intergranular β was produced by heat treating high in the $\alpha - \beta$ phase field (950°C for 8hrs) followed by furnace cooling. This resulted in an increase in average grain size as calculated from the mean linear intercept from 4.7 to 11.7 μm.

To examine the effects of orientation on fatigue crack propagation both long and short crack test pieces were cut in the orientations shown in Figure 2. For both heat treatments the predominant crystallographic texture was as shown with a prismatic plane pole coincident with the rolling direction.

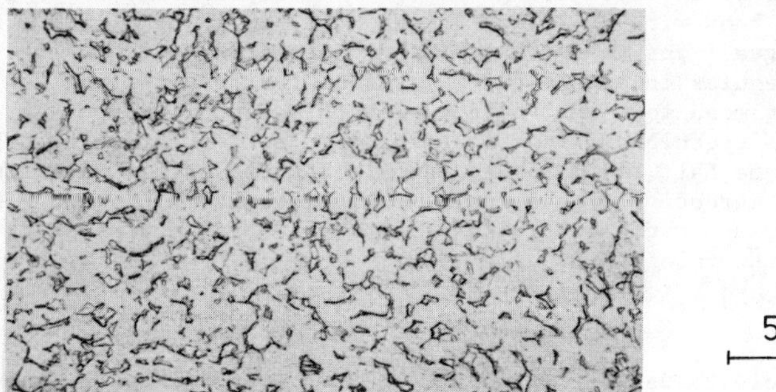

50 μm

Figure 1 Microstructure of as-received material

The long crack test pieces were of a standard four point bend loading design with the crack length being monitored continuously by a DC electrical potential drop technique. Loading was performed at a frequency of 70Hz with a ratio of minimum to maximum load (R) in the cycle of 0.2. Following crack initiation from the notch a step-down technique was employed to reduce the crack growth rates to threshold values before quantitative measurements of propagation rates were made under increasing K conditions. For this testing procedure, allowing for the electrical stability of the monitoring apparatus the minimum measurable crack growth rate and hence that at which the threshold was determined was 10^{-11} m/cycle.

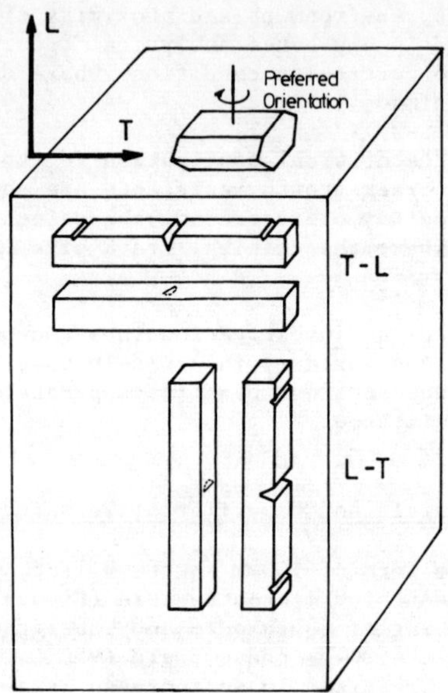

Figure 2. Orientations of test pieces as cut from the original bar. The schematic unit cell indicates the preferential crystallographic orientation induced by forging.

The short crack test pieces comprised a 10mm square bar, again cut in both orientations, with an electropolished and etched section on one face. With a monotonic yeild strength for this material of 900MPa loading in four point bend with $R(= \sigma_{min}/\sigma_{max}) = 0.2$ and $\sigma_{max} = 750$MPa resulted in total fatigue lives to failure of the order of 2×10^5 cycles. During this time, at regular intervals of $\sim 10^4$ cycles the load cycling was interrupted and held at mean load whilst crack length was monitored using a plastic replication technique (16). Assuming the depth of cracking below the surface to be half the measured surface length, stress intensities were calculated according to the analysis of Pickard (17) for pure bending. On average, under the conditions investigated for cracks up to 1mm deep this gave the relationship

$$\Delta K \text{ (MPa } \sqrt{(\text{m}}) \approx 0.7 \sqrt{a(\mu m)}$$

The threshold values for propagation of these short cracks were measured using two approaches:-

a) a load shedding method where reductions in amplitude of approximately 25% had to be made to cause crack arrest at short crack lengths.

b) a vacuum annealing technique where a pre-cracked bar containing a crack initiated at high stress amplitude was heated at 600°C for 30 minutes before cycling was recommenced at a low amplitude estimated to be near the threshold.

Both approaches differ greatly from the long crack technique and may lead to large inaccuracies in the values of threshold calculated.

Experimental Results

The fatigue crack growth rates in the as-received material for the T-L orientation are shown inFigure 3 for the long cracks and in Figure 4 for the short cracks. The dotted lines shown in these figures represent the best estimate mean lines through the data and are reproduced for comparison in Figures 6 to 9.

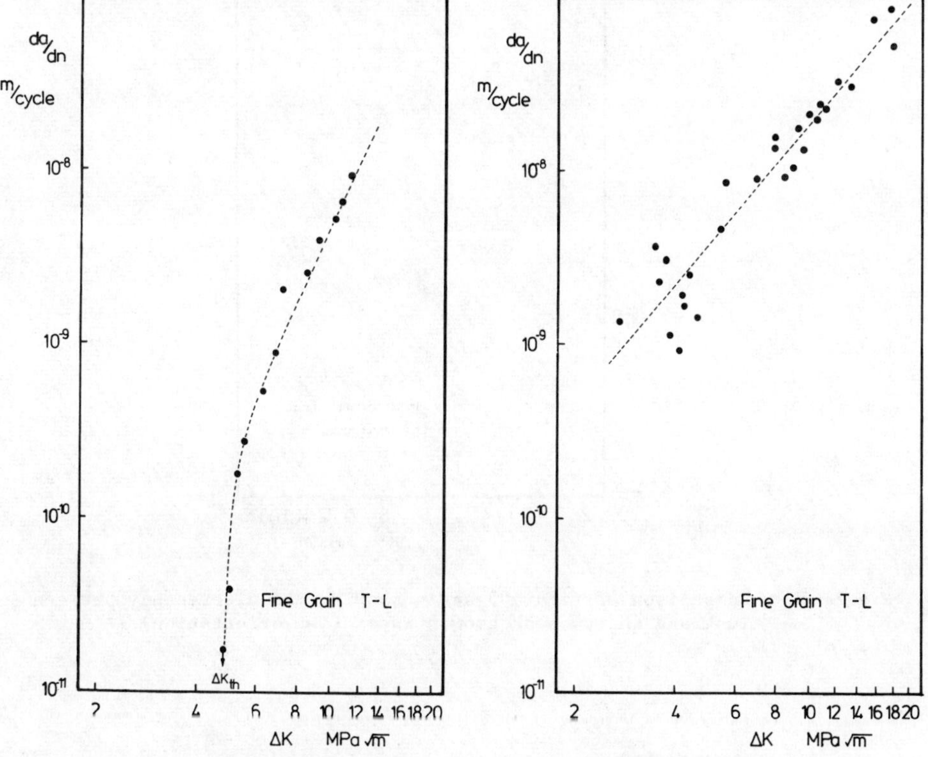

Figure 3. Near threshold fatigue crack growth rates for the fine grained material. T–L orientation – long crack

Figure 4 Crack growth rates for the fine grained material T–L orientation – short crack

It can be seen that the propagation of short cracks occurs at stress intensities well below the long crack threshold (Figure 5). The dotted line for the short crack case represents the simple power law relationship:-

$$\frac{da}{dn} = 10^{-10} \Delta K^{2.23}$$

where da/dn is in m/cycle

ΔK is in $MPa/\sqrt{m}$

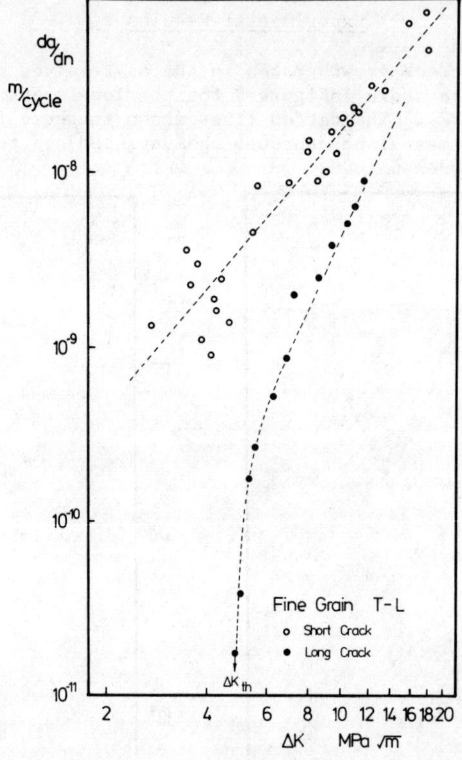

Figure 5. Comparison of figures 3 and 4 showing the discrepancy between long and short crack growth rates (T-L orientation)

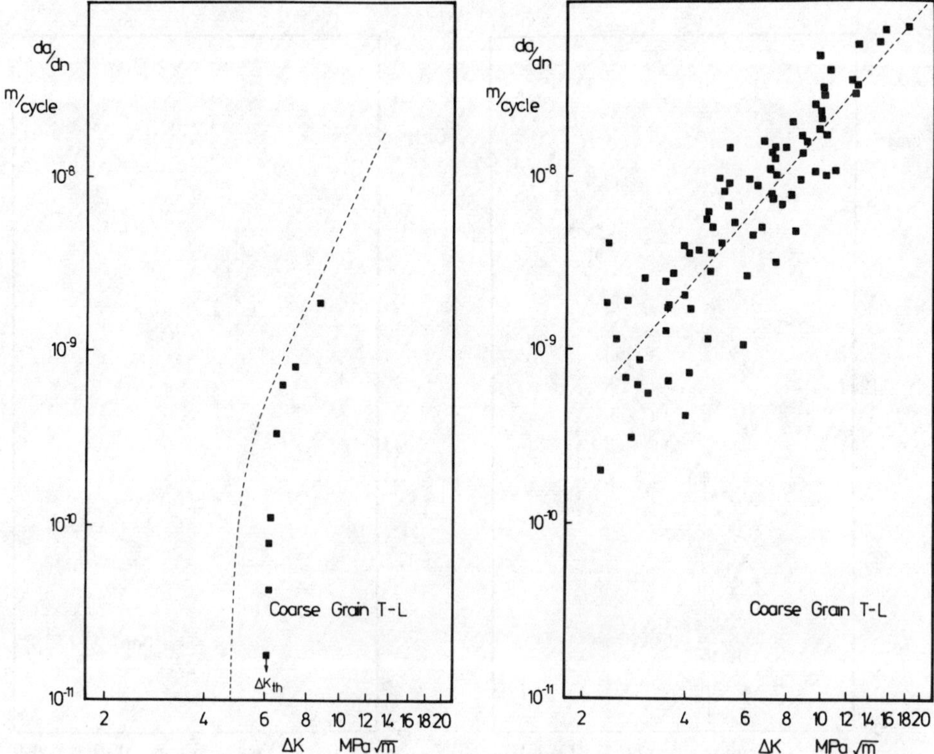

Figure 6 Near threshold fatigue crack growth rates for the coarse grained material T–L orientation – long crack

Figure 7 Crack growth rates for the coarse grained material. T–L orientation – short crack.

The corresponding behaviour of coarse grained material for both long and short cracks is reproduced in Figures 6 and 7. It is interesting to note that although the threshold is raised for this microstructure and the propagation rates are reduced for the long cracks, nevertheless the average behaviour of the short cracks is unaffected by the change in grain size. However, the scatter in the data is greater in the latter case and the possible relevance of this is discussed later.

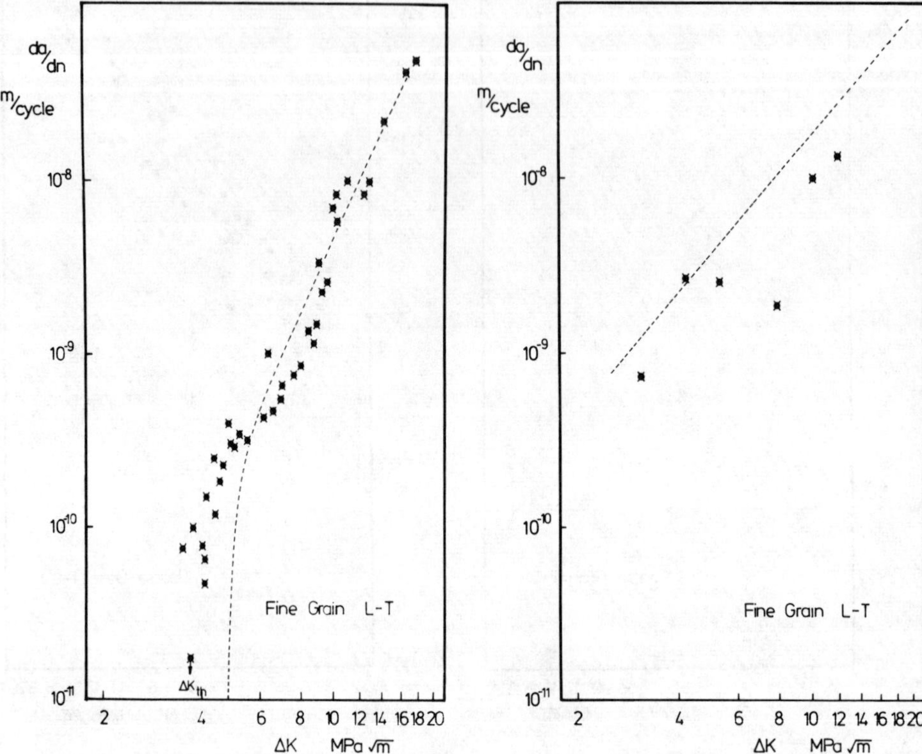

Figure 8 Near threshold fatigue crack growth rates for the fine grained material – L–T orientation – long crack

Figure 9 Crack growth rates for the fine grained material L–T orientation – short cracks.

A change to the L–T orientation using the fine grained microstructure shows a significant increase in near threshold growth rates, but at the same time, with the limited data collected, tends to indicate slower short crack propagation rates (figures 8 and 9)

The results of the threshold measurements for short cracks in all three cases are presented in figure 10. In detail those results with "error" bars represent load reduction tests were growth was achieved at a particular high value of stress intensity but not maintained at the subsequently reduced level, whilst those results with arrows are from annealing tests where growth was either achieved (arrow down) or not (arrow up) at the chosen K level for cycling after stress relief.

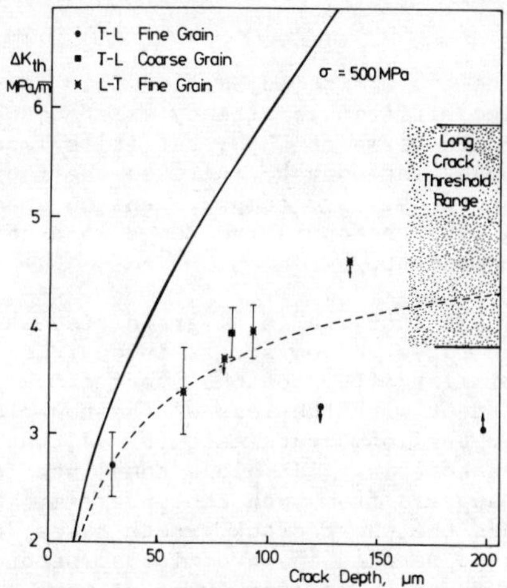

Figure 10. Estimates of the short crack growth threshold as a function of crack length for all three conditions.

As expected, a gread deal of scatter exists within the results making it impossible to draw quantitative conclusions or even to separate the different conditions tested. However, it is clear that all of the data is consistently below all the long crack values as well as the solid line representing the constant value of stress amplitude (500MPa) corresponding to the fatigue limit. The dotted line has been calculated using a slight modification to the analysis of El Haddad et al (18) assuming

$$\Delta K_{th} \text{ (short crack) } = \sqrt{\frac{a}{a+1}} \; \Delta K_{th} \text{ (long crack)}$$

where a is the crack depth and 1 is an extra "effective crack length" given by the equation

$$1 = \frac{1}{\pi Y^2} \; \frac{\Delta K_{th}^{\;2}}{\Delta \sigma_{fl}^{\;2}}$$

where ΔK_{th} is the long crack propagation threshold (5MPa $\sqrt{m}$)

$\Delta\sigma_{fl}$ is the smooth specimen fatigue limit (500 MPa)

Y is the compliance for a surface semi-circular crack.

By substitution of these experimental values 1 is calculated to be 69 μm.

Discussion

Both "long" and "short" fatigue crack growth rates have been successfully measured for different orientations of test piece from the initial billet and also different equiaxed α grain sizes for the same orientation.

These have highlighted the fact that changes in fatigue crack growth behaviour as measured in standard long crack test pieces are not necessarily reflected in the resistance to crack propagation in the early stages of the fatigue life when the crack size is small. This is perhaps not surprising when large differences already exist between long and short cracks when analysed in terms of LEFM, but it is important to note that microstructural changes intended to increase the fatigue resistance of a material as evidenced by, say, compact tension specimen data, may well accelerate cracking over the operative range in a service situation with a highly stressed component.

In particular, the increase in α grain size which resulted in reduced crack growth rates at low stress intensities for the long crack had apparently no similar effect on the short crack growth behaviour. This would be consistent with the loss of the non-closure of the crack faces postulated for the long crack case (8, 9), when the crack was too short for asperities to exist, let alone come into contact by crack tip shear. Indeed, it appears from both the indeterminate short crack threshold results and the short crack growth rates, that there is no difference in growth rate between the two microstructures. Importantly, this is in spite of there being over twice as many grain boundaries per unit length of crack extension in the fine grained case. Hence, ignoring all other effects, it would appear that over the range investigated, grain boundaries do not cause significant arrests in crack growth and that the principal contribution to the fatigue resistance arises from the intrinsic behaviour of the hexagonal matrix. However, this is not to say that significant arrests cannot occur, particularly at lower stress amplitudes (11) and moreover that the long crack improvement with a coarser grain size does not develop as the crack extends. Indeed it could well be the development of this long crack type roughness induced closure together with the overall coarseness of the structure that leads to the greater degree of scatter seen for the coarse grained material short crack growth rates over the fine-gained data.

A change in the orientation of the test pieces also produced different responses in the short and long crack regimes. Whereas a reduction in the fatigue crack propagation threshold occurred with the L-T orientation for long cracks a benefit, of a factor of two, in the mean line values was seen in the short crack growth resistance. The behaviour of the long cracks has been discussed previously (5), particularly in terms of the role of roughness induced crack closure arising from the coarse prior β grain size however, accepting the lack of an effect of grain size on short crack growth it is interesting to speculate on the apparent benefits achieved.

The initial difficulty is that the slower crack propagation rates are occurring when loading in the lower modulus direction contra to the general expectation (10). However, in discussing work on the effect of texture on the smooth specimen fatigue life Bowen (19), proposed that the changing Poisson's ratio as well as modulus permitted greater shear strains to be achieved when loading along the 'C' axis of the unit cell. This involved some degree of strain rather than stress control in the loading cycle to overcome the higher modulus in this direction reducing the overall levels of strain.

When observing the same effect of the highest modulus directions giving the lowest resistance to fatigue crack propagation Wanhill (7) however, favoured an explanation based on fracture mechanism. By changing environment from dry argon to 3.5% aqueous NaCl he produced an increased proportion of cleavage type features on the fatigue fracture surface and moreover increased the growth rate by over an order of magnitude. By comparing the orientation of the facets produced with the crystallographic texture of his material he inferred them to be associated with failure paralell to the basal plane of the hexagonal unit cell. Assuming laboratory air to be an aggressive environment for the material tested here, the T-L orientation would be more prone to such basal plane cracking than the L-T due to the crystallographic texture present (Figure 2), consistent with the faster propagation rates seen in this case.

When this increased proportion of facetted growth is translated to the long crack case it would result in a microstructurally rougher fracture surface. This would raise the level at which non-closure occurs due to asperity contact and thereby reduce the effective crack tip stress intensity range and hence growth rate as observed. These suggestions are consistent with the ideas of Beevers (20) that long crack growth behaviour is determined by two factors; firstly, the inherent resistance of the microstructure to crack propagation and secondly, the effect of the growth mode on the subsequent stress intensity range experienced at the crack tip.

With regard to the threshold measured for short cracks it is impossible to differentiate between the separate conditions but it is interesting to note the apparent success of the approach of El Haddad et al (18) in predicting their variation with crack length. This prediction relies on determining a value of effective crack length which, for this material is calculated to be 69 μm. This figure directly implies that true long crack behaviour is not even approached until cracks are longer than at least 500 μm. This is in direct contrast to the prediction of Taylor (1) which even for the larger grain size of 11.7 μm would expect true long crack behaviour with crack lengths greater than 120 μm. However, re-examination of the microstructure of both heat treatments under polarised light (Figure 11) revealed that much larger areas than single α grains showed identical contrast. This suggests similar crystallographic orientations for neighbouring α grains and therefore the possibility of co-operative deformation within such areas. Measuring these similarly oriented colonies to be approximately 300 μm across then indicates, according to Taylor (1) that short crack growth could indeed still be operative up to crack lengths of 3mm, consistent with our observations.

Figure 11 Similarly oriented colony size - fine grained material T-L - (polarised light)

Such co-operative deformation would also be consistent with the apparent inability of most grain boundaries in the fine grained material to cause significant crack arrests, as seen in the short crack growth behaviour, with the only effective boundaries being those at the colony perimeter. Importantly such behaviour might always be expected from microstructures which arise directly as a crystallographic transformation product from a single phase stable at the solution treatment temperature.

Conclusions

1. The influence of changes in grain size and test piece orientation on fatigue crack growth resistance are different in the long and short crack regimes. These differences are suggested to arise predominantly from crack growth mechanisms that give rise to roughness induced crack closure.

2. Short crack growth effects are maintained up to unexpectedly long crack lengths due to correlations between the orientations of adjacent α grains.

Acknowledgements

The authors would like to thank Professor R W Honeycombe FRS for the provision of Laboratory facilities at Cambridge University and the SERC for providing financial support. They would also like to thank Mrs J A Norris for her typing of the manuscript, and Mr T A Byrne for his help with the photographic work.

REFERENCES

1. D. Taylor and J. F. Knott, "Fatigue Crack Propagation Behaviour of Short Cracks. The Effect of Microstructure", Fatigue of Engineering Materials and Structures 4 (1981) pp147 - 155.

2. K.J. Miller, "The Short Crack Problem", Fatigue of Engineering Materials and Structures 5 (1982) pp223 - 232.

3. J. Schijve, "Differences between the Growth of Small and Large Fatigue Cracks : The Relation to Threshold K Values" pp881 - 908 in Fatigue Thresholds (2) J. Backlund, A Blom and C.J. Beevers, eds.; EMAS, Warley, UK 1981.

4. G.R. Yoder, L.A. Cooley and T.W. Crooker, "50-fold Difference in Region II Fatigue Crack Propagation Resistance of Titanium Alloys. A Grain Size Effect", Journal of Engineering Materials Technology 101 (1979) pp86 - 90.

5. C. W. Brown and G. C. Smith, "The Effect of Microstructure and Texture on the Fatigue Crack Growth Threshold in Ti-6Al-4V" pp329 - 343 in Fatigue Threshold (1) J. Backlund, A Blom and C. J. Beevers, eds; EMAS, Warley, UK., 1981.

6. A. W. Bowen, "The Relationship between Fatigue Crack Growth Rate and Texture in Ti-6Al-4V", pp446 - 450 in The Microstructure and Design of Alloys (1), Institute of Metals, London, 1973.

7. R. J. H. Wanhill, "Environmental Fatigue Crack Propagation in Ti-6Al-4V Sheet", Metallurgical Transactions 7A (1976) pp1365 - 1373

8. M. A. Hicks and J. E. King, "Temperature Effects on Fatigue Thresholds and Structure Sensitive Crack Growth in a Nickel Base Superalloy", International Journal of Fatigue 5 (1983) pp67 - 74.

9. R. O. Ritchie and S. Suresh, "Some Considerations on Fatigue Crack Closure at Near-Threshold Stress Intensities due to Fracture Surface Morphology", Metallurgical Transactions 13A (1982) pp937 - 940

10. A. R. Rosenfield, "An Analysis of Reported Fatigue Crack Growth Rate Data with Special Reference to Ti-6Al-4V", Engineering Fracture Mechanics 9 (1977) pp509 - 520.

11. C. W. Brown and M. A. Hicks, "A Study of Short Fatigue Crack Growth Behaviour in Titanium Alloy IMI685", Fatigue of Engineering Materials and Structures 6 (1983) pp67 - 76.

12. W. L. Morris, "The Non-Continuum Crack Tip Deformation Behaviour of Surface Microcracks", Metallurgical Transactions 11A (1980) pp1117 - 1123

13. J. Lankford, "The Growth of Small Fatigue Cracks in 7075-T6 Aluminium" Fatigue of Engineering Materials and Structures 5 (1982) pp233 - 248

14. W. L. Morris and M. R. James, "A Simple Model of Stress Intensity Range Threshold and Crack Closure Stress", Engineering Fracture Mechanics 18 (1983) pp871 - 877

15. M. R. James and W. L. Morris, "Effect of Fracture Surface Roughness on Growth of Short Fatigue Cracks", Metallurgical Transactions 14A (1983) pp153 - 155.

16. C. W. Brown and G. C. Smith "A Two-Stage Plastic Replication Technique for Monitoring Fatigue Crack Initiation and Early Fatigue Crack Growth" pp41 - 52 in Advances in Crack Length Measurement, C. J. Beevers, ed.; EMAS, Warley, UK., 1982

17. A. C. Pickard, "Stress Intensity Factors for Cracks with Circular and Elliptic Crack Fronts Determined by 3D Finite Element Methods" paper presented at 2nd Int. Conf on Numerical Methods in Fracture Mechanics Swansea, 1980

18. M. H. El Haddad, K. N. Smith and T. H. Topper, "Fatigue Crack Propagation of Short Cracks" Journal of Engineering Materials Technology 101 (1979) pp42 - 46

19. A. W. Bowen, "The Effect of Testing Direction on the Fatigue and Tensile Properties of Ti-6Al-4V Bar", pp1271 - 1281 in Titanium Science and Technology (2) R. I. Jaffee and H. M. Burte, eds.; Plenum Press, 1972.

20. C. J. Beevers, "Some Aspects of the Influence of Microstructure and Environment on ΔK Thresholds, pp257 - 276 in Fatigue Thresholds (1), J. Backlund, A. Blom and C. J. Beevers, eds.; EMAS, Warley, UK., 1981.

NEAR-THRESHOLD CRACK TIP STRAIN AND CRACK OPENING FOR LARGE AND SMALL FATIGUE CRACKS

J. Lankford and D. L. Davidson

Southwest Research Institute
6220 Culebra Road
San Antonio, Texas 78284, USA

The advantage of correlating fatigue crack growth rates for large cracks using concepts of linear-elastic fracture mechanics is that stress and crack length are replaced by a single unifying parameter--the cyclic stress intensity factor, ΔK. When ΔK is computed for small cracks, however, growth rates are found to be much faster than those for large cracks, and small cracks grow below the threshold ΔK for large cracks. The research reported here addresses the reasons for these differences in crack growth rate, the origin of the threshold ΔK for large cracks, and questions the use of ΔK in correlating the growth of fatigue microcracks. Crack tip strains and crack opening displacement have been measured for small cracks ranging from 35 to 200 μm, and the results compared with similar measurements from large cracks. The similitude between these parameters, which is found for large cracks of different length, is not maintained for small cracks. Threshold ΔK for large cracks is found to correspond to the stress intensity for which crack tip strain is reduced to the elastic limit.

Acknowledgement

The authors are grateful for the support of the U.S. Army Research Office, under contract number DAAG29-80-K-0034.

Introduction

Within the last few years, the "small crack problem" has come to be appreciated (1) as a significant factor in the prediction of fatigue lifetimes using linear-elastic fracture mechanics. The anomalous behavior of "small" cracks has been observed in a wide range of metal systems (aluminum (2-4), titanium (5), and ferrous (6) alloys), and can be summarized as follows.

First, cracks are often considered small when they are on the same scale as either a relevant microstructural element or the plastic zone within which they reside. They also may simply be physically small, i.e., $\lesssim$ 0.5-1 mm (1). The authors consider a fatigue microcrack* "small" when its rate of crack growth, under elastic net-section stress conditions, does not correlate, through ΔK, with that predicted by experiments using conventional through-crack specimens subject to nominally identical loading conditions. For example, for 7075 aluminum alloy, surface cracks of less than 300 μm have been found (3,4) to fit this definition. The behavior of such flaws is compared to the growth of a single-edge-notch through crack in Figure 1, which is based upon extensive crack growth measurements (as documented elsewhere (3,4)) for several dozen individual microcracks. It is apparent that small cracks and large cracks in this material differ in several important respects. The major difference is that small cracks grow at cyclic stress intensities below the stress intensity threshold (ΔK_{Th}) measured for large cracks. Furthermore, small cracks grow at much faster rates than do equivalent through cracks within the transition from the threshold to the Paris region. At a characteristic ΔK, hence characteristic crack length for a constant cyclic stress test, crack growth may experience transient deceleration, or even arrest. Throughout this region, and at lower stress intensities as well, the rate of growth is very high, and frequently independent of ΔK, corresponding to the apparent absence of a threshold stress intensity.

Considerable progress has recently been made in establishing the basis for some of these trends. For example, it has been determined that the characteristic average crack length for crack arrest in 7075-T651 (3) correlates quite well with the minimum grain dimension, suggesting grain boundary/crack tip interaction as the source of crack retardation/arrest. Further, an analysis (7) based on the alteration of crack tip slip due to resolved shear stress differences in neighboring grains has proven successful in modeling the varying degrees of retardation shown in Figure 1.

Recent measurements of crack opening loads in aluminum alloys have shown (8,9) that for large cracks, the ratio of opening load to maximum cyclic load, P_{op}/P_{max}, approaches unity as ΔK_{Th} is approached. This virtual absence of opening essentially explains the existence of the conventional, large crack ΔK_{Th}. The opening of small cracks, on the other hand, has recently been observed directly by the present authors (8), using a special cyclic loading stage within the SEM (10); under these conditions, it was found that microcracks in 7075-T651 consistently opened at $P_{op}/P_{max} \simeq$ 0.6 over a range in ΔK extending to well below the large crack ΔK_{Th}.

The latter observation affords one possible explanation of the observed ΔK-independence of, and lack of a threshold for, cyclic microcrack

---

*In this paper, "small crack" and "microcrack" will be considered synonymous, while "large" crack refers to a through-crack whose length is of the same order of size as the specimen thickness.

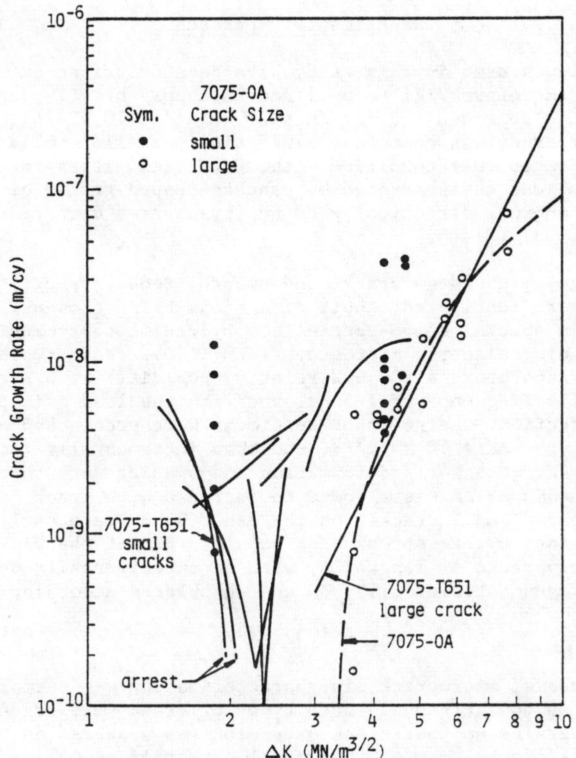

Figure 1. Growth of microcracks compared with large cracks, for 7075-OA and 7075-T651.

extension. However, this alone does <u>not</u> answer the critical question as to why the average rate of growth of small cracks is so fast. Other (8) recent measurements by the authors have shown that the crack tip openings of small cracks are much greater than those of large cracks at equivalent stress intensities. In addition, efforts to "fix" ΔK empirically, and thereby bring the large crack and small crack growth data into coincidence were unsuccessful (8). These experiences suggest that small scale yielding conditions may break down for the small crack, and that crack tip yielding (but not necessarily crack extension) may be fundamentally different for large cracks versus small ones. This likelihood is supported by recent measurements of plastic zones attending microcracks at low stress intensities in 7075-T651. Results (8) indicate that the ratio of plastic zone radius to crack size is nearly unity, far above the minimum (~0.02) usually considered valid (11) for small scale yielding approximations such as ΔK.

The purpose of the present work is to determine experimentally the relationships between crack tip opening and crack tip strain for large and small fatigue cracks. Effort is focussed on the near-threshold ΔK region of crack growth, and interest is centered on material behavior within a few microns of the actual crack tip. Results are interpreted in terms of 1) requirements for modeling microcrack growth, 2) the origin of the crack growth threshold, and 3) the applicability of the threshold concept.

Experimental Approach

The techniques used in this study have been described extensively elsewhere, and therefore will be outlined here only briefly; appropriate references should be consulted for further details (10,12-14). The material chosen for study was commercial 7075 aluminum alloy rolled plate material in the overaged (OA) condition. The 0.2% yield strength was 439 MPa, and the material was characterized by pancake-shaped grains of average dimensions 18 μm (normal direction) x 80 μm (transverse direction) x 150 μm (rolling direction) (3).

Single-edge-notch large crack, and smooth, reduced section small crack specimens (3) were fabricated, their dimensions being chosen so as to allow cycling within a special servo-controlled, hydraulic test machine situated within the scanning electron microscope (SEM) (10). Both types of specimen were tested in laboratory air (~60% relative humidity) at a stress ratio of ~0.1, at a cyclic frequency of 1-5 Hz, and with the load axis parallel to the rolling direction. Large crack specimens were precracked at a cyclic stress intensity of $\Delta K = 10\ MN/m^{3/2}$, and then incrementally load shed to near-threshold ΔK. Crack tip deformation and opening were then periodically observed in the SEM as ΔK was allowed to increase with crack length at constant cyclic load. Small cracks, on the other hand, were nucleated and grown at a constant cyclic stress ($\Delta\sigma$) equal to 80% of the yield strength. For surface microcracks of length 2a, with an experimentally-determined half-length to depth ratio of 1.15, ΔK was calculated according to (12,13)

$$\Delta K = 1.32\ \Delta\sigma \sqrt{a} \qquad (1)$$

As for large cracks, microcrack tip characterization was performed in the SEM, as crack length increased. Both types of crack were oriented relative to the pancake grains so that crack extension was measured in the transverse direction; small cracks grew in the normal direction as well.

Crack tip characterization was achieved by means of stereoimaging analysis (13) of high magnification photomicrographs of crack tip regions under loaded and unloaded conditions. Quantitative measurements (14) taken from these photographs yielded both crack opening displacement and the crack tip displacement field; differentiation of the latter gave the corresponding crack tip strain field. Stereoimaging observation of cracks at various fractions of the load cycle provided accurate values of crack opening loads.

Results

Data derived from the near-crack tip measurements are listed parametrically in Tables I-III. Various aspects of large and small crack tip behavior are compared graphically in Figures 2-10. Crack growth rates for the microcracks studied in this research, as well as corresponding large crack data, are presented in Figure 1. For comparative purposes, more complete crack growth rate data for 7075-T651 (3) also are shown. It is evident that for both ΔK levels studied here, small cracks are growing at much faster rates than would large cracks at the same value of ΔK. Furthermore, the average rate of microcrack growth at $\Delta K = 1.9\ MN/m^{3/2}$ is almost equal to that at $\Delta K = 4.2\ MN/m^{3/2}$, suggesting the absence of a crack growth threshold.

The displacements for a surface microcrack and a large crack (Figure 2) are observed to be qualitatively similar in that there is a large Mode I crack opening component (parallel to the load axis) for both. However, the large crack evidences more Mode II opening than does the small crack. The large crack displacement diagram which is shown was chosen for this comparison because it has a value of crack tip opening displacement approximately the same as for the small crack, although the significantly greater ΔK for the large crack should be noted.

The distribution of maximum shear strain near the tips of both a large and a small crack are compared in Figure 3. Computed ΔK values for both are nearly the same. As may be seen, the strain value at the tip of the microcrack (.132) is much larger than the corresponding value (.063) for the large crack. Also, the strain distribution for the microcrack decreases rapidly ahead of the crack tip, but drops off more gradually to the sides. While decreasing the length of the microcrack diminishes the spatial extent of the high strain region, it does not significantly decrease the local crack tip shear strain (.098), as is shown in Figure 4. In marked contrast, a typical strain distribution for a large crack at a higher ΔK is shown in Figure 5. In this case, the strain peaks strongly at the crack tip and decreases in all directions (compare with Figures 3 and 4 for microcracks).

The distribution of strain ahead of the crack tip is plotted for both a large crack and an equivalent microcrack in Figure 6. The value of strain at the crack tip was used, in addition to the data shown, to fit the following equation:

$$\Delta\varepsilon_t^{eff} = \frac{A}{(B+r)^2} \tag{2}$$

Table I. Parameters for Small Cracks; 7075-OA Grown in Air

Crack Length μm	ΔK $MN/m^{3/2}$	Crack Tip Strain $\Delta\varepsilon_t^{eff}(0)$	A	B	r_p^o μm	C_o μm	p
168	4.2	.077	38.2	22.3	52	.418	.408
206	4.65	.113	50.3	21.7	64	.454	.412
208	4.65	.082	65.7	28.3	69	.165	.640
168	4.2	.090	66.3	26.4	72	.525	.354
206	4.65	.10	27.7	16.8	47	.406	.391
36	1.9	.148	13.0	12.0	31	.100	.402
36	1.9	.112	30.1	17.5	48	.162	.504
36	1.9	.079	7.5	9.7	23	.097	.471
36	1.9	.104	11.6	10.5	30	.228	.272

r_p^o = value of r in Eq. (2) when $\Delta\varepsilon_t^{eff} = .0069$

$COD = C_o|-y|^p$

Table II. Parameters for Large Cracks; 7075-OA Grown in Air

ΔK $MN/m^{3/2}$	Crack Tip Strain $\Delta\varepsilon_t^{eff}(0)$	A	B	r_p^o µm	C_o µm	p
10	.23	47.0	14.3	68	.691	.354
10	.20	19.8	9.9	44	.565	.318
8	.098	15.4	12.6	35	.200	.473
6	.10	14.5	12.0	34	.185	.481
6	.08	14.9	13.7	33	.246	.412
5	.090	5.1	7.5	20	.179	.541
5	.06	5.1	9.2	18	.118	.308

Table III. Equations Fit by Least-Squares to the Data of Tables I and II; 7075-OA Grown in Air

Eq. No.	Large Cracks	Small Cracks
1	$\Delta\varepsilon_t^{eff}(0) = 6.54 \times 10^{-3}\ \Delta K^{1.464}$	No correlation
2	$\Delta\varepsilon_p^{eff}(0) = 5.1 \times 10^{-3}\ \Delta K^{1.56}$	No correlation
3	$C_o = 6.42 \times 10^{-9}\ \Delta K^{1.92}$	$C_o = 7.00 \times 10^{-8}\ \Delta K^{1.09}$
4	$\Delta\varepsilon_p^{eff}(0) = 1.26 \times 10^{4}\ C_o^{.772}$	No correlation
5	$r_p^s(0) = 2.43 \times 10^{-6}\ \Delta K^{1.35}$	$r_p^s(0) = 2.00 \times 10^{-5}\ \Delta K^{.74}$
6	$p = 6.83 \times 10^{-1}\ \Delta K^{-.273}$	No correlation
7	$\frac{da}{dN} = 3.99 \times 10^{-9}\ \Delta K^{1.64}$	Complex

ΔK in $MN/m^{3/2}$, C_o in m, r_p^c in m

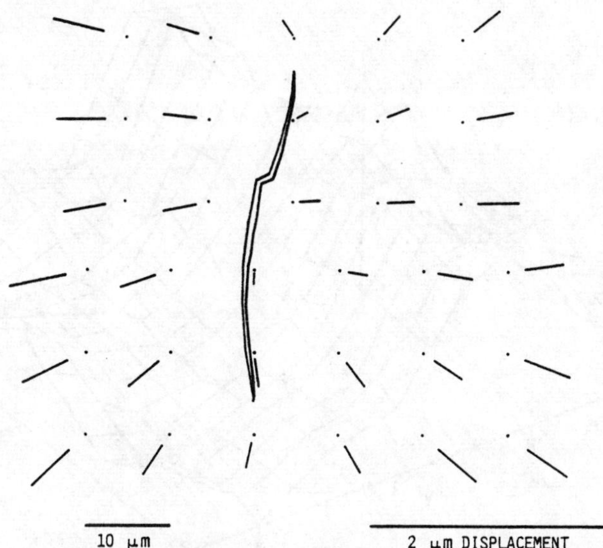

Figure 2a. Displacements around a small crack 36 μm in length loaded to $\Delta K = 1.9\ MN/m^{3/2}$. Note the difference between the spatial scale and the displacement scale.

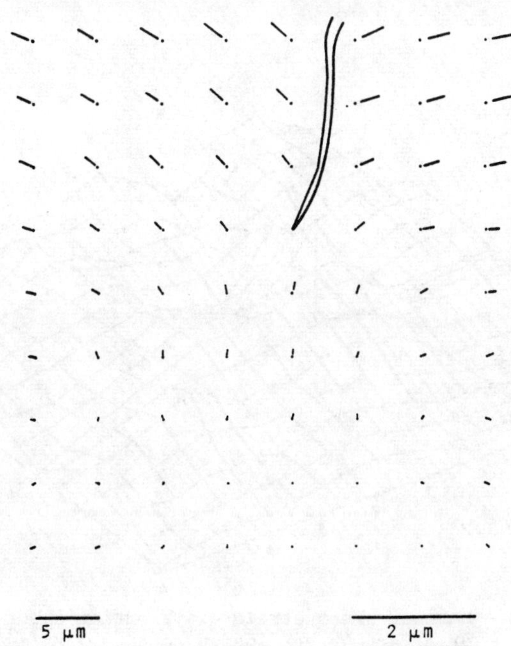

Figure 2b. Displacements around a large crack tip loaded to $\Delta K = 5\ MN/m^{3/2}$.

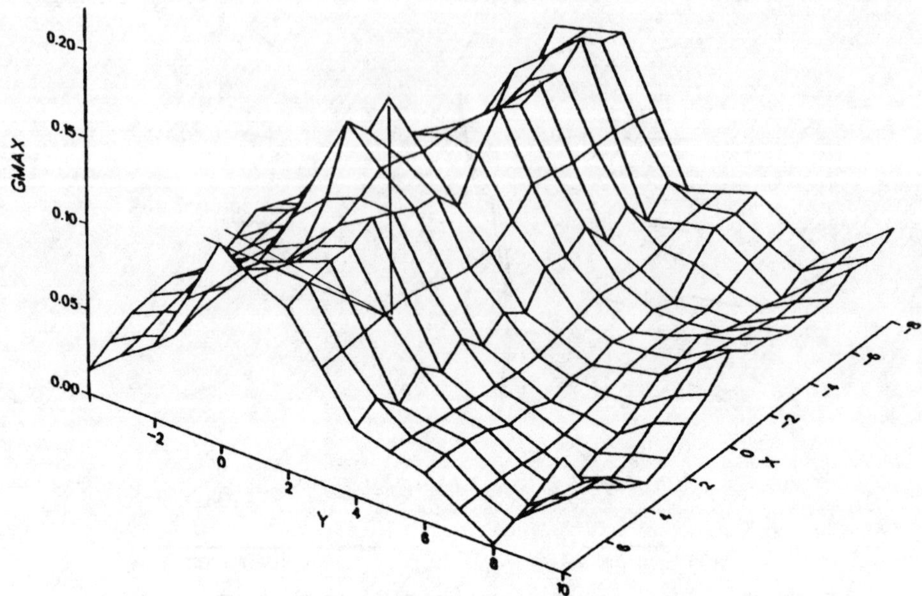

Figure 3a. Maximum shear strain distribution about the tip of a small crack 206 μm in length loaded to $\Delta K = 4.65\ MN/m^{3/2}$.

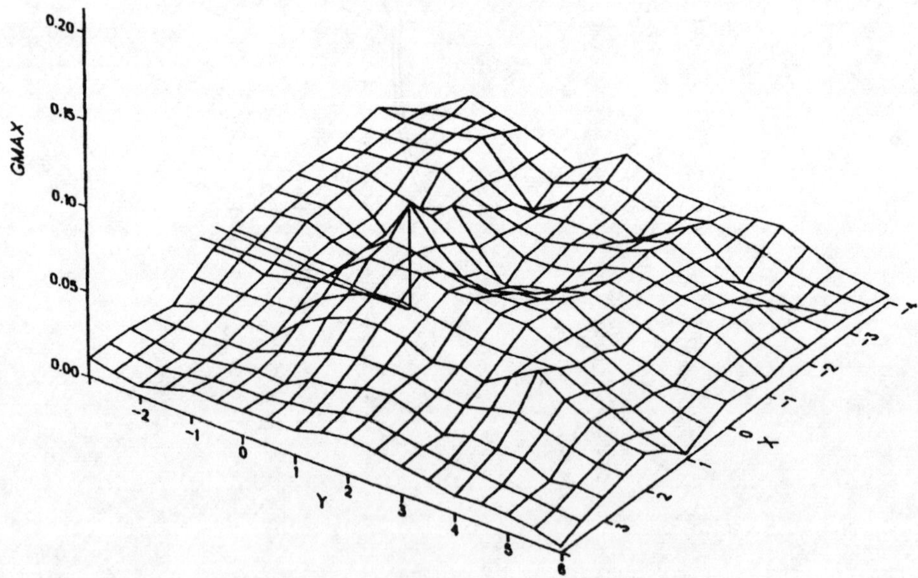

Figure 3b. Maximum shear strain distribution about the tip of a large crack loaded to $\Delta K = 5\ MN/m^{3/2}$.

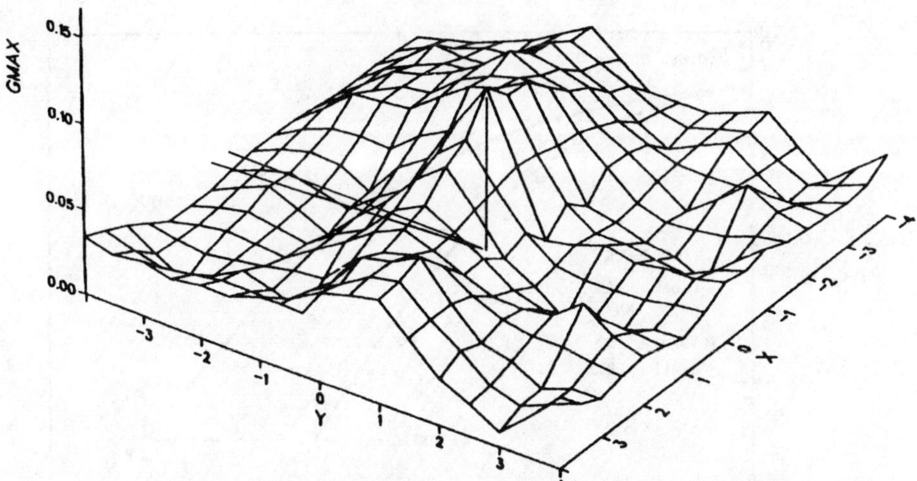

Figure 4. Maximum shear strain distribution about the tip of a small crack 36 μm in length loaded to $\Delta K = 1.9\ MN/m^{3/2}$.

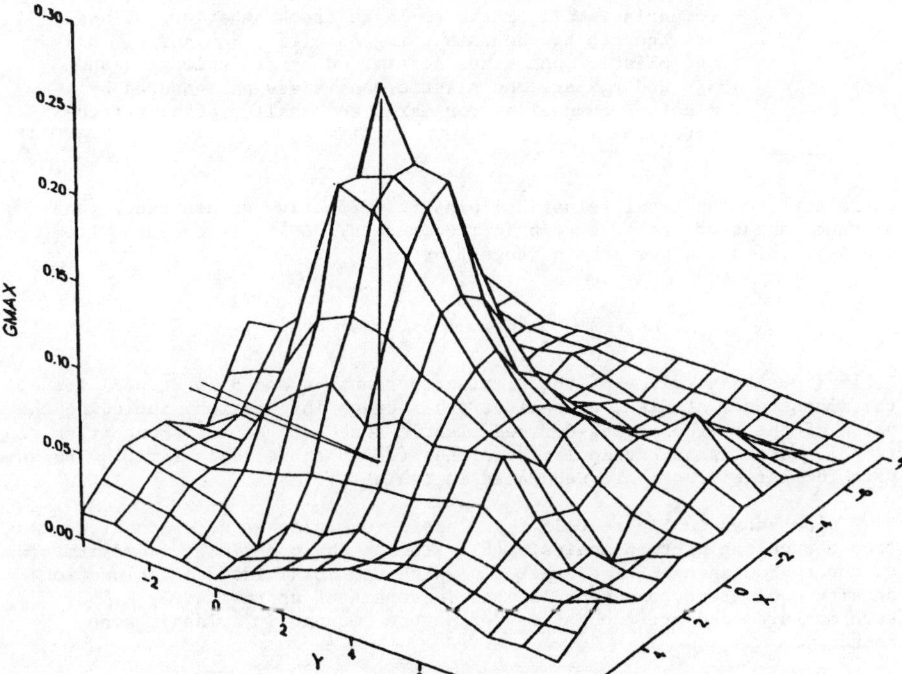

Figure 5. Maximum shear strain distribution about the tip of a large crack tip loaded to $\Delta K = 10\ MN/m^{3/2}$.

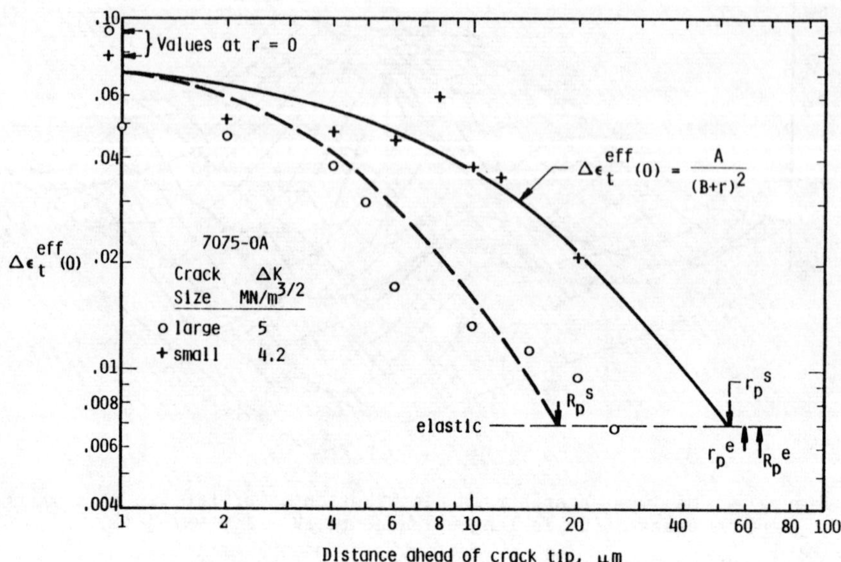

Figure 6. Distribution of effective strain ahead of the crack tip, comparing small cracks to large crack behavior. Lines are the fit to the data using Eq. (2). R_p^s and r_p^s are the plastic zone sizes determined by stereoimaging and R_p^e and r_p^e are the plastic zone sizes as measured by electron channeling for large and small cracks, respectively.

where $\Delta\varepsilon_t^{eff}$ = the total (elastic + plastic) effective strain range, and r = distance ahead of the crack tip in micrometers. $\Delta\varepsilon_t^{eff}$ is computed from $\Delta\varepsilon_1$ and $\Delta\varepsilon_2$, the principal strain ranges, by

$$\Delta\varepsilon_t^{eff} = \frac{2}{\sqrt{3}} (\Delta\varepsilon_1^2 + \Delta\varepsilon_1\Delta\varepsilon_2 + \Delta\varepsilon_2^2)^{1/2} \tag{3}$$

Table I presents values of the fitting constants A and B in Eq. (2) for all the microcracks studied, as well as other crack information, including the value of the stereoimaging-derived plastic zone size ahead of the crack tip $r_p^s(0)$, which is an extrapolation of Eq. (2) to the elastic limit. The same data for large cracks are tabulated in Table II.

Also shown in Figure 6 is the plastic zone size as determined by electron channeling pattern analysis (8). This technique yields a maximum size of the region ahead of the crack tip which has accumulated a dislocation density equivalent to a tensile plastic strain of approximately 0.003 (15); such measurements are thought to define the "monotonic" plastic zone boundary.

Strain at the crack tip is correlated with ΔK in Figure 7, which indicates that for small cracks, there basically is no correlation, while for large cracks correlation is reasonably good.

Crack opening displacement for cracks of both sizes may be described by

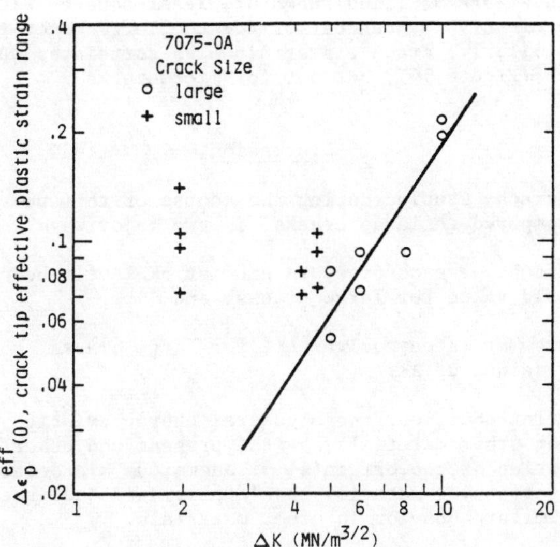

Figure 7. Effective plastic strain range at the crack tip as a function of ΔK. The line shown is a least-squares fit through the large crack data.

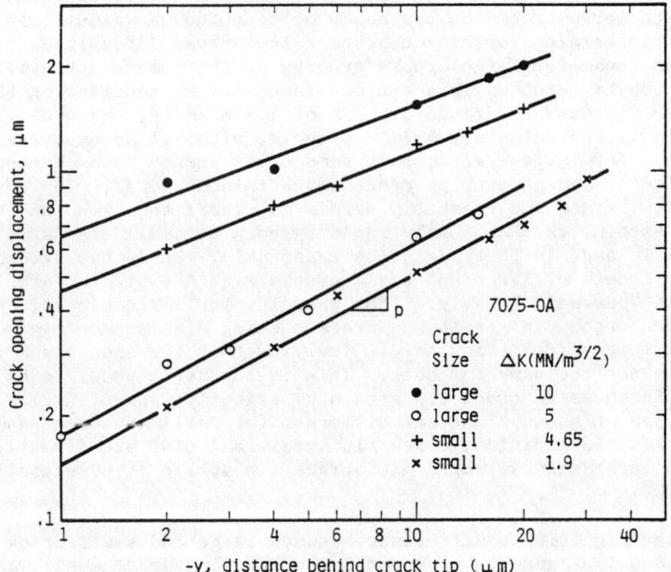

Figure 8. Crack opening displacement change with distance behind the crack tip. Slope is p and intercept is C_o in Eq. (4).

$$COD = C_o|-y|^p \tag{4}$$

where COD = the resultant value ($\sqrt{COD_x^2 + COD_y^2}$), and -y = the distance behind the crack tip. The crack tip opening displacement, CTOD (=C_o), is plotted vs ΔK in Figure 9; lines shown are least-squares fits to the data. The correlation clearly is much better for the large crack than for the small crack. Similarly, crack tip strain also correlates well with CTOD for large cracks (Figure 10), but not for microcracks.

Discussion

The small cracks studied during the course of this research behaved unusually, as compared to large cracks, in two major ways:

1) microcracks were observed to grow at ΔK levels below the threshold value for large cracks, and

2) they grew at rates equivalent, for large cracks, to much larger values of ΔK.

As noted in the Introduction, these general characteristics of behavior have been observed for other materials, by the present and other investigators. Thus, an explanation of the origin(s) of anomalous microcrack growth behavior in this particular material should provide a qualitative basis for understanding similar behavior in other materials.

For large cracks, we have shown that the crack growth rate is a direct function of the crack tip strain, and that ΔK provides a good correlation factor for linking crack tip plasticity parameters to crack growth rate. For small cracks, the measured values of crack tip strain, Figure 6, and opening displacement, Figure 8, are larger than would be expected by comparison with large cracks on the basis of computed ΔK values. In particular, the crack tip strains for microcracks at low stress intensities are roughly the same as those for large cracks growing at the same crack growth rate, but at a much higher ΔK. This can be visualized by considering Figures 1 and 7. For microcracks driven at a ΔK of 1.9 $MN/m^{3/2}$, the average crack growth rate is approximately 4×10^{-9} m/cycle, with ~.1 as an average crack tip strain. For large cracks, this same crack growth rate corresponds to $\Delta K \simeq 5\ MN/m^{3/2}$, and an average crack tip strain of 0.062. The correlation between growth rate and crack tip strain is clear; thus, on the basis of crack tip strain, the "unusually rapid" growth of small cracks is not unusual at all, and, in fact, is quite reasonable. It is the attempt to correlate the growth of large and small cracks with ΔK which creates the illusion of "unusual" behavior. For example, for small cracks, the relation between crack tip strain and crack opening displacement is not the same as for the large crack (Figure 10), implying that the open crack tip profile probably is not the same for both. This is further emphasized by the qualitative differences in crack tip strain distribution shown in Figures 2 thru 5; the shapes of these plots are different for small cracks as compared to large cracks. In addition, crack tip strain and CTOD are directly proportional for large cracks, while microcrack tip strain is essentially independent of CTOD.

Another significant difference between large and small crack behavior is that shown in Figure 6. For small cracks, plastic zone size as determined by stereoimaging is a much larger fraction of the monotonic plastic zone, established by selected area electron channeling, than is found for

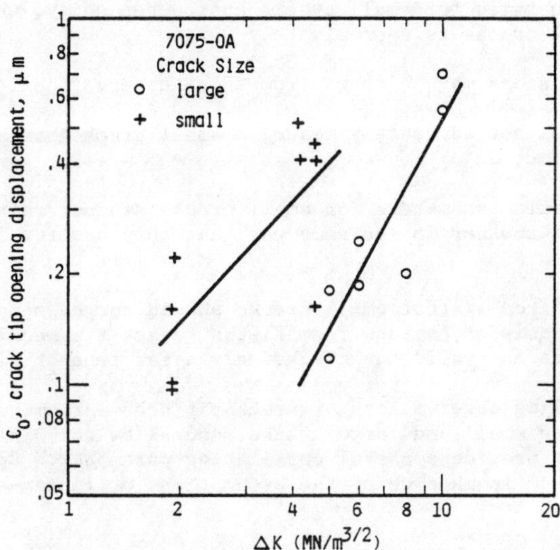

Figure 9. Crack tip opening displacement vs ΔK. Line shown is a least-squares fit through large crack data.

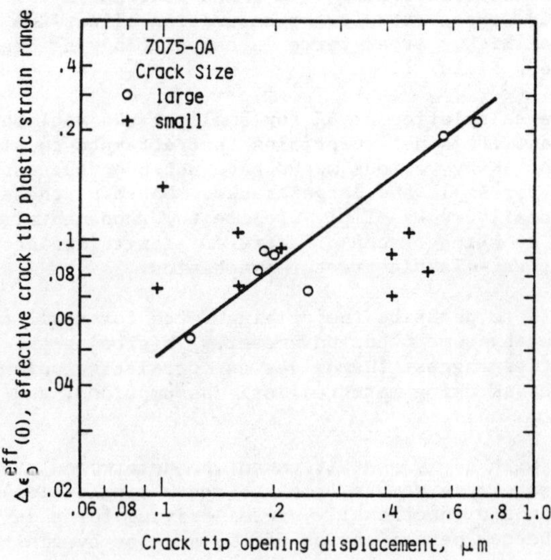

Figure 10. Crack tip effective plastic strain range vs CTOD. Line shown is through the large crack data.

large cracks at the same ΔK. Further, it is clear that the plastic zone size to crack length ratio for small cracks approaches unity, while the same ratio for large cracks is approximately 0.

Two conclusions may be drawn from these observations:

1) ΔK does not adequately describe small crack growth behavior;

2) crack tip parameters for small cracks are not related to one another in the same way that they are for large cracks.

It is apparent from 1) that small cracks should not be compared to large cracks on the basis of ΔK, and from 2) that crack tip mechanics analyses for large cracks are not valid for cracks only a few tens of micrometers long.

The preceding observations of crack tip behavior emphasize the differences between small and large cracks, and allow some insight regarding why ΔK does not provide a useful correlating parameter. Specifically, the computation of ΔK is made using the assumptions of linear-elastic fracture mechanics, which are:

1) that plasticity is well contained near the crack tip, so that the maximum extent of the plastic zone is small in comparison to:

   a) the remaining unbroken ligament of the body, and

   b) the overall crack length;

2) the gross section stress is a small fraction ($< .5$) of the yield stress (the latter assumption allows the deletion of higher order terms in calculating the crack tip stress field) (16).

In fact, the calculation of ΔK for small cracks violates both assumptions 1b and 2, and it is not surprising that attempts to empirically modify the computation of ΔK by various means have not been successful in unifying the growth rates for small and large cracks. Based on these considerations, it appears that small cracks simply violate the fundamental assumption of similitude (17), by which cracks of different length and geometry may be compared using linear-elastic fracture mechanics.

The use of ΔK to describe the driving force for small cracks evidently must be supplanted by some other parameter. This being the case, the following possibilities suggest themselves as correlating parameters: 1) ΔJ, 2) a derivation of ΔK using more realistic assumptions, and 3) gross section stress.

The problem with ΔJ is that it cannot be determined, for an elastic specimen, by measurements remote from the crack tip (at least at this time). Therefore, ΔJ is not a functionally useful driving force in the present investigation. A better determination of ΔK, such as by finite element calculations, does not appear to be possible either, because the basic concepts of linear-elastic fracture mechanics are violated. We have recently shown (8) that simple empirical corrections to ΔK are insufficient to correlate growth rates for large and small cracks. This apparently leaves only stress, the traditional parameter for defining the fatigue life of smooth specimens, as the crack driving force. It is interesting to note that Miller has suggested (18) that the fatigue limit corresponds not to cycles to nucleate a crack, but rather to a lifetime governed solely by the slow but steady growth of a single, dominant small crack.

It is possible to obtain an estimate of the growth rate of small cracks which is based on that for large cracks, by using the equations found in Table III in which crack tip strain can be used simply as a correlating response parameter. In particular, insertion of the crack tip strains of Table I into Eq. (1) of Table III permits the computation of an apparent ΔK. Using this ΔK in Eq. (7) then gives an estimate of the small crack growth rate. For the smallest cracks, this calculation is approximately 10 times too large, but for the larger microcracks (200 μm), the estimate is off, on average, by a factor of only 2. If crack tip strain somehow could be linked to gross section stress, then correlation of stress and microcrack growth rate might be possible.

The question as to the origin of ΔK_{Th} for large cracks still remains, but can be addressed by present results. If the equation for the total effective strain range at the crack tip, Eq. (1) of Table III, is extrapolated to the elastic limiting strain, $\Delta K = 1\ MN/m^{3/2}$ is derived. On the other hand, it will be recalled that COD $\alpha|-y|^p$, where p is a function of ΔK (Table III). Extrapolating p(ΔK) to its elastic value of 0.5 yields $\Delta K = 3\ MN/m^{3/2}$. This range of stress intensity (1-3 $MN/m^{3/2}$), derived on the basis of experimental characterization of crack tip parameters, effectively brackets the measured ΔK_{Th} range (Figure 1), and is a strong indication that ΔK_{Th} simply corresponds to a driving force which is no longer sufficient to produce plastic strain at the crack tip.

The physical basis for the approach of the crack tip strain to the elastic limit as ΔK decreases is not yet clear. We have observed (8), as have others (9), that P_{open} increases as ΔK is lowered, which means ΔK_{eff} decreases to 0 as P_o approaches P_{max}. This observation has been attributed (1) to both residual compressive stresses at the crack tip, and to residual displacement in the wake of the crack. Present results do not differentiate between these mechanisms.

Regardless of the specific bases for the large crack threshold, it is clear that these mechanisms do not manifest themselves in the same way for small cracks. That is, the mechanisms may affect the average rate of growth, but they do not produce a "threshold" for microcrack extension. The present results suggest in particular that for small cracks, the concept of a threshold ΔK is meaningless.

Summary and Conclusions

Summary of Results for 7075-OA Aluminum Alloy

1) Small cracks grow faster than large fatigue cracks when compared at the same ΔK.

2) Small cracks grow at ΔK values below the ΔK threshold for large cracks.

3) Crack tip strain and crack tip opening displacement are greater for small cracks than for large cracks at the same ΔK.

4) The distribution of strain at the tips of small cracks is different than that for a large crack.

5) Crack tip strain correlates well with crack tip opening displacement for large cracks but not for small cracks.

Conclusions

The similarity between crack tip opening displacement, strain, and plastic zone size which allows large crack growth data to be correlated using linear-elastic fracture mechanics ΔK calculations no longer exists for small cracks. Therefore, crack growth rates for small and large cracks cannot rationally be compared on the basis of ΔK. The magnitude of strain at the crack tip is found to provide an approximate method for computing the growth rate of small cracks from large crack growth data. Other definable macroscopic parameters (ΔJ or modified ΔK) are not likely to be useful as a substitute for ΔK in correlating crack growth rate; gross section stress may provide the best possibility.

References

1. S. Suresh and R. O. Ritchie, "The Propagation of Short Fatigue Cracks," Int. Metall. Rev. (in press).

2. S. Pearson, "Initiation of Fatigue Cracks in Commercial Aluminum Alloys and the Subsequent Propagation of Very Short Cracks," Eng. Frac. Mech. 7 (1975) pp. 235-247.

3. J. Lankford, "The Growth of Small Fatigue Cracks in 7075-T6 Aluminum," Fat. Eng. Mats. Structs. 5 (1982) pp. 233-248.

4. J. Lankford, "The Effect of Environment on the Growth of Small Fatigue Cracks," Fat. Eng. Mats. Structs. 6 (1983) pp. 15-31.

5. C. W. Brown and M. A. Hicks, "A Study of Short Fatigue Crack Growth Behavior in Titanium Alloy IMI 685," Fat. Eng. Mats. Structs. 6 (1983) pp. 67-76.

6. K. Tanaka, M. Hojo, and Y. Nakai, "Fatigue Crack Initiation and Early Propagation in 3% Silicon Iron," pp. 207-232 in Fatigue Mechanisms: Advances in Quantitative Measurement of Physical Damage, ASTM STP 811, American Society for Testing and Materials, Philadelphia, 1983.

7. K. S. Chan and J. Lankford, "A Crack-Tip Strain Model for the Growth of Small Fatigue Cracks," Scripta Met. 17 (1983) pp. 529-532.

8. J. Lankford, D. L. Davidson, and K. S. Chan, "The Influence of Crack Tip Plasticity in the Growth of Small Fatigue Cracks," Met. Trans. (submitted).

9. A. J. McEvily, "The Fatigue of Powder Metallurgy Alloys," Air Force Office of Scientific Research Annual Report, Grant No. AFOSR 81-0046, January, 1983.

10. D. L. Davidson and A. Nagy, "A Low Frequency Cyclic Loading Stage for the SEM," Jour. Phys. E 11 (1978) pp. 207-210.

11. R. A. Smith, "On the Short Crack Limitations of Fracture Mechanics," Int. Jour. Fract. 13 (1977) pp. 717-720.

12. T. A. Cruse, G. J. Meyers, and R. B. Wilson, "Fatigue Growth of Surface Cracks," pp. 174-189 in Flaw Grow and Fracture, ASTM STP 631, American Society for Testing and Materials, Philadelphia, 1977.

13. D. R. Williams, D. L. Davidson, and J. Lankford, "Fatigue Crack Tip Strains by the Stereoimaging Technique," Exp. Mech. 20 (1980) pp. 134-139.

14. D. L. Davidson and J. Lankford, "Fatigue Crack Tip Strains in 7075-T6 Aluminum Alloy by Stereoimaging and Their Use in Crack Growth Models," pp. 371-399 in Fatigue Mechanisms: Advances in Quantitative Measurement of Physical Damage, ASTM STP 811, American Society for Testing and Materials, Philadelphia, 1983.

15. D. L. Davidson and J. Lankford, "Fatigue Crack Tip Plastic Strain in High-Strength Aluminum Alloys," Fat. Eng. Mats. Structs. 3 (1980) pp. 289-303.

16. J. R. Rice, "Mechanics of Crack Tip Deformation and Extension by Fatigue," pp. 247-311 in Fatigue Crack Propagation, ASTM STP 415, American Society for Testing and Materials, Philadelphia, 1967.

17. R. O. Ritchie and S. Suresh, "The Fracture Mechanics Similitude Concept: Questions Concerning its Application to the Behavior of Short Fatigue Cracks," Mat. Sci. Eng. 57 (1983) pp. L27-L30.

18. K. J. Miller, "The Short Crack Problem," Fat. Eng. Mat. Struct. 5 (1982) pp. 223-232.

PROPAGATION OF SMALL SURFACE CRACKS IN Ti-ALLOYS

C. Gerdes, A. Gysler, and G. Lütjering

Technische Universität Hamburg-Harburg
2100 Hamburg 90
W.-Germany

The fatigue crack propagation behavior of surface cracks was studied on age-hardened Ti-8.6 Al and Ti-6Al-4V alloys, and compared with the propagation behavior of long through-cracks. The parameters of grain size, R-ratio, and environmental condition were varied in this study. It was found that in vacuum and in laboratory air, the small semi-elliptical surface cracks propagted faster and below the near-threshold stress intensity factors of long through-cracks. Propagation rates of surface cracks from low R-ratio tests approached those of through-cracks tested at high R-ratios as the surface crack depths increased. In vacuum, the merging of surface and through-crack propagation curves occurred at about four to five times the average grain size. Comparing small and large grain size material of Ti-8.6 Al at the same propagation rates of the surface cracks, higher ΔK-values were obtained for the material with a small grain size. In air the propagation rates of surface cracks were higher than in vacuum.

Introduction

The discrepancy between the fatigue crack propagation behavior of small surface cracks and long through-cracks in the same material has been recognized for several years [1]. It is generally accepted and well documented in the literature that fatigue crack propagation of small surface cracks occurs at stress intensities well below the threshold values observed for large through-cracks. The experimental results published with regard to this subject cover a rather wide spectrum of materials, including Al-alloys [1-5], steels [6-8], Ni-base alloys [9,10], and Ti-alloys [11].

Some of the fatigue crack propagation curves for small cracks reported in the literature show a very complex behavior: following high propagation rates after crack nucleation, the growth rates decline, pass through a minimum and then accelerate [5,9]. However this behavior has apparently not been observed in other studies [1,6,8]. There exists agreement that with increasing length the propagation rates of surface cracks approach those of through-cracks.

The experimental results reported in this study were aimed to contribute to the understanding of the different fatigue crack propagation behavior of small surface cracks and long through-cracks by studying the effects of variations in grain size, R-ratio, and environmental conditions on the crack propagation behavior in age-hardened Ti-alloys.

Experimental Procedure

The tests were performed on a binary Ti-8.6Al alloy and on a commercial Ti-6Al-4V alloy. The chemical compositions of these alloy are shown in Table I. Both alloys were unidirectionally hot rolled in the α- and (α+β)-phase field, respectively. Blanks of the rolled material were recrystallized, quenched in water and aged at 500°C to half of the final aging time. After machining, the specimens were then aged to obtain the total aging time of 10 h (Ti-8.6Al)

Table I. Chemical Compositions of the Alloys (wt.%)

Alloy	Al	V	O	N	H	C	Fe
Ti-8.6Al	8.6	-	0.07	0.01	0.017	0.02	0.07
Ti-6Al-4V	6.3	4.2	0.20	0.017	0.005	0.02	0.145

and 24h (Ti-6Al-4V) in order to remove internal stresses resulting from machining. The thermomechanical treatments applied in this work resulted in equiaxed grain sizes of about 20 μm and 100 μm for Ti-8.6Al and about 6 μm for the α-phase of Ti-6Al-4V. The texture of the binary alloy was a strong basal pole texture normal to the rolling plane [12] and a mixed basal/transverse texture for the (α+β) alloy. Details regarding the thermomechanical treatments and the texture pole figure determination of these alloy conditions can be found elsewhere [13,14]. Tensile tests were carried out at room temperature on round specimens with a diameter of 3 mm and a gage length of 25 mm, applying an initial strain rate of $7 \cdot 10^{-4} s^{-1}$. The tensile properties are summarized in Table II.

Table II. Tensile Properties

Alloy	Grain Size (μm)	E (GPa)	$\sigma_{0.2}$ (MPa)	σ_F (MPa)	$\varepsilon_F = \ln(A_o/A_F)$
Ti-8.6 Al	20	132	860	1055	0.24
	100	132	777	973	0.08
Ti-6Al-4V	6	113	1063	1590	0.69

The through-crack experiments of the binary alloy were performed on pin-loaded single-edge-notched (SEN) specimens, the gage section being 2 mm thick, 10 mm wide, and 30 mm long. The through-crack curves of the Ti-6Al-4V alloy were determined on compact tension (CT) specimens with a thickness of 7 mm and a width of 31 mm. Surface cracks were studied on slightly hour-glass shaped conventional S-N specimens with a midsection diameter of 2 mm. The loading axis for all specimen types was parallel to the rolling direction. The surfaces of the SEN and CT specimens were mechanically ground and polished, while the S-N specimens in addition were electrolytically polished to remove a surface layer of about 100 μm using the standard low temperature polishing technique.

All fatigue crack propagation tests were carried out on a servohydraulic testing machine under load control with a sinusoidal waveform at 30 Hz. The tests were done at room temperature in vacuum ($<10^{-4}$Pa) and in laboratory air (∿ 60% relative humidity). Crack propagation curves for SEN and CT specimens were established by applying conventional load shedding techniques until propagation rates of 10^{-9} m/cycle were reached. From this condition the cracks were allowed to extend at a constant load amplitude, and incremental growth rates were measured with a travelling microscope. Surface cracks were nucleated and propagated in the smooth specimens by cycling them to a maximum stress of about 85 % of the macroscopic yield stress of each particular microstructure (Table II). These tests also were performed at constant load amplitudes. It should be noted that these stress amplitudes resulted in specimen failure after about $5x10^5$ cycles. To monitor the surface crack length, the fatigue tests were interrupted at defined load cycles. The specimens were then transferred to a scanning electron microscope (SEM) equipped with a specially designed stage which permitted the specimen to be rotated in the SEM while under load at a predetermined stress level. The length (2ℓ) of surface cracks was obtained by projecting the surface crack traces from SEM micrographs on the plane normal to the stress axis.

The cyclic stress intensity factors (ΔK) for the through-cracks were calculated from the measured crack length using the equation of Srawley [15] for SEN specimens and the ASTM equation for CT specimens. The ΔK-values of the surface cracks were calculated for the crack tip in the interior of the specimen, applying the equation of Newman [16]. The necessary crack depth (a) was obtained by converting the projected surface crack length (2ℓ) into the depth (a) according to the experimentally verified relation $a/2\ell \sim 0.45$. Fig. 1 shows an axample of the semi-elliptical crack front in Ti-6Al-4V, which also was observed in the binary alloy.

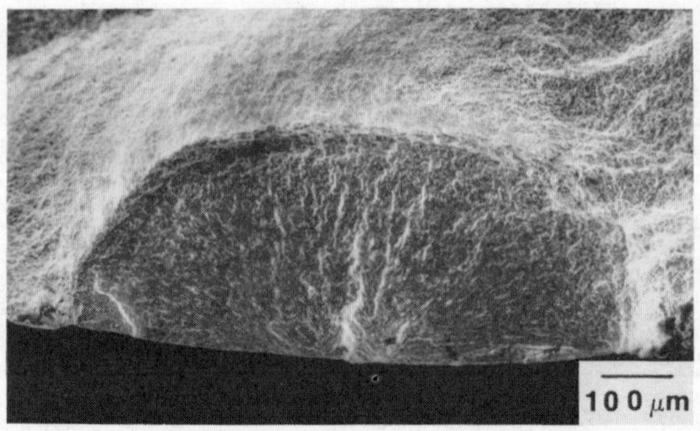

Figure 1 - Semi-elliptical surface crack in Ti-6Al-4V obtained in air (SEM).

Results and Discussion

Effect of Grain Size and R-Ratio on Crack Propagation Behavior

Fatigue crack propagation rates da/dN as a function of the cyclic stress intensity factor ΔK of surface and through-cracks in vacuum are shown in Fig.2 for Ti-8.6Al with grain sizes of 20 μm and 100 μm. As already reported earlier [14] at low R-ratios through-cracks in a fine grain size material propagated significiantly faster in comparison to a coarser microstructure. However, by increasing the R-ratio to 0.6, the da/dN vs ΔK curves not only shifted to lower ΔK-values, but the differences in propagation rates almost disappeared between the two grain sizes (Fig. 2). The decreasing effect of a grain size variation on da/dN with increasing R-ratio was also observed in other alloys (see for example [17]), and might be explained with decreasing contributions of roughness induced crack closure and slip reversibility at higher R-values. As can be seen from Fig. 2 variations of these effects are influencing especially the da/dN vs ΔK curves of the coarser microstructure.

The crack propagation behavior of surface cracks in both microstructures was found to be much more complex in comparison to through-cracks, as can be seen in Fig. 2. In all specimens tested in vacuum, always more than one surface crack was detected. These cracks were nucleated along grain boundaries by intersecting slip bands, or within slip bands. In both microstructures after crack nucleation high initial crack propagation rates (greater than 10^{-9} m/cycle) were observed at cyclic stress intensities of 4 $MPa \cdot m^{1/2}$ or less. However with further cycling da/dN decreased for most of the cracks, reached a minimum, and increased again. It should be noted that some of the small cracks did not propagate further after the propagation rates declined to a low value, they stopped. Similar observations were also made by Lankford [5] on an Al-alloy. Those cracks which were able to propagate further exhibited growth rates which gradually approached the curves obtained for through-cracks with the high R-ratio of 0.6 instead of those with R = 0.2 (Fig. 2). This behavior indicates that the role of roughness induced crack closure seems to be much less pronounced with regard to the propagation behavior of surface cracks in comparison to that of through-cracks.

Although the overall crack propagation behavior of surface cracks in the fine and coarse microstructures was very similar (Fig. 2) there exist some differences. The initial da/dN values after crack nucleation for the fine

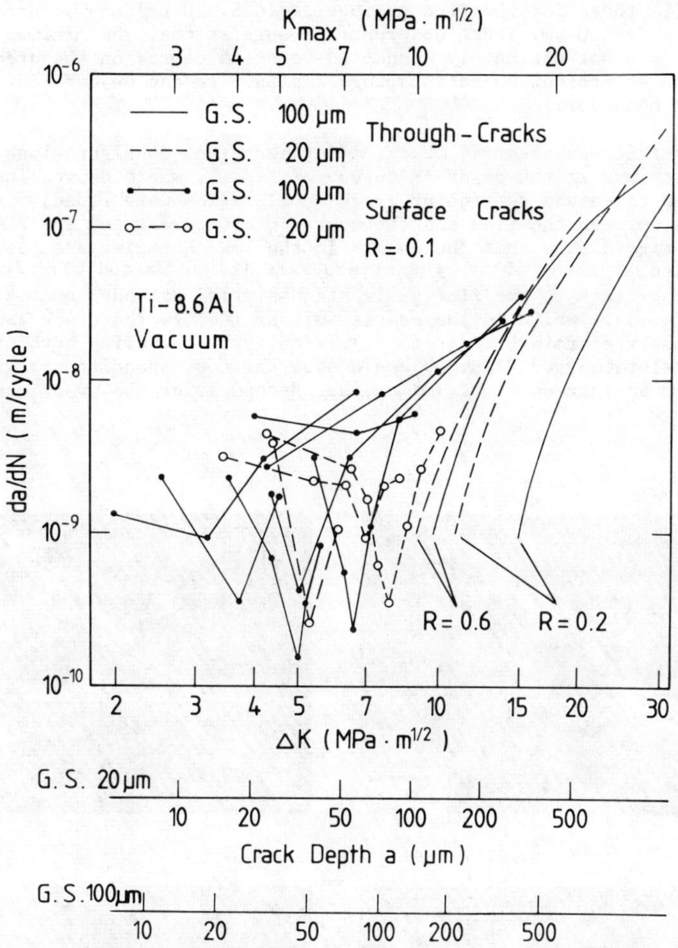

Figure 2 - Vacuum fatigue crack propagation curves of surface and through-cracks in Ti-8.6Al with two different grain sizes.

grain size material were somewhat higher and occurred at higher ΔK-values in comparison to the coarser microstructure. This seems to be due to the higher stress amplitude of 700 MPa needed to nucleate cracks in the fine grain size material as compared to a stress amplitude of only 600 MPa for the coarse microstructure. To demonstrate further differences, two additional lines are drawn below the abscissa in Fig. 2 which indicate the crack depth for each microstructure with regard to the ΔK-axis. The crack depth-ΔK correlation can be done since the tests were performed under constant stress amplitudes. Although there exists a fair amount of scatter in ΔK at which the minimum da/dN values were observed for each microstructure, the results at least show that the minima for the coarse microstructure (G.S. 100 μm) focus around about 50 μm, while those for the fine grain size (G.S. 20 μm) are occurring even slighthly above 50 μm. These observations suggest that the minimum propagation rates are not ultimately connected to crack depths on the order of the grain size. At present no satisfactory explanation can be offered for this unexpected behavior.

However it was observed that the surface crack configurations, belonging to those regimes in the da/dN-ΔK curves of Fig. 2, where decreasing crack propagation rates and subsequent crack acceleration were found, exhibited similarities between the fine and the coarse microstructures. The SEM micrographs in Fig. 3 show that the cracks in the low ΔK regime are rather straight and inclined at about 45° to the stress axis (Figs. 3a and c). It should be noted that the crack in the fine grain size material extended over several grains (Fig. 3a), while in the coarser microstructure the crack length was smaller than the grain size (Fig. 3c). With further cycling both of these cracks accelerated and Figs. 3b and d show the corresponding crack configurations after an increament of 10^4 cycles. By comparing the crack configurations

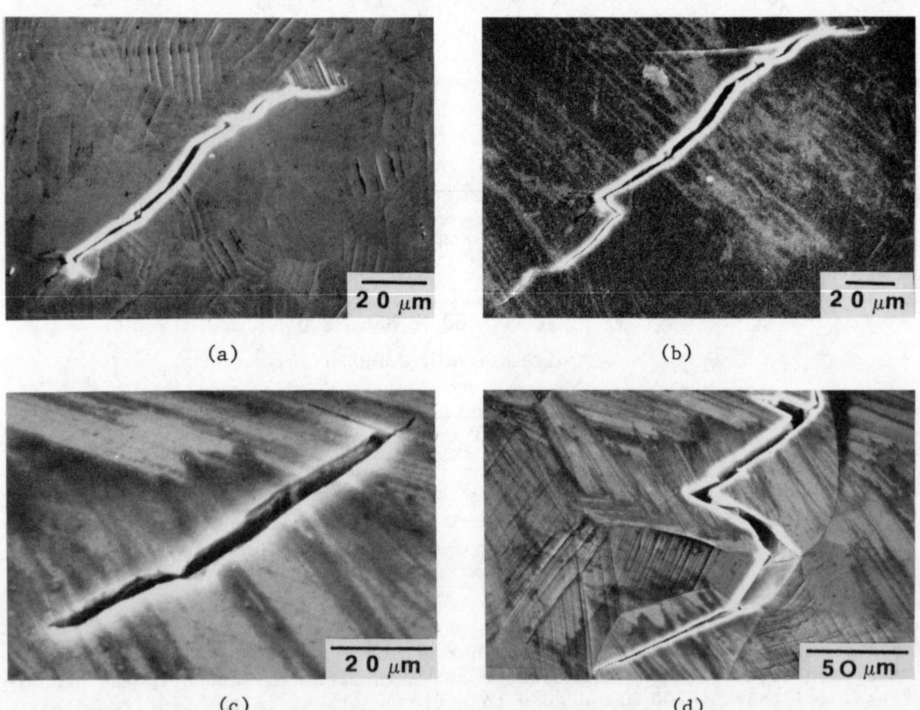

(a) (b)

(c) (d)

Figure 3 - Surface cracks in fine (a), (b) and coarse (c), (d) grain size Ti-8.6Al (SEM). Crack configurations in (a) and (c) belong to minimum propagation rates, and in (b) and (d) to crack acceleration. Stress axis is horizontal.

before (Figs. 3a and c) and after the cracks accelerated (Figs. 3b and d), it seems quite obvious that the minimum propagation rates are intimately connected to the fact that the original straight cracks in both microstructures changed their propagation directions completely. After reaching the configurations and length in Figs. 3b and d, both cracks exhibited a continuous increase in da/dN. Lankford [5] reported a similar behavior of surface cracks in Al-7075, where he observed that in vacuum, cracks started growing fast at $\sim 45°$, and then slowed or stopped when they had to change the direction.

The crack propagation rates at comparable ΔK-values were always lower for the fine as compared to the coarse microstructure in Fig. 2. This is thought to be a consequence of the larger number of grain boundaries in the fine grain size material at which the cracks had to change the propagation direction because of orientation differences between neighbouring grains.

Another, but rather tentative possibility for explaning the crack propagation minima in Fig. 2 could be due to the occurrence of a hardened surface layer as a consequence of fatigue cycling. Recently it was shown by X-ray data that fatigued specimens of a Ti-6Al-4V alloy exhibited a higher amount of lattice misalignment and therefore a higher dislocation density in a thin surface layer of about 25 μm in comparison to the specimen interior [18]. This also could account for the high initial propagation rates in Fig. 2, since it is known from previous results that pre-deformed specimens promoted higher fatigue crack propagation rates [19,20]. The minima could then occur if the cracks passed through such a hardened surface layer and subsequently grow with the lower propagation rates observed for cracks in a previously undeformed material.

Fracture surface studies of specimens which were broken after several surface cracks had propagated beyond the crack depths where the minimum da/dN-values occurred in Fig. 2, showed some evidence for a transition in fracture mode after the cracks reached a length of about 50 μm. Fig. 4 shows an example for a Ti-8.6Al specimen with a grain size of 100 μm. As can be seen from these micrographs the surface crack seemed to grow essentially along one slip system per grain to a depth of approximately 50 μm. Further crack propagation then occurred along at least two slip systems, which normally is also observed in this alloy in the near-threshold regime of through-cracks [14]. A similar transition from crack propagation along a single slip plane (stage I) near the crack initiation site towards an alignment of the crack plane normal to the stress axis with increasing crack length was also reported for a nickel-base superalloy single crystal by Duquette et al. [21].

Another observation which can be seen in Fig. 2 also seems to be of interest with regard to a correlation of surface and through-crack propagation behavior. From Fig. 2, crack depth values can be estimated for fine and coarse microstructures where the surface crack propagation rates approached those of through-cracks. Within experimental error, surface cracks tended to propagate at approximately the same rate as through-cracks if the surface crack depths reached about four to five times the particular average grain size.

Fatigue crack propagation curves of surface cracks in vacuum at $R = 0.6$ for the coarse microstructure of Ti-8.6Al are shown in Fig. 5 along with the corresponding curve for through-cracks at the same R-ratio. The overall crack propagation behavior of surface cracks at high (Fig. 5) and low (Fig. 2) R-ratios was found to be similar. However the da/dN-ΔK curves at the high R-ratio are shifted to very low ΔK-values in comparison to those obtained at a low R-ratio (compare Fig. 2). The crack depth values corresponding to the minimum propagation rates again were observed at about 50 μm, similar to the

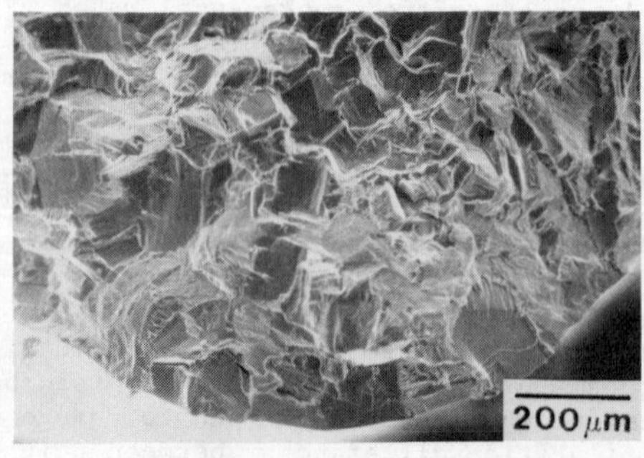

(a)

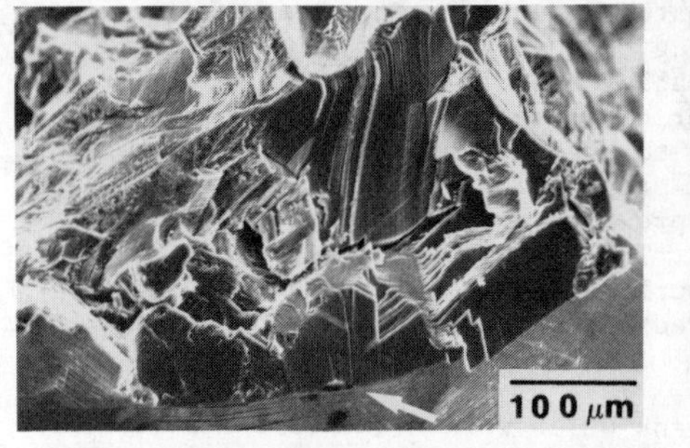

(b)

(c)

Figure 4 - Vacuum fatigue fracture surface of Ti-8.6Al with a grain size of ∿ 100 μm containing a surface crack (SEM). Crack initiation site marked by arrow. (a) viewed parallel to stress axis, (b), (c) slightly tilted in comparison to (a).

low R-ratio curves in Fig. 2. It will be noted that the crack propagation rates and the occurrence of the minima for the surface cracks, tested at low and high R-ratios, correlated reasonably well with K_{max} (compare Figs.2 and 5).

Effect of Environment on Propagation Rates of Surface and Through-Cracks in Ti-8.6Al

The fatigue crack propagation behavior of surface and through-cracks in laboratory air is shown in Fig. 6 for a coarse microstructure. In contrast to specimens fatigued in vacuum, where always several cracks were nucleated, in air only one crack per specimen was found at this particular stress amplitude. As can be seen by comparing Figs. 2 and 6 the initial propagation rates after crack nucleation are higher in air than in vacuum. So far it was not possible to interrupt a fatigue test at the right moment to detect surface cracks smaller than about 150 μm (crack depth 70 μm), although the number of load cycles between inspection intervals were comparable for the tests in air and in vacuum. From the data points in Fig. 6 it can be seen that in air only a slight discontinuity occurred in the da/dN-ΔK-curves of surface cracks,

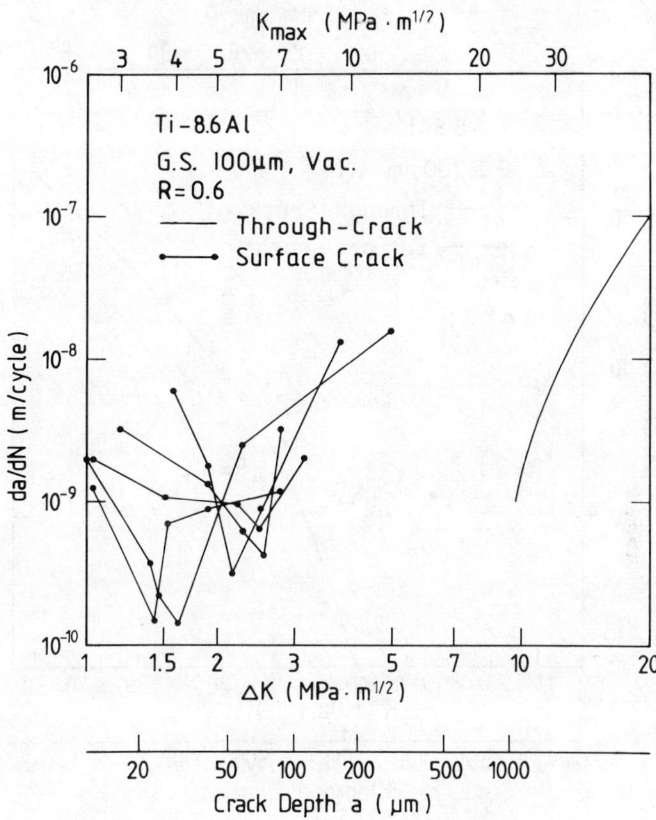

Figure 5 - Vacuum fatigue crack propagation curves of surface and through-cracks in Ti-8.6Al at R = 0.6 (Grain size ∿ 100 µm).

much less pronounced than in vacuum (Fig. 2). The concomitant crack depth of about 200 µm, belonging to the slight minimum in Fig. 6, was much longer in comparison to those observed in vacuum (about 50 µm), despite the fact that cracks in specimens tested in air also changed their direction after reaching a depth of about 50 µm. This can be seen from the SEM micrographs in Fig. 7.

Propagation of Surface and Through Cracks in Ti 6Al-4V

Fatigue crack propagation curves in laboratory air for the (α+β) Ti-alloy are shown in Fig. 8 for surface and through-cracks at different R-ratios. Crack nucleation and propagation of surface cracks in this microstructure occurred within the α-phase and along α/β-phase boundaries. In order to extend the da/dN-ΔK curves for surface cracks to higher propagation rates, some specimens were cycled to failure and propagation rates per cycle were determined from striations observed on the fracture surfaces. As can be seen in Fig. 8

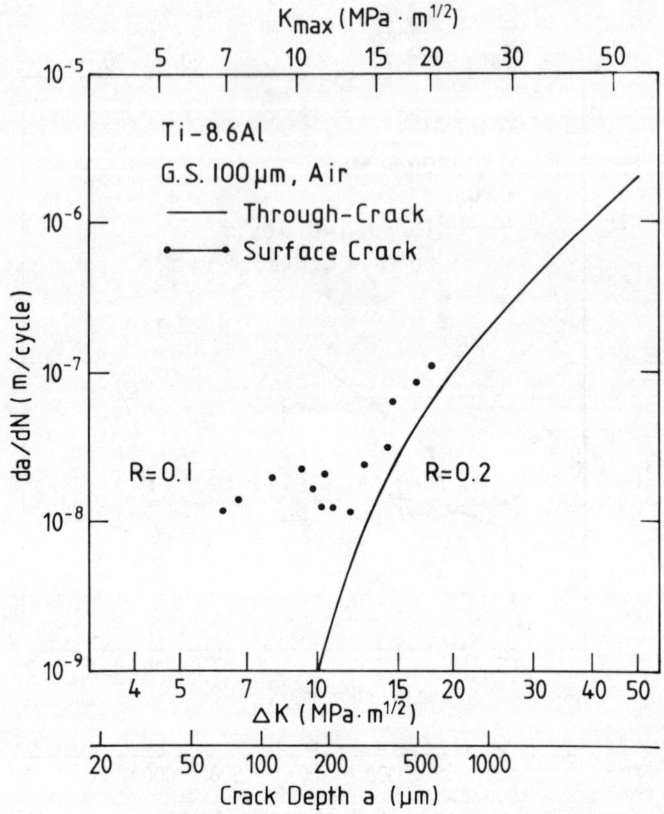

Figure 6 - Fatigue crack propagation curves in air of surface cracks (R = 0.1) and through-cracks (R = 0.2) in Ti-8.6Al (Grain size ∿ 100 µm).

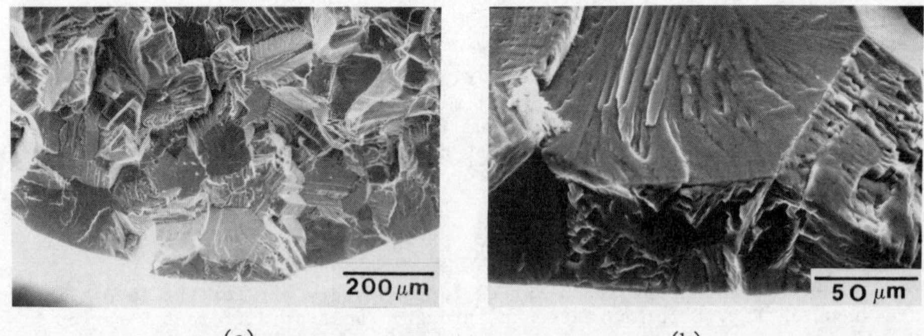

(a) (b)

Figure 7 - Fatigue fracture surface in laboratory air of Ti-8.6 Al with a grain size of ∿ 100 µm, containing a surface crack (SEM). Micrograph (b) is a higher magnification of (a).

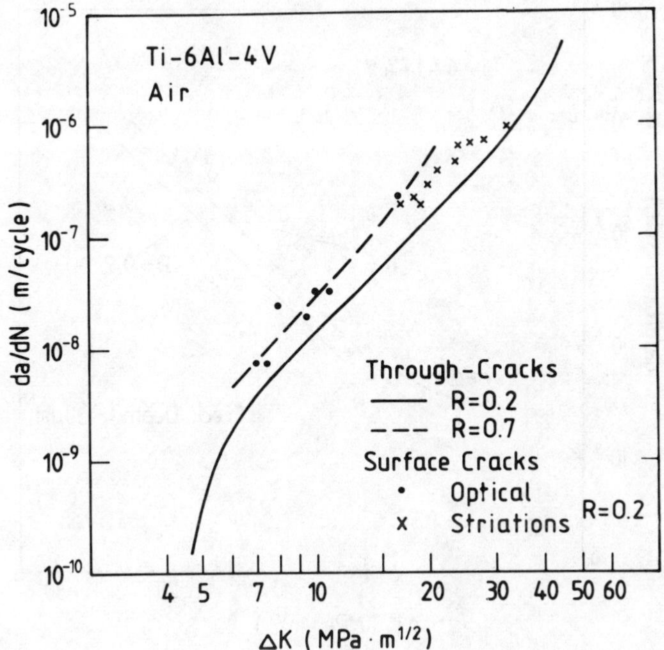

Figure 8 - Fatigue crack propagation curves in air of surface cracks (R = 0.2) and through-cracks (R = 0.2 and 0.7) in Ti-6Al-4V.

· determined from surface crack lengths

x determined from striation distances on fracture surfaces.

the data points determined from surface crack length measurements and from striation distances all fall on a smooth curve. The propagation rates of surface cracks with R = 0.2 were similar to those obtained from through-cracks with R = 0.7. This is in accordance with the results obtained for Ti-Al, and indicates also that crack closure effects do not play a significant role for the surface crack propagation behavior.

The effect of R-ratio on the propagation behavior of surface cracks in laboratory air is shown in Fig. 9 by tension-tension (R = 0.2) and push-pull (R = -1) loading conditions. Surface cracks with crack depths smaller than about 30 μm propagated under push-pull loading much faster than would be expected from an extrapolation of the straight curve obtained for longer surface cracks in Fig. 9. This behavior is similar to that observed for the cracks in the Ti-8.6Al alloy (see Fig. 2). Very short cracks in Ti-6Al-4V, which exhibited the high propagation rates, are again rather straight and inclined at about 45° to the stress axis (Fig. 10), in accordance with the observation made for the short cracks in Ti-8.6Al (compare Figs. 10 and 3a and c).

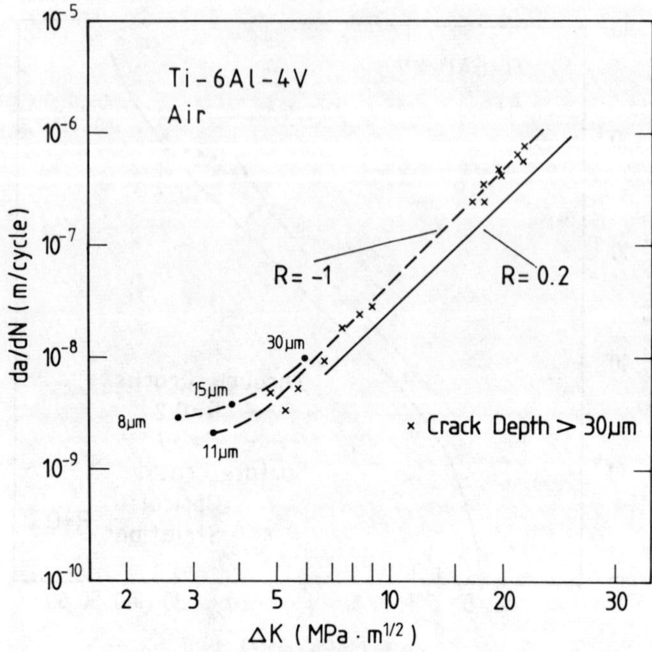

Figure 9 - Fatigue crack propagation curves in air of surface cracks in Ti-6Al-4V at R = 0.2 and R = -1.

· for indicated crack depth values
x crack depths > 30 μm

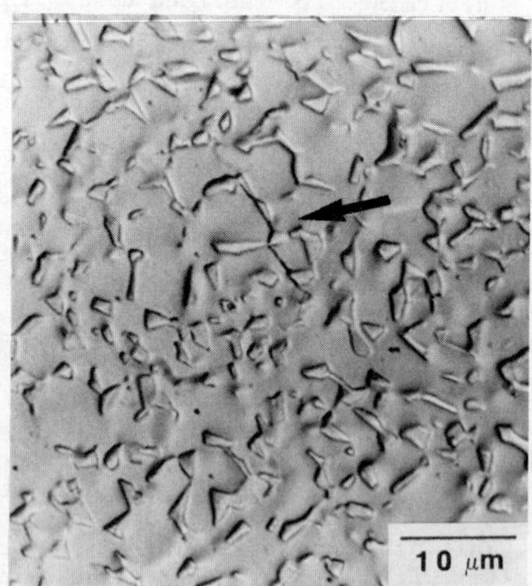

Figure 10 - Small surface crack in Ti-6Al-4V fatigued in laboratory air (LM). Stress axis is horizontal.

Summary

1. Small semi-elliptical surface cracks in Ti-alloys propagated faster than and below the cyclic stress intensity thresholds of through-cracks in both vacuum and laboratory air.

2. The fatigue crack propagation curves of surface cracks at low R-ratios approached those of through-cracks at high R-ratios, leading to the conclusion that crack closure effects do not play a significant role in surface crack propagation behavior.

3. In vacuum, the fatigue crack propagation behavior of surface and through-cracks became similar, if the surface cracks reached crack depths of four to five times the grain size.

4. Surface cracks which exhibited high initial propagation rates were generally straight and inclined 45° to the stress axis. Minimum propagation rates were observed as soon as the straight cracks had to change their direction.

5. Comparing small and large grain size microstructures of Ti-8.6Al at the same propagation rates of the surface cracks, higher ΔK-values were obtained in the material with the small grain size.

Acknowledgment

This work was supported by the Deutsche Forschungsgemeinschaft.

References

1. S. Pearson, "Initiation of Fatigue Cracks in Commercial Aluminium Alloys and the Subsequent Propagation of Very Short Cracks" Engineering Fracture Mechanics, 7 (1975) pp. 235-247.

2. W.L. Morris, O. Buck and H.L. Marcus, "Fatigue Crack Initiation and Early Propagation in Al 2219-T851", Metallurgical Transactions, 7 A (1976) pp. 1161-1165.

3. W.L. Morris, "The Early Stage of Fatigue Crack Propagation in Al 2048", Metallurgical Transactions, 8 A (1977) pp. 589-596.

4. J. Lankford, "The Growth of Small Fatigue Cracks in 7075 - T6 Aluminium", Fatigue of Engineering Materials and Structures, 5 (1982) pp. 233-248.

5. J. Lankford, "The Effect of Environment on the Growth of Small Fatigue Cracks", Fatigue of Engineering Materials and Structures, 6 (1983) pp. 15-31.

6. H. Kitagawa, and S. Takahashi, "Applicability of Fracture Mechanics to Very Small Cracks or the Cracks in the Early Stage", pp. 627-631 in Proc. 2nd Intern. Conf. on the Mechanical Behavior of Materials, American Society for Metals, Metals Park, OH, 1976.

7. M. H. El Haddad, T. H. Topper, and K. N. Smith, "Prediction of Non Propagating Cracks", Engineering Fracture Mechanics, 11 (1979) pp. 573-584.

8. K. Tanaka, Y. Nakai, and M. Yamashita, "Fatigue Growth Threshold of Small Cracks", International Journal of Fracture, 17 (1981) pp. 519-533.

9. D. Taylor and J. F. Knott, "Fatigue Crack Propagation Behavior of Short Cracks; The Effect of Microstructure", Fatigue of Engineering Materials and Structures, 4 (1981) pp. 147-155.

10. J. F. McCarver and R. O. Ritchie, "Fatigue Crack Propagation Thresholds for Long and Short Cracks in René 95 Nickel-Base Super alloy", Materials Science and Engineering, 55 (1982) pp. 63-67.

11. C. W. Brown and M. A. Hicks, "A Study of Short Fatigue Crack Growth Behavior in Titanium Alloy IMI 685", Fatigue of Engineering Materials and Structures, 6 (1983) pp. 67-76.

12. N. E. Paton, J. C. Williams, and G.P. Rauscher, "The Deformation of α-Phase Titanium", pp. 1049-1069 in Titanium Science and Technology, R. I. Jaffee and H. M Burte, eds; Plenum Press, New York, 1973.

13. M. Peters and G. Lütjering, "Controll of Microstructure and Texture in Ti-6Al-4V", pp. 925-935 in Titanium '80 Science and Technology, H. Kimura, O. Izumi, eds.; AIME, New York.

14. J. Lindigkeit, G. Terlinde, A. Gysler and G. Lütjering, "The Effect of Grain Size on the Fatigue Crack Propagation Behavior of Age-Hardened Alloys in Inert and Corrosive Environment", Acta Metallurgica, 27 (1979) pp. 1717-1726.

15. J. E. Srawley, "Wide Range Stress Intensity Factor Expressions for ASTM E 399 Standard Fracture Toughness Specimens", International Journal of Fracture, 12 (1976) p. 475.

16. J. C. Newman, "A Review and Assessment of the Stress-Intensity Factors for Surface Cracks", pp. 16-42 in ASTM-STP 687, American Society for Testing and Materials, 1979.

17. G. T. Gray, III, J. C. Williams and A. W. Thompson, "Roughness-Induced Crack Closure: An Explanation for Microstructurally Sensitive Fatigue Crack Growth", Metallurgical Transactions, 14 A (1983) pp. 421-433.

18. T. Takemoto, K. L. Jing, T. Tsakalakos, S. Weissmann and I. R. Kramer, "The Importance of Surface Layer on Fatigue Behavior of a Ti-6Al-4V Alloy", Metallurgical Transactions, 14 A (1983) pp. 127-132.

19. K. Schulte, H. Nowack and G. Lütjering, "Influence of Monotonic and Cyclic Predeformation on Fatigue Crack Propagation of High-Strength Aluminium Alloys", Engineering Fracture Mechanics, 13 (1980) pp. 1009-1021.

20. J. Lindigkeit, A. Gysler and G. Lütjering, "The Effect of Pre-Deformation on Fatigue Crack Propagation Behavior of an Al-Zn-Mg-Cu Alloy in Inert and Corrosive Environment", Zeitschrift für Metallkunde, 72 (1981) pp. 322-328.

21. D.J. Duquette, M. Gell, and J. W. Pieto, "A Fractographic Study of Stage I Fatigue Cracking in a Nickel-Base Superalloy Single Crystal", Metallurgical Transactions, 1 A (1970) pp. 3107-3115.

INVESTIGATION OF THE GROWTH THRESHOLD FOR SHORT CRACKS

W. L. Morris and M. R. James

Rockwell International Science Center
Thousand Oaks, California 91360

The role of the local stresses which form during fatigue in the surface of aluminum alloys (Al 2219-T851 and Al 7075-T6) in near threshold growth of short surface cracks is examined. These arise as the result of microplastic slip within individual grains and can sometimes be comparable in magnitude to the maximum externally applied stress. So long as the stress amplitude is below the cyclic yield strength, fatigue induced changes in residual stress primarily accompany loading about a non-zero mean. In tension-tension fatigue, long range compressive surface stresses develop which reduce growth rates. Stochastic grain-to-grain variations in this macroscopic stress are determined experimentally and are too small to explain the statistical variations seen in the propagation rate of the individual cracks examined. However, a simple model suggests that, on occasion, unusually small local residuals can cause relative accelerations in growth rate. For the alloys studied, stochastic growth rate variations found experimentally are attributed to closure stress phenomena and to the reduced stress intensity which accompanies irregularities in the crack path.

Introduction

Rates of crack growth depart substantially from those predicted by linear elastic fracture mechanics (LEFM), if either the alloy grain size or the plastic zone size is an appreciable fraction of the crack length. The growth of cracks of grain size dimensions is associated with large stochastic growth rate variations for individual cracks. These fluctuations decrease with crack length, typically disappearing at lengths of 5-15 grain diameters. Fractographic studies (1) show that cracks which span many grains experience local variations in rate within individual grains along the front, despite uniform average growth. Apparently, then, an important difference between long and short cracks is the absence of averaging of rates along the crack front in the latter. Clearly, the absolute crack size is unimportant and behavior is determined by the ratio of microstructural scale to crack length. Sited as causes of microstructurally related rate variations for short cracks are microstructurally caused variations in crack depth (2,3), crack path (4-7) and growth mode (5,8,9), and in Mode I closure stress. For the special case of planar Mode I cracks in Al 2219-T851, Morris et al (5) find that much of the several orders of magnitude scatter in growth of grain sized cracks is attributable to local closure stress variations.

A concensus is developing that a major factor in the acceleration of short crack growth rate due to elastic-plastic enhancement of ΔK is the ratio of plastic zone size to crack length. Smith (10) has used 10% as the ratio at which important departures from LEFM should be visible in an analysis for short cracks. If continuum analysis pertains (i.e., microstructural effects are minimum), the absolute crack length is again unimportant and both the ratio and the applicability of fracture mechanics are determined by the stress range. Supporting this thesis are the observations of many, beginning with Kitagawa et al (11), that LEFM works adequately even for quite short cracks if the fatigue limit is not exceeded. Also, on the high stress side Dowling (12) reports for short cracks in an alloy steel, growth accelerations are stress dependent in that convergence to LEFM results does not occur with increasing crack length.

Most short crack experiments have been done at constant stress amplitudes between the fatigue limit and yield strength. Here, the propagation of the shortest cracks is apparently faster than LEFM predictions. With increasing length, the rates converge to LEFM values. Lankford (3), and Chan and Lankford (13) suggest this length dependence stems from a fundamental variation with length of the plastic deformation about short cracks. Zurek et al (14) and James et al (15) find that a crack length dependence of both the closure stress and in the degree of blockage of growth by grain boundaries also contributes to the short crack effect. A possibility examined further here is that some of the acceleration in early growth is only apparent and an artifact of the large stochastic variations caused by alloy microstructure, especially if the stress amplitude range is low.

We want to make clear that rapid crack growth associated with particular microstructures is real and intrinsically important in its own right, because it leads to early failure. However, it is equally important to learn under what circumstances the early growth accelerations are simply stochastic, in which case an appropriate average of the rates will be predicted by LEFM; and when a fundamental length dependence of the growth rate is involved. Studies by James et al (15) for Al 2219-T851 and Zurek et al (14) for Al 7075-T6 indicate that LEFM describes the average rates of ensembles of surface cracks reasonably well for stress ranges near the fatigue limit. In apparent contradiction, recent results by Lankford (3) for an Al 7075-T6 show accelerated short crack growth at a total stress range close to that used by Zurek. A potentially important difference between the experiments is that the Lankford work was entirely in tension-tension loading compared to fully reversed loading for Zurek et al. Load ratio is very important for short cracks because at nonzero mean stress residual surface stresses can develop in aluminum alloys by microplastic surface creep (16).

Several experiments have been done to reconcile the observations of Lankford and Zurek and to further elucidate the mechanics of accelerated growth of short cracks. Lankford's published data for 7075 have been reprocessed to permit averaging based upon equal intervals in fatigue cycles for each rate measurement. The resulting average gives heavier weight to the shortest growing cracks, and the growth rate is found to be independent of stress amplitude in comparison to new data taken at $R = 0$. The consequence of grain-to-grain variations in the local residual stress to early growth statistics has also been examined by measuring the stresses by a compliance technique. These variations contribute to stochastic variations, but no evidence is found of an important crack length dependence of the stresses. The only departure from LEFM predictions is a dip in early growth rate seen at crack lengths of 20-30 μm in both Lankford's and the latest data.

Experimental Procedure

Data are presented for two aluminum alloys. The Al 2219-T851 has been characterized previously (15); the alloy has a yield strength of 360 MPa and an average grain size of 60 μm transverse to the rolling direction. The Al 7075-T6 alloy, described in Ref. (14), was prepared in several grain sizes by a thermomechanical treatment (17) and had yield strengths ranging from 500 to 510 MPa. The grains were pancake shaped with a mean depth of 12 μm and mean cross section at the surface of 12 μm and 130 μm.

Tapered cantilever beam specimens were prepared from the materials with the principle stress axis in the rolling direction. A careful machining and polishing schedule was used that left a mirror finish and minimized the surface residual stresses prior to fatigue. Specimens were fatigued in flexure using stroke control in the elastic regime. Data were obtained for loading in laboratory air (45-55% relative humidity) at 5 Hz and at either fully reversed or positive load ratios. Changes in surface crack length at intervals in fatigue were measured by optical microscopy using a loading jig to apply a tensile stress (of less than 75% of the fatigue amplitude) to the surface to open the short cracks for improved visability. This technique does not seem to introduce hold time affects in aluminum alloys studied at room temperature. The distribution in growth rates for ensembles of cracks are analyzed in a manner described previously (14).

A similar fixture was used in a scanning electron microscope (SEM) to load specimens for measurement of the local residual stresses at the surface tips of individual cracks, using a compliance technique. This was accomplished by measuring the crack tip opening displacement (measured ~5 μm behind the tip) as a function of applied surface stress as shown in Fig. 1. If the applied stress exceeds a closure stress, the opening-displacement curve is linearly elastic (18) for short cracks. In the absence of a residual surface stress, the load-displacement line passes through the zero load-zero displacement intercept as in Fig. 1a. In the presence of a compressive residual stress, the intercept shifts to the applied tensile stress required to return the local stress at the point of opening to zero, as in Fig. 1b. This tensile stress is simply that necessary to compensate for the local compressive residual stress at the crack tip and provides a convenient method of obtaining that stress. Thus, as shown in Fig. 1, crack tip compliance measurements can be used to measure both the crack tip closure stress and the residual stress in the grain at the crack tip. These residual stresses oscillate about the macroscopic value and we assume are due to the inhomogeneous microplasticity that induces the macroscopic value. These local residual stresses in the grain are apparently not significantly changed by the crack tip plastic zone.

Macroscopic residual stresses were measured using x-ray diffraction. Chromium radiation was employed and diffraction data from (311) planes were obtained for eight incident angles for each specimen, using a computerized data acquisition system. The samples were oscillated at ±2° at each tilt angle to increase the number of grains contributing to the diffraction profile. The irradiated area on the sample was 3 mm high by 2 mm wide with an approximate penetration depth of 10 μm. Thus, the x-ray technique averages the surface residual stress over many grains.

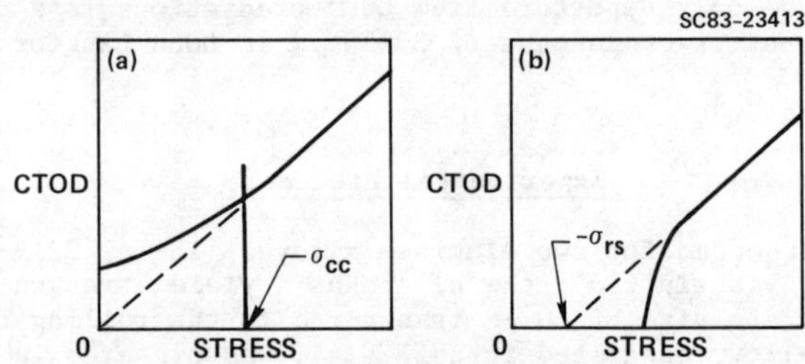

Fig. 1. Crack tip opening displacement (CTOD) versus surface stress analyzed to obtain residual surface stress: (a) For zero residual stress the elastic unloading line passes through the zero opening zero stress point, (b) a compressive residual stress shifts the intercept to an applied tensile stress.

Results

During fatigue at R = 0, a compressive residual stress develops on the surface of aluminum alloys prepared in an initially stress free condition (2,14). Data for Al 7075-T6 in Fig. 2 are the macroscopic average over many grains obtained by x-ray diffraction, and show that the surface of a 130 μm grain size alloy is more microplastic and develops larger residual stresses than does a 12 μm alloy. The results are for a peak tensile stress amplitude (σ_{max}) of 90% of the alloy yield strength. Eventually, the residual stress in the 130 μm alloy equilibrates at a value of $\approxeq -0.5\ \sigma_{max}$, so that at the surface the true stress (applied plus residual) is fully reversed. If the initial σ_{max} is decreased, the rate of change in residual stress diminishes as shown for an Al 2219-T851 alloy (Fig. 3). Notice that for comparable grain size and similar ratio of σ_{max} to yield strength, the 2219 alloy appears to be less microplastic than 7075 of comparable grain size (40 μm) in that a lower compressive residual stress is formed. Local variations in these stresses occur and we consider the magnitude of these fluctuations and their consequence to crack growth.

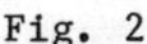

Fig. 2

Compressive residual stresses develop on Al 7075-T6 fatigued at R = 0 at a rate which increases with alloy grain size. σ_{max} = 453 MPa.

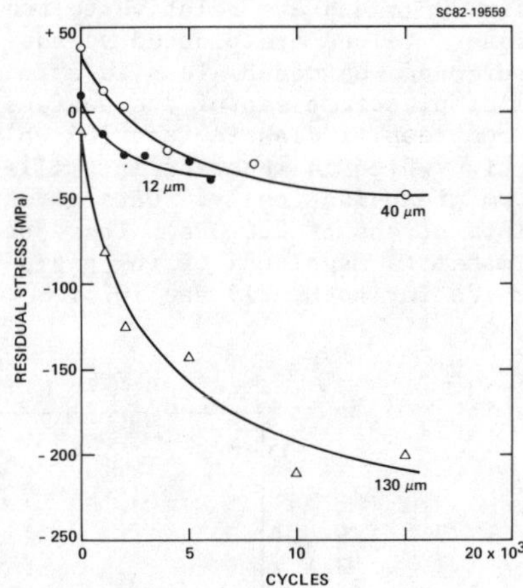

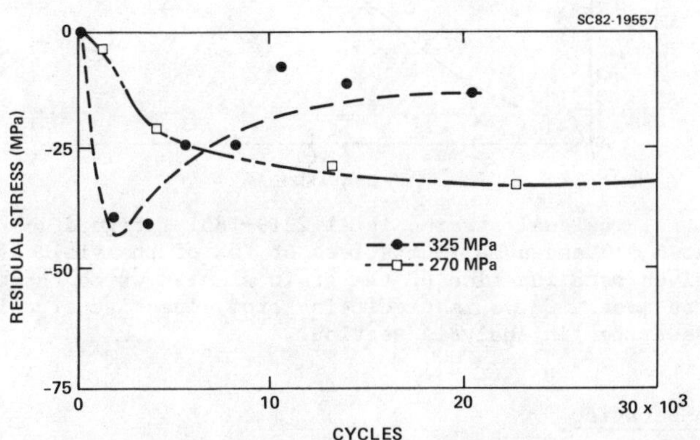

Fig. 3 The rate of formation of compressive surface stress in an Al 2219-T851 alloy increases with peak stress in R = 0 loading. Reversal at the higher amplitude is attributed to surface hardening and accompanying subsurface deformation.

Local Residual Stresses

Statistical variation of the residual stress from grain-to-grain can be anticipated. It must arise from the distribution in grain sizes and crystallographic orientations at the surface of an alloy. If these changes in local stress from grain-to-grain were comparable in magnitude to the average differences found versus grain size for 7075-T6 (Fig. 2), they would significantly affect early growth. To study this process, residual stresses in the path of a crack (shown later) are determined at the crack tip using the compliance technique illustrated by Fig. 1. In addition, insight into the statistical variation in the residual stress in many grains in a crack-free surface is obtained from a model which relates the local stresses to measured values of fatigue induced surface microplastic strain. The strains are determined using a reference gauge technique (19) for an Al 2219-T851 alloy fatigued in tension-tension loading. The surface strain data given in Fig. 4 were determined over gauge lengths of approximately 20 μm and are relative to the undeformed material prior to fatigue. Values are plotted versus the size of the grain in which the measurement was made. Tensile strains in the larger grains are the result of microplastic creep of the surface in each grain. These strains obscure the compressive elastic reaction which results from this deformation, a reaction which is apparent in smaller grains which do not significantly deform microplastically. Data given in Fig. 4 are for 10^4 cycles at a maximum stress of 270 MPa. These and similar results are used later to estimate the magnitude of the grain-to-grain variation in local residual stresses for both 2219 and 7075.

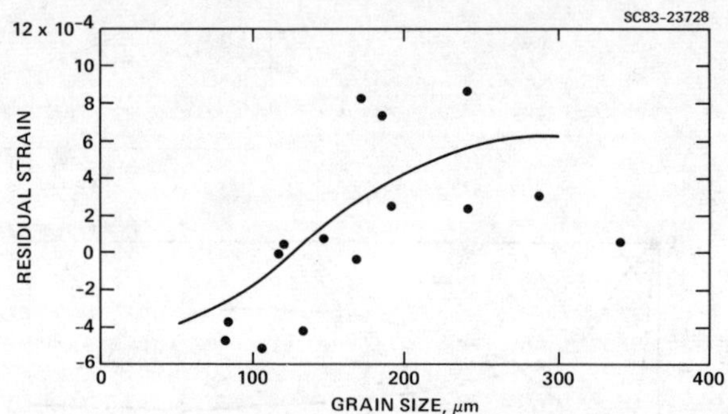

Fig. 4 Local residual strains in Al 2219-T851 fatigued 10^4 cycles at R = 0 and a maximum stress of 75% of the yield strength, given as a function of the grain size in which the measurement was made. Curve is prediction from linear superposition model described in Analysis section.

Early Growth Behavior

To assess the effect of fatigue induced residual stresses on the fluctuation in growth of individual cracks, growth rates and corresponding values of residual stress (measured by compliance) were obtained for propagation in the 130 μm Al 7075-T6 alloy. In Fig. 5, the local residual surface stress values are given for a crack whose length and growth rate are plotted in Fig. 6. The local residual stresses were measured at both surface tips at several intervals in fatigue. A statistical variation is seen

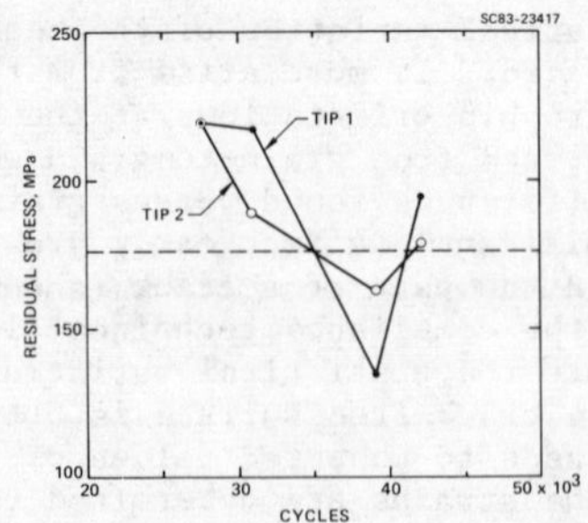

Fig. 5

Residual stress at the two surface tips of a short crack in Al 7075-T6 at several fatigue increments. Corresponding growth rates are in Fig. 6.

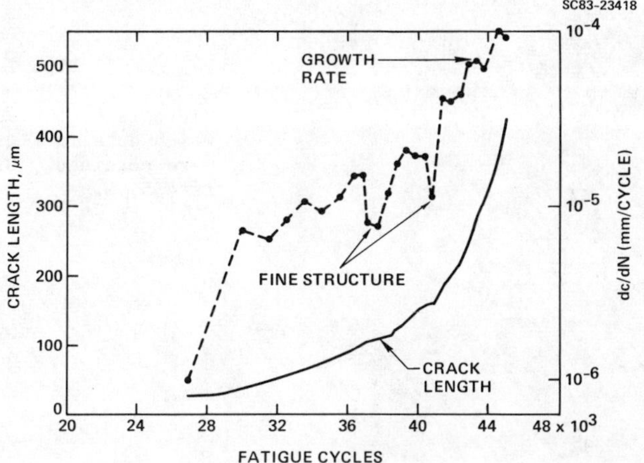

Fig. 6 Crack length and growth rate versus fatigue cycles.

which oscillates about the x-ray diffraction determined macroscopic value (dashed line). After fatigue, the surface was chemically etched so that the prior location of the crack path relative to grain boundaries, and the location of crack branching points could be determined. Numbers on the micrograph in Fig. 7 correspond to numbered features in growth rate data shown later. Analysis of the fracture surface for this crack (Fig. 8) (by rapid fracture of the sample) shows that the fatigue induced residual stress caused the crack to propagate more rapidly into the depth than along the surface producing a depth/length ratio of 0.8. Surface cracks in a stress free surface typically have an aspect ratio of 0.4-0.5 (3,14).

Results given in Figs. 9 and 10 are the average growth rate of ensembles of short surface cracks in the 130 μm material. Typically, 10-15 isolated non-interacting cracks are followed on a single sample. There is a very slight trend for the average growth rate to decrease with increased cyclic stress amplitude for small values of positive R ratio (Fig. 9) which we attribute to the increased residual surface stresses on the specimens fatigued at higher σ_{max}. The average growth rates (dc/dN) at the highest amplitude are reconstructed from results published by Lankford (3) by giving equal weight to each increment in growth over 2000 cycle intervals. This is the same cycle interval over which the additional data we took were averaged. Dashed lines are two sets of growth rate data for long cracks in Al 7075-T6 available in the literature (20,21). In Fig. 10 we illustrate, for constant σ_{max}, the pronounced difference in growth rates between loading at R = 0, and R = -1, which results from the absence of residual stress in the later case.

Analysis

It appears that the magnitude of the stochastic variations in fatigue induced residual surface stresses from grain-to-grain are too small to be the dominant source of the statistical fluctuations observed in the growth rate of short surface cracks in aluminum alloys. The magnitude of the grain-to-grain variations are small (Fig. 5), and there is evidence that

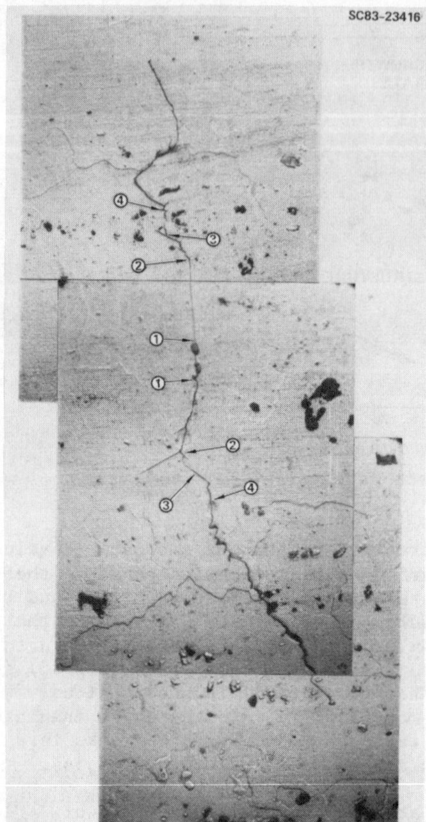

Fig. 7

Micrograph of crack from which data in Figs. 5 and 6 were obtained. Crack length is 435 μm.

Fig. 8 Fracture surface of the Fig. 7 crack overloaded after the experiment was complete. Bar at left is 100 μm.

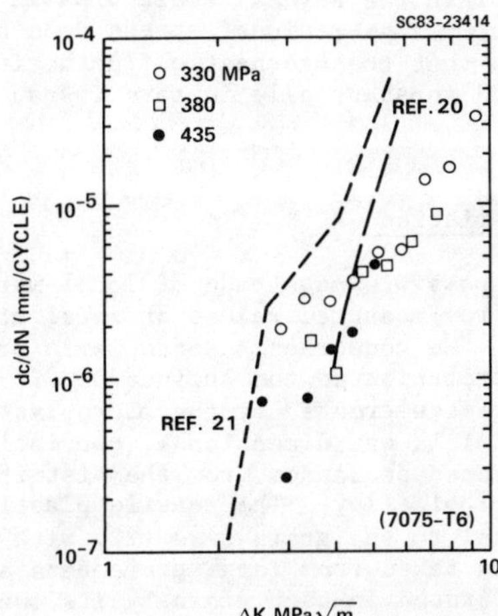

Fig. 9 Growth rate in Al 7075-T6, at positive load ratios for several cyclic stress amplitudes. Closed circles (435 MPa) are averaged from Lankford's results (Ref. 3). Others are for R = 0. Dashes are long crack data.

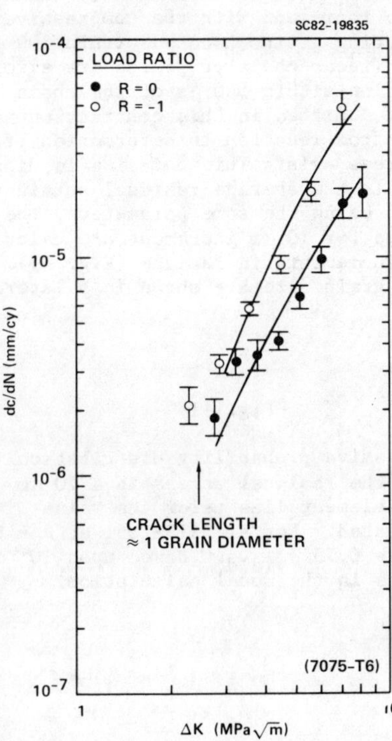

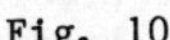

Fig. 10

Growth rate is faster in R = -1 loading by an amount attributable to the absence of substantial residual stresses in the fully reversed loading case. Curves are from a Monte Carlo calculation of growth rate (14). Maximum applied stress is 300 MPa for both cases.

other mechanisms can explain the several small dips in growth rate. Nevertheless, the macroscopic average residual stress does have important consequences and it appears that the stochastic fluctuations are of sufficient magnitude to effect dc/dN substantially in rare instances. This is shown next.

Local Residual Stress Variation

An estimate of the possible magnitude of local variations in residual surface stress is made from measured values of local microplastic strain such as given in Fig. 4. We construct a Monte Carlo procedure to calculate the residual stress distribution at the surface of 2219, by making a linear superposition of the stresses created by the microplastic strain in each grain. The model material is one-dimensional, consisting of a linear array of grains of length selected at random from the distribution in grain size determined for the 2219-T851 alloy. The tensile plastic strain (ε_p) is assumed to be proportional to the grain size (22) with a proportionality and statistical variation taken from large grain data as in Fig. 4. The internal stress (σ_i) generated in each grain by its own deformation is compressive and is taken as $\sigma_i = -|\varepsilon_p|E$, where E is Young's modulus. The net strain developed in each grain by its own plastic deformation is tensile, and the reaction strain and stress in neighboring grains is taken to be compressive. After Timoshenko and Goodier (23), we assume that the stress produced by each deformed grain produces a reaction within its neighbors which decreases inversely with the second power of distance from the deforming source. The local stresses and local strains are calculated over increments of 10 μm in length by summing the self-induced stress or strain in each increment with the compressive contribution arising from the neighboring grains located within 500 μm. Inaccuracies due to end effects in the linear chain of grains are eliminated by discarding data in the simulation within 500 μm of the chain end after the superposition is completed. Omitted in this contruction are tensile hoop stresses which would result from reaction to deformation of grains on a line perpendicular to the stress axis. This omission is discussed later. The curve shown in Fig. 4 is the average residual strain versus grain size predicted by this model. Using the same parameters, the distribution in local stresses averaged per 10 μm increment are calculated for 2219 fatigued at R = 0 for three increments in fatigue (Fig. 11). Predictions for Al 7075-T6 of 130 μm grain size are shown in a later section.

Fig. 11

Cumulative probability distribution that the residual stress in a 10 μm long element lies below the value indicated. For Al 2219-T851 at R = 0 $\sigma_{max} = 0.75\ \sigma_{yield}$. Based upon 10^4 grains in the model calculation.

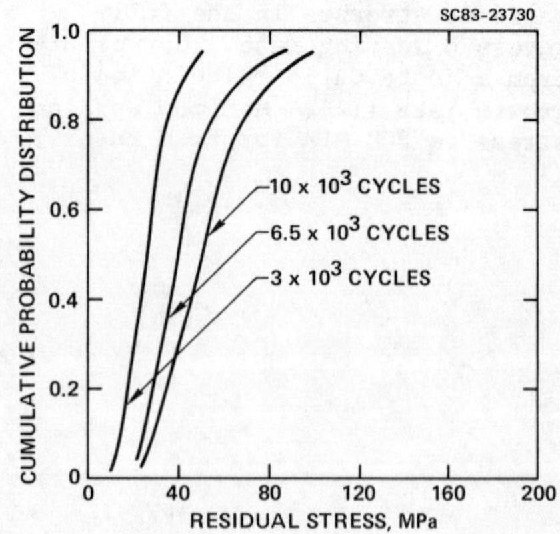

Individual Crack Behavior

In Figs. 12 and 13 we compare the growth rate data from Fig. 6 to several simple models. Solid curves in 12 are of the form dc/dN $\propto (\Delta\sigma\sqrt{2c})^m$ where 2c is the projection of the surface crack length perpendicular to the surface stress axis. Experimental values of 2c are used to calculate the rates shown and the proportionality is selected so that all predicted values converge to a common point at dc/dN = 10^{-4} mm/cycle. Results are given for m = 2 and 3, for the stress range $\Delta\sigma$ assumed equal to the applied. For the m = 2 case, we also show the result of reducing the peak tensile stress by the magnitude of the experimentally measured residual surface stress given in Fig. 5. The general trend in growth rates appears to be determined by a cyclic stress intensity ($\sigma\sqrt{2c}$). Fluctuations in residual stress found at the surface are far too small to account for the fine structure seen in the growth rate data.

Next, we examine the influence of microstructure on dc/dN. With reference to the micrograph in Fig. 7, we relate the crack length and crack growth data in Fig. 6 to microstructural features. Prior to position 1 in

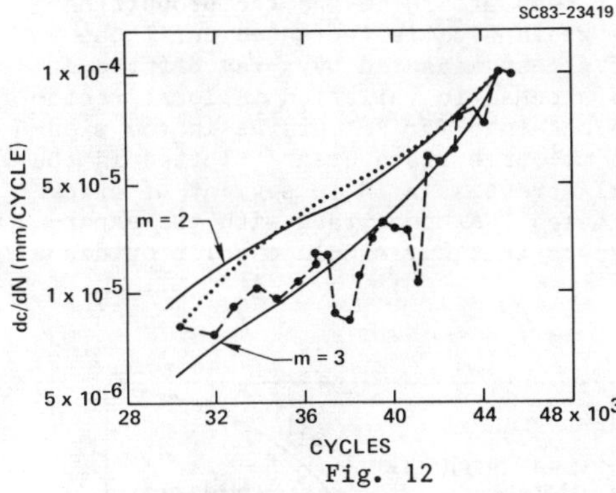

Fig. 12

Comparison of rate of propagation of crack in Fig. 7 to several simple models: constant exponents at powers of m = 2 and 3 solid lines, dotted line is m = 2 model corrected for measured residual stress.

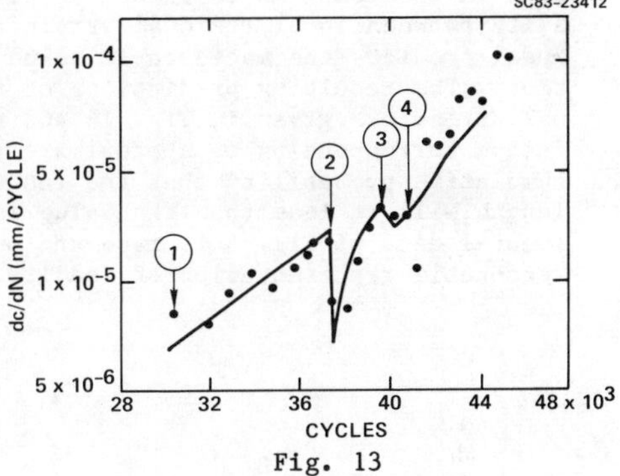

Fig. 13

Comparison of rate of propagation of the Fig. 7 crack to a model which incorporates crack closure stress.

Fig. 13, the crack was less than 25 μm long and was within several microns of the two small surface pits at which it initiated. Between 24 × 10^3 cycles, when the crack was first found, and 30 × 10^3 cycles, the crack grew only 6 μm, displaying a dip in growth rate which was only poorly resolved. More accurate length data were not taken at this point because the crack was still embedded within the zone of initiation. Between points 1 and 2, both surface tips were propagating, apparently in a crystallographic mode. At 2 the upper tip reached a grain boundary and the lower tip branched. At 3 the top tip branched and at 4 the bottom tip reached a grain boundary. In the regions 1-2 and 2-3, the rates increased progressively as a tip traversed the grain. The increase was more rapid from 2-3 for which the path of the upper tip was not planar. The predicted growth rate (Fig. 13) is obtained from a model which corrects the cyclic stress intensity range for crack closure stress giving $dc/dN \propto [(\sigma_{max}-\sigma_{cc})\sqrt{2c}]^2$. The closure stress $\sigma_{cc} = \alpha z \sigma_{max}/2c$. Variable z is the distance of a surface crack tip to the next grain boundary in its path. For the 7075 alloy, material parameter $\alpha = 0$ for crystallographic growth and $\alpha = 0.5$ for noncrystallographic growth (14). The rates plotted are averages of the rates at the two surface tips. At a point of branching, we have arbitrarily set $dc/dN = 0$ at a branched tip for 1000 cycles. While crack closure accounts for much of the fine structure in growth rate, notice that at 4 there is a kink in the crack path and the predicted rates exceed the measured; a result anticipated by Suresh (4).

Discussion

For a randomly selected element of material ahead of a crack tip, the chance of finding a large deviation from the average macroscopic residual stress in the case considered for Al 2219-T851 (Fig. 11) is small. Predictions commensurate with the highest cyclic stress loading conditions considered for Al 7075-T6 are made by making minor adjustments to our one-dimensional residual stress model. The scale of the grain size distribution used in the model is increased to give a mean of 130 μm and the proportionality between local residual strain and grain size is increased until the model predicts the macroscopic value of stress measured by x-ray diffraction. The resulting predictions of the stochastic variation in local residual stress are given in Fig. 14 and are obtained for 10^4 grains in the simulation corresponding to approximately 1 cm^2 of surface area. Plotted is the cumulative probability that the residual stress in a 10 μm segment of crack length will be less than the value indicated. A comparison with the experimental data of Fig. 5 is made and suggests that our simple model provides a reasonable representation of reality.

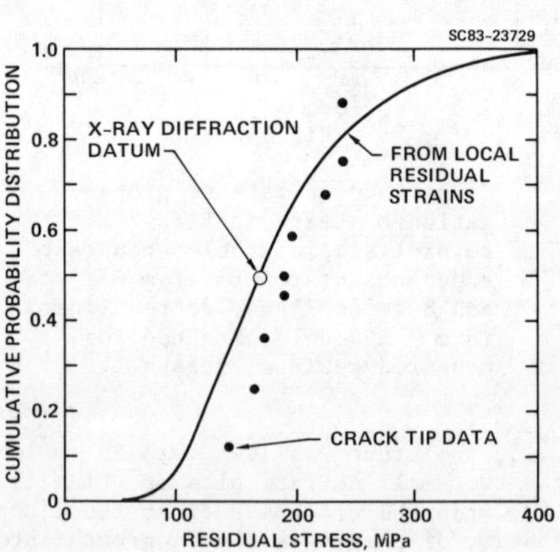

Fig. 14

Cumulative probability that residual stress in a 10 μm long crack segment will fall below the indicated value for Al 7075-T6 (130 μm) with a macroscopic residual stress of 160 MPa. Curve is from superposition model. Data taken from Fig. 5.

The local residual stress must fall below 100 MPa for the fluctuation to have an obvious effect on dc/dN for an individual crack. The model predicts that the probability of occurrence of stresses in this range is small. This assessment might change if the local stresses in some of the grains were tensile even though the aggregate was compressive. It is known that later during fatigue macroscopic tensile stresses can develop in aluminum alloys fatigued at R = 0 (16). The source of this behavior is not entirely clear. One possibility is that the surface cyclically hardens and continued plastic tensile deformation of the subsurface causes a tensile reaction in the surface. Another possibility is that tensile hoop stresses around a deformed grain may cancel the compressive radial stresses of neighboring grains. Hoop stresses have not been included in our one-dimensional model, in part because no evidence of local elastic tensile stresses have been found by our reference gauge measurements. We do not know if this is because of the small sample size, or if there are fundamental constraints which prevent the formation of local tensile stresses in an essentially compressive field.

Before this study was begun, we thought that the dip in early growth rate reported by Lankford (3) might result from the progressive development of a macroscopic residual stress after crack initiation. Cracks would propagate progressively more slowly as the compressive residual stress formed. It is now clear that, for R = 0 loading of 7075-T6, cracks do not initiate until the macroscopic residuals are essentially established. Residual stresses cannot therefore be responsible for the observed dip in growth rate. Actually, we do not see an early dip in growth rate in the average growth rate data (Fig. 9), probably because our observations are confined to cracks longer than 30 μm. A dip can be inferred, however, from the short time to crack initiation. In fatigue at positive R ratios, cracks initiate only at the largest surface discontinuities and cracks shorter than 20 μm are typically embedded in this zone. In fully reversed loading even 2 μm wide inclusions may initiate cracks (22), and we have previously thought that the rapid average growth rates from these must stem from the absence of nearby grain boundary obstacles in the crack path.

Yet another residual stress anomaly should be noted. At high R ratios, data for 2219 (15) show that creep within the plastic zone causes the formation of a compressive residual stress at the tips of short surface cracks. The local stress intensity at the crack tip is consequently less than that calculated from the applied stress. This residual stress forms at cyclic surface stress amplitudes so small that no surface macroplastic deformation is apparent in the crack free surface. No macroscopic residual stresses form in this circumstance and the residuals are localized to and unique to each crack. The influence of grain size at the crack tip on the magnitude of the residuals is unknown (15). This mechanism would produce a crack length dependence of the crack tip residual stress which has not been sought in Al 7075-T6.

The essential question of what is the correct way to prepare an average of short crack rates which are naturally stochastic for comparison to mechanistically averaged stochastic variations within a single long crack still remains. In calculating the short crack average, we have chosen to give equal weight to units of growth over equal periods of fatigue cycles. This gives a higher weight in the average to the slower growing cracks. This procedure can be rationalized since slower growing segments within a large crack must have an important retarding effect on the entire crack front. Support is found in the observation that ΔK usually provides a good description of the average short crack growth rates when this averaging technique is followed (14). The most direct test has been in the measurement of rates at several different cyclic stress amplitudes in 2219 (15) and 7075 (14) alloys. A stress dependence in addition to ΔK has been found only when $\Delta\sigma$ was above a critical value, and even this was apparent only for very short cracks (15). Another option of course is to compare short crack results directly to fracture mechanics data. Short cracks are less susceptible to retardation from closure stresses encountered during load shedding, however, so such comparisons are best made to the most recent data wherein this potential problem has been recognized.

A sensitive method to evaluate short crack growth phenomena is to compare rates averaged for an ensemble of cracks. Statistical fluctuations are minimized and trends in growth rate with stress, crack length and grain size become visible. Compared in Fig. 10 are averaged growth rates of approximately 15 cracks per specimen in coarse grained (130 μm) Al 7075-T6. In fully reversed loading, no surface residual stress was evident while at $R = 0$, a fatigued induced compressive residual stress of -135 MPa was measured by x-ray diffraction. This was essentially constant during crack propagation. The abscissa has not been corrected for the residual stress with the result that longer cracks propagate more slowly at $R = 0$ than at $R = -1$. The solid curves in Fig. 10 are from a Monte Carlo procedure which involves an algebraic summing of the applied and residual stress in calculating an effective ΔK (14). Interestingly, the apparent threshold for crack growth is almost unchanged by the residual stress. This is because the threshold occurs at a crack length of about one average grain diameter and this corresponds to the same calculated value of ΔK for both cases. We wonder if this same consideration (compressive residual stress) contributes to the departure of long from short crack growth rates at high ΔK in Fig. 9.

The major stochastic effects of microstructure on the near threshold growth rate of short surface cracks in 7075-T6 appears to be associated with grain boundaries and crack branching. In the absence of any irregularities in crack path, growth rates tend to accelerate as a crack traverses a grain until the crack stops suddenly at or just before it reaches the next boundary in its path. The acceleration is more exaggerated if the mode of propagation is transgranular and noncrystallographic. This behavior can be rationalized by a closure stress model in which the plasticity induced closure stress is largest just as a crack enters a new grain. A major elastic-plastic contribution to ΔK would cause deceleration of growth across a grain in which plastic deformation was constrained by the boundaries (13). Since we do not know the true power dependence of the growth rate, we cannot be certain that $m = 2$ (Fig. 12) and that no elastic-plastic effect is present. But, clearly it is not large in this case.

Perspectives and Conclusions

To place these results in perspective, we refer to Fig. 15 which schematically illustrates an observation of James et al. (15) based on averaged growth rates of ensembles of cracks. It is the total cyclic stress range ($\Delta\sigma$) which determines the applicability of linear elastic fracture mechanics. If $\Delta\sigma$ is high, ΔK_{th} disappears, i.e., short cracks grow below ΔK_{th}. This stress dependent transition in the dc/dN data is another manifestation of the reduction in ΔK_{th} with increasing stress first described by Kitagawa and Takahashi (11). Apparently, the transition stems from surface microplastic deformation at high $\Delta\sigma$, and so there is the resulting connection between alloy grain size, yield strength and transition stress described by El Haddad et al. (24) and by Ritchie (25). Reduced grain size increases the $\Delta\sigma$ at which the transition occurs. The mechanics of the interaction of surface plasticity induced by the net section stress with small surface cracks is of continuing interest. It is thought to directly accelerate the propagation of moving cracks (13) and to reduce the duration of those arrested by obstacles such as grain boundaries (15). There is evidence that the maximum rather than the cyclic stress intensity is important in the latter process (15). Other mechanisms which for short cracks are

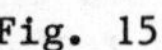

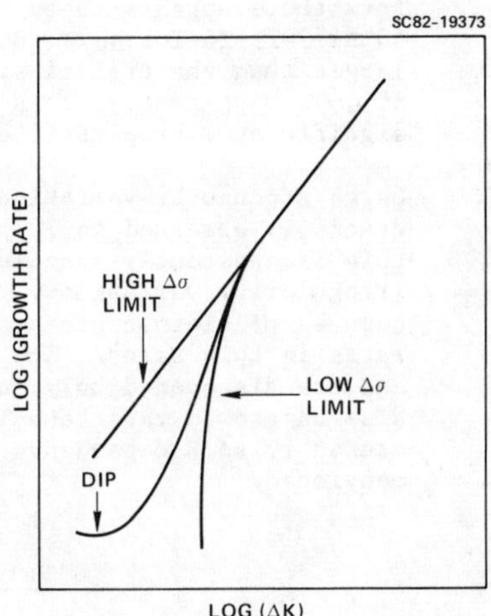

Fig. 15

Schematic representation of the effect of cyclic stress range on near threshold crack growth.

directly sensitive to crack length, such as closure stress, further complicate this picture. Stochastic effects of several of these have been described in this paper. Their consequence to average rates are becoming clear only as the sensitivity of early growth to the ratio of crack length to grain size is studied (14).

Fatigue at positive R ratios involves a lower ratio of stress range to maximum stress. This means that in comparison to experiments done in fully reversed loading, dc/dN results at R = 0 are more likely to fall on the low $\Delta\sigma$ branch of Fig. 15. The source of dip in dc/dN at intermediate stress is of special interest (3,26,27). In smooth bar specimens it is found only when the cracks are very short and usually less than one grain diameter in length. It is reminiscent of crack growth from notches (28), and hence it may result from the microscopic notch which constitutes the crack initiation site, such as an inclusion.

Several explanations for the apparent disparity between fully reversed and tension-tension loading have been considered.

1. Compressive residual surface stresses of substantial magnitude are induced by fatigue of Al 7075-T6. These are found to develop too rapidly to produce an observable dip in average rates of short cracks in smooth surface specimens fatigued at constant stress amplitude. Grain-to-grain variations in these local residuals are also found. A simple model suggests that only in exceptional cases will they be of sufficient magnitude to produce an anomaly in the growth rate of individual cracks that is distinguishable from stochastic variations arising from crack closure and crack path irregularity.

2. Perception of the crack length dependence of early growth rates is apparently sensitive to the manner in which the rate data are analyzed. A long crack front averages diverse local rates of growth naturally, but for comparison with short crack data to find stress/length dependences an average rate must be calculated for an ensemble of short cracks. By giving equal weight to increments of growth per fatigue cycle we have prepared averages

which give heavier weight to the slower growing cracks. In this form there appears to be no stress dependence of the growth rate in Al 7075-T6 for $\Delta\sigma \leq 400$ MPa, for cracks which are somewhat larger than the typical sites of initiation (i.e., longer than 30 μm). Apparently, this value of $\Delta\sigma$ is below that required for significant microplastic deformation of the surface.

3. Large stochastic variation in the growth rate of individual cracks is observed in 7075-T6. Beyond the region of initiation, this is apparently associated with crack closure stress and path irregularity variations. Zurek et al. (14) have discussed consequences of microscopic growth behavior to the average growth rates in this alloy. The effects on the average are not large and are discovered only in an assessment of the effect of grain size on growth rate behavior. The bias in early growth behavior caused by an R dependence of the crack initiation habit is mentioned.

References

1. C. Q. Bowles, The Role of Environment, Frequency and Wave Shape During Fatigue Crack Growth in Aluminum Alloys, Ph.D. Thesis, Technische Hogeschool Delft, p. 35, 1978.
2. W. L. Morris and O. Buck, "Crack Closure Measurements for Microcracks Developed During Fatigue of Al 2219-T851," Met. Trans. 8A, 597-601, 1977.
3. J. Lankford, "The Growth of Small Fatigue Cracks in 7075-T6 Aluminum," Fatigue Engrg. Mater. Struct. 5, 233-248, 1982.
4. S. Suresh, "Crack Deflection: Implications for the Growth of Long and Short Fatigue Cracks," Met. Trans. A, 14A, 2375-2385, 1983.
5. W. L. Morris, M. R. James, O. Buck, "Growth Rate Models for Short Surface Cracks in Al 2219-T851," Met. Trans. A, 12A, 57-64, 1981.
6. M. R. James, and W. L. Morris, "The Effect of Fracture Surface Roughness on Growth of Short Fatigue Cracks," Met. Trans. 14A, 153-155, 1983.
7. T. Kunio, and K. Yamada, "Microstructural Aspects of the Threshold Condition for Nonpropagating Fatigue Cracks in Martensitic-Ferritic Structures", pp. 342-370 in Fatigue Mechanism (edited by J. J. Fong), ASTM STP 675; American Society for Testing and Materials, Philadelphia, Pa, 1979.
8. W. L. Morris, "A Comparison of Microcrack Closure Load Development for Stage I and II Cracking Events for Al 7075-T651," Met. Trans. 8A, 1087-1093, 1977.
9. K. Tanaka, M. Hojo and Y. Nakai, "Fatigue Crack Initiation and Early Propagation in 3% Silicon Iron," pp. 207-232, Fatigue Mechanisms: Advances in Quantitative Measurement of Physical Damage, ASTM STP 811, American Society for Testing and Materials, Philadelphia, Pa, 1983.
10. R. A. Smith, "On the Short Crack Limitations of Fracture Mechanics," Int. J. Fract. 13, 717-720, 1977.
11. H. Kitagawa and S. Takahashi, "Applicability of Fracture Mechanics to Very Small Cracks," pp. 627-631, Proc. 2nd Int. Conf. on Mech. Behavior of Materials, Boston, MA, Aug. 1976.
12. N. E. Dowling, "Crack Growth During Low Cycle Fatigue of Smooth Axial Specimens", pp. 97-121, Cyclic Stress Strain and Plastic Deformation Aspects of Fatigue Crack Growth, ASTM STP 637; American Society for Testing and Materials, Philadelphia, Pa, 1977.
13. K. S. Chan, and J. Lankford, "A Crack-Tip Strain Model for the Growth of Small Fatigue Cracks," Scripta Met. 17, 529-532, 1983.

14. A. K. Zurek, M. R. James and W. L. Morris, "The Effect of Grain Size on Fatigue Growth of Short Cracks," Met. Trans. 14A, 1697-1705, 1983.
15. M. R. James, W. L. Morris, A. K. Zurek, "On the Transition from Near-Threshold to Intermediate Growth Rates in Fatigue," Fatigue Engrg. Mater. Struct., 6, 293-305, 1983
16. M. R. James, and W. L. Morris, "Fatigue Induced Changes in Surface Residual Stress," Scripta Met, 17, 1101-1104, 1983.
17. J. A. Wert, N. E. Paton, C. H. Hamilton, and M. W. Mahoney, "Grain Refinement in 7075 Aluminum by Thermomechanical Processing," Met. Trans. 12A, 1267-1176, 1981.
18. W. L. Morris, "The Noncontinuum Crack Tip Deformation Behavior of Surface Microcracks," Met. Trans. 11A, 1117-1123, 1980.
19. W. L. Morris, R. V. Inman, and M. R. James, "Measurement of Fatigue Induced Plasticity," J. Matls. Sci. 17, 1413-1419, 1982.
20. W. G. Truckner, A. B. Thakker and B. J. Bucci, Research on Investigation of Metallurgical Factors on the Growth Rate of High Strength Aluminum Alloys, U.S. Air Force Materials Laboratory, Contract F33615-74-C-5079, Technical Report, May, 1975.
21. P. E. Bretz, A. K. Vasudevan, R. J. Bucci, and R. C. Malcolm, Effect of Microstructure on 7XXX Aluminum Alloy Fatigue Crack Growth Behavior Down to Near-Threshold Rates, NADC Contract N00019-79-C-0258, Final Report, Oct. 1981,
22. M. R. James, and W. L. Morris, "The Fracture of Constituent Particles During Fatigue," Matls. Sci. and Engrg. 56, 63-71, 1982.
23. S. P. Timoshenko, and J. N. Goodier, Theory of Elasticity, 3rd Edition, p. 68; McGraw-Hill, New York, N.Y., 1970.
24. M. H. El Haddad, K. N. Smith and T. H. Topper, "Fatigue Crack Propagation of Short Cracks," Trans. ASME - J. Engrg. Matls. and Tech. 101, 42-46, 1979.
25. R. O. Ritchie, "Near Threshold Fatigue-Crack Propagation in Steels," Int. Metals Review 24, 205-230, 1979.
26. R. G. deLange, "Plastic-Replica Methods Applied to Study of Fatigue Crack Propagation in Steel 35 CD4 and 16 St. Aluminum Alloy," Trans. Met. Soc. AIME 230, 644-648, 1964.
27. M. M. Hammouda, R. A. Smith and K. J. Miller, "Elastic-Plastic Fracture Mechanics for Limitation and Propagation of Notch Fatigue Cracks," Fatigue Engr. Matls. and Structures 2, 139-154, 1979.
28. D. Taylor and T. F. Knott, "Fatigue Crack Propagation Behavior of Short Cracks; The Effect of Microstructure," Fatigue Engrg. Matls. and Struct. 4, 147-155, 1981.

MECHANICS OF GROWTH THRESHOLD

OF SMALL FATIGUE CRACKS

K. Tanaka
Department of Engineering Science
Kyoto University
Kyoto 606, Japan

and Y. Nakai
Department of Mechanical Engineering
Osaka University
Suita 565, Japan

The effect of crack length on the threshold condition for fatigue crack growth is predicted based on the blocked slip band model combined with the semi-empirical equation for the change of crack closure with crack length. The experimental data required for the model are the total range and effective range of the threshold stress intensity for a large crack, and the fatigue limit of smooth specimens. The above threshold model is applied to the non-propagation of a small fatigue crack at the notch tip. The fatigue limits of notched components predicted as a function of notch geometry agree with the experimental data. The change of crack closure with crack length in smooth and notched specimens is analysed theoretically by using Budiansky-Hutchinson's crack closure model. The results agree fairly well with the formula obtained semi-empirically.

Introduction

Fracture mechanics has been successfully applied to fatigue crack propagation. Under gross elastic condition, the rate of fatigue crack growth uniquely determined by the stress intensity factor (SIF). The threshold condition for fatigue crack growth is expressed by the threshold value of the SIF range. The SIF characterization is, however, questioned when the crack length is small. The plastic deformation at the crack tip, which is responsible for crack growth, is no longer contained within the SIF singularity field. The rate-SIF relation established for large cracks is no longer applicable to small cracks.

The threshold value of the SIF range is known to decrease with decreasing crack length and the threshold value of the applied stress approaches the fatigue limit of smooth specimens at very short cracks [1-5]. Therefore, the theoretical modeling for the threshold condition for crack growth is required to yield the model for the fatigue limit of smooth specimens at a limit of vanishing crack length. The crack closure is known to be pronounced near the threshold of large cracks [6-8]. The amount of crack closure changes with crack length [5, 9]. The model for the threshold should also include the change of crack closure level with crack length.

In the present paper, previous models for the growth threshold of small cracks in smooth specimen will first be reviewed, and a semi-empirical formula for the variation of crack closure with crack length will be proposed. The threshold model is then applied to the non-propagation of small cracks at the tip of sharp notches. Finally, the effect of crack length on crack closure is theoretically analysed by using Budiansky-Hutchinson's model and is compared with a semi-experimental formula.

Growth Threshold of Small Fatigue Cracks

According to the experimental results, the threshold value of the SIF range, ΔK_{th}, tends to decrease as a crack becomes shorter, approaching the value corresponding to the constant stress range equal to the fatigue limit to smooth specimens. This transition behavior was first modeled by Haddad et al. [3]. They assumed that the threshold condition for crack growth was determined by the constant value of the SIF range ($=\Delta K_{th\infty}$) for a fictitious crack whose length was the actual crack length plus the intrinsic crack length. The intrinsic crack length a_o is defined by

$$a_o = (\Delta K_{th\infty}/\Delta\sigma_{wo})^2/\pi \quad (1)$$

where $\Delta\sigma_{wo}$ is the stress range at the fatigue limit of smooth specimens. For an isolated crack of length a, Haddad condition yields the threshold stress range as

$$\Delta K_{th\infty} = \Delta\sigma_{th}[\pi(a + a_o)]^{1/2} \quad (2)$$

Using Eq.(1), Eq.(2) can be rewritten as

$$\Delta\sigma_{th} = \Delta K_{th\infty}/[\pi(a+a_o)]^{1/2} = \Delta\sigma_{wo}(1+a/a_o)^{-1/2} \quad (3)$$

Since $\Delta K_{th} = \Delta\sigma_{th}(\pi a)^{1/2}$, we have

$$\Delta K_{th} = \Delta K_{th\infty}(1 + a_o/a)^{-1/2} \quad (4)$$

Equations (3) and (4) have been confirmed to agree well with the experimental data for various metals [3-5].

The authors [5], combining their micromechanical model for crack growth threshold with crack closure, gave a different interpretation of Eqs.(3) and (4). They started with a model of crack-tip slip band blocked by the grain boundary as shown in Fig.1(a), and obtained a model for the fatigue limit of smooth specimens by reducing the crack length to zero as shown in Fig.1(b). When shear incompatibility is modeled by infinitesimal (super-) dislocations distributed continously on the slip plane, the density of pileup dislocations at the tip of the slip band is infinite and the elastic stress field has a singularity like the crack-tip stress field. The intensity of the field is characterized by the microscopic stress intensity factor (MSIF). The authors have assumed that the threshold condition for fatigue crack growth is determined by the condition whether the slip band near the crack tip propagates into an adjacent grain or not, which is expressed by the critical value of MSIF. The threshold value derived through this model was found to give the effective component of the applied threshold stress, i.e. the maximum stress minus the crack opening stress.

For mathematical simplicity, the coplaner slip band shown in Fig.2(a) was analysed previously. The effective component of the applied stress $\Delta\sigma_{effth}$ is given by

$$\Delta\sigma_{effth} = K^{m}{}_{c}/(\pi b)^{1/2} + (2/\pi)\sigma_{fr}{}^{*}\cos^{-1}(a/b) \tag{5}$$

where $K^{m}{}_{c}$ is the critical value of MSIF, $\sigma_{fr}{}^{*}$ is the friction stress, and b is the crack length a plus the size of the blocked slip band zone ℓ. The corresponding effective value of SIF, ΔK_{effth}, is

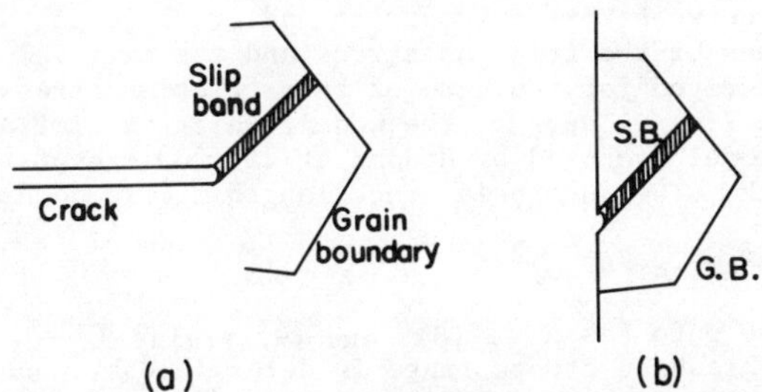

Figure 1 - Models of fatigue thresholds.
(a) Threshold for fatigue crack propagation.
(b) Threshold for fatigue crack initiation.

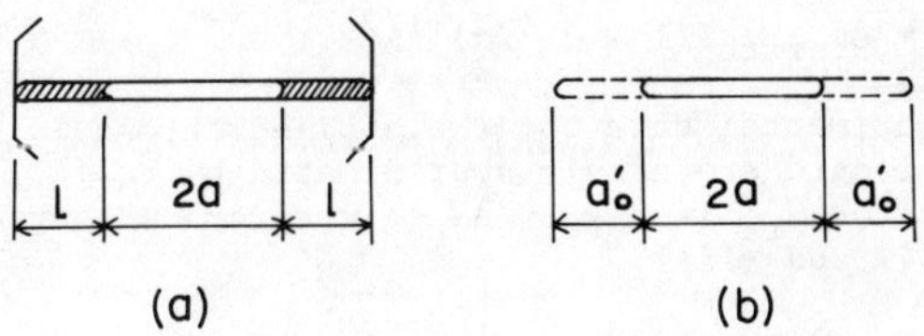

Figure 2 - Simplified models.
(a) Blocked slip band model.
(b) Intrinsic crack model.

$$\Delta K_{effth} = \Delta\sigma_{effth}(\pi a)^{1/2} = K^m{}_c(a/b)^{1/2} + 2(a/\pi)^{1/2}\sigma_{fr}{}^*\cos^{-1}(a/b) \quad (6)$$

For long cracks, the threshold SIF, $\Delta K_{effth\infty}$, can be obtained by taking a limit of a small ℓ as

$$\Delta K_{effth\infty} = K^m{}_c + 2(2/\pi)^{1/2}\sigma_{fr}{}^*\sqrt{\ell} \quad (7)$$

The fatigue limit of smooth specimen is obtained by substituting $a = 0$ into Eq.(5) as

$$\Delta\sigma_{wo} = \sigma_{fr}{}^* + K^m{}_c/(\pi\ell)^{1/2} \quad (8)$$

Since the size of the slip band zone is an order of the grain size, the threshold SIF increases and the fatigue limit decreases with increasing grain size. Equations (7) and (8) agree with the experimental data with mild steels [4, 8, 10].

Tanaka and Mura [11] used an alternate criterion for the threshold. They assumed the density of dislocation strain energy within the slip band was a critical parameter. The functional relation of the threshold condition with grain size and crack length is quite similar to that derived from the MSIF criterion.

By using the Petch type relation obtained between $\Delta\sigma_{wo}$ and d $(=2\ell)$, Eq.(8), the values of $\sigma_{fr}{}^*$ and $K^m{}_c$ were determined for mild steel as $\sigma_{fr}{}^*=$ 114 MPa and $K^m{}_c=0.318$ MPa$\sqrt{m}$. The values of $\Delta K_{effth\infty}$ given by substituting $\sigma_{fr}{}^*$ and $K^m{}_c$ into Eq.(7) were found to agree with the experimentally measured values of $\Delta K_{effth\infty}$ [5].

The relation between threshold stress and crack length does not vary much with the values of the friction stress and the critical MSIF. Simple equations can be derived for the case of zero friction stress within the slip band. Figure 2(b) illustrate the model, which is similar to the fictitious crack model proposed by Haddad et al. [3] except the length of the intrinsic crack. The intrinsic crack length $a_o{}'$ is determined by

$$a_o{}' = (\Delta K_{effth\infty}/\Delta\sigma_{wo})^2/\pi \quad (9)$$

Substituting $\sigma_{fr}{}^*=0$ into Eqs. (7), (8), and (9) yields $a_o{}'=\ell$. The effective component of the threshold stress range is determined by equating the SIF range at the fictitious crack tip to $\Delta K_{effth\infty}$ $(=K^m{}_c$ for $\sigma_{fr}{}^*=0)$ as

$$\Delta\sigma_{effth}=\Delta K_{effth\infty}/[\pi(a+a_o{}')]^{1/2}=\Delta\sigma_{wo}(1+a/a_o{}')^{-1/2} \quad (10)$$

The above equation gives $\Delta\sigma_{effth}=\Delta\sigma_{wo}$ for $a=0$. By using $\Delta K_{effth}=$ $\Delta\sigma_{effth}(\pi a)^{1/2}$ and Eq.(10), we have

$$\Delta K_{effth} = \Delta K_{effth\infty}(1 + a_o{}'/a)^{-1/2} \quad (11)$$

Since the experimental data for the applied threshold SIF range as a function of crack length are nicely approximated by Eq.(4), the variation of the effective fraction U_{th} $(=\Delta K_{effth}/\Delta K_{th})$ with crack length can be given as follows from Eqs.(4) and (11):

$$U_{th} = U_{th\infty}[(a + a_o)/(a + a_o{}')]^{1/2} \quad (12)$$

where

$$U_{th\infty} = \Delta K_{effth\infty}/\Delta K_{th\infty} = (a_o{}'/a_o)^{1/2} \quad (13)$$

(Eqs.(1) and (9) are used). The value of U_{th} is one at $a = 0$ and is reduced down to $U_{th\infty}$ with increasing crack length.

In the above equations, ΔK and $\Delta\sigma$ are the values corresponding to the nominal tensile cycle; $\Delta K = K_{max}$ and $\Delta\sigma = \sigma_{max}$ for the cases of negative R values [5, 12]. The same notation will be used throughout the paper.

Figure 3 shows the changes of the ratios of $\Delta\sigma_{effth}/\Delta\sigma_{wo}$, $\Delta\sigma_{th}/\Delta\sigma_{wo}$, and U_{th} as a function of crack length a for a low-carbon steel (0.20C-0.92Mn -0.26Si-0.11P-0.15S-0.009N) with a ferrite grain size of 55μm [4, 5, 10]. The curves are drawn based on Eqs.(3), (10) and (12) for the case of stress ratio R=-1 by using the data: $\Delta K_{th\infty}$=6.20 MPa$\sqrt{m}$, $\Delta K_{effth\infty}$=1.30 MPa$\sqrt{m}$, and $\Delta\sigma_{wo}$=163 MPa. The effective fraction at the threshold U_{th} is unity for a very small crack, and decreases toward $U_{th\infty}$=0.21 with increasing crack length. The experimental data plotted in the figure were obtained for a corner crack or surface crack in plane bending fatigue [4, 5]. The length of those cracks is converted to that of an isolated through-crack by using the SIF-equality equation. The data points lie close to the theoretical curves.

By analysing the data for various other metals, the a_o value is about ten or twenty times of the grain size and the a_o' value is about the grain size [5]. Both values tend to decrease with increasing stress ratio [5,12].

Figure 3 is important to judge the growth behavior of a crack under cyclic loading. The diagram of stress versus crack length can be divided into three regions: Region I which is below the $\Delta\sigma_{effth}$ line, Region II which is above the $\Delta\sigma_{th}$ line, and Region III lying in between Regions I and II. Region I indicates the non-propagation of a fatigue crack, and Region II means a propagating crack. Region III corresponds the crack arrest region. For a given stress, a crack whose length is in Region III will grow. The corresponding point in Fig.3 moves horizontally. The crack decelerates and eventually stops at the crack length corresponding to the boundary between Regions III and II.

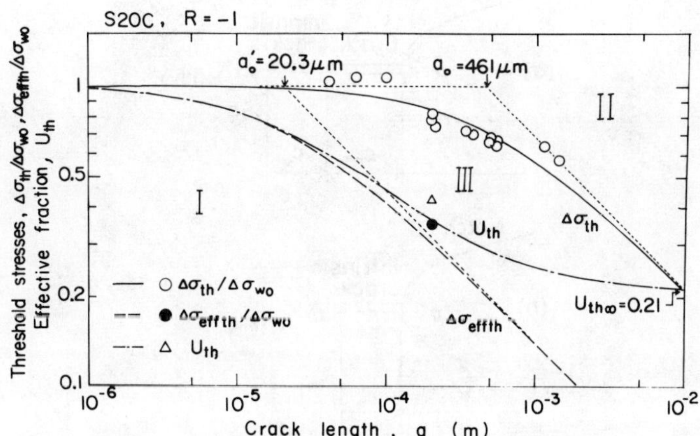

Figure 3 - Changes of threshold stresses and crack closure with crack length.

Non-Propagating Small Cracks in Notch Fatigue

At the root of sharp notches, a crack can decelerate as it grows, and become a non-propagating crack depending on the applied stress level [13]. The stress intensity range for a crack of length c formed at the notch tip can be expressed as

$$\Delta K = \Delta\sigma \cdot \sqrt{\pi c} \cdot F(c) \tag{14}$$

where $\Delta\sigma$ is the applied nominal stress range and $F(c)$ is the geometrical correction factor [12]. Since ΔK is an increasing function of c at a constant $\Delta\sigma$ [14, 15], the conventional fracture mechanics fails to predict the decreasing growth rate of a small fatigue crack near the notch tip.

The model of blocked slip band or intrinsic crack described in the preceding section can be applied to the non-propagation of a short crack at the notch tip. For simplicity, the intrinsic crack model will be used. Figure 4(a) illustrates a non-propagating crack of length c_{np} formed at the notch. When c_{np} diminishes to zero, we have the model for the threshold for crack initiation as shown in Fig.4(b). Because of geometrical difference, the length of intrinsic crack is defined in a slightly different way [12, 16]. The length c_o' is determined from the condition that the fatigue limit of smooth specimens $\Delta\sigma_{wo}$ is to be derived from the non-propagation criterion for an edge crack c_o' in a semi-infinite plate:

$$c_o' = (\Delta K_{effth\infty}/1.122\Delta\sigma_{wo})^2/\pi \tag{15}$$

where 1.122 is the correction factor due to the free surface. From Eqs. (9) and (15), $c_o'=0.794\,a_o'$.

The condition that the SIF range at the tip of a fictitious crack of length c_{np} plus c_o' equals $\Delta K_{effth\infty}$ yields

$$\Delta K_{effth\infty}=\Delta\sigma_{effth}[\pi(c_{np}+c_o')]^{1/2}\cdot F(c_{np}+c_o') \tag{16}$$

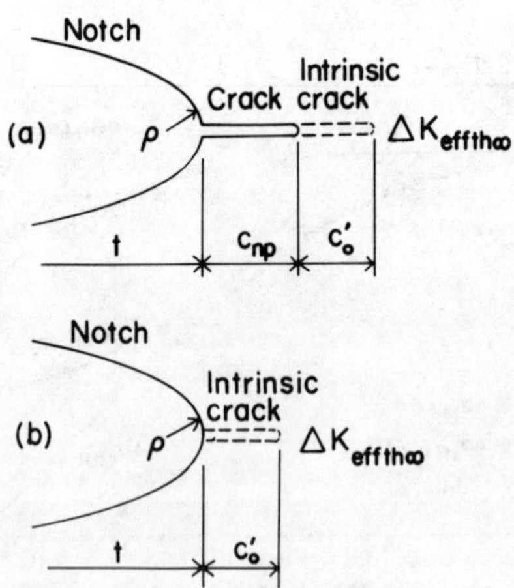

Figure 4 - Models of thresholds of notch fatigue.
(a) Crack propagation threshold.
(b) Crack initiation threshold.

The value of ΔK_{effth} for a small crack at the notch tip is calculated from $\Delta\sigma_{effth}$ and c_{np} as $\Delta K_{effth}=\Delta\sigma_{effth}(\pi c_{np})^{1/2}\cdot F(c_{np})$. Then, we have

$$\frac{\Delta K_{effth}}{\Delta K_{effth\infty}} = \left[\frac{c_{np}}{c_{np}+c_o'}\right]^{1/2} \cdot \frac{F(c_{np})}{F(c_{np}+c_o')} \tag{17}$$

Since the crack closure comes from the touching of crack faces, it may depend on the distance of the crack face made by fatigue. It seems reasonable to use Eq.(12) where a is substituted by c_{np}

$$U_{th}/U_{th\infty} = [(c_{np} + a_o)/(c_{np} + a_o')]^{1/2} \tag{18}$$

The threshold stress intensity range ΔK_{th} is given by

$$\Delta K_{th}/\Delta K_{th\infty}=(\Delta K_{effth}/\Delta K_{effth\infty})(U_{th}/U_{th\infty})^{-1} \tag{19}$$

The corresponding stress range $\Delta\sigma_{th}$ is calculated from ΔK_{th} through

$$\Delta\sigma_{th} = \Delta K_{th}/[(\pi c_{np})^{1/2} \cdot F(c_{np})] \tag{20}$$

When $c_{np}=0$, Eq.(18) gives $U_{th}=1$, i.e. $\Delta\sigma_{th}=\Delta\sigma_{effth}$. The fatigue limit for crack initiation, $\Delta\sigma_{wl}$, is determined from Eq.(15) with $c_{np}=0$ as

$$\Delta\sigma_{wl} = \Delta K_{effth\infty}/[(\pi c_o')^{1/2} \cdot F(c_o')] \tag{21}$$

To compute the variation of ΔK_{th} with crack length c by using Eqs.(17)-(19), the material properties required are $\Delta K_{th\infty}$, $\Delta K_{effth\infty}$ and $\Delta\sigma_{wo}$. The F-term in those equations accounts for the influence of notch geometry. The variation of ΔK_{th} with crack length is calculated for a center-notched plate (width=45mm, thickness=4mm) having a notch with tip radius 0.16mm and length 6mm [12]. The stress ratio R is -1. The materials is a structural low-carbon steel (0.17C-0.79Mn-0.19Si-0.016P-0.020S) with a ferrite grain size of 64μm. The experimental data used are $\Delta K_{th\infty}$=6.11 MPa$\sqrt{m}$, $\Delta K_{effth\infty}$=3.06

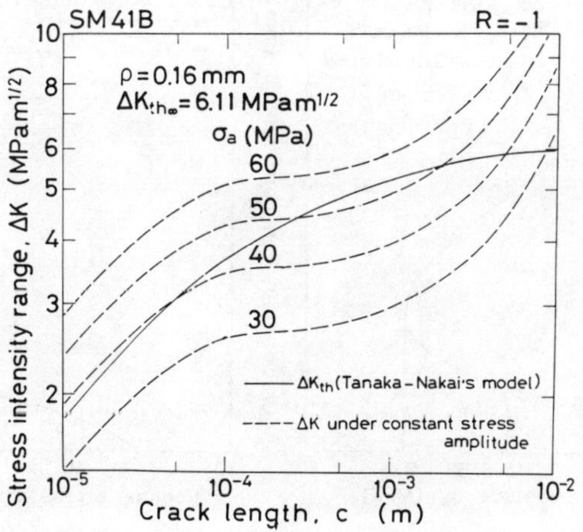

Figure 5 - Change of threshold stress intensity range with crack length in notch fatigue.

MPa$\sqrt{m}$, and $\Delta\sigma_{wo}$=163 MPa. The values of c_o', a_o' and a_o are calculated as 0.089, 0.112 and 0.447mm, respectively. Figure 5 shows the variation of ΔK_{th}, together with the change in ΔK for several applied stress amplitudes. Since ΔK increases monotonically with c, a constant ΔK_{th} criterion does not explain the non-propagation of a fatigue crack. A fatigue crack can propagate as far as the applied ΔK value is above the threshold value. Under a given amplitude, the crack becomes non-propagating when it grows up to the length at the intersection of ΔK-curve and ΔK_{th}-curve.

By using Eqs.(17)-(20), the threshold stress range $\Delta\sigma_{th}$ (=amplitude for R=-1) for the same material is calculated as a function of crack length for the cases of several notch-tip radii. The relation is drawn in a diagram of crack length c versus nominal stress amplitude σ_a (Fig.6). As indicated in Fig.6(b) (for $\rho = 0.16$mm), the diagram can be divided into three regions. The stress corresponding to the intersection of the curve with the abscissa is the fatigue limit σ_{w1} (stress amplitude) for crack initiation. When a certain combination (c, σ_a) lies in Region I, a crack does not propagate. If it is in Region III, the crack propagates and eventually is arrested at the boundary between Regions I and III. In Region II, a crack continues to propagate. The stress at the nose is the fatigue limit σ_{w2} (stress amplitude) for fracture. Even at σ_a below σ_{w2}, a crack can cause fracture if its length is long enough to be in Region II. As seen in Fig.6(a), σ_{w1} is larger than σ_{w2} for the case of $\rho = 2.0$mm, which means that a crack propagates once nucleated.

From the calculation of σ_{w1} and σ_{w2} for different values of ρ, the diagram of the fatigue limits, σ_{w1} and σ_{w2}, against the elastic stress concentration factor K_t is drawn in Fig.7, together with the experimental data. The experimental data well agree with the predicted lines. The length of the non-propagating crack as a function of the applied stress

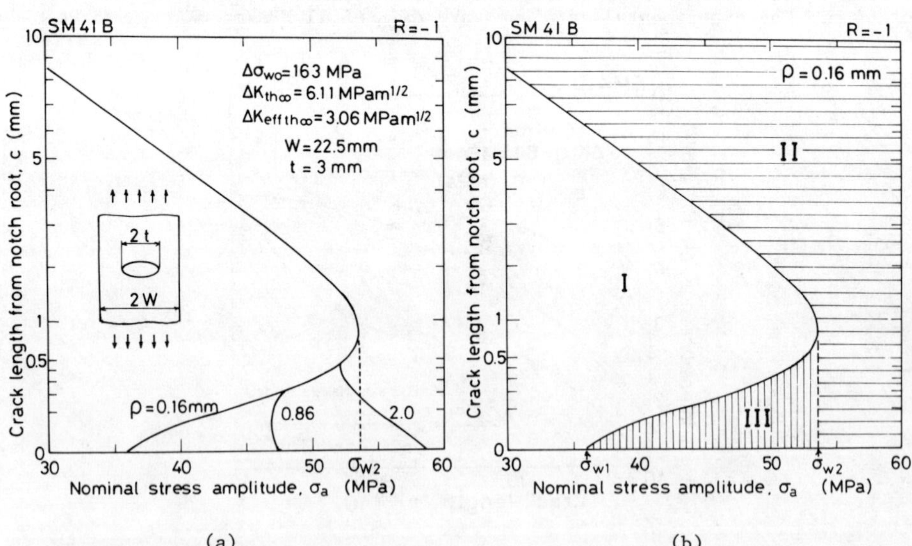

(a) (b)

Figure 6 - Limiting curves for non-propagation of fatigue crack, (a) $\rho = 0.16$, 0.86 and 2.0mm, (b) $\rho = 0.16$mm.

amplitude predicted as above was also found to agree with the experimental result.

Equations (17)-(19) can be directly compared with measurements on the crack closure for non-propagating cracks. By using the values of c_o, a_o' and a_o, the ratios of $\Delta K_{effth}/\Delta K_{effth\infty}$, $U_{th}/U_{th\infty}$ and $\Delta K_{th}/\Delta K_{th\infty}$ are calculated as a function of c_{np}. Figure 8 shows the prediction and experimental data. For a long non-propagating crack, ΔK_{effth} and ΔK_{th} are nearly constant, equal to $\Delta K_{effth\infty}$ and $\Delta K_{th\infty}$, respectively. For a short crack, the ratios of $\Delta K_{effth}/\Delta K_{effth\infty}$ and $\Delta K_{th}/\Delta K_{th\infty}$ decrease approaching

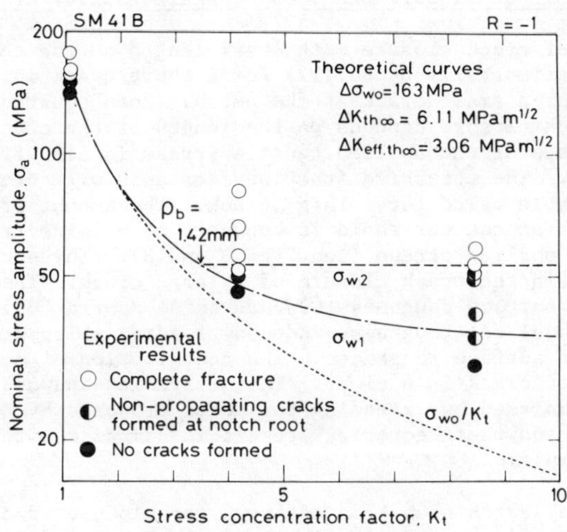

Figure 7 - Fatigue strength reduction as a function of stress concentration factor.

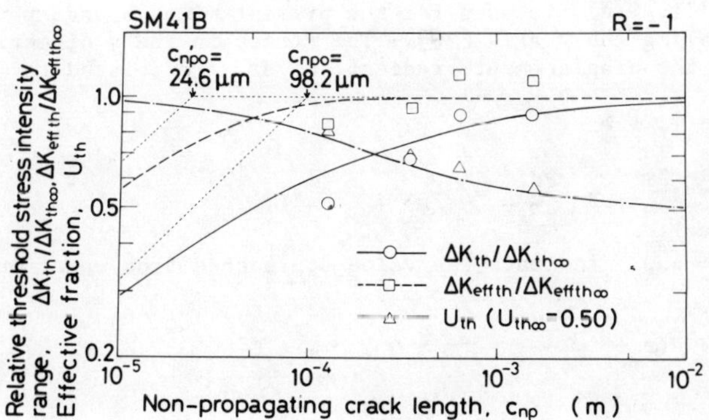

Figure 8 - Change of threshold stress intensity ranges and crack closure with crack length in notch fatigue.

the dotted line with slope -0.5 which corresponds to $\Delta\sigma_{th}=\Delta\sigma_{wl}$ (calculated). The transition crack lengths c_{npo}' and c_{npo} shown in the figure are determined by

$$\left.\begin{aligned}\Delta K_{effth\infty} &= 1.122K_t\Delta\sigma_{wl}(\pi c_{npo}')^{1/2}\\ \Delta K_{th\infty} &= 1.122K_t\Delta\sigma_{wl}(\pi c_{npo})^{1/2}\end{aligned}\right\} \tag{22}$$

where the geometrical correction factor F=1.122K_t is used. The experimental data lie close to the predicted lines.

Analysis of Closure Behavior of Small Fatigue Cracks

The change of crack closure with crack length at the threshold is expressed semi-experimentally by Eq.(12) for a short crack in smooth specimens and by Eq.(18) for a small crack at the notch. Both equations indicate that the level of crack closure depends on the length of the crack wake made by fatigue. The range of the applied tensile stress is all effective for a very small crack. The effective fraction decreases with crack growth, approaching a stable value for a large crack. The amount of crack closure for a large crack at the threshold is known to be a large fraction, 50 to 80 per cent, of the applied stress (see Figs.3 and 8). Three causes have been proposed to explain the crack closure of a large crack. They are residual plasticity [17], surface roughness [7], and oxide debris [6]. Morris et al. [9] proposed a model for roughness-induced closure, which utilized the measured value of surface roughness. The contribution of oxide-induced closure to a short crack is unclear. The plasticity-induced closure for a short crack is analysed by extending the model proposed by Budiansky and Hutchinson [18], and the theoretical result is compared with the above semi-experimental formula.

Budiansky and Hutchinson [18] analysed the closure of a semi-infinite crack propagating under steady cyclic loading. They used Dugdale model [19] in the small scale yielding situation. Their model is extended to a small crack under large scale yielding. The residual extension attached to the crack face is assumed to be uniform as in Budiansky-Hutchinson's model. The analysis is conducted for the case of stress ratio R=-1, where the crack is completely closed at the minimum load [18]. The complex stress function of Muskhelishivilli, $\Phi(z)$, is used for the present mixed boundary value problem with a crack along the x axis [20]. The stress in the y direction on the x axis, σ_y, and the displacement gradient, $d\delta/dx$, are related to $\Phi(z)$ by

$$\sigma_y = \Phi_+ + \Phi_- \tag{23}$$

$$\frac{d\delta}{dx} = \left(\frac{\partial v}{\partial x}\right)_+ - \left(\frac{\partial v}{\partial x}\right)_- = \frac{4}{iE}(\Phi_+ - \Phi_-) \tag{24}$$

where suffix + and - indicate the value approached from upper and lower planes as

$$f_\pm \equiv f(x \pm iy) \qquad y > 0\ , \quad y \to 0 \tag{25}$$

Figure 9 illustrates the opening and closure of a crack with length $2a$ in an infinite plate. (a) indicates the opening of a crack under the maximum stress $\sigma=\sigma_{max}$. The size of the maximum plastic zone and the opening displacement, δ_M, are the same as those in the Dugdale crack without

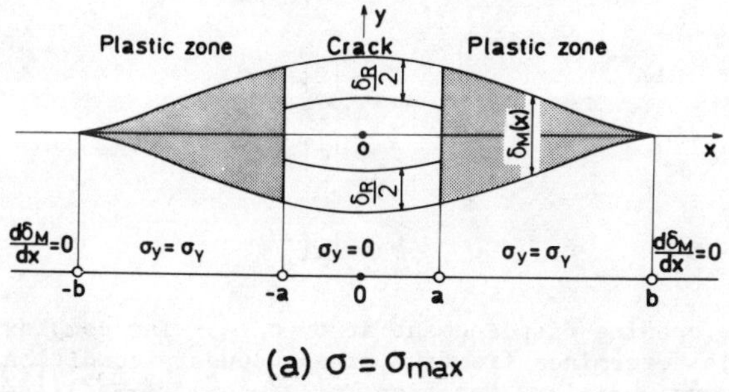

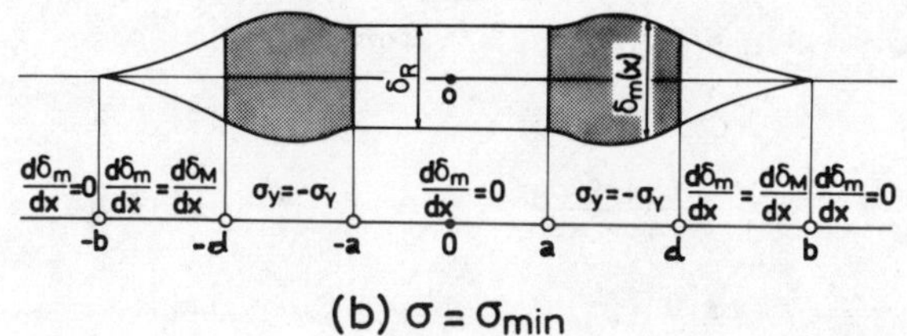

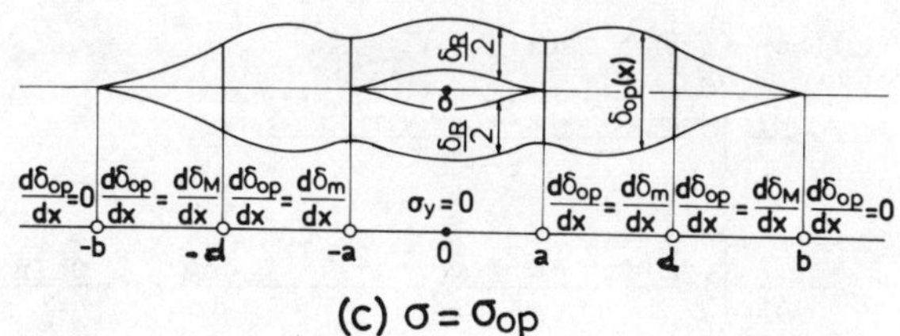

Figure 9 - Crack closure model for small fatigue crack.

residual extension. The plastic zone size is determined by

$$\frac{a}{b} = \cos\left(\frac{\pi}{2} \cdot \frac{\sigma max}{\sigma_Y}\right) \tag{26}$$

and the derivative of δ_M is

$$\frac{d\delta_M}{dx} = -\frac{4\sigma_Y}{\pi E}\,\ell n \left| \frac{x(b^2-a^2)^{1/2} + a(b^2-x^2)^{1/2}}{x(b^2-a^2)^{1/2} - a(b^2-x^2)^{1/2}} \right| \tag{27}$$

where b is the sum of the crack length a and the plastic zone size ω, σ_Y is the yield strength, and E is Young's modulus. The residual extension is assumed to be a constant value of δ_R. At the minimum stress $\sigma = \sigma_{min}$, the crack is closed and the reversed plastic zone spreads over ω^* $(=d-a)$. The boundary conditions are

$$\frac{d\delta_m}{dx} = 0 \qquad |x| < a$$

$$\sigma_y = -\sigma_Y \qquad a < |x| < d$$

$$\frac{d\delta_m}{dx} = \frac{d\delta_M}{dx} \qquad d < |x| < b \tag{28}$$

$$\frac{d\delta_m}{dx} = 0 \qquad b < |x|$$

where δ_m is the opening displacement at $\sigma = \sigma_{min}$. The complex stress function $\Phi(z)$ is determined from the mixed boundary conditions, which are expressed by using auxiliary function $\chi = \{(z^2 - a^2)(z^2 - d^2)\}^{1/2}$ ($\chi \to z^2$ for $z \to \infty$) as

$$(\chi\Phi)_+ - (\chi\Phi)_- = 0 \qquad \text{for} \qquad |x| < a$$

$$= -\chi_+ \sigma_Y \qquad a < |x| < d$$

$$= \frac{iE\chi}{4} \bullet \frac{d\delta_M}{dx} \qquad d < |x| < b \tag{29}$$

$$= 0 \qquad b < |x|$$

The general solution is obtained by using Plemelj's formula [20] as

$$\{(z^2 - a^2)(z^2 - d^2)\}^{1/2}\Phi(z)$$

$$= -\frac{\sigma_Y}{\pi} \int_a^d \frac{x\{(d^2 - x^2)(x^2 - a^2)\}^{1/2}}{x^2 - z^2}\, dx$$

$$+ \frac{E}{4\pi} \int_d^b \frac{x\{(x^2 - d^2)(x^2 - a^2)\}^{1/2}}{x^2 - z^2} \frac{d\delta_M}{dx}\, dx + \frac{\sigma_{min}}{2} z^2 + A \tag{30}$$

For $\Phi(z)$ to be finite at $z = \pm a$ and $\pm d$, the right hand side of Eq.(28) needs to be zero:

$$-\frac{\sigma_Y}{\pi} \int_a^d x\left(\frac{x^2 - a^2}{d^2 - x^2}\right)^{1/2} dx + \frac{E}{4\pi} \int_d^b x\left(\frac{x^2 - a^2}{x^2 - d^2}\right)^{1/2} \bullet \frac{d\delta_M}{dx}\, dx$$

$$+ \frac{\sigma_{min}}{2} a^2 + A = 0 \tag{31}$$

$$\frac{\sigma_Y}{\pi} \int_a^d x\left(\frac{x^2 - a^2}{d^2 - x^2}\right)^{1/2} dx + \frac{E}{4\pi} \int_d^b x\left(\frac{x^2 - a^2}{x^2 - d^2}\right)^{1/2} \bullet \frac{d\delta_M}{dx}\, dx$$

$$+ \frac{\sigma_{min}}{2} d^2 + A = 0 \tag{32}$$

Eliminating A from Eqs.(29) and (30) yields

$$\frac{E}{2\pi}\int_d^b \frac{x}{(x^2-d^2)^{1/2}(x^2-a^2)^{1/2}}\frac{d\delta_M}{dx}\,dx + \sigma_Y + \sigma_{min} = 0 \qquad (33)$$

The opening point of a crack at the tip is calculated from the condition shown in Fig.9(c). Denoting the crack opening displacement at $\sigma=\sigma_{op}$ (the opening stress) by δ_{op}, the boundary conditions are

$$\begin{aligned}
&\sigma_y = 0 && \text{for} \quad |x| < a \\
&\frac{d\delta_{op}}{dx} = \frac{d\delta_m}{dx} && a < |x| < d \\
&\frac{d\delta_{op}}{dx} = \frac{d\delta_M}{dx} && d < |x| < b \\
&\frac{d\delta_{op}}{dx} = 0 && b < |x|
\end{aligned} \qquad (34)$$

By using the auxiliary function $\chi = (z^2-a^2)^{1/2}$ and Plemelj's formula, the stress function is determined as

$$(z^2-a^2)^{1/2}\Phi(z) = \frac{zE}{4\pi}\left\{\int_a^b \frac{(x^2-a^2)^{1/2}}{x^2-z^2}\frac{d\delta_m}{dx}\,dx\right.$$

$$\left. + \int_d^b \frac{(x^2-a^2)^{1/2}}{x^2-z^2}\left(\frac{d^2-x^2}{x^2-a^2}\right)^{1/2} \cdot \frac{d\delta_M}{dx}\,dx\right\} + \frac{\sigma_{op}}{2}z \qquad (35)$$

where $d\delta_m/dx$ is

$$\frac{d\delta_m}{dz} = -\frac{8}{E}\left(\frac{z^2-a^2}{d^2-z^2}\right)^{1/2}\left\{-\frac{\sigma_Y}{\pi}\int_a^d \frac{x}{x^2-z^2}\left(\frac{d^2-x^2}{x^2-a^2}\right)^{1/2} dx\right.$$

$$\left. + \frac{E}{4\pi}\int_d^b \frac{x}{x^2-z^2}\left(\frac{d^2-x^2}{x^2-a^2}\right)\frac{d\delta_M}{dx}\,dx + \frac{\sigma_{min}}{2}\right\} \qquad (36)$$

The opening stress σ_{op} is given from the condition that $\Phi(z)$ is bounded at $z = \pm a$ as

$$\sigma_{op} = \frac{2}{\pi}(\sigma_Y + \sigma_{min})\cos^{-1}\left(\frac{a}{d}\right)$$

$$-\frac{E}{2\pi}\int_d^b \left\{1 - \frac{2}{\pi}\tan^{-1}\left(\frac{x}{a}\left(\frac{d^2-a^2}{x^2-d^2}\right)^{1/2}\right)\right\}\frac{d\delta_M}{dx}\frac{1}{(x^2-a^2)^{1/2}}\,dx \qquad (37)$$

At a limit of ω, $\omega^* \ll a$, d, the above equations are reduced to those given by Budiansky and Hutchinson [18].

If the values of σ_{max}/σ_Y and σ_{min}/σ_Y are given, the maximum plastic zone size relative to crack length, $\omega/a = (b-a)/a$, and the reversed plastic zone size relative to crack length, $\omega^*/a = (d-a)/a$, are determined by Eqs. (26) and (33), respectively. Substitution of those values into Eq.(37) yields the opening stress σ_{op}/σ_Y. Then the effective stress range is determined through

$$\Delta\sigma_{eff} = \sigma_{max} - \sigma_{op} \tag{38}$$

Since Eqs.(33) and (37) can not be solved analytically, numerical integration is used. The solution for d/a of Eq.(33) is obtained through a trial and error method.

At the threshold for crack growth, the crack-tip plastic zone is confined within a grain as described before. We can assume zero MSIF value in the blocked slip band model for simplicity, because the relation between the effective threshold stress and the crack length does not vary much with the value of the critical MSIF. Then, the threshold stress can be regarded as the stress required to produce a constant ω^* value of the grain size order. The yield strength σ_Y used in the above equations should be interpreted as the friction stress σ_{fr}^* at the threshold.

The calculation is carried out for the case of $R=-1$, i.e. $\sigma_{max} = -\sigma_{min} = \sigma_a$. For a given value of σ_a/σ_Y, the values of ω^*/a and $\Delta\sigma_{eff}/\sigma_Y$ are calculated. Figure 10 shows the relations of σ_a/σ_Y and $\Delta\sigma_{eff}/\sigma_Y$ against a/ω^*. The figure can be regarded to show the changes of σ_a and $\Delta\sigma_{eff}$ with crack length for a constant value of ω^*. It is worthy to note resemblance between Figs.3 and 10. As a/ω^* becomes large, the ratios of $\sigma_a\sqrt{a}/\sigma_Y\sqrt{\omega^*}$ and $\Delta\sigma_{eff}\sqrt{a}/\sigma_Y\sqrt{\omega^*}$ approach constant values, indicating the constant values of ΔK and ΔK_{eff}. On the other hand, σ_a/σ_Y and $\Delta\sigma_{eff}/\sigma_Y$ becomes unity as a/ω^* tends to zero. At the threshold, the limiting amplitudes should be the fatigue limit of smooth specimens. Therefore, we make $\sigma_{wo} = \sigma_Y$ $(=\sigma_{fr}^*)$. The crack lengths, a_o' and a_o, determined from Eqs.(1) and (9) are a_o=8.73 ω^* and a_o'=1.71 ω^*. a_o is about five times larger than a_o'. From Eq.(13), the effective fraction for a large crack is 0.443 which is the same value obtained by Budiansky and Hutchinson for $R=0$ [18].

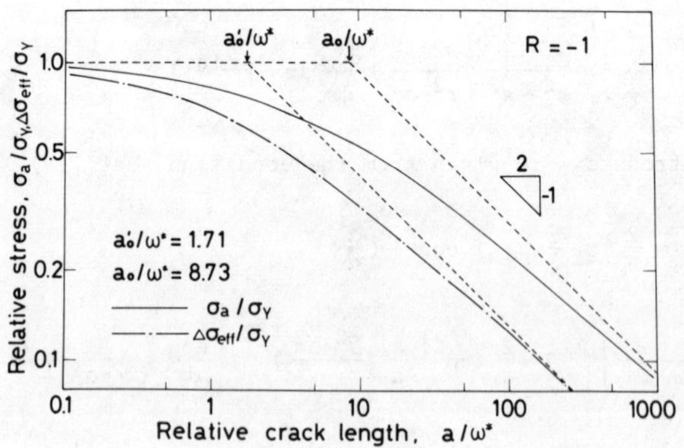

Figure 10 - Change of threshold stresses with crack length.

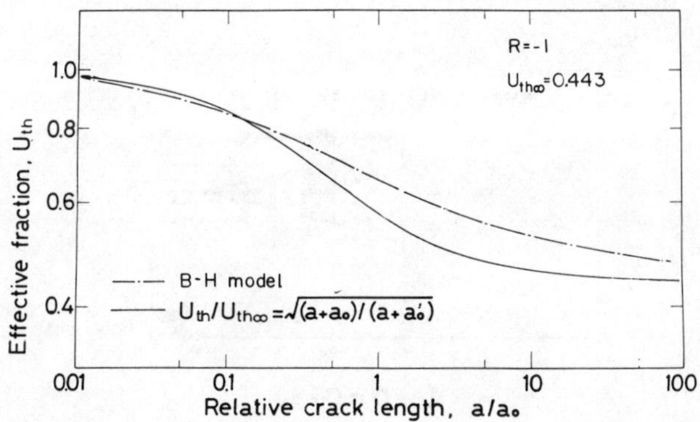

Figure 11 - Change of effective fraction of applied load with crack length.

The change of the effective fraction U ($=\Delta\sigma_{eff}/\sigma_a$ for $R=-1$) with crack length is shown in Fig.11. The theoretical calculation gives $U=1$ at $a=0$ and $U=0.443$ for a large crack. In the figure, the semi-empirical formula, Eq.(12), is drawn by using $a_o=8.73\ \omega^*$ and $a_o'=1.71\ \omega^*$. The theoretical trend of the change of the effective fraction with crack length is similar to the semi-experimental relation, although the theoretical value is slightly higher than the semi-experimental relation.

The present model can be applied to a small crack made at the notch. Figure 12 shows the model. For simplicity, the notch tip radius is assumed to be zero. The distributions of stress and displacement are the same as those for a crack whose length is the sum of the real crack length and the notch depth. The notch part is assumed to be open even at the minimum stress. The analysis is conducted when the notch depth is large compared with the plastic zone at the crack tip and when $R=-1$. The crack has a length of c and a uniform residual elongation δ_R as shown in Fig.12. The crack is closed at the minimum load. The stress intensity factor at the crack tip opening K_{op} is determined in a similar way. The detail is given elsewhere [21].

Figure 13 shows the variation of the effective fraction with the crack length in notch fatigue, where the crack length is normalized by the reversed plastic zone size. The dashed line is the result for a crack without notch given previously where c means the half of the total crack length. When the crack length from the notch tip is above the reversed plastic zone size, the fraction is about the same for cracks with and without notch. This results support the substitution of a by c_{np} to derive Eq.(17). When the crack length is small, the value of U_{th} is different between cracks with and without a notch. For a small crack without notch, $\sigma_{op}=0$ and $U_{th}=1$. On the other hand, $\sigma_{op}=\sigma_{min}=-\sigma_{max}$ for a notch-tip crack, and then $U_{th}=\Delta\sigma_{effth}/\Delta\sigma_{th}=(\sigma_{max}-\sigma_{min})/\sigma_{max}=2$.

The theoretical analyses of the plasticity-induced crack closure described above support the semi-experimental formulation of the crack length effect on crack closure. Further refinement of analysis of small crack closure is necessary to include the stress ratio effect and the

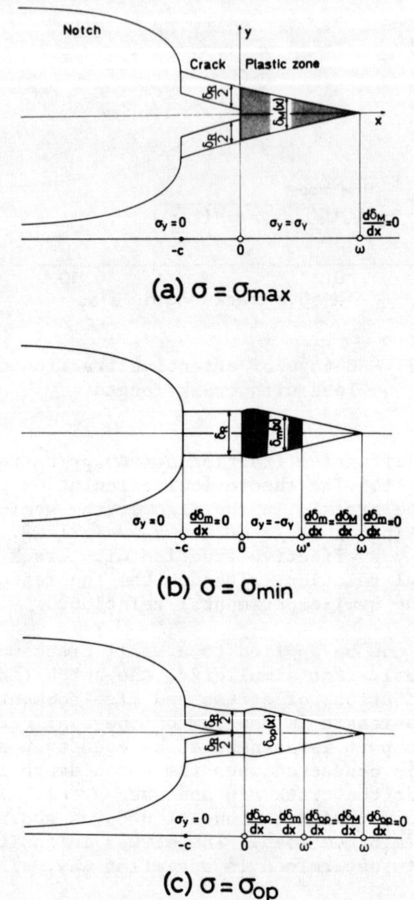

Figure 12 - Crack closure model for small fatigue crack formed at a notch.

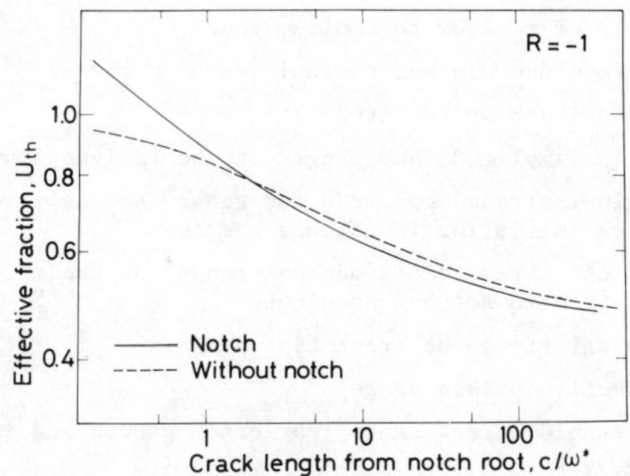

Figure 13 - Change of effective fraction of applied load with crack length for cracks formed at a notch.

interaction of slip band with the grain boundary often observed for a small crack near the threshold [4, 22, 33]. The contribution of surface roughness to crack closure can be substantial for small cracks in some metals [9].

Conclusions

(1) The relation between the effective component of the threshold stress and the crack length is predicted by assuming the critical value of the microscopic stress intensity factor for the crack-tip slip band blocked by the grain boundary. The model can be simplified by assuming zero friction stress within the slip band.

(2) The changes of the threshold stress and the crack closure with crack length are formulated by combining the simplified model with crack closure behavior. The experimental data required for the formulation are the total and effective ranges of the threshold stress intensity factor for a large crack, and the fatigue limit of smooth specimens.

(3) The threshold model for a small fatigue crack is applied to the nonpropagation of a small fatigue crack at the notch tip. The fatigue limits of notched components predicted as a function of notch geometry agree with the experimental data. A similar semi-experimental formula for the change of crack closure of the notch-tip crack is obtained.

(4) The crack closure model proposed by Budiansky and Hutchinson for a large fatigue crack is extended to analyse the closure of a small fatigue crack formed in a smooth specimen or at the notch. The theoretical results on the change of crack closure with crack length agree fairly well with the formula obtained semi-experimentally.

Nomenclature

R	ratio of minimum to maximum load
σ_{max}, σ_{min}	maximum and minimum nominal (net-) stresses
σ_a, $\Delta\sigma$	nominal stress amplitude and range*
σ_{wo}, $\Delta\sigma_{wo}$	stress amplitude and range* at the fatigue limit
σ_{w1}, $\Delta\sigma_{w1}$	nominal stress amplitude and range* at the threshold for crack initiation in notched specimen
σ_{w2}, $\Delta\sigma_{w2}$	nominal stress amplitude and range* at the threshold for fracture in notched specimen
σ_{op}	nominal stress at crack tip opening
$\Delta\sigma_{eff}$	effective stress range
$\Delta\sigma_{th}$, $\Delta\sigma_{effth}$	threshold stress range* for crack growth and its effective component
σ_Y	yield strength
σ_{fr}^*	friction stress
E	Young's modulus
ρ	notch tip radius
a	crack length
c, c_{np}	crack length from notch tip
a_o', a_o	intrinsic crack length for center-cracked specimen
c_o', c_o	intrinsic crack length for notched specimen
ℓ	size of blocked slip band
ω	size of monotonic plastic zone
ω^*	size of reversed plastic zone
K_t	elastic stress concentration factor
K	stress intensity factor (SIF)
F	correction factor for SIF
K_{max}, K_{min}	maximum and minimum SIF values
K_{op}	SIF at crack opening
ΔK, ΔK_{eff}	SIF range* and effective SIF range
ΔK_{th}, ΔK_{effth}	SIF range* and effective SIF range at the threshold (suffix ∞ indicates the value for a long crack)
U_{th}, $U_{th\infty}$	effective fractions for a short crack and for a long crack
K^m_c	critical value of microscopic SIF

* The range of each quantity is taken for the nominal tensile portion only; the range equals its maximum value for the case of negative R.

References

[1] H.Kitagawa and S. Takahashi, "Application of Fracture Mechanics to Very Small Cracks or the Cracks in the Early Stage", *Proc. 2nd Inter. Conf. Mech. Behav. Mater.*, Boston, ASM (1976) pp. 627-631.

[2] S. Usami and S. Shida, "Elastic-Plastic Analysis of the Fatigue Limit for a Material with Small Flaws", *Fatigue Engng Mater. Struct.* 1 (1979) pp. 471-481.

[3] M.H. El Haddad, K.N. Smith, and T.H. Topper, "Fatigue Crack Propagation of Short Cracks", *J. Engng Mater. Technol., Trans. ASME* 101 (1979) pp. 42-46.

[4] Y. Nakai and K. Tanaka, "Grain Size Effect on Growth Threshold for Small Surface-Cracks and Long Through-Cracks under Cyclic Loading", *Proc. 23rd Japan Cong. Mater. Res.* (1980) pp. 106-112.

[5] K. Tanaka, Y. Nakai, and M. Yamashita, "Fatigue Growth Threshold of Small Cracks", *Int. J. Fract.* 17 (1981) pp. 519-533.

[6] R.O. Ritchie, S. Suresh, and C.M. Moss, "Near-Threshold Fatigue Crack Growth in 2•1/4 Cr-Mo Pressure Vessel Steel in Air and Hydrogen", *J. Engng Mater. Technol., Trans. ASME* 102 (1980) pp. 293-299.

[7] K. Minakawa and A.J. McEvily, Jr., "On Closure in the Near-Threshold Region", *Scripta Metall.* 15 (1981) pp. 633-636.

[8] Y. Nakai, K. Tanaka, and T. Nakanishi, "The Effects of Stress Ratio and Grain Size on Near-Threshold Fatigue Crack Propagation in Low-Carbon Steel", *Engng Fract. Mech.* 15 (1981) pp. 291-302.

[9] W.L. Morris, M.R. James, and O. Buck, "A Simple Model of Stress Intensity Range Threshold and Crack Closure Stress", *Engng Fract. Mech.* 18 (1983) pp. 871-877.

[10] S. Taira, K. Tanaka, and M. Hoshina, "Grain Size Effect on Crack Nucleation and Growth in Long-Life Fatigue of Low-Carbon Steel", *ASTM STP 675* (1979) pp. 135-173.

[11] K. Tanaka and T. Mura, "Fatigue Crack Growth along Planer Slip Bands", *Acta Metall.* (submitted).

[12] K. Tanaka and Y. Nakai, "Propagation and Non-Propagation of Short Fatigue Cracks at a Sharp Notch", *Fatigue Engng Mater. Struct.* (in press)

[13] N.E. Frost, K.I. Marsh, and L.P. Pook, Metal Fatigue, pp. 130-201; Oxford Univ. Press, London, 1974.

[14] J.C. Newman, Jr., "An Improved Method of Collocation for Stress Analyois of Cracked Plates with Various Shaped Boundaries", *NASA Technical Note* D-6376 (1971).

[15] P. Lukas and M. Klesnil, "Fatigue Limit of Notched Bodies", *Mater. Sci. Engng* 34 (1978) pp. 61-66.

[16] K. Tanaka and Y. Nakai, "Prediction of Fatigue Threshold of Notched Components", *J. Engng Mater. Technol., Trans. ASME* (submitted).

[17] W. Elber, "The Significance of Fatigue Crack Closure", *ASTM STP 486* (1971) pp. 230-242.

[18] B. Budiansky and J.W. Hutchinson, "Analysis of Crack Closure in Fatigue Crack Growth", *J. Appl. Mech., Trans. ASME* 45 (1978) pp. 267-276.

[19] D.S. Dugdale, "Yielding of Steel Sheets Containing Slits", *J. Mech. Phys. Solids* 8 (1960) pp. 100-104.

[20] N.I. Muskhelishvili, Some Basic Problems of the Mathematical Theory of Elasticity, Translated by J.R.M. Radok, pp. 425-503; Noordhoff, Groningen-Holland, 1953.

[21] Y. Nakai, K. Tanaka, and M. Yamashita, "Analysis of Closure Behavior of Small Fatigue Cracks", *J. Soci. Mater. Sci. Jpn* 32 (1983) pp. 19-25.

[22] W.L. Morris, "The Noncontinuum Crack Tip Deformation Behavior of Surface Microcracks", *Metall. Trans.* 11A (1980) pp. 1117-1123.

[23] J. Lankford, "The Growth of Small Fatigue Cracks in 7075-T6 Aluminum", *Fatigue Engng Mater. Struct.* 5 (1982) pp. 233-248.

CRACK CLOSURE AND THE CONDITIONS FOR FATIGUE CRACK PROPAGATION

A. J. McEvily and K. Minakawa

Metallurgy Department
University of Connecticut
Storrs, Connecticut 06268
USA

Abstract

This paper deals with the relationship between the nature of the crack closure process and the conditions for fatigue crack propagation. It is shown that important closure events affecting the threshold level occur within one millimeter of the crack tip, and in the absence of closure it is conclusively shown that there is no mean stress (R) dependency of the threshold level. The incorporation of closure concepts into the analysis of the growth of cracks from notches leads to a clear understanding of matters such as the non-propagation of cracks, notch size effects and notch sensitivity. The application of the findings to design is also considered.

Introduction

Recent studies have shown that the threshold level for fatigue crack propagation is associated with the crack closure process. It is therefore important that we understand just what the closure process entails. We know it is a complex process being influenced by deformation mode, i.e., Mode I or Mode II, the environment, and microstructure, and many of these aspects are discussed in this symposium. The purpose of this paper is to describe some experimental work intended to define in a particular alloy just how far behind the crack tip does the closure process occur which brings about the development of the threshold. In addition the relationship between crack closure and the R-dependency of the threshold level will be considered. Finally, concepts developed as a result of closure considerations will be used to consider the arrest of cracks (non-propagating cracks), notch size effects, notch sensitivity, the growth of short and long cracks, and design based upon threshold.

Crack Closure at Threshold

In relating crack closure to the threshold level a question arises as to just where in the relationship to the crack tip does the important closure event occur. For example, it has been suggested (1) that at threshold the tip might in fact be open and that closure might be occurring at some distance well behind the crack tip. In order to determine if such a circum-

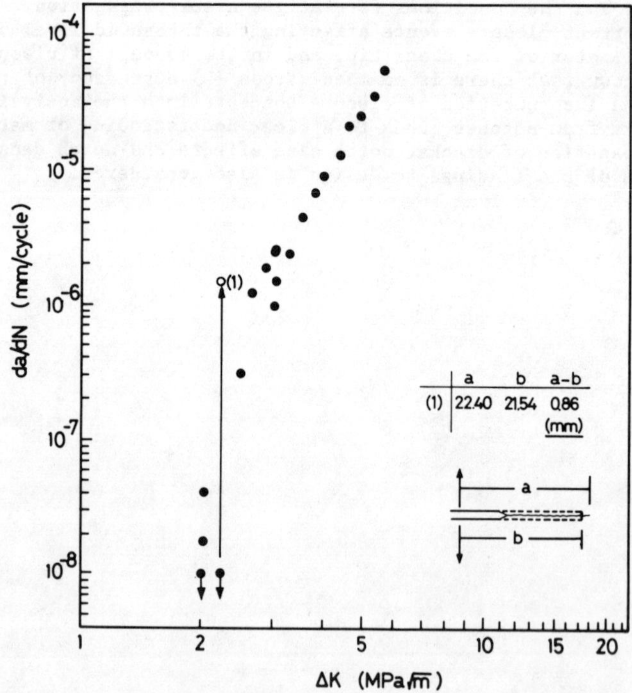

Figure 1 - Fatigue crack growth rate, da/dN, as a function of ΔK. The open symbol refers to the condition immediately after removal of the material.

stance existed in a given material a series of crack growth experiments were carried out using compact specimens of the high strength (580 MPa yield strength), P/M aluminum alloy 7090-T6 (2). In this experiment a fatigue crack was grown under decreasing ΔK conditions at R = 0.05 until the threshold level was reached, and the crack closure behavior as determined by a clip gauge mounted at the mouth of the specimen was determined as a function of ΔK. After the threshold level had been reached the material behind the crack tip was removed by an electro-discharge process to within 1 mm of the tip itself, as shown in Fig. 1. The specimen was then retested at the previously determined threshold ΔK level and the crack growth and closure behavior were observed.

Fig. 1 shows the result of this experiment. It is seen that after removal of the material from behind the crack tip that the closure level decreases markedly, Fig. 2, and the crack growth rate increases correspondingly. However, as the crack advances in this stage of the test the extent of closure increases as shown in Fig. 2. A new threshold was determined and was found to be at 2.0 MPa$\sqrt{m}$ as compared to the original value of 2.2 MPa$\sqrt{m}$. The crack growth increment after metal removal was only 0.8 mm, and it is therefore concluded that in this alloy the primary closure process affecting the threshold level occurs quite close to the crack tip, i.e., within 1 mm of the tip. From the increase in closure level after metal removal we also conclude that the closure level builds up over a correspondingly short distance, i.e., of the order of 1 mm in this alloy.

Effect of R-ratio and Closure on Threshold

A number of studies have shown that dependency of the threshold level on R can be related to the change in closure level with R ratio. (3-6) Since the closure level itself has been related to structure size it is of interest to study the variation of threshold as a function of R in materials of different structural size. The alloys selected for these experiments

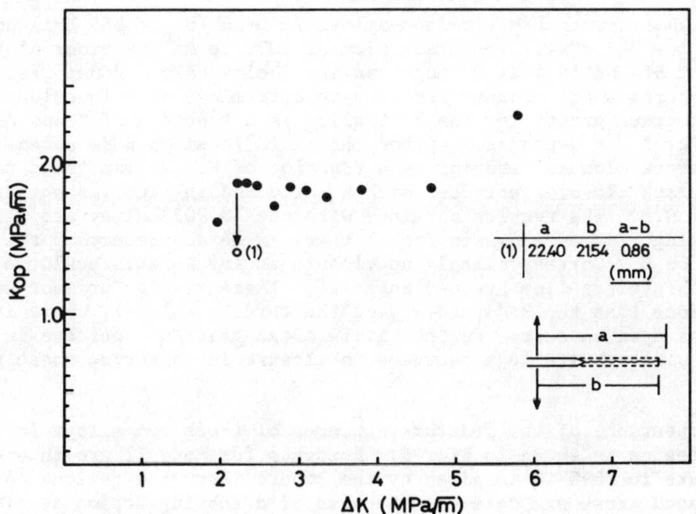

Figure 2 - Crack opening level, K_{op}, as a function of ΔK. The open symbol refers to the condition immediately after removal of the material.

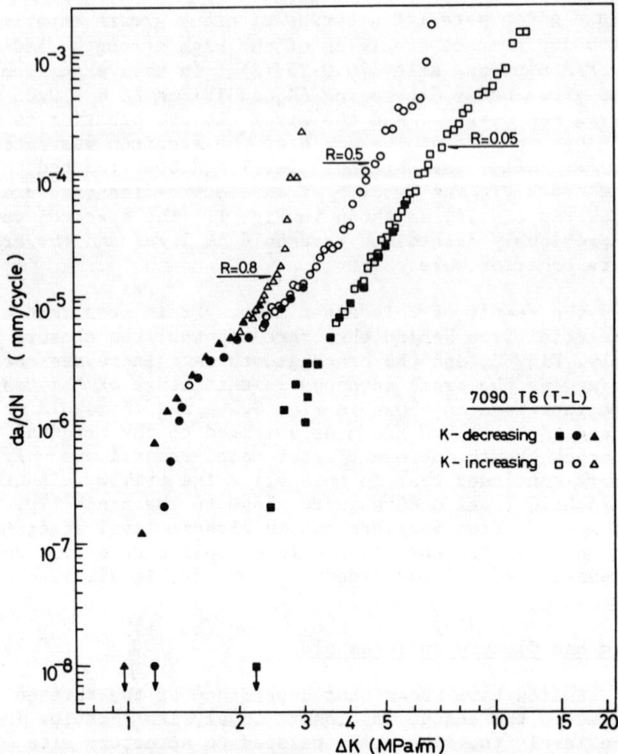

Figure 3 - Fatigue crack growth rate, da/dN, as a function of ΔK for the 7090-T6 aluminum alloy at three R ratios.

were the high strength P/M aluminum alloys 7090-T6 (σ_{ys} = 580 MPa) and IN 9021-T4 (σ_{ys} = 523 MPa). The grain size of 7090 is of the order of 3 μm whereas that of the IN 9021 is much smaller, being only 0.2 μm. Fatigue crack growth rates and closure levels were determined as a function of R. The rate of crack growth for the 7090 alloy as a function of R and ΔK is shown in Fig. 3. A dependency of the threshold level on R is noted. Fig. 4 shows the crack closure behavior as a function of R. It was found that the extent of crack closure decreased with R ratio and in fact was not detectable at R = 0.8. The results obtained with the IN 9021 alloy are particularly striking, for as shown in Fig. 5 there is no dependency of the threshold level on R. Correspondingly no closure at any R-ratio could be detected in this ultra-fine grained material. These results present unequivocal evidence that the R dependency of the threshold level, where it exists, is dependent upon closure. Further it is clear that the decrease in microstructural size results in a decrease in closure in comparing these two alloys.

The appearance of the fracture surfaces of these two alloys in the threshold region is shown in Fig. 6. Evidence for Mode II growth and associated closure in 7090-T6 is given by the smooth, serrated regions in Fig. 6. The smooth areas are developed because of a rubbing action as closure develops. On the other hand, the fracture surface of the IN 9021 alloy is devoid of features which might be linked to a closure process.

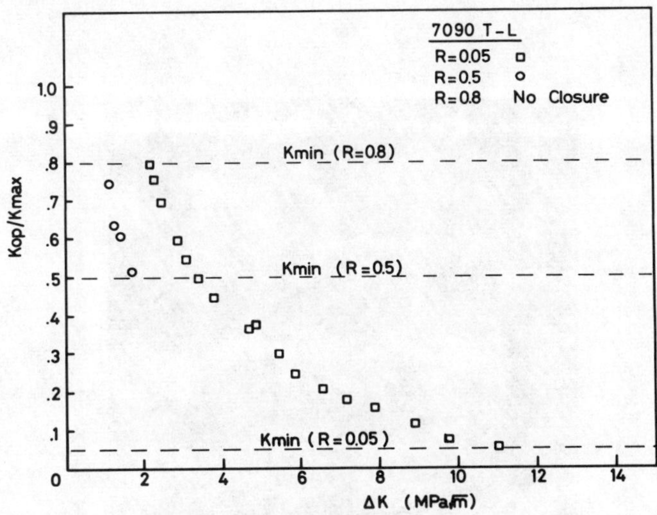

Figure 4 - Crack opening level in terms of the ratio of K_{op} to K_{max} as a function of ΔK for the 7090-T6 aluminum alloy at three R ratios.

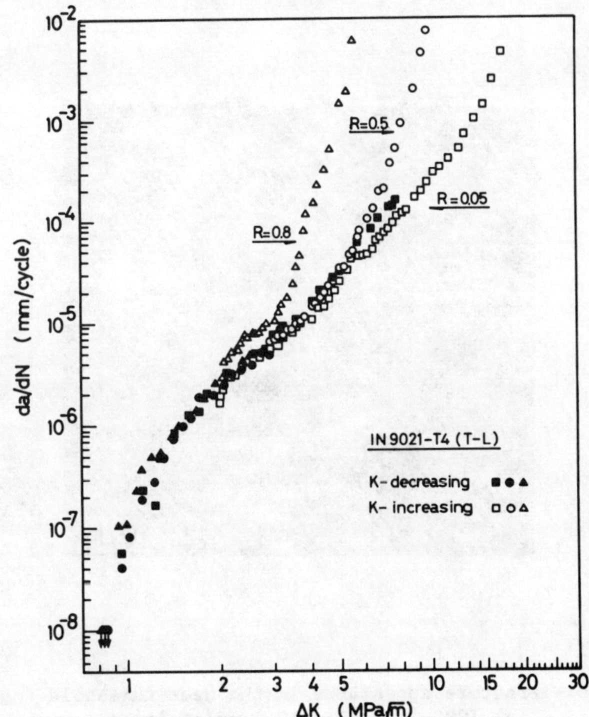

Figure 5 - Fatigue crack growth rate, da/dN, as a function of ΔK for the IN 9021-T4 aluminum alloy at three R ratios.

(a)

(b)

Figure 6 - Fracture appearance in the near-threshold region:
(a) 7090-T6 at 1×10^{-6} mm/cycle
(b) IN 9021-T4 at 1×10^{-6} mm/cycle

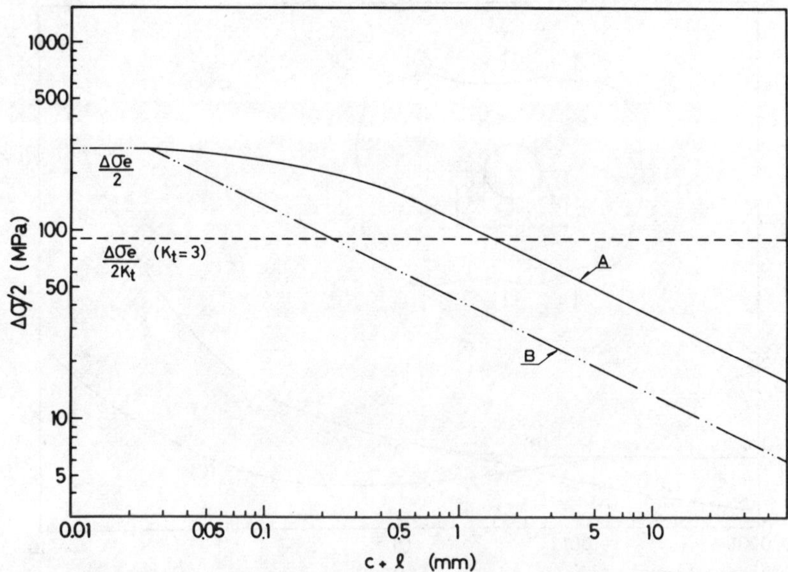

Figure 7 - Crack propagation conditions as a function of stress amplitude and total crack length, c + ℓ, where c is radius of a circular hole and ℓ is crack length. Fully reversed loading assumed.

Growth of Short and Long Cracks

In this section we consider the effect of closure on the stress level needed to propagate a crack as a function of crack length, a. In the case of long cracks the stress amplitude for propagation is directly related in simple form to the condition that the product of the tensile stress range and the square root of the crack length be equal to a constant. On a plot of log $\Delta\sigma$ against log a this relationship is a straight line of slope -1/2. However, as has been shown (7), this relationship does not hold for short cracks and instead there is a gradual transition to the endurance limit as indicated by curve A in Fig. 7. In this transition there is also a change in the crack closure level. It is fully developed in the case of the long crack, but as recent work has indicated it decreases to zero for an extremely small crack which has just initiated (2,8,9). If there were no closure in the case of a long crack, then a lower value of stress would be required for propagation as compared to a long crack with fully developed closure. Curve B in Fig. 7 represents the stress needed to propagate a crack in the absence of closure. Again, there will be a transition to short crack behavior. The tensile portion of a loading cycle is considered to be important with respect to crack growth.

In the treatment of the growth of short cracks from notches or defects an expression for the dependence of the stress intensity factor is needed. This dependency of K on crack length is shown by curve A in Fig. 8 for the case of a crack initiated at a circular notch (10). However, in dealing with short cracks another factor has to be taken into account, namely that the size of the plastic zone is large with respect to crack length. In this case the rate of growth would be higher than predicted by linear elas-

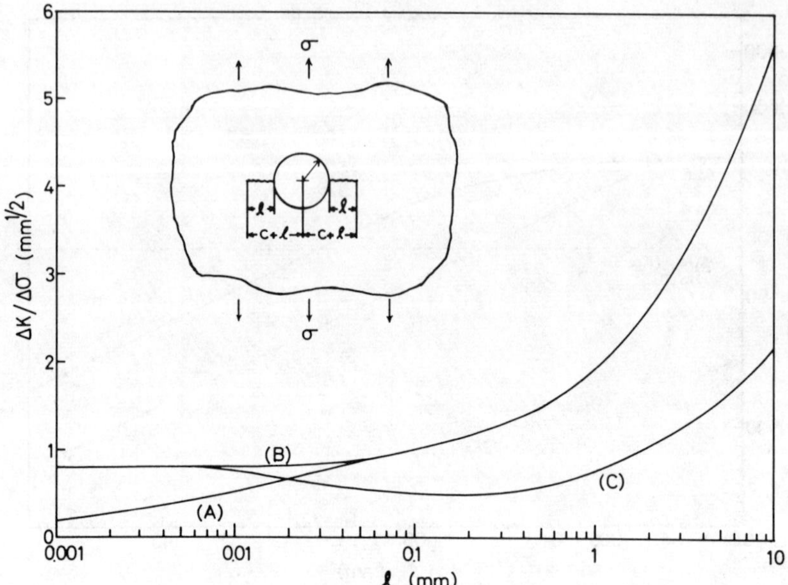

Figure 8 - Variation of $\Delta K/\Delta\sigma$ as a function of crack length, ℓ for three cases: (A) based on short crack, ℓ, from a hole of radius c equal to 0.2 mm; (B) based on a crack of length, ℓ + c; and (C) based on crack of length, ℓ + c, as influenced by the development of closure.

tic fracture mechanics (11). We therefore assume that if curve B, Fig. 8 is used in place of curve A, this feature will be accounted for. Curve B is obtained simply by taking the effective crack length to be c + ℓ where c is the radius of the hole and ℓ is the actual length, and we will use this method for determining the stress intensity factor in this paper.

Next we consider the influence of crack closure on the stress requirement for crack growth. For illustrative purposes the material is taken to be a medium-strength steel. Based upon our observations (2) and those of Morris et al. (8) we assume that closure in this steel develops in the wake of a crack over a distance of 0.5mm as shown in Fig. 9. The maximum value of K_{op} is taken to be 3.8 $MPa\sqrt{m}$ to correspond to a value of K_{op}/K_{max} at threshold (R=0) of 0.6 which is typical of a medium-strength steel. Although some additional closure may develop as the crack extends beyond 0.5 mm, as a simplification in the present analysis this additional closure will be considered to be negligible. It is also assumed that the maximum closure value is independent of ΔK level although in some cases there is evidence that the closure level actually decreases above the threshold (12). If we assume that a constant value of ΔK_{eff} ($\Delta K_{eff} = K_{max} - K_{op}$) is needed to just keep a crack propagating at a rate corresponding to the threshold level (taken to correspond to a growth rate of 10^{-8} mm/cycle) then as a short crack increases in length the stress requirement for propagation must be increased to maintain a constant value of ΔK_{eff} while crack closure is developing. This stress requirement is shown in Fig. 10 for three hole sizes in a medium- strength steel, curves C, D, and E. This stress requirement goes through a maximum before falling off due to increase

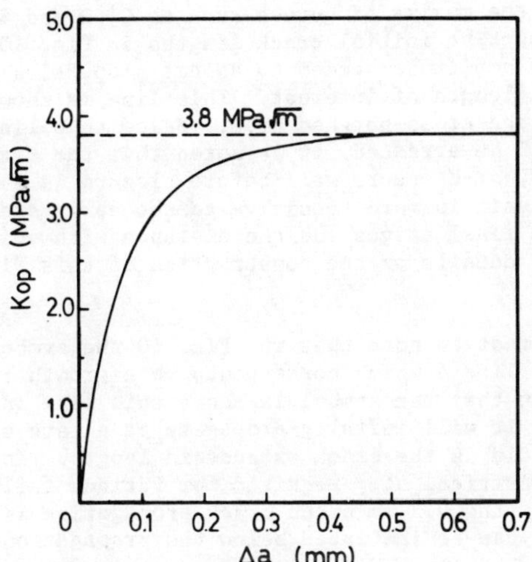

Figure 9 - Assumed development of closure as a short crack propagates from zero length to a length of 0.5 mm.

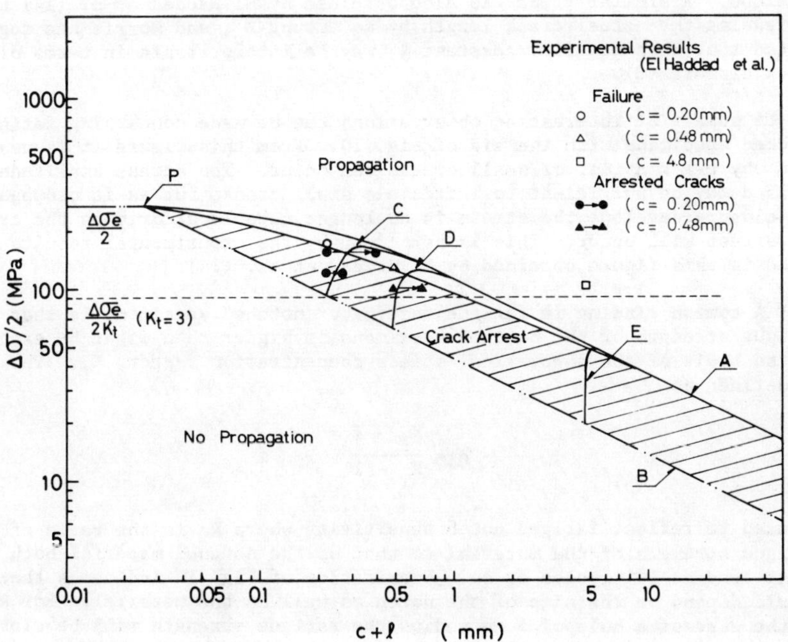

Figure 10 - Influence of crack closure on the stress required to propagate fatigue cracks as a function of notch or flaw size.

in crack length and merges with the line representing the threshold condition for long cracks. The stress amplitude required for propagation of a crack is given by the maxima of curves such as C, D and E. If we plot the maxima at the respective initial crack lengths in Fig. 10 we then have a plot of the stress amplitude needed to assume properties of a crack of any particular initial length of interest. This line is shown in Fig. 10 as the upper band to the cross-hatched area. Below this line if cracks are initiated they will be arrested. It is noted that the maximum point in either curves C, D, or E occurs well before closure is fully developed. Therefore the analysis is more sensitive to the early stages of closure rather than to the final stages and the distance with which closure fully develops. Further details on the construction of this figure are available in Ref. 9.

It is of interest to note that in Fig. 10 the symbol for the 4.8 mm notch is above the line A which corresponds to a growth rate of 10^{-8} mm/cycle. The fact that the symbol is above this line indicates that once a crack is started it will initally propagate at a rate above 10^{-8} mm/cycle and even more rapidly as the crack extends in length. In such a case crack initiation is the critical step required for fatigue failure. On the other hand in the case of the 0.2 mm notch crack propagation is the critical step, since cracks can be initiated below the propagation limit.

The variation of $\Delta K/\Delta\sigma$ as a function of crack length and crack closure (taken to be 60% of K_{max} to correspond to closure at threshold level in a steel of medium strength (14)) is also shown in Fig. 8, curve C. For a given applied level of $\Delta\sigma$, the ΔK value initially decreases as closure develops. A similar trend was also obtained by El Haddad et al (14) by increasing the actual crack length by an amount ℓ_o, and Morris has suggested that this material constant ℓ_o may be interpretable in terms of crack closure (8).

A number of interesting observations can be made concerning fatigue of notched specimens with the aid of Fig. 10. From this figure it is readily seen why crack arrest of small cracks can occur. The stress amplitude applied may be sufficient to initiate a small crack, but as it propagates and closure develops the stress is no longer capable of driving the crack and arrest will occur. This is borne out by the experimental results indicated in this figure obtained by El Haddad et al. (14).

A common finding in fatigue testing of notched specimens is that the fatigue strength of the notched specimens is higher than might be expected on the basis of the theoretical stress concentration factor, K_T. The index q, defined as

$$q = \frac{K_F - 1}{K_T - 1} \qquad (1)$$

is used to reflect fatigue notch sensitivity where K_F is the ratio of the fatigue strength of the material to that of the notched material both being determined at 10^7 cycles or so. Examination of Fig. 10 indicates that q should depend on the size of the notch as well as the material. For example, in the case of a hole of 5 mm radius the fatigue strength will be close to $\Delta\sigma_e/2K_T$ and q will be close to unity. On the other hand as the radius is made smaller then the notched fatigue strength given as the maximum of the curves marked C and D increases as the notch radius decreases. For curve D the value of the notched fatigue strength is 125 MPa and for curve C it is

170 MPa. Corresponding values of K_F are 2.16 and 1.59 respectively. The value of q is tabulated as

ρ	K_F	q
5 mm	3.0	1.0
.5 mm	2.16	0.58
.2 mm	1.59	0.30

Clearly the fatigue notch sensitivity is strongly related to crack closure characteristics.

Such results are also sometimes interpreted in terms of a notch size effect, expressed as

$$K_F = 1 + \frac{K_T - 1}{1 + \sqrt{\rho'/\rho}} \quad (2)$$

where ρ' is a material constant (15,16), or

$$K_F = 1 + \frac{K_T - 1}{1 + \frac{a_o}{\rho}} \quad (3)$$

where a_o is also a material constant (17). The predicted value of K_F based on these relations are plotted in Fig. 11 together with the K_F values tabu-

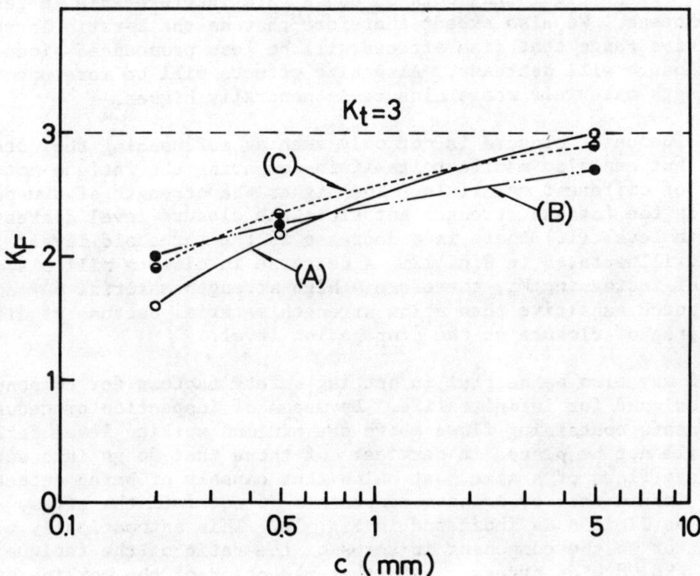

Figure 11 - Variation of K_F as a function of hole radius for three cases; (A) based on Fig. 10; (B) based on Eq. 2; and (C) based on Eq. 3. ($\rho' = 0.4$ mm, $a_o = 0.25$ mm).

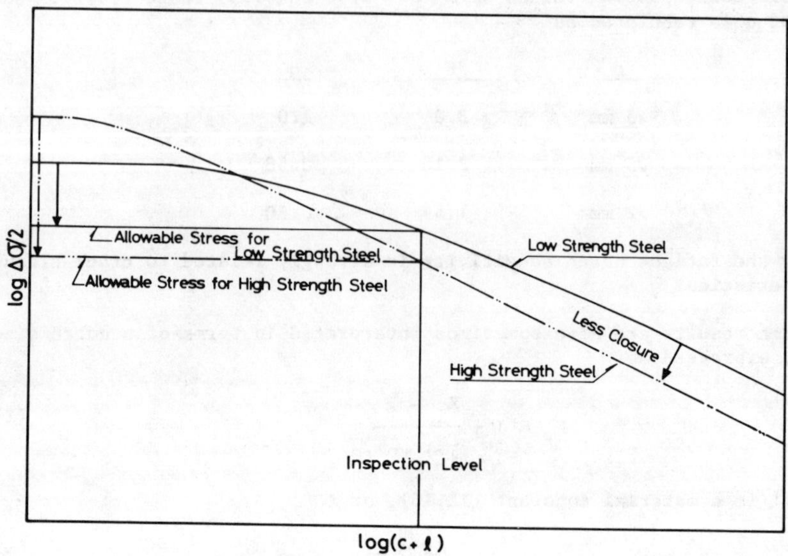

Figure 12 - Effect of strength level and crack closure on the allowable stress from non-propagation of a crack as influenced by inspection capability.

lated above. It is clear that both ρ' and a_o are interpretable in terms of closure phenomena. We also expect therefore that as the R-ratio increases in the positive range that size effects will be less pronounced since the extent of closure will decrease. Also size effects will be more pronounced in low strength materials where closure is generally higher.

The influence of closure is not only seen as influencing the notch size effect but can also manifest itself in comparing the fatigue notch sensitivity of different materials. The higher the strength of a material the higher is the fatigue strength but since the closure level decreases with strength level (18) there is a decrease in the threshold level. This situation is illustrated in Fig. 12. A decrease in closure will also have the effect of increasing K_F, therefore a high strength material may appear to be more notch sensitive than a low strength material because of differences in degree of closure at the propagation level.

Fig. 12 may also be helpful in setting safety factors for components which are designed for infinite life. By means of inspection procedures those components containing flaws above the minimum working level for flaw detection will not be placed in service. Of those that do go into service some will have flaws of a size just below that capable of being detected. In order to insure that even these components do not fail the stress amplitude should be limited as indicated in Fig. 12. This automatically places a safety factor on the component in terms of the ratio of the fatigue strength to the working stress. A further reduction of the working stress to account for uncertainty in the loading spectrum, environmental effects, and scatter would also have the benefit of increasing the allowable flaw size. It is noted that at the same flaw size detection level the allowable stress level may be lower in the case of a material of higher strength, and if so, a material of lower strength may be more suited to the application.

Conclusions

1. In the high strength P/M aluminum alloy 7090-T6 the important closure event affecting the threshold level occurs within 1 mm behind the crack tip.

2. In the absence of crack closure there is no R effect on the threshold level.

3. The arrest of cracks initiated at notches (non-propagating cracks) is due to the development of crack closure.

4. Notch size effects and notch sensitivity are understandable on the basis of crack closure considerations.

5. Safety factors and maximum permissible flaw sizes can be determined on a rational basis through consideration of the fatigue strength and the propagation limit for the growth of flaws.

Acknowledgement

The authors are pleased to acknowledge the support of this research by the Air Force Office of Scientific Research. Helpful discussions with Dr. H. Nakamura are also acknowledged.

References

1. J. C. Newman, Jr., presented at the AGARD Specialist's Meeting on Behavior of Short Cracks in Airframe Components, Toronto, September (1982).

2. K. Minakawa, J. C. Newman, Jr. and A. J. McEvily, to be published in Fatigue of Engng. Mater. and Struct., V. 6, pp. 359-365 (1983).

3. R. A. Schmidt and P. C. Paris, ASTM STP 536, 79 (1973).

4. A. T. Stewart, Engng. Frac. Mech., 13, 463 (1980).

5. P. K. Liaw, V. P. Swaminathan, T. R. Leax and J. K. Donald, Scripta Met., 16, 871 (1982).

6. S. Suresh and R. O. Ritchie, Engng. Fract. Mech., 18, 785 (1983).

7. H. Kitagawa and S. Takahashi, 2nd Intl. Conf. on Mech. Beh. of Mater., Boston, 627 (1976).

8. W. L. Morris and M. R. James, Engng. Frac. Mech. 18, 871 (1983).

9. K. Tanaka and Y. Nakai, to be published in "Fatigue Crack Growth Threshold Concepts", Proc. of Intl. Symposium, AIME, Philadelphia, (1983).

10. H. Tada, P. C. Paris and G. R. Irwin, The Stress Analysis of Cracks Handbook, Del Research Corp., Hellertown, PA (1973).

11. R. A. Smith and K. J. Miller, Intl. J. of Mech. Engng. Sci., 19 (1977).

12. K. Minakawa, Y. Matsuo and A. J. McEvily, Met. Trans. A, 13A, 439 (1982).

13. A. J. McEvily and K. Minakawa, to be published in Scripta Metallurgica, January (1984).

14. M. H. El Haddad, T. H. Topper and K. N. Smith, Engng. Frac. Mech. 11, 573 (1979).

15. H. Neuber, Kerbspannungslehre, Springer, (1958).

16. P. Kuhn and H. F. Hardrath, NACA Tech. Note 2805, (1952).

17. R. E. Peterson, Stress Concentration Factors, John Wiley and Sons, New York, (1974).

18. K. Minakawa and A. J. McEvily, in "Fatigue Thresholds", 1st Intl. Conf., Stockholm, J. Backlund, A. Blom and C. J. Beevers, eds., EMAS Publ. Ltd., Warley, U.K., 1, 373 (1982).

DISLOCATION MODELS FOR THRESHOLD FATIGUE CRACK GROWTH

T. Mura* and J. Weertman+

Department of Civil Engineering*
Department of Materials Science and Engineering+
and
Materials Research Center*,+
Northwestern University
Evanston, IL 60201

This paper reviews dislocation models that have been applied to the near-threshold stress intensity factor region. Theoretical work on dislocation theories of fatigue crack growth in the near-threshold region is still in its infancy. Because of the spareness of existing theory this region of the fatigue crack growth curve remains ill understood.

General Considerations

There are two "natural" physical reasons why a threshold cyclic stress intensity factor might exist. If a fatigue crack propagates each stress cycle it might be expected that no crack growth occurs once the crack growth rate approaches an interatomic distance per cycle. However, average growth rates smaller than an atomic distance per cycle have been measured. On the other hand, crack growth should not occur if the stress intensity factor never exceeds that for propagating or blunting a Griffith crack in a perfectly elastic solid. In this section a brief review is given of what are the main experimental findings that bear on these "natural" explanations of a threshold and thus what it is that must be accounted for by any dislocation model.

Below a threshold cyclic stress intensity factor ΔK_{th} a fatigue crack presumably never propagates. (In this paper the cyclic stress intensity factor is defined for a crack in an infinite plate as $\Delta K = K_{max}-K_{min}$ where $K_{max} = \sigma_{max}(\pi a)^{1/2}$ and $K_{min} = \sigma_{min}(\pi a)^{1/2}$. Here σ_{max} and σ_{min} are the maximum and minimum values of the applied stress and a is the half length of the crack. The R-ratio is defined to be $R = K_{min}/K_{max}$.) For ΔK greater than ΔK_{th} but K_{max} smaller than the critical stress intensity factor K_c for fracture the crack growth rate da/dN, where N is the number of stress cycles, is given by the Paris equation

$$da/dN = C(\Delta K)^n \tag{1}$$

where C and n are constants. The value of n typically lies in the range $2 < n < 4$. An interesting form of this equation was proposed by Pearson (1) as

$$da/dN = C_p(\Delta K/E)^{3.6} \tag{2}$$

where E is Young's modulus and C_p is another constant. Pearson and later workers (1-5) have shown that equation (1) gives a reasonable account of crack growth data for a number of metals and alloys although, admittedly, not of the changes in da/dN produced by altering such things as the microstructure.

A generalization of equation (1) that gives a reasonable account of literature data (see Table I) is

$$da/dN = (\alpha/b^{(n-2)/2})(\Delta K/nE)^n \tag{3}$$

where b is the interatomic distance and α is a dimensionless constant of order of $\alpha \simeq 10$. (In an active environment, such as moist air, that accelerates the crack growth rate, the value of $\alpha \simeq 30$ to 100.)

A fatigue crack cannot grow in a cleavage mode if K_{max} is smaller than the critical stress intensity factor K_{cb} of a Griffith crack in an elastic solid. The value of K_{cb} is given by

$$K_{cb} = [2E\gamma/(1-\nu^2)]^{1/2} \tag{4}$$

where γ is the surface energy of the metal and ν is the Poisson's ratio. If a fatigue crack grows in a crack tip blunting mode it cannot grow if K_{max} is smaller than the critical stress intensity factor gK_{cb} for the onset of crack tip blunting through dislocation emission from the crack tip (14). The constant g must be smaller than 1 to have crack tip blunting instead of cleavage propagation. For ductile material $g \simeq 0.6$ (14).

Since a crack cannot be propagated when $K_{max} < K_{cb}$ or gK_{cb} a lower limit to the threshold cyclic stress intensity factor ΔK_{th} is

$$\Delta K_{th} = (1-R)K_{cb} \text{ or } (1-R)gK_{cb} \tag{5}$$

Table II lists values of $\Delta K_{th} = K_{cb}$ (for R=0) obtained from equations (4) and (5) for a number of metals.

The crack growth rate given by equation (3) when ΔK is very slightly larger than ΔK_{th} given by equation (5) is

$$(da/dN)_{th} = (\alpha/b^{(n-2)/2})(1-R)^n(gK_{cb}/nE)^n \tag{6}$$

where g is set equal to 1 if propagation occurs in a cleavage mode.

It should be noted that K_{cb}, when no environment lowers the value of the surface energy γ, is of the order of $(1/3)Eb^{1/2}$ (see Table II). Thus equation (5) can be approximated by the equation

$$(da/dN)_{th} \simeq \alpha b(g/3n)^n(1-R)^n \tag{7}$$

For $\alpha = 10$, $g = 1$ and $R > 0$ the predicted growth rate is smaller than an atomic distance when $n > 1.6$. (For $g = 1$, $n = 3.5$ and $R = 0$ the growth rate per cycle da/dN given by this last equation is 0.0027b.)

Table I. Crack Growth Rate da/dN at $\Delta K = 10$ MPa $m^{1/2}$

Metal	Reference	$(da/dN)_{exp}$ nm	$(da/dN)^{+}_{Eq. 3}$ nm	n	R
Aluminum alloys					
7050-T76	(6)	58	61	3.3	0.05
7XXX series	(7)	300*	180*	3.5	0.33
2XXX series	(7)	150*	170*	3.8	0.1
Steels					
300-M	(8)	10*	12*	2.54	0.05
4Cr-0.35C	(8)	9*	11*	2.6	0.05
4140, 550°C, temper	(9)	6	2.8	2.9	0.05
4140, 200°C temper	(9)	5	1.2	3.7	0.05
Nickel Base					
Mar-M-200 and U-700	(10)	≃ 9*	7.6*	3**	-1
Cu and Cu alloys	(11)	≃ 6++	5	4	≃0
"	(11)	≃60*	16*	4	≃0
Ti-6Al-4V	(12)	8	13	3.5+*	0.12
"	(13)	1+++	2.5	5.5+++	0.05

+E and b given in Table II; $\alpha = 10$ unless otherwise indicated; room temperature data unless otherwise indicated.

*Tested in non-inert environment; $\alpha = 30$.

**Our estimate from data plot.

++Liquid nitrogen temperature.

+*Our estimate from data plot

+++Our extrapolated value from data of ΔK greater than 10 MPa $m^{1/2}$. Actual data shows threshold-like drop off in crack growth rate at 10 MPa $m^{1/2}$ to $da/dN < 0.1$ nm.

Table II. Theoretical Threshold Stress Intensity Factor

$\Delta K_{th} = K_{cb}$ (R=0)

Metal	E 10^{10}Pa	ν	b nm	γ J/m^2	γ* J/m^2	$Eb^{1/2}/3$ MPa m$^{1/2}$	$\Delta K_{th}=K_{cb}$ MPa m$^{1/2}$
Al	7.1	0.356	0.286	0.98	0.83	0.40	0.40
Cu	13	0.34	0.255	1.65	1.36	0.69	0.70
Fe	21	0.30	0.248	2.0	2.1	1.05	0.96
Au	8	0.41	0.288	1.35	0.94	0.45	0.51
Ni	21	0.30	0.249	2.3	2.1	1.1	1.0
Pb	2.4	0.40	0.350		0.34	0.14	0.14
Ag	8	0.36	0.289	1.13	0.94	0.45	0.46
W	39	0.28	0.274	3.0	4.4	2.2	1.6
Pt	13	0.39	0.278	3.0	1.46	0.72	0.96
Pd	17	0.38	0.275		1.92	0.94	0.87
Ti	11	0.32	0.289		1.3	0.62	0.57
Zn	10	0.24	0.266	0.1	1.2	0.54	0.046

E and ν are polycrystalline averaged values from Simmons and Yang compilation (15); b, the interatomic spacing, from Table A-6 of Barrett and Massalski (16); experimental values of γ from Table 1.1 of Kelly (17) or from Table A-1 of Hirth and Lothe (18); surface energy γ* is estimate from the relationship $\gamma^* = Ea_o/20$ where a_o is the largest atomic plane spacing ($a_o = a/3^{1/2}$ for fcc, $a_o = a/2^{1/2}$ for bcc and $a_o = c/2$ for hcp); K_{cb} is calculated from equation (4) using γ where available, otherwise with γ*.

If equation (5) gives a good estimate of the true cyclic threshhold stress intensity factor, in general, the crack growth rate in the near threshold region is smaller than an interatomic distance per cycle. Table III lists experimental values of ΔK_{th} and the crack growth rate $(da/dN)_{th}$ which is defined here as the crack growth rate at a cyclic stress intensity factor slightly greater than ΔK_{th} (see Fig. 1). What should be noted in Tables II and III is that for tests conducted in an inert environment that the measured values of ΔK_{th} are a factor of 5 to 10 larger than the theoretical value given by equation (5). Moreover, the crack growth rate $(da/dN)_{th}$ at the threshold is of the order of or larger than an interatomic distance per cycle. Only on the very steep part of the crack growth curve for values of ΔK very close to ΔK_{th} are measured crack growth rates smaller than an interatomic distance per cycle. (For the materials tested in more active environments only tests on a complex nickel alloy in Table III produced data not in conflict with equation (5).)

From this brief review it appears that dislocation models of the near-threshold region should account for threshold stress intensity factors that are appreciably larger than the critical one of a Griffith crack. Moreover, dislocation models should be able to explain how average crack growth rates smaller than an interatomic distance per cycle come about. The fact that threshold occurs when the crack growth rate $(da/dN)_{th}$ is of the order of one to ten times the interatomic distance suggests that this "natural" explanation of the threshold has an element of truth to it. If so, dislocation models should show explicitly why the crack growth curve on a da/dN versus ΔK plot drops so precipitously below a crack growth rate of a few interatomic distances per cycle rather than continuing in a normal fashion to the Griffith crack critical stress intensity factor (for either cleavage or blunting) and then dropping off. This paper will not give the answers to these questions, only the start that has been made towards obtaining them.

Threshold and Crack Closure

One reason that the measured ΔK_{th} may be larger than ΔK_{th} given by equation (5) (pointed out by R. O. Ritchie to one of us (JW), private conversation, 1981) is that the crack closure phenomenon reduces the effective value of ΔK. In fact, Suresh et al (22) have shown that an excess of corrosion products on the crack faces increases the magnitude of the crack closure effect and further reduces the effective ΔK.

Since an active environment can also reduce the surface energy of a metal an environment can produce opposite changes in the measured value of ΔK_{th}. From equation (5) if γ is reduced then ΔK_{th} is reduced. But if crack closure is increased because of corrosion products on the crack surface the measured ΔK_{th} is increased.

Crack closure also occurs in the absence of an active environment. Both Fine and his students (20,21) and Suresh and Ritchie (23,24) have invoked plastic deformation as well as the roughness it produces to show that closure reduces the effect of ΔK in the threshold region. Thus even in an inert atmosphere the effective values of ΔK_{th} are smaller than the values listed in Table III and are closer to the theoretical values of equation (5). For the case of iron (0.025%C) Lin and Fine (21) showed that whereas the ΔK_{th} as measured ranged from 7.8 to 9.9 MPa $m^{1/2}$ the effective ΔK_{th} were all approximately equal to 3.5 MPa $m^{1/2}$. Park and Fine (21) found for pure iron that the measured ΔK_{th} is of the order of 5 to 6 MPa $m^{1/2}$ but the effective ΔK_{th} is 3.8 MPa $m^{1/2}$ and for Al-3Mg at liquid nitrogen the value of ΔK_{th} given in Table III is reduced for the effective

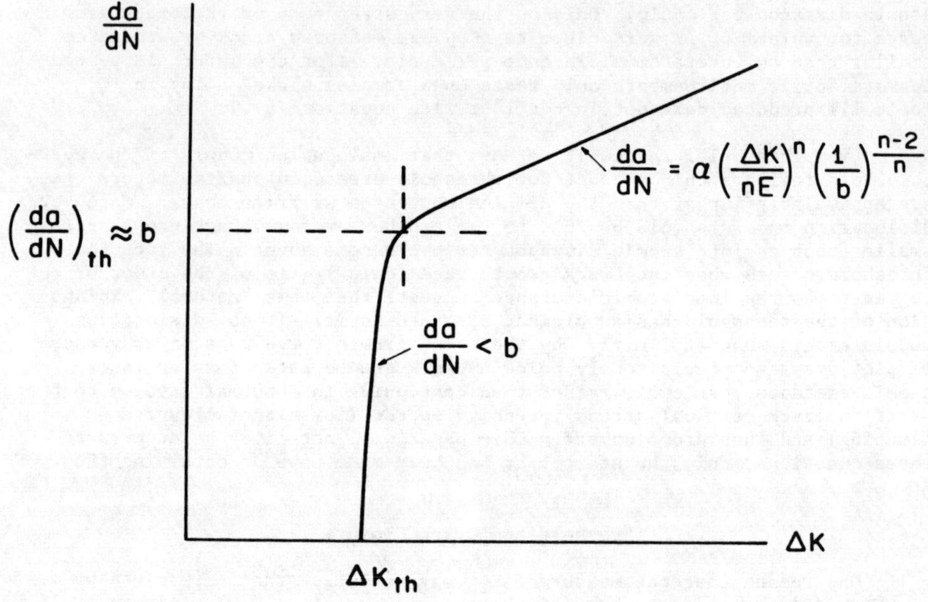

Figure 1. Schematic log-log plot of da/dN versus ΔK in the threshold region that shows ΔK_{th} and $(da/dN)_{th}$.

ΔK_{th} to 2.5 MPa $m^{1/2}$. (The effective ΔK is equal to the nominal ΔK minus the ΔK at which crack closure no longer exists. Taira et al (25) and Stanzl and Tschegg (26) have shown that the fatigue crack growth curve in the steep part of the threshold region is made into a normal Paris law curve when an effective ΔK is used rather than the nominal ΔK.) Thus even with the crack closure correction the threshold stress intensity factor is significantly larger than the value given by equation (5). But perhaps it is more impressive that the corrected threshold stress intensity factor is, nevertheless, within an order of magnitude of the equation (5) value. However, the value of ΔK_{th} is known to depend upon metallurgical factors such as grain size. Equation (5) cannot account even qualitatively for variations in ΔK_{th} produced by metallurgical factors.

Fine (21,27) has proposed that ΔK_{th} is determined by the activation of dislocation sources close to the fatigue crack tip. In particular, he suggests that the equation

$$K_{th} = \alpha'\sigma_s(\ell)^{1/2} \tag{8}$$

gives the threshold stress intensity factor. Here α' is a dimensionless parameter whose value depends on environment and the strength of the metal, σ_s is the stress required to activate a dislocation source, and ℓ is the distance a source is from the crack tip. This model has not yet been given a quantitative treatment.

The Yokoboris and co-workers (28-31) have used dislocation emission from the crack tip as a criterion of ΔK_{th}. This criterion was later used by one of us (JW) too for ductile material (6). But this criterion is simply equation (5) with $g < 1$. As it stands it cannot account for variations of ΔK_{th} produced by altering a metals microstructure. The Yokoboris and co-workers add a dislocation dynamic law (dislocation velocity proportional to stress acting on a dislocation raised to an arbitrary power m) in an attempt to obtain a dependence of ΔK_{th} on factors such as grain size. They find (31)

$$\Delta K_{th} = \Delta K_i + \sigma_s\lambda^{1/2} + M(\rho d/Nb)^{(1+\beta)/2m\beta} \tag{9}$$

where m is the power just mentioned, ρ is the grown in dislocation density, β is the cyclic strain hardening exponent, d is the grain size. ΔK_i is the stress intensity factor required to have dislocation emission, N is the number of dislocations emitted at the start of yielding, σ_s has the same meaning as in equation (8), λ is a distance from the crack tip over which the stress is averaged, and M is a constant which is a complicated function of many physical parameters. The model of the Yokoboris et al, as we understand it, can be summarized as follows. Equation (5) with $g < 1$ gives the condition when dislocation move easily in the crystal lattice once they are created. But if they do not move freely a larger value of ΔK is required. Thus ΔK_{th} is greater than the value given by equation (5).

Increasing the R-ratio decreases ΔK_{th} (see data on pure Fe and Al-3Mg of Park and Fine in Table III). Since increasing R increases K_{max} at a given value of ΔK this is a not unexpected result. Increasing R eliminates crack closure and thus must decrease ΔK_{th}. Moreover, the ΔK_{th} of equation (5) is reduced when R is increased. However, Park and Fine (20) found no decrease in the effective ΔK_{th} (the threshold stress intensity factor corrected for crack closure) with increase in R for Al-3Mg tested at room temperature and at liquid nitrogen temperature in pure Fe at room

Table III. Measured Values of ΔK_{th} and $(da/dN)_{th}$

Metal	Reference	R	ΔK_{th} MPa $m^{1/2}$	$(da/dN)_{th}$ nm
Aluminum alloys				
2124-T4	(19)	0.05	6++	0.2
2024-T4	(19)	0.05	3++	0.3
Al-3Mg	(20)	0.05	3.4	≃1
"	(20)	0.5	2.3	≃1
"	(20)	0.05	4.2++	≃1
"	(20)	0.5	2.8++	≃1
7XXX series	(7)	0.33	3	2
7XXX	(7)	0.33	2*	1
2XXX	(7)	0.1	3*	6
Fe	(20)	0.05	5.9	≃1
"	(20)	0.3	5.5	≃1
"	(20)	0.5	5.0	≃1
Fe(0.024¢C)	(21)	0.05	7.8 to 9.9	0.2
Steels				
300-M	(8)	0.05	3*	1
300-M	(8)	0.7	2.3*	1
4Cr-0.35C	(8)	0.05	3* to 4*	1
Nickel base				
Mar-M-20	(10)	-1	<1+,*	<0.01+,*

+Threshold not attained at lowest $\Delta K = 1$ MPa $m^{1/2}$ used.

++Tested at liquid nitrogen; other tests in table all at room temperature.

*Tested in a not inert environment.

temperature (R = 0.05 and R = 0.5). Liaw et al (21,22) also find (for tests carried out in air) that the effective value of ΔK_{th} is virtually independent of R for a CrMoV steel, a NiCrMoV steel (tested at 93°C), 2219 Al and for copper for R-ratios of 0.1 and 0.5 to 0.8. These results suggest that threshold might be determined by a "natural" interatomic distance criterion rather than by some kind of stress intensity factor criterion.

Growth Rate in the Near-Threshold Region

Consider next the problem of determining the growth rate in the near-threshold region. Two theories will be considered. The first is that developed by Tanaka and Mura and further developed here. The second is a theory that is discussed here of the double slip plane crack model of Lin, Thomson and Weertman.

Tanaka-Mura Dislocation Ratcheting Model

According to experimental observations in low-carbon steels (34-36) and Ti-alloys (37,38), the growth path of a long fatigue crack with a very low growth rate near the threshold is in shear bands developed at the crack tip. The slip band ahead of the crack tip is small compared with the crack length and it is blocked by grain boundaries (36,39) as shown in Fig. 2a. The slip band is inclined by θ from the crack plane. If a crack grows along the slip band, the new increment of crack surface is subjected to a mixed mode loading (Mode I and Mode II).

Tanaka and Mura (40) have characterized the crack growth near the threshold as a crack growth under the mixed mode. The following is a slight modification of their theory. For analytical simplicity, the geometrical configuration shown in Fig. 2a is replaced by that in Fig. 2b, or in Fig. 3. The crack and slip band is subjected to the two stress intensity factors $\Delta K_1'$ and $\Delta K_2'$. The slip band consists of layer I and layer II dislocation pileups. Forward loading causes the pileup of negative dislocations on layer II. Under the assumption of irreversible dislocation motion, ratcheting dislocation accumulation takes place through the help of the back stress produced by the dislocations of the preceding half cycle (the back stress from the dislocations on layer II helps produce the pileup of dislocations on layer I and vice versa).

In the mixed mode situation shown in Fig. 3, the normal stress σ opens the crack, the shear stress τ produces the shear band. The propagation of a fatigue crack along the slip band is nothing more than repetition of microcrack initiation ahead of the main crack tip. Thus the theory of Tanaka and Mura (41-44) for initiation of fatigue crack can be applied to the propagation. This model probably is adequate for accounting for Stage I crack growth near threshold.

The accumulated damage after ΔN cycles is considered to be the dislocation dipole density or its associate accumulated strain energy. When the energy density attains a critical value $2W_s^*$, a part of damage zone of length λ becomes a new crack. The process is repeated and the crack grows. The energy density per unit length of the slip band ℓ is $\Delta N \Delta U/\ell$, where ΔU is an increment of the self-energy of dislocations in each reversal of loading. The crack propagation law that is obtained is

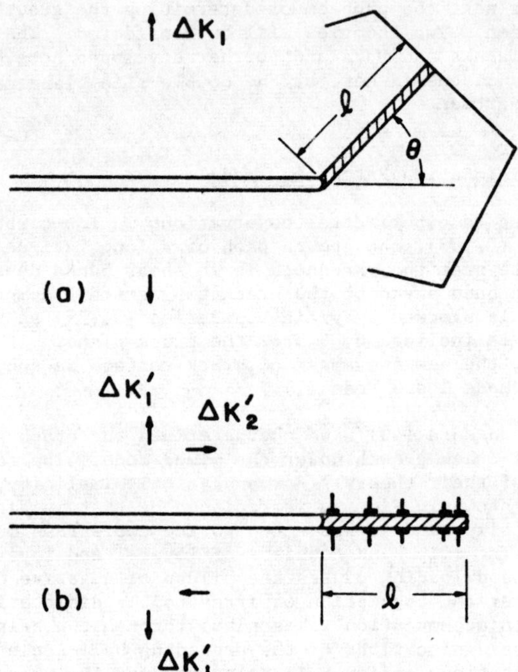

Figure 2

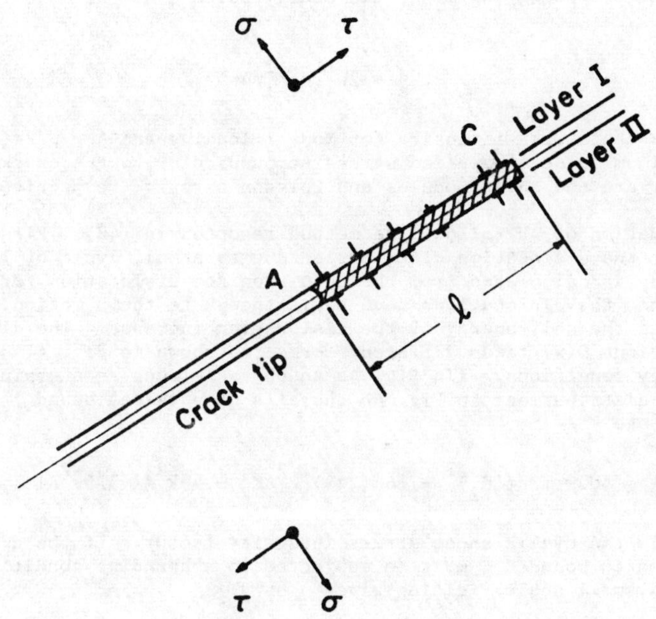

Figure 3

$$\frac{da}{dN} = \lambda/\Delta N = (\lambda/\ell)(\Delta U/2W_s^*). \tag{10}$$

The energy ΔU for a mixed mode loading is obtained through a simple superposition of strain energies of the edge and screw dislocations (the dislocations of the two kinds are not interactive). The near-tip stress due to Mode I loading is not relaxed by slip deformation of Mode II and provides an additional energy release when the crack extends. Hence W* must be the real fracture surface energy W_s minus the energy release due to Mode I crack growth:

$$2W_s^* = 2W_s - K_1^2/4\pi A , \tag{11}$$

where K_1 is the stress intensity for Mode I loading and $A = \mu/2\pi(1-\nu)$. σ_1 is the maximum value of applied stress perpendicular to the crack surface, and μ and ν are the shear modulus and Poisson's ratio, respectively.

Evaluation of ΔU follows the method reported in (42). First, an increase of the dislocation distribution due to a half cycle of loading or unloading is calculated from the condition for dislocation force balance assuming that the dislocations have resistance k to their motion. ΔU is evaluated as the self-energy of the dislocation increase. The distribution of dislocations D(x) takes different forms, as shown in Fig. 4, depending on the boundary conditions. (In Fig. 4a an obstacle such as a grain boundary bounds the distribution; in Fig. 4b there is no physical bound.) For the unbounded case

$$\Delta U = \{3\pi(\Delta K_2)^2 - 16k(2\pi\ell)^{1/2}\Delta K_2 + 48k^2\ell\}(\ell/6\pi^2 A) \tag{12}$$

where ΔK_2 is the cyclic shear stress intensity factor. If the dislocation distribution is bounded ℓ must be subjected to a bounding condition (45) which for a small scale yielding gives

$$\ell \simeq (\pi/8)(\Delta K_2/2k)^2 \equiv \omega . \tag{13}$$

When equation (13) is substituted into equation (12)

$$\Delta U = k^2(\Delta K_2/2k)^4/24A . \tag{14}$$

It is found that the crack growth rate is

$$\frac{da}{dN} = \frac{2\lambda}{3\pi} \; \frac{3\pi(\Delta K_2)^2 - 16k(2\pi\ell)^{1/2}\Delta K_2 + 48k^2\ell}{8\pi A W_s - K_1^2} \tag{15}$$

when the dislocation distribution is blocked, and

$$\frac{da}{dN} = \frac{\lambda}{3} \; \frac{(\Delta K_2)^2}{8\pi A W_s - K_1^2} \tag{16}$$

when the dislocation distribution is unblocked. The condition for the blocked dislocation distribution leads to

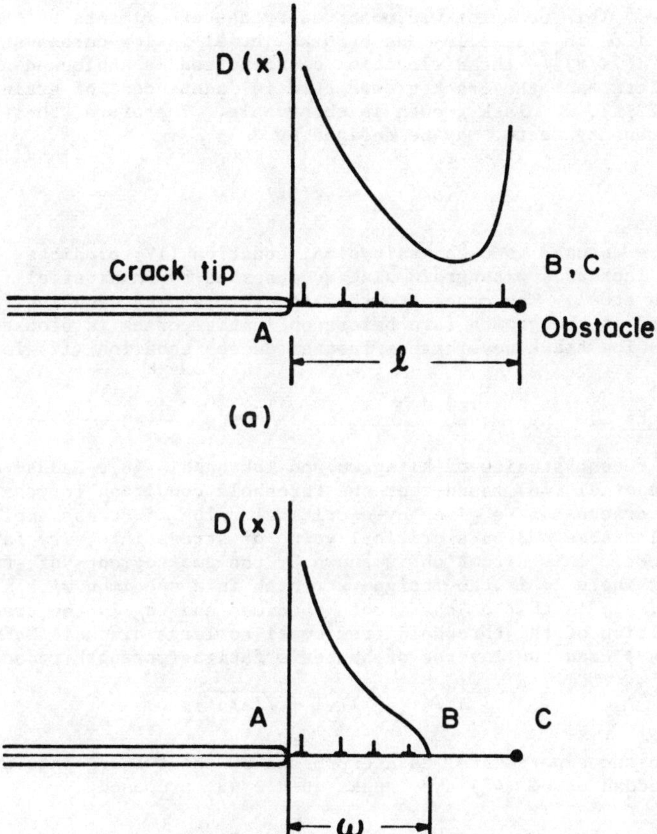

Figure 4

$$\Delta K_2 > 4k(2\ell/\pi)^{1/2} \qquad (17a)$$

When ΔK_2 satisfies condition (17) and ℓ is a grain size, the crack growth rate predicts from equation (15) is a rate which is slower the larger is the grain size. This behavior was observed by the experiments by Chakrabortty and Stark (38) in a Ti-alloy and by Nakai et al in low-carbon steel (36). If $\Delta K_2 = 4k(2\ell/\pi)^{1/2}$ the dislocation distribution is unblocked and equation (16) predicts that the crack growth rate is independent of grain size. When $\Delta K_2 < 4k\sqrt{(2\ell/\pi)}$, no crack growth is observable. Therefore, the threshold stress intensity factor may be defined by

$$(\Delta K_2)_{th} = 4k(2\ell/\pi)^{1/2} \ . \qquad (17b)$$

If ℓ can be assumed to be a grain size, equation (17) predicts that the threshold increases with grain size as observed by Nakai et al (36) in a low-carbon steel. The crack growth rate expressed by equation (16) may be interpreted as the growth rate before an initial crack is blocked by a grain boundary. The crack never grows further unless equation (17) is satisfied.

Small Cracks

The recent studies of Kitagawa and Takahashi (46), Haddad et al (47), and Tanaka et al (48) found that the threshold condition for the fatigue growth of cracks can be given by a critical value of stress amplitude for very small cracks and by a critical value of stress intensity factor for large cracks. This situation is shown by the two segments of straight lines in Fig. 5, where $\Delta\tau$ is the fatigue strength in a specimen with a crack length 2a, $\Delta\tau_o$ is that of the smooth specimen and $2a_o$ is the crack length in the transition of the threshold from small to large cracks. McEvily and Groeger (49) used the inverse of Neuber's fatigue strength reduction (50),

$$\Delta\tau/\Delta\tau_o = 1/\{1 + \sqrt{(a/a_o)}\} \qquad (18)$$

to explain the experimental data expressed by the two segments of straight lines. Haddad et al (47) and Tanaka et al (48) proposed

$$\Delta\tau/\Delta\tau_o = (1 + a/a_o)^{-1/2}. \qquad (19)$$

These formulae curves are plotted in Fig. 5. Tanaka and Mura (42) have considered theoretically this short crack behavior. Since the fatigue crack growth near the threshold can be regarded as repetition of a microcrack initiation at the main crack tip they were able to put

$$\Delta U = \Delta U_o \qquad (20)$$

where $U_o = (\Delta\tau_o - 2k)^2\ell^2/4A$, and U is given by equation (12). Equation (20) provides a relation between $\Delta\tau/\Delta\tau_o$ and a/a_o for a given value of $\Delta\tau_o/2k$ as shown in Fig. 5. When $\Delta\tau_o/2k = 5$, equation (20) reduces to equation (19).

Double Slip Plane Model

The double slip plane (DSP) crack model of Lin, Thomson and Weertman (51) can be used to obtain fatigue crack growth laws (52). The

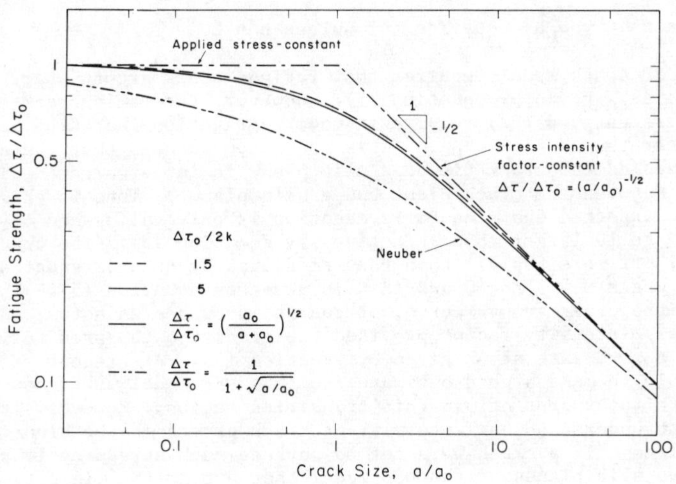

Figure 5. The effect of crack length on the threshold stress range for fatigue crack growth.

calculations as they stand (52) predict crack growth down to ΔK_{th} given by equation (5). In the model the crack is assumed to grow each cycle. Thus if the parameters in the DSP model are chosen so that the derived crack growth equations predict quantitatively the same crack growth rates as given by the empirical equation (3) not only is the value of ΔK_{th} smaller than observed but the crack growth rate is smaller than an interatomic distance per cycle at the lowest values of ΔK. In the derivation (52) of fatigue crack growth laws two modes of crack propagation were considered. These are propagation by a cleavage mode and propagation through crack tip blunting. For the case of propagation by a cleavage mode there is no real physical difficulty should the average crack fall below the level of an atomic distance per cycle. The crack can still be considered to advance each stress cycle. The position of the crack tip in this situation is analagous to the position of a dislocation line in a low stress amplitude internal friction experiment. In such an experiment the average shift in position of the dislocation line may be considerably less than b. The average shift by less than an atomic position can be accomplished by the movement of kinks along the dislocation line by distances larger than b. Similarly for the position of the crack tip an average shift of less than b can be accomplished by movements of "jogs" at the crack front by distances larger than b. Thus if crack propagation is in a cleavage mode an advance per cycle by an average distance smaller than b makes physical sense.

For fatigue crack growth by crack tip blunting through dislocation emission (6,52) the situation is different. The smallest physically meaningful crack advance is of the order of b which is the smallest "radius of curvature" that the blunted crack tip can have. Thus once da/dN falls below b crack growth only can occur intermittently if the advance is through blunting of the crack tip. The value of ΔK_{th} at $(da/dN)_{th}$ must be larger than the value given by equation (5) unless $n \simeq 2$.

The DSP crack model requires that fatigue crack growth start when ΔK exceeds the ΔK_{th} given by equation (5). However, the Paris crack growth laws that are derived (52) require in their derivation that K_{max} is appreciably larger than K_{cb} or gK_{cb}. (This condition insures that the active slip zone behind the crack tip on a slip plane is large compared with the separation between the crack plane and a slip plane.) Thus in this theory it is to be expected that the Paris equation is only valid when ΔK_{th} is a factor of 5 to 10 larger than ΔK_{th} given by equation (5). The crack rate should fall off more rapidly than that predicted by a Paris equation when $\Delta K < M\Delta K_{th}$ where $M \simeq 3$ to 10 and ΔK_{th} is given by equation (5). This break in the fatigue crack growth curve, of course, produces an actual threshold cyclic stress intensity factor provided the crack growth rates fall very rapidly to a zero rate at ΔK given by equation (5). At present we have not yet been able to make a good estimate from the DSP model as to how fast the crack growth rates drop off in this transition region. However, it is not difficult to understand how intermittent crack growth in the blunting mode can take place. If a crack were not to advance each stress cycle the regions on two slip planes of the DSP model that are in the vicinity of the crack tip work harden more and more. The more these regions work harden the easier it becomes for the crack to advance by a distance b. (Crack advance in the DSP model occurs more easily the more difficult it is to produce slip on the two slip planes.) Thus if ΔK is larger than ΔK_{th} given by equation (5) a stationary crack eventually will grow by an increment after many stress cycles in material that work hardens.

Acknowledgement

This work was supported by the National Science Foundation through the Materials Research Center of Northwestern University under Grant Number DMR82-16972.

References

1. S. Pearson, "Fatigue Crack Propagation in Metals," Nature, 211 (1966) pp. 1077-1078.
2. M. O. Speidel, "Corrosion Fatigue in Fe-Ni-Cr Alloys," pp. 1071-1094 in Stress Corrosion Cracking and Hydrogen Embrittlement of Iron Base Alloys, National Association of Corrosion Engineers, Houston, TX, 1977.
3. M. O. Speidel, "The Resistance to Fatigue Crack Growth of the Platinum Metals," Platinum Metals Rev, 25 (1981) pp. 24-31.
4. N. E. Frost, L. P. Pook, and K. Denton, "A Fracture Mechanics Analysis of Fatigue Crack Growth Data for Various Materials," Eng. Fracture Mech., 3 (1971) pp. 109-126.
5. S. E. Stanzl, "Fatigue Crack Growth (Physical Principles, Models and Experimental Results)," Res. Mechanica, 5 (1982) pp. 241-292.
6. R. D. Brown and J. Weertman, "Mean Propagation Effects on Crack Propagation Rate and Crack Closure in 7050-T76 Aluminum Alloy," Eng. Fracture Mechs., 10 (1978) pp. 757-771.
7. A. K. Vasudevan and S. Suresh, "Influence of Corrosion Deposits on Near-Threshold Fatigue Crack Growth Behavior in 2XXX and 3XXX Series Aluminum Alloys," Met. Trans., 13A (1982) pp. 2271-2280.
8. R. O. Ritchie, "Near-Threshold Fatigue-Crack Propagation in Steels," International Metals Revs., 20 (1979) pp. 205-230.

9. P. N. Thielen and M. E. Fine, "Fatigue Crack Propagation in 4140 Steel," Met. Trans., 6A (1975) pp. 2133-2141.
10. S. Purushothaman and J. K. Tien, "Slow Crystallographic Fatigue Crack Growth in a Nickel-Base Alloy," Met. Trans., 9A (1978) pp. 351-355.
11. H. Ishii and J. Weertman, "Fatigue Crack Propagation in Copper and Cu-Al Single Crystals," Met. Trans., 2 (1971) pp. 3441-3452.
12. P. E. Irving and C. J. Beevers, "The Effect of Air and Vacuum Environments on Fatigue Crack Growth Rates in Ti-6Al-4V," Met. Trans., 5 (1974) pp. 391-398.
13. R. J. H. Wanhill and H. Döker, "Vacuum Fatigue Fracture of Ti-6Al-4V," pp. 799-810 in Titanium and Titanium Alloys, Volume 2, J. C. Williams and A. G. Belov, eds.; Plenum Press, New York, NY, 1982.
14. J. Weertman, "Fatigue Crack Growth in Ductile Material," pp. 11-19 in Mechanics of Fatigue AMD-Vol. 47, T. Mura, ed.; ASME, New York, NY, 1981.
15. G. Simmons and H. Wang, Single Crystal Elastic Constants and Calculated Aggregate Properties: A Handbook, Second Edition, M.I.T. Press, Cambridge, MA, 1971.
16. C. S. Barrett and T. B. Massalski, Structure of Metals, Third Revised Edition, Pergamon Press, Oxford, 1980.
17. A. Kelly, Strong Solids, Second Edition, Clarendon Press, Oxford, 1973.
18. J. P. Hirth and J. Lothe, Theory of Dislocations, Second Edition, John Wiley, New York, NY, 1982.
19. J. McKittrick, P. K. Liaw, S. I. Kwun, and M. E. Fine, "Threshold for Fatigue Macrocrack Propagation in some Aluminum Alloys," Met. Trans., 12A (1981) pp. 1535-1539.
20. D. H. Park and M. E. Fine, "Near-Threshold Fatigue Crack Propagation in Fe and Al-3% Mg," (Abstract) p. 36, Fall Meeting Program, AIME, 1983 J. Metals (August 1983) and unpublished data.
21. G. M. Lin and M. E. Fine, "Effect of Grain Size and Cold Work on the Near Threshold Fatigue Crack Propagation Rate and Crack Closure in Iron," Scripta Met., 16 (1982) pp. 1249-1254.
22. S. Suresh, G. F. Zamiski, and R. O. Ritchie, "Oxide-Induced Crack Closure: An Explanation for the Near-Threshold Corrosion Fatigue Crack Growth Behavior," Met. Trans., 12A (1981) pp. 1435-1443.
23. R. O. Ritchie and S. Suresh, "Some Considerations on Fatigue Crack Closure at Near-Threshold Stress Intensities due to Fracture Surface Morphology," Met. Trans., 13A (1982) pp. 937-940.
24. S. Suresh and R. O. Ritchie, "A Geometric Model for Fatigue Crack Closure Induced by Fracture Surface Roughness," Met. Trans., 13A (1982) pp. 1627-1631.
25. S. Taira, K. Tanaka, and M. Hoshina, "Grain Size Effect on Crack Nucleation and Growth in Long-Life Fatigue of Low-Carbon Steel," pp. 135-173 in Fatigue Mechanisms, STP 675, J. T. Fong, ed.; ASTM, Philadelphia, PA, 1979.
26. S. Stanzl and E. Tschegg, "Fatigue Crack Growth and Threshold Measured at Very High Frequencies (20 kHz)," Metal Sci., 14 (1980) 137-143.
27. M. E. Fine, "Threshold for Fatigue Failure," Bull. JIM,20 (1981) pp. 668-677 [in Japanese].
28. T. Yokobori, A. T. Yokobori, Jr., and A. Kamei, "Dislocation Dynamics Theory for Fatigue Crack Growth," Int. J. Fracture, 11 (1975) pp. 781-788.
29. T. Yokobori, S. Konosu, and A. T. Yokobori, Jr., "Micro and Macro Fracture Mechanics Approach to Brittle Fracture and Fatigue Crack Growth," pp. 665-682 in Fracture 1977, (ICF 4) Volume 1, D. M. R. Taplin, ed.; University of Waterloo Press, Waterloo, Ontario, 1977.

30. A. T. Yokobori, Jr. and T. Yokobori, "On Micro- and Macro-Mechanics of Fatigue Thresholds," pp. 171-189 in Fatigue Thresholds, Volume II, Engineering Materials Advisory Services, Ltd. (EMAS), Warley, UK, 1982.
31. A. T. Yokobori, Jr. and T. Yokobori, "A Criterion for Threshold Stress Intensity Factor in Fatigue Crack Growth," pp. 1373-1380 in Advances in Fracture Research, Volume, 3, (Fracture 81, ICF 5), D. Francois, ed.; Pergamon Press, Oxford, 1982.
32. P. K. Liaw, T. R. Leax, V. P. Swaminathan, and J. K. Donale, "Influence of Load Ratio on Near-Threshold Fatigue Crack Propagation Behavior," Scripta Met., 16 (1982) pp. 871-876.
33. P. K. Liaw, T. R. Leax, R. S. Williams, and M. G. Peck, "Influence of Oxide-Induced Crack Closure on Near-Threshold Fatigue Crack Growth Behavior," Acta Met., 30 (1982) p. 2071-2078.
34. A. Otsuka, K. Mori, and T. Miyata, "The Condition of Fatigue Crack Growth in Mixed Mode Condition," Eng. Fract. Mech., 7 (1975) pp. 429-439.
35. A. Otsuka, K. Mori, and Y. K. Kawamura, "Relationship Among Crack Growth Rate, Growth Mode and ΔK in Mild Steel," Trans. Jap. Soc. Mech. Engng. A45 (1979) pp. 1312-1322.
36. Y. Nakai, K. Tanaka, and T. Nakanishi, "The Effects of Stress Ratio and Grain Size on Near-Threshold Fatigue Crack Propagation in Low-Carbon Steel," Engng. Fract. Mech., 15 (1981) pp. 291-302.
37. G. Yoder, L. A. Cooley, and T. W. Crookers, "Observations on Microstructurally Sensitive Fatigue Crack Growth in a Widmanstätten Ti-6Al-4V Alloy," Met. Trans. 8A (1977) pp. 1737-1743.
38. S. B. Chakrabortty and E. A. Starke, Jr., "Fatigue Crack Propagation of Metastable Beta Titanium-Vanadium Alloys," Met. Trans. 10A (1979) pp. 1901-1911.
39. K. Katagiri, A. Omura, K. Koyanagi, J. Awatani, T. Shiraishi, and H. Kaneshiro, "Early Stage Crack Tip Dislocation Morphology in Fatigue Copper," Met. Trans. 8A (1977) pp. 1769-1773.
40. K. Tanaka and T. Mura, "Fatigue Crack Growth Along Planar Slip Bands," Acta Met. (in press).
41. K. Tanaka and T. Mura, "A Dislocation Model for Fatigue Crack Initiation," J. Appl. Mech. 103 (1981) pp. 97-103.
42. K. Tanaka and T. Mura, "A Micromechanical Theory of Fatigue Crack Initiation from Notches," Mech. Materials, 1 (1981) pp. 63-73.
43. K. Tanaka and T. Mura, "A Theory of Fatigue Crack Initiation at Inclusions," Met. Trans. 13A (1982) pp. 117-123.
44. T. Mura and K. Tanaka, "Dislocation Dipole Models for Fatigue Crack Initiation," pp. 111-131 in Mech. Fatigue, AMD 47, T. Mura, ed.; ASME, New York, NY, 1981.
45. A. K. Head and N. Louat, "The Distribution of Dislocations in Linear Arrays," Austral. J. Phys., 8 (1955) pp. 1-7.
46. H. Kitagawa and S. Takahashi, "Applicability of Fracture Mechanics to Very Small Cracks or the Cracks in the Early Stage," pp. 627-631 in Proceedings 2nd International Conference on Mechanical Behavior of Materials, Boston, MA, 1976.
47. E. M. H. Haddad, K. N. Smith and T. H. Topper, "Fatigue Crack Propagation of Short Cracks," Trans. ASME, J. Eng. Mat. Tech., 101 (1979) pp. 42-46.
48. K. Tanaka, Y. Nakai, and M. Yamashita, "Fatigue Growth Threshold of Small Cracks," Int. J. Fracture, 17 (1981) pp. 519-533.
49. A. J. McEvily and J. Groeger," On the Threshold for Fatigue Crack Growth," Fracture 1977 (ICF4), Volume 2, D. M. R. Taplin, ed.; pp. 1293-1297, University of Waterloo Press, Waterloo, Ontario, 1977.
50. H. Neuber, Kerbspannungslehre, Springer-Verlag, 1958.

51. J. Weertman, I.-H. Lin, and R. Thomson, "Double Slip Plane Crack Model," Acta Met., 31 (1983) pp. 473-482.
52. J. Weertman, "Crack Growth for the Double Slip Plane and the Modified DSP Crack Model: Part II, Fatigue Crack Growth with and without Blunting," submitted to Acta Met.

Concluding Remarks

Introduction

At the end of the scheduled papers presented during the symposium, a General Discussion of the current status of research on fatigue thresholds was held and invited session summaries were presented. The purpose of these remarks was to (1) assess as succinctly as possible what had been learned, (2) determine some of the remaining unanswered questions concerning thresholds, and (3) speculate on what should be done next. Two written summaries are presented in the following pages. We appreciate the efforts of the summary speakers as well as the authors and the participants in the discussion session.

D. L. Davidson and S. Suresh
Editors

CONCLUDING REMARKS

R. O. Ritchie

Department of Materials Science and Mineral Engineering
University of California
Berkeley, California 94720
USA

In providing one of the summaries for this conference, I am reminded of my concluding remarks to the last international meeting on Fatigue Thresholds, held in Stockholm, Sweden, in June 1981. In comparing these two meetings, it is clear that, in two short years, the interest and participation in near-threshold fatigue has increased significantly. This is perhaps most obvious in two areas, namely the analysis of small cracks and particularly in the adoption of the fatigue crack closure concept (which now appears to be the primary factor in the majority of papers on near-threshold fatigue crack growth!). Based on my impressions of the meeting in 1981, I listed five principal themes where I considered significant advances had been made or where future studies, in my opinion, could be directed. With a single exception, these themes remain as important today, and provide a similar basis to summarize the present conference.

Physical Mechanisms

The notion that crack closure can be induced by a number of mechanisms, such as corrosion debris, fracture surface roughness, etc., rather than simply cyclic plasticity is now well accepted. Conceptually, several papers both prior to and in this meeting, have shown the significance of such closure to the roles of load ratio, microstructure, environment, strength level, crack size and so forth in influencing the fatigue threshold. Clearly the phenomenon is real and can dominate behavior below $\sim 10^{-6}$ mm/cycle, at least for "long" cracks at low load ratios. Furthermore, the influence of such closure on the very existence of a threshold, in the analysis of short cracks and in rationalizing variable amplitude behavior has now been established. However, although the concept of various origins of closure appears feasible, and their occurrence can provide, at least phenomenalogically, a description of behavior, aside from a few simple micro-mechanical models the precise physical and mechanical nature of the various closure mechanisms is not well understood. In particular, the exact locations of fracture surface contact with respect to the crack tip, relationships between opening and closing loads, and so forth are still essentially unanswered questions, although certain studies have begun to address these points. Furthermore, the underlying fundamental mechanism of fatigue crack advance still remains largely a mystery despite renewed efforts to apply classical mechanics analyses based on instantaneous crack tip displacement or damage accumulation ideas.

Measurement Techniques

One of the major problem areas associated with near-threshold fatigue remains the question of measurement techniques. Not only is there still the need to standardize techniques to measure near-threshold growth rates and the value of the threshold, but more importantly there is an urgent need to arrive at a concensus on methods to estimate the extent of crack closure. Numerous authors now "correct for closure" to derive a quantitative measure of the effective ΔK, using techniques as diverse as crack mouth displacement gauges, back-face strain gauges, electrical potential measurements ultrasonics, **in situ** microscopy to name but a few. Yet there has been no comparative study to date of the relative merits of these techniques to assess near-threshold closure, and no attempts at standardization. In particular, the validity of the commonly used global compliance measurements using external strain gauges to detect closure, which is presumed to be within 10-100 μm from the crack tip, must be examined, especially where such procedures are applied to short cracks. Clearly, this problem restricts our current understanding of near-threshold fatigue by questioning any comparison of closure data from different investigators and furthermore restricting our ability to develop **quantitative** analyses of closure. The efforts of the ASTM and other standards committees should be encouraged in this regard.

Specificity of Phenomenon

Back in 1981, I commented that although 10 to 15 years ago the majority of basic research on fatigue crack propagation was performed on aluminum alloys, most current data on near-threshold fatigue had been determined for steels. Although the large majority of such data still pertains to ferrous materials, in the past two years the question of whether the near-threshold and closure mechanisms, first documented for steels, were specific to other alloys has been answered by numerous studies on aluminum alloys, titanium alloys, copper, superalloys and so forth. Clearly many of the mechanisms and behavioral patterns are common to all these materials, but as one might expect differences exist, and these have been summarized in several papers in this conference.

Short Cracks

The question of the short, or small, crack remains, and likely will remain for some time, the major limitation to the practical relevance of threshold data. As has been shown previously and in this conference, cracks which are small compared to microstructural or local plasticity size-scales can propagate below the "long crack" threshold at anomalously high growth rates, challenging both our fracture mechanics-based notions of similitude and our assumptions of conservativism in present day defect-tolerant design. As reviewed in detail by Dr. Lankford in the accompanying summary, significant progress, both in experimental observations and interpretation, has emerged since 1981, and specifically has highlighted differences in the crack tip fields and the related question of plasticity and closure effects in the wake of the crack tip. It is my own view that, aside from the microstructural and environmental effects which differ between long and short crack behavior, the problem will become increasingly more difficult until we have a realistic mechanics-based "characterizing parameter" to describe crack advance. This, however, will not be achieved merely by replacing our linear elastic fracture mechanics parameter, ΔK, by one based on non-linear elastic behavior, e.g., ΔJ, since although the latter approach can allow plastic zones on the order of the crack size, it still does not tolerate crack extension and cannot take into account prior cyclic plasticity in the wake of the crack tip. In this regard, it is perhaps necessary to depart from stationary crack analyses, **since fatigue cracks clearly do move,** and consider characterizing parameters based on crack tip fields for non-stationary cracks. Recent observations of the $\ell n(1/r)$ strain singularity ahead of fatigue cracks, rather than fields varying with $r^{-1/2}$ or $r^{-(1/(n+1))}$, certainly suggest that this approach may be viable.

Practical Relevance of Threshold Data

Somewhat surprisingly little emphasis was devoted in this meeting to the practical application of threshold fatigue data in engineering design and life prediction procedures. Although much of our work is inspired by scientific curiosity in the relatively unchartered area of near-threshold behavior, there always remains the question of the usefulness of such data in real world applications, particularly in view of the non-uniqueness of threshold values at small crack sizes. However, our knowledge of near-threshold crack growth and closure mechanisms has provided a far greater insight into the general nature of fatigue, has helped interpret such practically significant topics as variable amplitude behavior, has suggested guidelines for the design of new alloys with improved resistance to fatigue, and in my mind will provide a key to the unification of classical S-N/low cycle fatigue and fracture mechanics approaches to fatigue. Due in no small part to the organizing and editing efforts of our chairmen, Drs. Davidson and Suresh, the present state-of-art of such knowledge is amply surveyed in the current proceedings.

CONCLUDING REMARKS - SMALL CRACKS

J. Lankford

Department of Materials Sciences
Southwest Research Institute
San Antonio, Texas
USA

In assessing the small crack fatigue research which has been reported at this symposium, it seems worthwhile to consider the subject from two perspectives: (1) What do we think we have learned? (2) Where do we go from here? The following observations are necessarily subjective, but may help provoke the thoughts of others who bring different backgrounds to bear on the problem.

What Did We Learn?

It was noted that there may actually be two classes of small cracks, namely, those in which all dimensions of the crack are small, and those in which every dimension but one is small. In other words, through cracks which are very short in length are not necessarily small cracks (microcracks)--they are instead short cracks, and may behave differently from both nominally equivalent (same length) "small" half-penny surface cracks, and "long" through cracks. Although "small" cracks in Al and Ti, and "short" cracks in Ti, all grow faster than equivalent long cracks, the "short" cracks seem not to exhibit a dip in da/dN at low ΔK, as do the "small" ones in both alloy systems.

It was previously proposed by several workers that the low-ΔK crack growth rate deceleration for microcracks was due to grain boundary interactions. However, the experiments on small cracks in Ti by Gerdes, Gysler, and Lutjering suggest that some other explanation may be required. The Al work by Morris and James likewise implies, however, that this explanation does not involve residual stresses.

Previous experiments have indicated that for steels, the large crack threshold merely shifts to lower cyclic stress intensities as ΔK decreases, and it has been possible to collapse the curves together by means of relatively simple corrections to ΔK. However, for Al alloys, small cracks seem to lack a threshold altogether, so that no such simple "fix" in terms of ΔK is possible. Although the absence of crack closure at low ΔK can explain the absence of ΔK_{Th} for small cracks, it does <u>not</u> explain why they grow so fast. Further, it is doubtful if considering a short or small crack as a long crack with a negligible wake will explain the absence of closure in the former at low ΔK. Recent work has shown that contact due to the wake is most important just a few micrometers behind the crack tip, so that the remainder of the long crack wake is unimportant.

The rapid growth of small cracks seems to be related to their extraordinarily large crack tip strains and crack tip openings versus those of nominally equivalent large cracks. These observations imply that small cracks and large cracks (at least for Al alloys) do not share similitude in ΔK. Thus, we may require a driving force term other than ΔK (or even ΔJ) for small cracks.

Where Do We Go From Here?

It is evident that we need to characterize the influence of crystallography on the growth of small cracks, especially in terms of the dip in da/dN. Why aren't small cracks retarded as they cross all grain boundaries? Are crack tip slip bands really blocked by grain boundaries, as has been suggested by several researchers? If we could understand physically the da/dN dip for small cracks, we might be in a position to design alloys in which microcracks grew only with difficulty. However, if it turns out that all small cracks cannot be stopped (as is probably the case), then it is necessary to accept the presence of growing microcracks, and to learn how to predict the growth rates of the fastest ones.

At some minimum ΔK, small and large cracks begin to grow at the same rate. We need to determine what parameters (r_p/a, r_p/D, a/D, $(a+r_p)/D$, P_{op}/P_{max}) correlate with this. Further, we need to determine and not simply infer or suppose the origin of crack tip behavior. For example, if the all-encompassing term "crack closure" is invoked, we must try to establish its basis (residual displacement; asperity contact; mixed mode opening; oxide debris; crack tip residual stress; others sure to come). In short, we require micromechanical models based on experimentally verifiable crack tip and material parameters.

Subject Index

Adsorption of water vapor, 106
Alloys, 302, 303
  aluminum, 3, 25, 28, 31, 33, 34, 37, 50, 53, 66, 137, 147, 155, 163, 164, 263, 265, 334, 361, 381, 404, 447, 471, 479, 517, 521, 533
  brass, 66
  copper, 533
  copper-zinc, 404
  ferrous, 135, 137, 234, 240, 245, 251, 281, 336, 422, 500, 533
    $CrMoV_2$, 206
    dual-phase steel, 116, 118
    HSLA steel, 302, 303, 305, 308-319
    iron, 146, 155, 302, 303, 305, 308-319
    structural steel, 100
    4340 steel, 312-319
  iron-silicon, 302, 303, 308-312, 319
  nickel-base, 140, 185, 186, 199, 329, 330, 341, 533
  powder metallurgy, 163, 164, 165
  stainless steel, 263, 265
  titanium, 85, 137, 329, 433, 465, 533

Corrosion,
  active path, 411
  fatigue, 65, 101, 104, 106, 108, 403
Crack closure, 90, 164, 227, 242, 249, 299, 304, 321, 322, 537
  compliance, 103, 117, 157, 171, 266, 268, 270, 304, 346, 349, 518, 519
  computed, 266
  optical, 330-332
  oxide-induced, 5, 29, 64, 105, 178, 179, 196, 216, 218, 230, 233, 270, 276, 328, 334-336, 370, 412
  phase transformation-induced, 230, 241
  plasticity-induced, 64, 79, 103, 155, 157, 229, 230, 266, 369, 443, 448, 481, 492, 505, 506-511
  roughness-induced, 64, 124, 125, 136, 152, 155, 157, 177, 179, 217, 218, 230, 235, 237, 266-276, 316, 321, 328, 334-336, 343, 352, 354, 411, 414, 442, 468, 518, 519
Crack deviation, 164, 179, 182
Crack growth,
  branching, 135, 164, 179, 182
  continuous, 475, 476
  crystallographic, 236
  discontinuous, 48, 78, 408, 413, 449, 469, 474, 485, 489
  intergranular, 77, 92, 271-274, 307, 309, 311-313, 315, 408, 413
  non-linear, 361, 364
  retardation, 364, 374, 382, 390-394
  shear mode, 148, 152, 153, 157
  transgranular, 92, 152, 157, 171, 271, 307, 309, 311-313, 347, 407, 408, 486
Crack initiation, 468, 500, 505
Crack opening,
  displacement, 48, 105, 165, 332, 451, 452, 457, 508
  load, 67, 75, 318, 322, 500, 509, 521
  mode, 3, 49, 106, 110, 125, 138, 236, 238, 331, 457, 518, 539
Crack path, 177, 367, 370
Crack size or shape, 433, 447, 468, 471, 473, 525
  long, 44, 100, 163, 404, 411, 417, 424, 433, 437, 468, 473, 500, 518-522, 531, 545
  short, 253, 408, 411, 424, 433, 438, 443, 491, 502, 506, 523
  small, 447, 498, 511, 544
  surface, 330, 447, 468, 473, 480, 486, 489
Crack tip blunting, 545
Crack tip micromechanics,
  plastic zone size, 46, 92, 135, 330, 352, 451, 452, 456, 510

strain, 46, 451, 452, 454, 456
stress, 481
Cyclic fatigue effects, 245
Cyclic material properties,
hardening, 44, 46, 66, 88, 135, 537
yield strength, 45, 51, 66, 88, 100, 141, 500

Damage characteristics,
dislocations, 499, 502
microstrains, 483
persistent slip bands, 44, 49, 354
slip lines, 190, 499, 502
strain, 46, 403, 451, 452, 454, 456
Damage measurement techniques,
electron channeling pattern, 456
interferometry, 420
stereo imaging, 450
Dislocations,
in slip bands, 347, 499, 500, 502, 531, 537, 545
models using, 531, 537, 539
pile ups, 499, 502, 540
source, 125

Environmental effects, 472, 535
argon, 67, 70, 119, 120
dry argon, 147, 155
helium, 250
high temperature air, 192, 194, 196, 209, 215, 216, 217, 252
hydrogen, 106, 108, 232, 244, 250, 300, 301, 319
liquid, 252
liquid nitrogen, 147, 157
oxide thickness, 8
salt water, 405
silicon oil, 67, 70, 405
species transport, 107
vacuum, 10, 28, 53, 58, 104, 106, 404, 410, 472
very dry air or nitrogen, 13, 53, 58, 101, 104, 106
water, 31, 251, 405
water vapor in air, 4, 12, 28, 53, 58, 67, 70, 101, 104, 165, 244, 250, 300, 301, 322, 403, 410, 475

Flaw,
size, 253, 497, 525
Fractography, 12, 16, 17, 50, 70, 73, 93, 148, 152, 155, 171, 174, 175, 176, 191, 195, 212, 271-274, 307, 311, 312, 314, 316, 348, 351, 372, 407, 409, 410, 472, 474, 486, 522
Fracture mechanics,
j-integral, 460
stress intensity factor, 2, 27, 31, 32, 33, 35, 47, 55, 56, 67, 69, 87, 90, 102, 119, 147, 149, 177, 178, 189, 192, 194, 209, 228, 240, 268, 270, 306, 308, 347, 350, 367, 386, 406, 419, 422, 425, 436, 437, 449, 451, 452, 457, 459, 469, 473, 487, 491, 498, 502, 518, 519, 532
Fracture surface roughness, 117, 123, 148, 154, 208, 214, 215

Hydrogen embrittlement, 38, 40, 65, 106, 108
internal hydrogen, 301, 323

Load interaction effects, 361, 367

Microstructural effects, 89, 132, 246, 248, 264, 266, 435
colony size, 94
crystallographic orientation, 136, 236, 255, 443
dispersoids, 7
grain boundaries, 136, 182, 449, 468, 470, 489, 492, 539
grain size, 35, 37, 49, 50, 52, 86, 125, 126, 132, 139, 164, 165, 167, 177, 182, 307, 309, 310, 319, 322, 329, 337, 344, 353, 382, 387, 392, 439, 440, 468, 470, 484, 499, 501, 513, 539
heat treatment, 7, 10, 13, 27, 28, 31, 53, 86, 117, 121, 164, 182, 363, 366, 36 367, 375
precipitate, 7
process zone, 387
stacking fault energy, 134
subgrain size, 387-396
Models,
Bilby-Cottrell-Swinden (BCS), 497
Chakrabortty, 93
closure, 328
Coffin-Manson/LCF, 46
critical strain, 132
damage accumulation, 5
deflection, 54, 368
double slip plane, 531, 540
hydrogen embrittlement, 21

plastic work to fracture, 20, 542
strain accumulation, 47
superposition, 15
Monte-Carlo techniques, 492

Notch effects, 502, 504, 511, 523, 525, 527

Oxide thickness measurement, 29, 105, 207, 214, 215, 371, 373

Plasticity, 492
irreversible, 48, 50, 482
reversible, 45, 49, 133

Slip, 132, 138, 248
cross, 7
distance, 47, 50
homogeneous, 182
multiple, 133
planar, 8, 30, 35, 141, 182, 190, 354, 471
wavy, 7, 30
Slip band,
band blocking, 132, 141, 499, 500, 502, 539, 542
band spacing, 132
S-N curve, 399
Strain,
microstrain, 45, 46, 47, 50
Stress,
closure, 138, 148, 155, 171, 172, 304, 316-319, 322, 328, 330-338, 355, 518, 519, 521
crack opening, 318, 322, 483, 490, 499, 503, 506, 509
critical shear, 45, 52
cyclic, 47
mean, 300, 301, 306, 314, 315, 321, 337
ratio, 28, 30, 34, 39, 70, 75, 90, 91, 101, 137, 148, 155, 157, 164, 169, 171, 177, 179, 189, 192, 198, 209, 211, 219, 232, 243, 267, 269, 270, 300, 301, 304, 306, 309, 314, 315, 319, 338, 347, 349, 350, 363, 372, 384, 386, 388, 419, 421, 425, 468, 476, 504, 520, 521, 538
residual, 483, 488, 490
shear, 539
state, 419

Threshold measurement technique, 254

Ultrasonic frequency fatigue, 400

Variable amplitude loading, 253, 361

Author Index

Bailon, J. P., 63
Beevers, C. J., 327
Bell, K., 327
Bignonnet, A., 99
Blom, A. F., 263
Bouchet, B., 99
Bretz, P. E., 25, 163
Brook, R., 281, 417
Brown, C. W., 433

Cadman, A. J., 281
Carlson, R. L., 327
Chen, R. T., 43
Chesnutt, J. C., 83

Dickson, J. I., 63
Davidson, D. L., 447, 553

Ebenberger, H. M.. 399
El Boujdaini, M., 63
Esaklul, K. A., 299

Fine, M. E., 115, 145

Gerberich, W. W., 299
Gerdes, C., 465
Gray, G. T., III, 131
Gysler, A., 465

Heikkenen, H. C., 43
Hertzberg, R. W., 379
Hicks, M. A., 341
Horng, J. L., 115

Jaccard, R., 379
James, M. R., 479

King, J. E., 341
Kwon, J. H., 99

Lankford, J., 447, 559
Liaw, P. K., 205
Lin, F. S., 43
Loison, D., 99
Lutjering, G., 465

McEvily, A. J., 517
Minakawa, K., 517
Morris, W. L., 479
Mura, T., 531

Nakai, Y., 497
Namdar-Irani, R., 99
Newton, C. H., 379
Nicholson, C. E., 281

Park, D. H., 145
Petit, J., 3, 99
Petit, J. I., 163

Ritchie, R. O., 227, 555
Roy, P., 185

Saxena, A., 205
Shih, T. T., 205
Stanzl, S. E., 399
Starke, E. A., Jr., 43
Suresh, S., 227, 361, 553
Swaminathan, V. P., 205

Tanaka, K., 497
Taylor, D., 433
Thompson, A. W., 131

Vasudevan, A. K., 25, 163, 361
Vecchio, R. S., 379
Venables, R. A., 341

Weertman, J., 531
Wert, J. A., 83
Williams, J. C., 131
Wright, A. G., 299

Yuen, J. L., 185